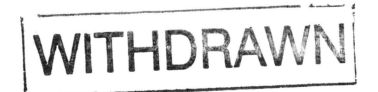

Land Restoration
Reclaiming Landscapes
for a Sustainable Future

Edited by

Ilan Chabay

Martin Frick

Jennifer Helgeson

AMSTERDAM • BOSTON • HEIDELBERG • LONDON
NEW YORK • OXFORD • PARIS • SAN DIEGO
SAN FRANCISCO • SINGAPORE • SYDNEY • TOKYO
Academic Press is an imprint of Elsevier

ELSEVIER

Academic Press is an imprint of Elsevier
225 Wyman Street, Waltham, MA 02451, USA
525 B Street, Suite 1800, San Diego, CA 92101–4495, USA
The Boulevard, Langford Lane, Kidlington, Oxford OX5 1GB, UK
125 London Wall, London, EC2Y 5AS, UK

Notices
Knowledge and best practice in this field are constantly changing. As new research and experience broaden
our understanding, changes in research methods, professional practices, or medical treatment may become
necessary.

Practitioners and researchers must always rely on their own experience and knowledge in evaluating and
using any information, methods, compounds, or experiments described herein. In using such information or
methods they should be mindful of their own safety and the safety of others, including parties for whom they
have a professional responsibility.

To the fullest extent of the law, neither the Publisher nor the authors, contributors, or editors, assume any liability
for any injury and/or damage to persons or property as a matter of products liability, negligence or otherwise, or
from any use or operation of any methods, products, instructions, or ideas contained in the material herein.

Library of Congress Cataloging-in-Publication Data
A catalog record for this book is available from the Library of Congress

British Library Cataloguing in Publication Data
A catalogue record for this book is available from the British Library

For information on all Academic Press publications
visit our website at http://store.elsevier.com/

ISBN: 978-0-12-801231-4

Disclaimer
The views expressed in this paper are those of the author and do not reflect the position of their associated
institutions.

Working together
to grow libraries in
developing countries

www.elsevier.com • www.bookaid.org

Dedication

We dedicate this text to all those who are committed to land restoration for the good of humanity. We hope that this text will assist in and inspire your continued efforts to restore degraded lands and in turn improve lives and encourage an ever more peaceful, equitable world.

Contents

PART 3 SOIL, WATER, AND ENERGY—THE RELATIONSHIP TO LAND RESTORATION

PART 4 ECONOMICS, POLICY, AND GOVERNANCE OF LAND RESTORATION

PART 5 THE COMMUNITY AS A RESOURCE FOR LAND RESTORATION

PART 6 GENDER IN THE CONTEXT OF LAND RESTORATION

PART 7 COMMUNITIES, RESTORATION, RESILIENCE

*WOCAT case studies are available in our companion web site at booksite.elsevier.com/9780128012314

Contributors

Erhan Akça
Adiyaman University, Technical Sciences, Vocational School, Adiyaman, Turkey

Sasha Alexander
United Nations Convention to Combat Desertification, Platz der Vereinten Nationen 1, 53113 Bonn, Germany, and University of Western Australia, 35 Stirling Highway, Crawley, Western Australia 6009, Australia

Pua Bar (Kutiel)
Department of Geography and Environmental Development, Ben-Gurion University, Beer-Sheva Israel

Monique Barbut
United Nations Convention to Combat Desertification, Platz der Vereinten Nationen 1, 53113 Bonn, Germany

Nir Becker
Department of Economics and Management, Tel-Hai College, Upper Galilee Israel

Hartmut Behrend
Bundeswehr Geoinformation Centre, German Armed Forces, Euskirchen, Germany

Julia Birch
Independent consultant, Melbourne, Australia

Lili Hernandez Boesen
The Humanitarian Water and Food Award

Dieter van den Broeck
Living Lands, Patensie, South Africa

Kathleen Buckingham
World Resources Institute, Washington, DC

Jared Buono
Ecohydrologist, Mumbai, India

Kees Burger
Development Economics Group, Wageningen University and Research Centre, Wageningen, The Netherlands

Delia C. Catacutan
World Agroforestry Centre (ICRAF-Vietnam), Hanoi, Vietnam

Ilan Chabay
Professor and Senior Fellow, Institute for Advanced Sustainability Studies, Potsdam, Germany and Chair of the Knowledge, Learning, and Societal Change Alliance (www.KLaSiCA.org)

Alan Channer
Initatives for Lands, Lives, and Peace, Geneva, Switzerland

Jonathan Davies
International Union for Conservation of Nature

Sean DeWitt
World Resources Institute, Washington, DC

Ton Dietz
African Studies Centre, Leiden, The Netherlands

Tom Duncan
Home Ecology & Brooklet Farm Education, Brooklet, Australia

Hannes Etter
Economics of Land Degradation Initiative, Bonn, Germany

Willem H. Ferwerda
Rotterdam School of Management, Erasmus University, Rotterdam, the Netherlands; IUCN Commission on Ecosystem Management, Gland, Switzerland, and Commonland Foundation, Amsterdam, the Netherlands

Lynn Finnegan
Quaker United Nations Office, Geneva

Rob Francis
World Vision Australia, Melbourne, Australia

Stephen Freeman
The Humanitarian Water and Food Award

Kees van der Geest
United Nations University Institute for Environment and Human Security, Bonn, Germany

Luc Gnacadja
Executive Secretary of the UN Convention to Combat Desertification (2007–2013)

Bradley T. Hiller
Freelance Consultant to the World Bank, Cambridge Institute for Sustainability Leadership, University of Cambridge, University of Anglia Ruskin

Stephen Hinton
The Humanitarian Water and Food Award

Dhanasree Jayaram
Manipal University, Manipal, Karnataka, India

Ednah Kang'ee
Armed Conflict and Peace Studies Program, Department of History and Archaeology, University of Nairobi, Kenya

Rhamis Kent
Permaculture Research Institute, Australia

Bettina Koelle
Indigo Development & Change, Nieuwoudtville, South Africa

Adam Koniuszewski
Green Cross International, Geneva, Switzerland

Lars Laestadius
World Resources Institute, Washington, DC

Rattan Lal
Carbon Management and Sequestration Center, Ohio State University, Columbus, OH, USA

Hanspeter Liniger
Centre for Development and Environment, University of Bern, Bern, Switzerland

John D. Liu
Director, Environmental Education Media Project; Visiting Fellow, Netherlands Institute of Ecology, and Ecosystems Ambassador, Commonland Foundation

Guy Lomax
The Nature Conservancy, Oxford, UK

Georgina McAllister
GardenAfrica, Burwash, East Sussex, UK

Luca Montanarella
European Commission, DG Joint Research Centre (JRC), Ispra, Italy

Mark Mulligan
King's College London, Strand, London, UK

Stela Nenova
Energy and Environmental Policy Consultant

Noel Oettle
Environmental Working Group, Nieuwoudtville, South Africa

Isabelle Providoli
Centre for Development and Environment, University of Bern, Bern, Switzerland

Jayashree Rao
Grampari, Maharashtra, India

Tony Rinaudo
World Vision Australia, Melbourne, Australia

Tetsu Sato
Research Institute for Humanity and Nature, Motoyama, Kamigamo, Kita-ku, Kyoto, Japan

Allan Savory
Savory Institute, Boulder, Colarado

Meira Segev
Department of Geography and Environmental Development, Ben-Gurion University, Beer-Sheva Israel

Rima Mekdaschi Studer
Centre for Development and Environment, University of Bern, Bern, Switzerland

Kume Takashi
Ehime University, Department of Rural Engineering, Faculty of Agriculture, Matsuyama City, Japan

Maura Talbot
Living Lands, Patensie, South Africa

Mary Thompson-Hall
Basque Centre for Climate Change, Bilbao, Spain

Grace B. Villamor
Center for Development Research, University of Bonn, Bonn, Germany

Peter Weston
World Vision International, Solomon Islands

Donald A. Wilhite
University of Nebraska, Lincoln, NE, USA

Augustine Yelfaanibe
Faculty of Integrated Development Studies (Wa Campus), University for Development Studies, Tamale, Ghana

Acknowledgments

This book is the product of tireless work and dedication by many people. It has been a rare and wonderful opportunity to work with a remarkable group of authors, spread around the globe and from such diverse disciplines, all working on aspects of land restoration. It was made possible through an extensive network of relationships mainly built at the yearly "Caux Dialogue for Land and Security" in Switzerland. The authors would like to thank all those who, as volunteers, commit to organize this important conference and maintain the network.

The authors in this volume represent a range of viewpoints on the topic of land restoration and its relationship with security in different dimensions. The expertise represented in this book spans a variety of professional positions and experience—practitioners, nongovernmental organizations (NGOs), university professors, and policy specialists. These authors' inputs complement one another in formulating a representative and holistic view of land restoration. We thank them for their time, collaboration, and contributions.

We would like to acknowledge the outstanding efforts of some volunteer text editors who went through select chapters and worked with some of the authors during the drafting process. Many heartfelt thanks to Meera Shah, Natassia Ciuriak, Scott Darby, Irina Fedorenko, Jane Feeney, Guy Lomax, Barb Smeltzer, Rachel Waggott, and Wessel van der Meulen. And our thanks also to Dr. Lori Adams Chabay for inspiring the inclusion of notes for educators and others on the uses of this text.

Foreword

Land matters. It is worth putting it as bluntly as that. It feeds and nourishes us. It provides fuel to warm us and the homes we live in. Its loss is a real-world driver and amplifier of instability. Humans will fight to defend their land and kill to grab enough land for their needs. It matters more than we recognize, and more than political decision makers dare to calculate. But let us be clear—the foundation of our homes is on shaky ground.

As the global population expands dramatically toward a total of more than 9.5 billion people by 2050, demand for the goods and services that the land provides will only get stronger. In the context of a changing climate, as demand starts to outstrip supply, competition for this increasingly scarce and valuable productive resource will only heat up. As the loss of productive land and soil continues at an alarming rate, we expect to see declining food production and more hunger. The natural end point of these trends is mass forced migration, radicalization and deadly conflict in climate change and land degradation hot-spots. The so-called developing world, where natural infrastructure and resource governance mechanisms are often weakest and most vulnerable, is at the greatest risk. It will be hit hardest. However, in an interconnected world, no region or community is immune. No country can rest on its laurels.

That is the reality. Yet, you will see that if decision makers are bold enough, there is a chance that we can put our house in some semblance of order and strengthen its foundations. Land and soil are set to be a fundamental part of the Sustainable Development Goals for post-2015 implementation. Achieving land degradation neutrality will be a crucial first step. From the community to the national level, we must stop the loss of healthy and productive land by avoiding degradation wherever we can and by rehabilitating already degraded land. More well-managed land means more food and water available, more of the ecosystems we enjoy, less forced environmental migration, and a greater chance of security and peace in an unstable world.

Work on securing our land resources also needs to start now because sustainable land and soil management can buy valuable time in the fight against climate change—time that is desperately needed. Land and soil is the second-largest carbon sink, after the oceans. Getting carbon back in the soil could buy the 30 years that we may need to move to a low-carbon economy. It could also get vulnerable populations, who are already experiencing climate change impacts and resource scarcity, time to adapt.

So this book could not have come at a more opportune moment. It brings together findings from various levels of governance and experiences from diverse stakeholder communities toward a viable solution to one of the most pressing issues of our time.

We see all around us how fragile peace and stability can be. It is fragile precisely because we have ripped the ground from underneath people's feet. By addressing the real world drivers of instability, we can build resilience to shocks of every sort. Sustainable land management can diffuse competition and conflict and put solid ground under the most vulnerable. It is a foundation for the future—if we are daring enough to just grab it.

Monique Barbut

Executive Secretary of the United Nations Convention to Combat Desertification (UNCCD)

Governing Land Restoration: Four Hypotheses

We are living in times of unprecedented environmental change. Speaking about soils alone, we are currently losing about 24 billion tons of fertile soil per year. Some soil types even become extinct because of human use. Due to atmospheric pollution, the nitrogen input to our soils alters soil properties on a global scale. All these are examples of what Paul Crutzen has dubbed the "Anthropocene": We are living in the age of humans. In his ground-breaking piece, Crutzen concludes, "A daunting task lies ahead for scientists and engineers to guide society towards environmentally sustainable management during the era of the Anthropocene" (Crutzen, 2002, p. 23).

There cannot be any doubt. Reversing these soil and land degradation trends requires technical and managerial solutions for land and soil rehabilitation. The *World Overview of Conservation Approaches and Technologies* (WOCAT, 2015) provides a valuable account of the experiences gained so far. We also have to invest in research on new technical solutions. Finding ways to respond to the lack of biomass at the farm level to increase the content of organic matter in soils is just one example among many. We need new technical responses to respond to these challenges.

Meta analyses of the adoption rates of sustainable land management emphasize the importance of social, political, and economic factors. Market access and land tenure serve as examples in this regard. This implies that attention has to be paid to the question of how soil and land restoration activities become adopted. There is another line of reasoning to add to this. Given the magnitude of the problem, we are likely to face calls for large-scale land and soil rehabilitation soon. This will affect the lives of many who are currently living on or off these lands. While we all depend on scientists and engineers to develop new responses to challenges such as land degradation—to return to Paul Crutzen—the question of which response is chosen (and which questions are asked by science) in open and democratic societies cannot be left to the "guidance" of scientists and engineers alone. Hence, there are close ties between the way we govern soil and land restoration and the way we do research in support of such transformation processes towards more sustainable futures. In our opening remarks, we would like to address this topic by means of introducing four hypotheses on governing land restoration and research in support of it:

- We need a multilateral response to land and soil degradation, most likely through a slightly altered multilateral system.
  - There are probably few resources that are more local than land and soils. Communities have strong cultural ties to specific regions, such as families handing over plots to the following generation. At the same time, soils and land provide globally relevant ecosystem services. Soils are the largest terrestrial carbon store. They store about 4 billion tons of carbon, which is 10 times as much as the world's forests. Humanity derives more than 95% of its food from soil and land. Anybody who doubts the global implications of this fact is encouraged to look at the population dynamics and projected impact of climate change on major staple crops in sub-Saharan Africa. The 2012 population projection by UN DESA (2013) foresees a doubling of the population in Africa from 2013 to 2050 and a more than quadrupling until 2100. The fifth Intergovernmental Panel on Climate Change (IPCC) assessment report finds that maize-based

systems—which are expanding in many parts of Africa—do belong to the most vulnerable crop systems in sub-Saharan Africa. Estimates for yield losses by 2050 for sub-Saharan Africa arrive at 22%. At the same time, many countries in Africa and on other continents are already heavily affected by land degradation. Hence, we do need multilateral approaches to respond to soil and land degradation and to restore degraded soils. At the same time, multilateral environmental governance did not, unfortunately, maintain the spirit of Rio 1992. More often than not, it is currently characterized by slow progress and stalemate. Furthermore, it is still an open question whether the multilateral system is always best equipped to deal with the new complexity of actors and their potential contributions to solving sustainability challenges. There is the need to strengthen the multilateral system. One approach—among several—could be to blend the normative power of the United Nations (UN) with more flexible approaches that are linked to, but independent of, the UN system. The achievements and potential challenges of the Global Water Partnership seem worthwhile to explore in this regard. On a side note, land and soil degradation become increasingly linked to security questions. From our point of view, the need for multilateral approaches, as well as the urgent necessity of dialogue on how to move the global agenda on land and restoration ahead, pose strict limitations on the use of the security narrative. Far too often, questions related to land are primarily treated as a national security topic. This might run counter to our joint endeavor to lift soil and land to the international policy agenda.

- Multistakeholder processes to identify pathways of change for soil and land restoration on landscape level hold great potential, but they need to include deliberate efforts to empower marginal groups.
  - Successful land and soil restoration measures need to go beyond the plot level. They need to be designed and implemented on the landscape level. First, there are various extra-local factors—social, economic, political, and ecological—concerning land use that influence land degradation patterns. Finding adapted responses to land degradation and identifying suitable land and soil restoration techniques demands taking these extra-local influences into account. Second, many restoration activities requires collective activities beyond the plot level to become effective. Third, thinking in terms of benefits of land restoration beyond the plot level also holds the potential to identify other support systems for land restoration. The famous upstream-downstream metaphor is just one example among many in this regard.
  - Developing these landscape-level soil and land restoration approaches is not a quick fix. Various interests, perceptions, and experiences come to interact at this level. Finding ways to adapt to these complexities requires a local search and negotiating process. This also requires a different type of external support to land restoration initiatives. For example, the Doing Development Differently (DDD) Manifesto holds some valuable considerations in this regard (DDD, 2014). While such a landscape-level approach is likely to be burdensome and to be characterized by back-and-forth discussions, it is difficult to imagine a different approach that might lead to successful land and soil restoration activities. The history of watershed development approaches holds valuable lessons on this topic.

    To be successful, landscape-level approaches need to be inclusive. Again, that is probably easier said than done. Land and soil restoration activities on a landscape level need to include deliberate measures to empower those who are marginal to bring their voices to the fore. To give just one example: How can we ensure that landless people, who tend to be among the poorest, benefit from land and soil restoration activities? Finding viable responses to this question depends on inclusive decision-making processes, which are likely to be successful only if they deliberately deploy means to empower those who are not used to, or are often prohibited from, voicing their opinion and concerns in public.

- For pro-poor land and soil restoration to take place, we need to secure the land rights of the most marginal groups in society.
  - The overwhelming majority of the rural poor live on degraded lands and on soils of low fertility.[1] Hence, soil and land restoration measures need to contribute to the progressive realization of the right to food. For marginal groups in society, land and soil restoration measures come with a risk: they increase the value of the land. There are examples from development cooperation in which attempts to increase the productivity of the land used by women led to them losing access to the land. Higher productivity turned "women's land" into high potential cash crop land and, hence, "men's land." Poverty and lack of political or social representation often go hand in hand. For land and soil restoration initiatives aimed at food security and poverty reduction, this implies that restoration activities need to be accompanied by activities to empower those who are most marginal in rural communities. Land and soil restoration need to be embedded in a broader change process. Securing the land rights of these groupings in society is an essential element of that process. Principles for such a human rights–based approach to responsible land governance are described in the Voluntary Guidelines on the Responsible Governance of Tenure of Land (VGGT). The principles of the VGGT need to form part and parcel of land and soil restoration activities.
- To support the necessary governance transformations for land and soil rehabilitation, we need transdisciplinary research that gets involved in the respective transformation processes.
  - New approaches to soil and land restoration require excellent disciplinary and interdisciplinary science. However, there are way too many findings that are relevant but are not taken up by decision makers. In this respect, it is useful to distinguish between research *for* transformations, *on* transformations, and *in* transformations. Research for transformations establishes an objective for research endeavors that support transformation. This can either be done by conducting research on transformation processes (e.g., analyses of pathways of earlier transformations) or research in transformation processes. The latter involves the engagement of researchers in the change process.
  - This engagement is essential, as research findings do not only lead to, foster, or support transformation processes ("change through knowledge") but also because change processes generate knowledge ("knowledge through change"). It is through the stepwise learning in actual change processes that a research programme is codeveloped (and implemented together) in support of change. Getting involved in governance transformations is a basis for reflecting on these processes. There is an urgent need for transdisciplinary research to complement well-established disciplinary or interdisciplinary research.
  - There are two additional remarks on this. First, experience with rethinking rural extension services and action research for sustainable land management shows that codeveloped knowledge is better adapted to the needs of those who shall benefit from land and soil restoration activities. It is a stepping stone to achieve sustainability of land and soil restoration measures.

---

[1]It is pivotal to point out that this statement does not imply attributing responsibility to the rural poor for the state of the land they farm. There are other processes that might lead to the rural poor inhabiting marginal land.

Second, inclusive and participatory ways of doing research hold a greater potential to contribute to the empowerment of vulnerable or marginal groups who ought to benefit first from land and soil restoration activities. In times of strong calls for "evidence-based policies," what counts as evidence or as relevant knowledge is highly contested. Transdisciplinary research in this understanding is, then, not only about creating robust knowledge, but also about developing shared framings or understandings of the problem at hand. It is an inclusive and emancipatory way of constructing the problems and responses to them.

In conclusion, an increasingly complex sustainability governance landscape, participatory ways of arriving at democratic decisions, and an increasingly diverse range of media channels offer various ways for transformations through transdisciplinarity. These developments call for engaged research. Knowledge-based organizations need to support actual transformation processes. This engagement is, of course, in need of scrutiny. Global Soil Week (www.globalsoilweek.org), to provide one example from our work program, rests on a particular theory of change, on the hypothesis that the multilateral governance system of soils and land would benefit from certain governance functions that such a flexible platform can offer. To what extent this hypothesis is true, and to what extent Global Soil Week actually leads to transformations through transdisciplinarity, then become a matter of research itself. There are several other systematic approaches to foster transdisciplinarity. The Future Earth program (2015) and the design of the new Horizon 2020 program by the European Commission (EC, 2015) need to be mentioned here. We hope that transdisciplinary work on land and soil rehabilitation will be one of the focal areas of these two programs, and we would like to offer Global Soil Week as a platform for reviewing the experiences gained so far.

We are convinced that this book is a valuable contribution toward this goal. It is a timely publication and provides a fertile learning ground for fresh thinking on governance transformations of soil and land rehabilitation. We do hope it will be widely read and the lessons learned widely applied.

Jes Weigelt

Co-Lead Sustainability Governance Programme, Global Soil Forum Coordinator,
Institute for Advanced Sustainability Studies, Potsdam, Germany

Alexander Müller

Senior Fellow, Sustainability Governance Programme,
Institute for Advanced Sustainability Studies, Potsdam, Germany

REFERENCES

Crutzen, P.J., 2002. Geology of mankind. Nature 415, 23.

Doing Development Differently (DDD), 2014. The Manifesto. from, www.doingdevelopmentdifferently.com (accessed 08.03.15.).

European Commission (EC), 2015. Horizon 2020 The EU Framework Programme for Research and Innovation. from, www.ec.europa.eu/programmes/horizon2020 (accessed 08.03.15.).

Future Earth, 2015. Future Earth: Research for Global Sustainability. from, www.futureearth.org (accessed 08.03.15.).

United Nations, Department of Economic and Social Affairs, Population Division (UNDESA), 2013. World Population Prospects: The 2012 Revision. United Nations, New York. www.esa.un.org/unpd/wpp/index.htm.

World Overview of Conservation Approaches and Technologies (WOCAT), 2015. Knowledge Base. from, https://www.wocat.net (accessed 08.03.15.).

Introduction

After Neil Armstrong set foot on the Moon, the excitement of that accomplishment yielded to the sobering experience of having conquered a place for humanity that was unfit for any form of life. The prevalent theory of the origin of the Moon says that a large meteorite hit the Earth billions of years ago and a part of the planet was shot into space but was caught by the Earth's gravitational pull.

Thus, the Moon and the Earth are made of the same material. Yet, while the Moon is a place on which people can only survive using most sophisticated life-support systems, our home planet is the most hospitable place in the known universe.

If we look at the giant life-support system that this planet provides us, we must acknowledge the central role played by soils. From the dawn of civilization, people have been aware of the central importance of soils to maintaining life. Many myths about the origin of humans revolve around earth and soil. In fact, the word *Adam* itself means "the one who is made out of clay." Many religions and traditions around the world feature similar myths and stories.

Strangely, as humans shape this planet on an unprecedented (indeed geological) scale, as is reflected in the growing recognition of the Anthropocene as a new geological age, land is strangely underrepresented in international discussions focused around the topical issues of climate, energy, water, and food. Yet, land is the nexus at which all these issues converge.

Most of our food is produced on land; functional soils and related ecosystems are key to the availability and management of drinking water; even energy from organic matter is being produced by agriculture at an increasing rate. But land is being appropriated for many uses other than food production. The pressure on arable land by rapid-growing urbanization is tremendous. Already more than half of the world's population resides in cities, and this share is likely to increase to a staggering two-thirds.

Even more serious than land loss due to urbanization are the large areas of land in rural areas that have become severely degraded, with consequent loss of productivity and diminished ecosystem services.

Today, 33 percent of land is moderately to highly degraded due to the erosion, salinization, compaction, acidification and chemical pollution of soils. Worldwide some 1500 million people depend on degrading land. Indeed, though most people are much less aware of it than of other critical crises (e.g. the climate crisis), we do have a soil crisis on a planetary scale: Every single year, an area thrice the size of Switzerland is lost for agriculture, exacerbating existing situations in climate change, water management, availability of food, and ultimately human security.

We believe that land and its ecosystem services is an issue of central importance for human well-being and development and thus must be broadly discussed. The core question is: How can an adequate area of land with sufficient quality of soil become and remain available to humanity worldwide, now and in the future?

Expanding agricultural land by deforestation is not a viable option, as forests are vital for carbon sequestration and water management. They represent, in fact, a precondition for agriculture. Preventing land degradation and desertification and reclaiming degraded land for agriculture and ecosystem services are much more feasible and sustaining approaches.

The good news, and a relatively little known fact, is that land degradation can be reversed, with sometimes spectacular results, at relatively low cost. Against conventional wisdom, land can be restored without artificial irrigation. What it takes is often simple, manual labor and, before that, addressing often complex and interdependent societal conditions.

This book intends to showcase examples of successful land restoration and experiences gained from the respective processes. Different methodologies exist, which proved to work in different settings. This book is not intended to advocate any specific technology or technique, but rather to make the case for the necessity and the feasibility of land restoration through different pathways. What holds true for different methodologies is also, and particularly, true for variations in social and political conditions. There is no one-size-fits-all method because the local societal and bio-geo-physical conditions are very different, as the examples in this book illustrate. However, it is safe to say that every successful land restoration project is, in the best sense of the word, "grounded" in the respective local context.

To reflect the complexity and interdependence of issues around land, this book tries to provide different perspectives on land. Our authors are political decision makers, scientists, practitioners, and consultants. Their stories often also double as accounts of personal journeys.

As the book gives voice to multiple disciplines, we hope that it adds to the interdisciplinary dialogue that is needed for successful land restoration projects. Different perspectives are based on different cultural, personal, and economic backgrounds and on individual values. This is reflected by the style, tone, and attitude of the chapters, which have been edited for clarity and accuracy while also intentionally and gratefully leaving the authors' voices to come through as much as possible.

Land management and, in particular, land restoration, needs to be approached in a holistic manner. Generally, technical questions are not the greatest obstacles, though the science is complex and, of course, continues to unfold. The real challenge is to get communities to work together and to shape an effective framework for action. Significant action can happen only if various stakeholders define and support the same goal. Building a community around land issues—from the bottom up, addressing stakeholders at multiple levels—is crucial. Silo thinking, whether by experts, policy makers, or practitioners, needs to be overcome.

A recurrent theme is competing problem-solving mechanisms from the realms between traditional and modern lawmaking. Land tenure is an excellent example for the need to balance modern tenure systems with traditional systems, particularly to protect the poorest among us.

It is also important to highlight the interdependence between global governance issues and localized decisions. In the same way that legal frameworks need to support local action, grass-roots action can provide inspiration for global decision- making.

The current international efforts to curb CO_2 emissions to keep climate change to a manageable level provide an excellent example. The carbon sequestration potential of land restoration is still undervalued in these negotiations. At the same time, land restoration can dramatically improve local conditions and contribute to the management global commons—climate and water management. As land restoration can address climate change and rural poverty simultaneously, it has serious potential to create support for the climate negotiations by a broad alliance of poorer countries.

In addition to the climate negotiations, security considerations should be a prime reason for international diplomacy to look closer at the issue of land. Pressure on land is often mirrored by security situations since land is directly relevant to water and energy supplies, producing food, and provision of income.

With growing urbanization, the pressure on land will increase. Only about 3% of the global surface is arable land. Land will not only be lost due to urbanization. Agricultural planning needs to take the "real footprint" of cities into account—i.e., how much land is needed for production of water, food, and energy, as well as for the disposal of waste from urban areas. Just as the footprint of cities is much larger than just the built structure, the real footprint of an entire country is often larger than the country itself. In this vein, seemingly distant problems are directly affecting industrialized countries and highly developed countries. In a globalized and interdependent world, the effects of land degradation might also materialize in other countries. A spike in bread prices in 2011 was one of the triggers for the Arab Spring rebellions. Global draughts and fires in Russia had a direct impact on the Arab peninsula, the biggest grain importer in the world.

Security experts agree that land can affect security in both, the traditional "hard" security sense (i.e., military) and in the definition of "human security" (i.e., well-being). Two-thirds of the world's active conflicts today are in the drylands. The global struggle for arable land is likely to become a further source of conflict. To mitigate the risk of future conflicts, we need to improve the quality of existing arable land, or land that has become unfit for agriculture.

This book builds on a series of on-going conferences held in Caux, Switzerland, since 2011. The so-called Mountain House in Caux, a beautiful turn-of-the-century former palace hotel uniquely situated 1000 m above Montreux and overlooking Lake Geneva, has a remarkable historic tradition. Since 1946, it has been a place for behind-the-scenes talks to promote peace and reconciliation. Famously, it helped bring the Germans and French together right after World War II to start the difficult, but highly successful, path to reconciliation with Germany after the bitter experiences of war and Nazi terror. Ever since, Caux has served as a place where former enemies would meet and forgiveness paves the way for future cooperation.

Initiated by Mohammed Sahnoun, former Algerian ambassador and United Nations (UN) diplomat, the Caux Forum on Human Security focused in annual conferences over 5 years, from 2008–2012 on the deeper causes of conflict. Environmental pressure was one of five focus areas. In 2008, Luc Gnacadja, then executive secretary of the UN Convention to Combat Desertification (UNCCD), was invited to speak about land and security in this context. His presentation was both alarming and inspiring at the same time. It led to the founding of the "Initiative for Land, Lives, Peace," which organized two subsequent conference days on the topic of land and security in 2011 and 2012. The broad resonance of interests in those days encouraged the conference's organization team to start a new conference entitled "The Caux Dialogue on Land and Security," which has been held every summer since 2013.

The Caux Dialogue is unique in having a wide range of stakeholders participating in the discussions—from the "hard" security communities to grass-roots activists, from UN organizations to academics. If this book stimulates your interest in the issue, please do find out more at www.landlivespeace.org.

We hope that you will enjoy using this book and that it will enrich your understanding of how multi-disciplinary (and even transdisciplinary) the issue of land restoration is. The foreword by Monique Barbut, executive secretary of the UNCCD, provides the context for this text and sets the scene for the importance of land restoration globally. Section 1 of this book sets the social context for land restoration. Section 2 reviews the concepts and methodologies for restoration and maintenance for land restoration efforts. The complex relationship between land restoration and water and energy is noted throughout section 3. Section 4 delves into the relationship between economics, policy, and governance for land restoration by drawing on theory and related case study examples. The significance of community as a backbone for land restoration is discussion in section 5. Section 6 offers a review of the relationship between gender and land restoration. The connection between communities, land restoration, and providing resilience more generally is discussed in section 7. Section 8 of the book rounds out the discussion through a series of case studies from various areas of the world that have implemented a variety of approaches. Suggestions for ways to use this book in your own research and work are provided in section 9. Finally, the conclusions and a look forward by Luc Gnacadja are provided as a means of reflecting upon the text.

It has been our pleasure to bring all these contributors together to create such a volume. We sincerely hope that it will serve to further connect and inspire those dealing with land restoration in any and all capacities.

Jennifer Helgeson

Environmental Economist, National Institute of Standards and Technology (NIST), USA; London School of Economics and Political Science, Department of Geography and Environment, The Grantham Research Institute on Climate Change and the Environment

Ilan Chabay

Professor and Senior Fellow, Institute for Advanced Sustainability Studies, Potsdam, Germany and Chair of the Knowledge, Learning, and Societal Change Alliance (www.KLaSiCA.org)

Martin Frick

Ambassador of Germany to the United Nations and other International Organisations based in country, currently Director, Climate Change, Energy and Tenure Division of the UN Food and Agricultural Organization

August 18, 2015

SOCIAL CONTEXTS OF LAND RESTORATION

LAND DEGRADATION AS A SECURITY THREAT AMPLIFIER: THE NEW GLOBAL FRONTLINE

1.1

Monique Barbut[1], Sasha Alexander[1,2]

United Nations Convention to Combat Desertification, Platz der Vereinten Nationen 1, 53113 Bonn, Germany.[1]
University of Western Australia, 35 Stirling Highway, Crawley, Western Australia 6009, Australia.[2]

1.1.1 INTRODUCTION

Land degradation is generally understood to be the reduction or loss of biological or economic productivity (UNCCD, 1994: Article 2) resulting in decreased yields, incomes, food security, and the loss of vital ecosystem services. These impacts, in turn, serve to undermine the peace and stability of land-dependent communities. Thus, there appears to be a demonstrable link between land degradation and human security, especially when we consider how poverty and hunger lead to migration and conflict.

The health and resilience of our land resources (e.g., soil, water, and biodiversity) are largely determined by our management practices, governance systems, and environmental changes. The conversion of natural ecosystems or the unsustainable use of fertilizers, pesticides, and irrigation for food production contributes not only to land degradation at the local level, but also to increased carbon emissions, reduced biodiversity, and diminished rainfall on regional and global scales (Sivakumar, 2007). Indeed, land degradation, biodiversity loss, and climate change are considered to be intertwined threats to human security that contribute to a downward spiral in the productivity and availability of land resources (D'Odorico et al., 2013; Mueller et al., 2014).

Demonstrating sufficient causation or the existence of feedback loops between land degradation, climate change, and human security has proved difficult due to a host of other contributing factors, such as political and economic instability and social fragmentation, which are often considered more proximate determinants (Thiesen et al., 2013; Koubi et al., 2014; Selby and Hofmann, 2014). While the political and economic influences on human security are often more visible and are often cited as causes, land degradation and climate change are recently being more clearly identified in empirical studies from multiple places in the world, most prominently in Africa and Asia (Paul and Roskaft, 2013; Taylor, 2013; Bond, 2014; van Schaik and Dinnissen, 2014).

In these cases, a strong association between land degradation, climate change, and human security is found when one critically examines the host of underlying drivers that contribute to poverty, hunger, migration, and conflict. Without adaptation strategies and resilience building devoted to responsibly managing and restoring our natural capital, land degradation, especially in developing countries, will continue to be a significant factor threatening rural livelihoods, triggering forced migration, and aggravating conflicts over limited natural resources (Barnett and Adger, 2007). In extreme cases of land degradation, desertification, and drought, entire communities are forced to migrate from their ancestral lands to areas where competition for scarce resources already exists and thus contributing to a higher risk of conflict (Kumssa and Jones, 2014).

Given the global scope of land degradation, considering both its causes and impacts, there is an environmental and a socio-economic imperative to address it as an underlying threat to peace and security, from the scale of local communities to that of entire continents. This chapter will briefly explore these linkages and potential responses through the lens of human security, setting the stage for the remainder of this book. The aim is to convince decision makers of the urgent need to halt and reverse land degradation trends, while at the same time highlighting practical and cost-effective solutions, including the adoption and scaling-up of sustainable land management (SLM) practices and ecosystem restoration activities.

1.1.2 THE HUMAN SECURITY LENS

The perceived causes of insecurity have increased and diversified considerably over the last few decades. While political and military issues remain key, perceptions of conflict and security have broadened to include economic and social threats, such as poverty, infectious diseases, and environmental degradation as significant contributing factors (Kingham, 2011). This new understanding of the diverse challenges to human security is now reflected in national and international policy debates.

National security has long been seen as the ability of a country to militarily defend itself against threats and aggression both foreign and domestic. In the 1970s, the notion of "environmental security" was introduced to broaden the national security perspective by focusing on the long-term sustainability of our climate system, the natural resource base, and its capacity to provide for future generations (Falk, 1972; Brown, 1977). In 1982, the Independent Commission on Security and Disarmament Issues advanced the concept of "common security," pushing national security assessments and strategies to adopt a more interdisciplinary approach while also recognizing the transboundary nature of environmental security (ICDSI, 1982).

The 1994 United Nations Human Development Report (UN HDR) makes it clear that national and environmental security can no longer be considered in isolation; it coined the term "human security" to encompass the less often considered threats to human well-being—namely, those related to food, economic, health, and environmental security (UNDP, 1994). In essence, human security is seen as the universal right of individuals and communities to enjoy peace (not fear) and stability (not want) in the pursuit of sustainable development; in broad strokes, it embodies freedom from conflict, access to resources, and the opportunity to improve the human condition (Biswas, 2011; Westing, 2013).

The human security perspective is provoking novel ways of thinking in defense ministries around the world, including the recent attention by the North Atlantic Treaty Organization (NATO) to "hybrid threats," which are seen as more than just the amalgamation of existing security challenges (NATO, 2009). While still evolving, the human security lens embraces complexity—namely, the interdependencies among constituent elements, a multiplicity of stakeholders with vested interests, and a dynamic security landscape where traditional military solutions may not be a key component (Aaronson et al., 2010). It also recognizes the importance of transnational cooperation and governance for the sustainable use of the stocks and flows of land resources that transcend political boundaries.

With respect to land resources, human security provides a multifaceted perspective conducive to the observation of the causal factors linking the trinity of land degradation, biodiversity loss, and climate change to increased human insecurity. In terms of its practical application, the lens must be fixed on addressing the conditions and processes that reduce the options available to deal with potential insecurities of any kind (Zografos et al., 2014). Most notably, the 1994 UN HDR emphasizes the role of early prevention and the cost effectiveness of reducing the underlying threats that increase the vulnerability of communities and ecosystems: the basic premise of this book.

1.1.3 LAND DEGRADATION CAN MAKE THINGS WORSE

The combined challenge of an increasing population (i.e., more demand for land-based resources) and the impacts of climate change (particularly drought) on food and water security will no doubt continue to jeopardize human security. Land degradation, even when not a direct cause, is a significant contributing factor or "threat amplifier" that acts in combination with other factors in a highly specific geographical location and socio-economic context (van Schaik and Dinnissen, 2014). A number of qualitative case studies indicate that environmental stress may, under specific circumstances, increase or amplify the risk of violent conflict, but not necessarily in a systematic or unconditional way (Bernauer et al., 2012).

As already stated, the term *land degradation* refers to a loss in productivity and availability of land resources, such as soil, water, and biodiversity, while recognizing that the condition of these resources are inextricably linked in nonlinear processes and are difficult to segregate (Sterzel et al., 2014). Globally, approximately 25% of all land is highly degraded, while 45% is considered stable or slightly/moderately degraded; remarkably, only 10% of all land is considered to be improving (FAO, 2011). Over 2 billion people worldwide depend on 500 million small-scale farmers for their food security; this accounts for 80% of the food consumed in Asia and sub-Saharan Africa (IFAD, 2013). In India, it is estimated that 296 million acres (approximately 70%) of the 417 million acres of land under cultivation are degraded, with over 200 million people dependent on this degraded land for their sustenance (ICAR, 2010).

In addition to lower yields and incomes, land degradation reduces water productivity and affects its availability, quality, and storage (Bossio et al., 2010). This often leads to a reduction in other important regulating services, such as soil stability and fertility, climate control, and carbon sequestration. Around 1.2 billion people (i.e., almost one-fifth of the world's population) live in areas experiencing water scarcity, and another 500 million people are fast approaching this situation (UNDP, 2006). Increased water scarcity is evident throughout much of the world as groundwater tables recede, rivers and lakes run dry, and rainfall lessens and becomes more erratic. In much of Africa and Asia, this means less arable land for farmers and the loss of grazing lands and freshwater sources for pastoralists. In northern Nigeria, a significant relationship exists between freshwater scarcity and conflicts among farmers and pastoralists: the availability and management of water sources were found to be the most potent predictors of conflict between sedentary farmers and nomadic herders (Audu, 2013).

As with climate change, biodiversity loss is both a cause and consequence of land degradation, contributing to its impact on many key elements of human security. In the last hundred years, forest cover in Haiti has been reduced from 60% to 2%. This has significantly increased the country's vulnerability to both rapid-onset disasters, such as landslides and flooding, and to slow-onset environmental degradation, such as drought, soil erosion, and the loss of productive land (Williams, 2011). Without economic support and effective governance mechanisms for the sustainable management of land resources, the Haitian rural economy will continue to suffer, while the incentives for migration as an adaptation strategy increase. The influx of migrants to urban areas, coupled with extreme poverty and a lack of political institutions, has led to an alarming rise in armed gangs and crime over the past decade (Alscher, 2011).

Land degradation directly affects the health, stability, and livelihoods of approximately 1.5 billion people. It is particularly acute in the world's drylands (i.e., arid, semi-arid, and dry sub-humid areas), which account for land on which one-third of the global population lives, up to 44% of all the world's cultivated systems, and about 50% of the world's livestock breeding and feeding grounds (MA, 2005). Climate change and drought, shifts in vegetation composition, accelerated soil erosion, and other functional disturbances caused by human activities have made these working landscapes susceptible to rapid land degradation, with observed feedbacks on regional climate patterns and desertification (Ravi et al., 2010). The increased levels of stress in dryland ecosystems, combined with weak economies and poor governance systems, often render them unable to withstand the pressures of population growth.

1.1.4 GLOBAL THREATS TO HUMAN SECURITY

Scarcity and uneven distribution generate additional pressures on natural resources, which put at risk the health and well-being of affected communities, often with wider implications for regional and global security. An increasing number of fragile states are unable to cope with the human consequences of environmental degradation or to provide much-needed assistance to communities experiencing land resource scarcities. While the causes of conflict in Darfur are many and complex, regional climate variability, water scarcity, and the steady loss of fertile land were found to be important underlying factors (UNEP, 2007). This can be measured in terms of economic assistance, Fatou Bensouda, chief prosecutor of the International Criminal Court, told the UN Security Council that the 10-year conflict has cost the UN and humanitarian aid organizations more than US$10.5 billion (UN, 2013).

While acknowledging that conflicts can rarely be characterized as purely resource-driven, competition over natural resources can intensify and exacerbate existing tensions, amplifying risks for intra- and interstate conflict, rural to urban and international migration, and potentially contributing to crime or extremism. Land degradation is more often the consequence of conflicts in many regions in Africa, the Middle East, and Asia that are particularly prone to relapse, in part as a result of continued poor governance and the failure to address land and water management issues in the post-conflict period (Brinkman and Hendrix, 2011; Weinthal et al., 2014).

In recent decades, international migration flows have been characterized by people moving from Asia and Africa to North America and Europe. This is expected to continue; for the period 2010–2050, the number of migrants heading to the developed countries is likely to reach 96 million (DESA, 2013). Once again, the dryland regions are a particular focal point: they account for about 40% of the Earth's total land surface, are home to more than 2 billion people (Safriel et al., 2005), and are expanding, with one study suggesting that an estimated 135 million people will be at risk of being displaced by desertification over the coming decades due to water shortages and reduced agricultural output (GHF, 2009). In sub-Saharan Africa alone, another assessment indicates that some 60 million people are expected to move from desertified or degraded areas to northern Africa and Europe by 2020, with this figure likely to increase until 2045 (UK MOD, 2014).

As mentioned, land degradation is one of many factors that cause internal displacement, rural to urban and international migration. For instance, legal uncertainty, especially regarding land tenure, can be an important social driver for multiple migrations. This is the case in some parts of Benin, where clear legal status would help to promote sustainable land management and avoid a perpetuation of environmental degradation (Doevenspeck, 2011). The lack of rural employment opportunities is another major factor contributing to these migration trends. The UK Ministry of Defense predicts that rural areas with larger youth populations, lack of employment opportunities, and poor governance are likely to suffer from instability, which could lead to unrest or conflict; at the same time, growing urban unrest could pose major security challenges, with the potential for countrywide repercussions (UK MOD, 2014).

The impacts of land degradation not only pose serious challenges to sustainable development, but they also amplify the underlying social, economic, and political weaknesses that exist at the local and national levels. Food shortages can lead to sharp price increases and result in instability in those areas unable to cope; according to the World Bank, rising food prices have caused 51 food riots in 37 countries since 2007 (World Bank, 2014). For countries in which social safety nets or alternative sources of income are lacking, victims become refugees; in some cases, internally displaced people and forced migrants turn to crime and extremism for survival and a sense of purpose.

1.1.5 SUSTAINABLE LAND MANAGEMENT AND RESTORATION

Practical and cost-effective solutions are at hand. Recent studies show the great potential for the large-scale restoration and rehabilitation of land resources in many parts of the world. For example, it is estimated that there are over 2 billion hectares that are suitable and available for forest landscape restoration (GPFLR, 2014). This analysis spurred the Bonn Challenge to mobilize financial pledges to restore 150 million hectares of degraded and deforested land by 2020, with nearly a dozen countries now participating (IUCN, 2014). The Bonn Challenge is not a new global commitment, but rather a vehicle for realizing existing international objectives, including the Convention on Biological Diversity (CBD)'s Aichi Biodiversity Targets; the UN Framework Convention on Climate Change (UNFCCC)'s REDD+ mechanism, which goes above and beyond the Reducing Emissions from Deforestation and Forest Degradation (REDD) principles; and the 2012 UN Conference on Sustainable Development (Rio+20) goal of land degradation neutrality.

Another global study estimated that there are up to 500 million hectares of abandoned land (crop and pasture) with the potential for recovering agricultural productivity and other services through appropriate land management and restoration (Campbell et al., 2008). This represents an important but neglected opportunity, as there are many proven and cost-effective conservation and restoration practices that could benefit both people and ecosystem functioning (Stavi and Lal, 2015). A concerted global effort is clearly needed to halt and reverse land degradation, restore degraded ecosystems, and sustainably manage land resources. The priority now is to tackle the immediate challenge of how to sustainably intensify the production of food, fuel, and fiber to meet future demand without further degrading our finite land resources (Aronson and Alexander, 2013).

SLM practices, such as agroforestry and conservation agriculture, can boost yields, improve food security, and prevent future land degradation (Branca et al., 2013). SLM practices include the integrated management of crops (trees), livestock, soil, water, nutrients, biodiversity, disease, and pests in order to optimize a range of ecosystem services (Liniger et al., 2011). As water is a limiting factor in most rain-fed regions, SLM practices can be tailored to improve water management and availability, particularly for small farmers, while also improving nature's functions and the livelihoods of the rural poor (Bossio et al., 2010). In terms of human security, the overall objective of SLM is to maximize provisioning services (e.g., food, water, and energy), while enhancing the resilience of land resources and the communities that depend on them.

The term *ecosystem* or *ecological restoration* refers to the process of assisting in the recovery of an ecosystem that has been degraded, damaged, or destroyed (SER, 2004), typically as a result of human activities. The science and practice of restoration have made significant advances in the past two decades, increasing our understanding and ability to better manage ecosystems and their connectivity in the wider landscape. Many restoration efforts have the objective of improving the delivery of ecosystem services of high value in supporting human livelihoods, including carbon storage, climate and water regulation, the provision of clean water, and the maintenance of soil fertility (Holl and Aide, 2011; Rey Benayes et al., 2009; Lamb, 2011). With traditional practices, such as farmer-managed natural regeneration (agroforestry), water harvesting, and the creation of windbreaks, farmers in Niger and throughout the Sahel have experienced significant income gains due to higher production values, with little expenditure other than that of additional labor (Haglund et al., 2011).

At the UN Conference on Environment and Development in Rio de Janeiro in 1992, the global community recognized that healthy and productive ecosystems are necessary for sustainable and equitable development. This conference gave birth to Agenda 21 and the Rio Conventions—namely, the CBD, the UNFCCC, and the UN Convention to Combat Desertification (UNCCD). These resulting forms of international governance set out to address the causes and impacts of land degradation, biodiversity loss, and climate change by addressing these global environmental challenges and concerns as key security issues (UNDP, 1994). Member states reiterated these commitments and common goals with even greater urgency 20 years later at Rio+20.

1.1.6 LAND DEGRADATION NEUTRALITY

At Rio+20, world leaders agreed on the urgent need to reverse land degradation and recognized that good land management provides significant social and economic benefits. The Rio+20 outcomes document, *The Future We Want*, set out a new level of ambition: "to strive to achieve a land-degradation neutral world" (UNGA, 2012). This heralds a new commitment to a world where all nations individually strive to achieve land degradation neutrality by (i) managing land more sustainably, which would reduce the rate of degradation; and (ii) increasing the rate of restoration of degraded land, so that the two trends converge to give a zero net rate of land degradation (Grainger, 2015).

Land degradation neutrality is a hybrid lay-scientific concept that is now being developed in parallel processes, so that scientific analysis leads to findings that will better inform decision makers (Grainger, 2010; UNCCD, 2013b). In many countries, achieving land degradation neutrality will require a paradigm shift in land stewardship: from "degrade-abandon-migrate" to "protect-sustain-restore" (UNCCD, 2013a). This means cooperation among sectors, including those concerned with human security, and national sustainable development plans that embrace complementary land management options: (i) adopting and scaling up sustainable land management policies and practices in order to minimize current, and avoid future, land degradation; and (ii) rehabilitating degraded and abandoned production lands, as well as restoring degraded natural and semi-natural ecosystems that provide vital (albeit indirect) benefits to people and working landscapes.

In the case of Brazil, a country rich in terrestrial carbon and biodiversity, agricultural production is forecast to increase significantly over the next 40 years. A recent study produced the first estimate of the carrying capacity of Brazil's 115 million hectares of cultivated pasturelands, where researchers investigated if the more sustainable use of these existing production lands could meet the expected increase in demand for meat, crops, wood, and biofuels. They found that current productivity is at 32%–34% of its potential and that sustainable intensification to 49%–52% would provide an adequate supply of these goods until at least 2040, without further land or ecosystem degradation and with significant carbon sequestration benefits (Strassburg et al., 2014).

While some of the scientific knowledge and many of the organizing principles needed currently exist to support land degradation neutrality as an important element in the Sustainable Development Goals and the post-2015 development agenda, significant challenges remain, which are addressed in this book, including:

- Scaling up locally relevant tools and technologies
- Overcoming social constraints and reforming economic incentives
- Creating enduring institutions and equitable governance systems
- Improving methods and indicators for monitoring, evaluation, and communication

Moving forward, there is an immediate need to identify existing projects suitable for testing the ambition of land degradation neutrality, as well as establishing new projects at the local, community, or landscape scale to further its effectiveness in implementation and monitoring.

1.1.7 CONCLUSIONS

Land degradation is a widespread crisis, destabilizing nations and communities on a global scale. To be clear, food will be less plentiful (and thus more expensive) unless responsible land management and restoration is given priority on the international political agenda. The commitment to halt and reverse land degradation will undoubtedly feature prominently in post-2015 development and climate agendas. If this helps to bring about a transformative shift in land management policies and practices, it will certainly contribute to achieving the global priorities of ensuring human security by eradicating poverty and hunger and by reducing migration and conflict. While safeguarding human security poses numerous challenges, to which we need to react, there are many proactive solutions—early interventions on the ground—that offer cost-effective opportunities for reducing risk and vulnerability at various scales.

The human security perspective allows us to focus on shared solutions to multiple challenges and to better assess the social, economic, and environmental dimensions of sustainable development in order to leverage integrated and mutually reinforcing approaches that are context specific. In this book, the authors and editors have provided valuable insights into a number of issues related to protecting and restoring our land resources for the benefit of current and future generations. Using case studies from around the world, they have outlined the diverse strategies and multiple benefits of a holistic land-based approach to reducing some of the underlying drivers of human insecurity. They conclude that investing in these practical nature-based solutions, which transform lives and reduce vulnerability, would be cheaper and more effective in many cases than investing in walls, wars, and relief.

REFERENCES

Aaronson, M., Diessen, S., de Kermabon, Y., Long, M.B., Miklaucic, M., 2010. NATO countering the hybrid threat. Prism 2 (4), 111–124.

Alscher, S., 2011. Environmental degradation and migration on Hispaniola Island. Intl. Mig. 49, e164–e188.

Aronson, J., Alexander, S., 2013. Ecosystem restoration is now a global priority: Time to roll up our sleeves. Restor. Ecol. 21, 293–296.

Audu, S.D., 2013. Conflicts among farmers and pastoralists in northern Nigeria induced by freshwater scarcity. Dev. Country. Stud. 3 (12), 25–32.

Barnett, J., Adger, W.N., 2007. Climate change, human security, and violent conflict. Polit. Geogr. 26, 639–655.

Bernauer, T., Bohmelt, T., Koubi, V., 2012. Environmental changes and violent conflict. Environ. Res. Lett. 7, 015601.

Biswas, N.R., 2011. Is the environment a security threat? Environmental security beyond securitization. Intl. Affairs Rev. 20(1).

Bond, J., 2014. Conflict, development, and security at the agro-pastoral-wildlife nexus: A case of Laikipia County. Kenya. J. Dev. Stud. 50, 991–1008.

Bossio, D., Geheb, K., Critchley, W., 2010. Managing water by managing land: Addressing land degradation to improve water productivity and rural livelihoods. Agr. Water Manag. 97, 536–542.

Branca, G., Lipper, L., McCarthy, N., Jolejole, M.C., 2013. Food security, climate change, and sustainable land management. A review. Agron. Sustainable Dev. 33 (4), 635–650.

Brinkman, H.J., Hendrix, C.S., 2011. Food insecurity and violent conflict: Causes, consequences, and addressing the challenges. Occasional Paper 24, World Food Programme, Rome.

Brown, L., 1977. Redefining national security. Paper 14, Worldwatch Institute, Washington DC.

Campbell, J.E., Lobell, D.B., Genova, R.C., Field, C.B., 2008. The global potential of bioenergy on abandoned agriculture lands. Environ. Sci. Tech. 42, 5791–5794.

D'Odorico, P., Bhattachan, A., Davis, K.F., Ravi, S., Runyan, C.W., 2013. Global desertification: Drivers and feedbacks. Adv. Water Res. 51, 326–344.

Doevenspeck, M., 2011. The thin line between choice and flight: Environment and migration in rural Benin. Intl. Mig. 49, e50–e68.

Falk, R.A., 1972. This Endangered Planet: Prospects and Proposals for Human Survival. Vintage Books, New York.

Food and Agriculture Organization (FAO), 2011. The State of the World's Land and Water Resources for Food and Agriculture (SOLAW): Managing Systems at Risk. FAO, Rome; and Earthscan, London.

Global Humanitarian Forum (GHF), 2009. The Anatomy of a Silent Crisis. Human Impact Report: Climate Change. GHF, Geneva.

Global Partnership on Forest Landscape Restoration (GPFLR), 2014. Accessed 01.09.14 from, http://www.forestlandscaperestoration.org/.

Government of the United Kingdom, Ministry of Defence (UK MOD), 2014. Strategic Trends Programme: Global Strategic Trends—Out to 2045, fifth ed. UK MOD, London.

Grainger, A., 2010. Reducing uncertainty about hybrid lay-scientific concepts. Curr. Opin. Environ. Sustain. 2, 444–451.

Grainger, A., 2015. Is land degradation neutrality feasible in dry areas? J. Arid Environ. 112A, 14–24.

Haglund, E., Ndjeunga, J., Snook, L., Pasternak, D., 2011. Dry land tree management for improved household livelihoods: Farmer-managed natural regeneration in Niger. J. Environ. Manage. 92, 1696–1705.

Holl, K.D., Aide, T.M., 2011. When and where to actively restore ecosystems? Forest Ecol. Manage. 261, 1558–1563.

Independent Commission on Disarmament and Security Issues (ICDSI), 1982. Independent Commission on Disarmament and Security Issues (ICDSI). Simon & Schuster, New York.

Indian Council of Agricultural Research (ICAR), 2010. Degraded and Wasteland of India: Status and Spatial Distribution. ICAR, New Delhi.

International Fund for Agricultural Development (IFAD), 2013. Smallholders, Food Security and the Environment. IFAD, Rome.

International Union for the Conservation of Nature (IUCN), 2014. The Bonn Challenge and Landscape Restoration. Accessed 01.09.14 from https://www.iucn.org/about/work/programmes/forest/fp_our_work/fp_our_work_thematic/fp_our_work_flr/more_on_flr/bonn_challenge.

Kingham, R.A., May 30, 2011. Discussion paper on economic and environmental confidence- and peace-building measures and the role of the OSCE. Presented at the OSCE Chairmanship Workshop on Economic and Environmental Activities as Confidence Building Measures, Vienna.

Koubi, V., Spilker, G., Bohmelt, T., Bernauer, T., 2014. Do natural resources matter for interstate and intrastate armed conflict? J. Peace Res. 51, 227–243.

Kumssa, A., Jones, J.F., 2014. Human security issues of Somali refugees and the host community in northeastern Kenya. J. Immig. Refugee Stud. 12, 27–46.

Lamb, D., 2011. Regreening the bare hills: Tropical forest restoration in the Asia-Pacific region. Springer, Dordrecht.

Liniger, H.P., Mekdaschi Studer, R., Hauert, C., Gurtner, M., 2011. Sustainable land management in practice: Guidelines and best practices for sub-Saharan Africa. Food and Agriculture Organization of the United Nations (accessed 01.09.14)., from http://www.fao.org/3/a-i1861e.pdf.

Millennium Ecosystem Assessment (MA), 2005. Dryland systems. In: Hassan, R., Scholes, R.J., Ash, N. (Eds.), Ecosystems and Human Well-Being: Current State and Trends. Earthscan, London, pp. 623–662.

Mueller, E.N., Wainwright, J., Parsons, A.J., Turnbull, L., 2014. Land degradation in drylands: An ecogeomorphological approach. In: Mueller, E.N., Wainwright, J., Parsons, A.J., Turnbull, L. (Eds.), Patterns of Land Degradation in the Drylands: Understanding Self-Organised Ecogeomorphic Systems. Springer, Dordrecht, pp. 1–9.

North Atlantic Treaty Organization (NATO), 2009. Navigating Towards 2030: Final Report of the Allied Command Transformation (ACT) Multiple Futures Project (MFP). NATO, Brussels.

Paul, A.J., Røskaft, E., 2013. Environmental degradation and loss of traditional agriculture as two causes of conflicts in shrimp farming in southwestern coastal Bangladesh: Present status and probable solutions. Ocean Coast. Manage. 85, 19–28.

Ravi, S., Breshears, D.D., Huxman, T.E., D'Odorico, P., 2010. Land degradation in drylands: Interactions among hydrologic-aeolian erosion and vegetation dynamics. Geomorph. 116, 236–245.

Rey Benayas, J.M., Newton, A.C., Diaz, A., Bullock, J.M., 2009. Enhancement of biodiversity and ecosystem services by ecological restoration: A meta-analysis. Science 325, 1121–1124.

Safriel, U., Adeel, Z., Niemeijer, D., Puigdefabregas, J., White, R., Lal, R., Winslow, M., Ziedler, J., Prince, S., Archner, E., King, C., 2005. Dryland systems. In: Hassan, R., Scholes, R.J., Ash, N. (Eds.), Ecosystems and Human Well-Being. In: Findings of the Conditions Trends Working Group of the Millennium Ecosystem Assessment Vol. 1. Island Press, Washington DC, pp. 623–662.

Selby, J., Hofmann, C., 2014. Beyond scarcity: Rethinking water, climate change and conflict in the Sudans. Glob. Environ. Chang. 29, 360–370.

Sivakumar, M.V.K., 2007. Interactions between climate and desertification. Agr. Forest. Meteorol. 142, 143–155.

Society for Ecological Restoration (SER), 2004. SER Primer on Ecological Restoration. Society for Ecological Restoration, Science and Policy Working Group, Washington, DC.

Stavi, I., Lal, R., 2015. Achieving zero net land degradation: Challenges and opportunities. J. Arid Environ. 112, 44–51.

Sterzel, T., Lüdeke, M., Kok, M., Walther, C., Sietz, D., de Soysa, I., Lucas, P., Janssen, P., 2014. Armed conflict distribution in global drylands through the lens of a typology of socio-ecological vulnerability. Reg. Environ. Change 14, 1419–1435.

Strassburg, B.B.N., Latawiec, A.E., Barioni, L.G., Nobre, C.A., da Silva, V.P., Valentim, J.F., Vianna, M., Assadal, E.D., 2014. When enough should be enough: Improving the use of current agricultural lands could meet production demands and spare natural habitats in Brazil. Glob. Environ. Chang. 28, 84–97.

Taylor, M., 2013. Climate change, relational vulnerability, and human security: rethinking sustainable adaptation in agrarian environments. Clim. Dev. 5, 318–327.

Theisen, O.M., Gleditsch, N.P., Buhaug, H., 2013. Is climate change a driver of armed conflict? Clim. Change 117, 613–625.

United Nations (UN), 2013. Security Council hears criticism over "inaction and paralysis" in Darfur crisis. UN News Centre, December 11. Accessed 01.09.14 from, http://www.un.org/apps/news/story.asp?NewsID=46722#.VAeTbjbD-cw.

United Nations Convention to Combat Desertification (UNCCD), 1994. Elaboration of an International Convention to Combat Desertification in Countries Experiencing Serious Drought and/or Desertification, Particularly in Africa. Final text of the Convention, UN, New York.

United Nations Convention to Combat Desertification (UNCCD), 2013a. A Stronger UNCCD for a Land Degradation Neutral World. Issue Brief, UNCCD, Bonn, Germany.

United Nations Convention to Combat Desertification (UNCCD), 2013b. Follow-up to the outcomes of the United Nations Conference on Sustainable Development (Rio+20). Decision 8/COP11.

United Nations Department of Economic and Social Affairs (DESA), 2013. World Population Prospects: The 2012 Revision, Key Findings, and Advance Tables. Working Paper ESA/P/WP.227.

United Nations Development Programme (UNDP), 1994. Human Development Report. Oxford University Press, New York.

United Nations Development Programme (UNDP), 2006. Human Development Report: Beyond Scarcity: Power, Poverty, and the Global Water Crisis. UNDP, New York.

United Nations Environment Programme (UNEP), 2007. Sudan: Post-Conflict Environmental Assessment. UNEP, Nairobi.

United Nations General Assembly (UNGA), 2012. The Future We Want. Resolution 66/288.

van Schaik, L., Dinnissen, R., 2014. Terra incognita: Land degradation as underestimated threat amplifier. Netherlands Institute of International Relations. Clingendael, The Hague.

Weinthal, E., Troell, J.J., Nakayama, M. (Eds.), 2014. Water and Post-Conflict Peacebuilding. Routledge, London.

Westing, A.H., 2013. From Environmental to Comprehensive Security. Springer Briefs on Pioneers in Science and Practice 13, Springer, Dordrecht.

Williams, V.J., 2011. A case study of desertification in Haiti. J. Dev. Stud. 4 (3), 20–31.

World Bank, 2014. Food Price Watch, May 2014. World Bank Poverty Reduction and Equity Department, Washington, DC.

Zografos, C., Goulden, M.C., Kallis, G., 2014. Sources of human insecurity in the face of hydro-climatic change. Glob. Environ. Chang. 29, 327–336.

LAND DEGRADATION AND ITS IMPACT ON SECURITY

1.2

Hartmut Behrend

Bundeswehr Geoinformation Centre, German Armed Forces, Euskirchen, Germany

1.2.1 INTRODUCTION

The prevailing discourse surrounding conflicts and human insecurity focuses on religious or ethnic divides and power politics as causes. While these may be important contributing factors, other factors should not be overlooked. This section analyzes the relationship between land degradation and security, while also taking into account the impact of climate change, as climate change and land degradation are closely linked. It focuses on the effects these two factors have on human security and, thus, on the potential for violent conflicts. Climate change and, to a much lesser extent, land degradation have both long been recognized as threat multipliers by such world actors as the North Atlantic Treaty Organization (NATO) and the European Union (EU). However, their ongoing and potential contribution to insecurity is still underestimated. They are not sufficiently incorporated into mainstream means of conflict analysis, early warning mechanisms, and risk assessment. A better understanding of how land degradation and climate change contribute to insecurity could help solve or prevent conflicts. This text first scrutinizes the security-related impacts of land degradation and climate change on a global scale. It then takes a closer look at those regions where these environmental problems are projected to increase conflict potential the most. It concludes by offering mitigation strategies, of which the most important is to use sustainable agriculture to restore degraded land.

1.2.2 THE RECOGNITION OF LAND DEGRADATION AND CLIMATE CHANGE AS SECURITY INFLUENCES

On an international scientific and political level, climate change was the first global environmental problem to be addressed as a serious risk to human security and, thus, a source of influence for violent conflicts: in April 2007, the United Nations (UN) Security Council discussed this issue for the first time; a few months later, the German Advisory Council on Global Change (Wissenschaftlicher Beirat Globale für Umweltveränderungen; WBGU) submitted the first comprehensive analysis of the impacts of climate change on security (Schubert et al., 2008); the EU acknowledged climate change as a threat multiplier in its 2008 update of the European Security Strategy (European Council, 2008); and, in November 2010, NATO included climate change in its new Strategic Concept as one factor that will have a significant impact on future security within its area of interest (NATO, 2010: Article 15). Even

Land Restoration. http://dx.doi.org/10.1016/B978-0-12-801231-4.00004-5

the *5th Assessment Report* of the Intergovernmental Panel on Climate Change (IPCC, 2014a) has outlined the impacts of climate change on human security.

The impact of land degradation on human security is addressed by the UN Convention to Combat Desertification (UNCCD), which first recognized and noted the link between land degradation and insecurity in 2009 (Brauch and Spring, 2009). Land degradation's contribution to insecurity has also been underlined in many reports (e.g., Van Schaik and Dinnissen, 2014) and at the Caux Dialogue on Land and Security organized by the UN (Caux Initiatives of Change, 2014).

Land degradation poses a direct threat to security by reducing agricultural productivity through the overexploitation of land in drylands. The most important human activities causing this overexploitation are the following:

- **Exploiting wood for fuel:** The most important source of land degradation along desert margins
- **Cultivating increasingly marginal land:** The most important source in the less arid regions of the drylands
- **Irrigating for agriculture:** Leads to soil salinization
- **Livestock farming:** Mostly applied by nomads in the more arid regions of the drylands

The reduction in agricultural productivity caused by land degradation, combined with population growth, leads to increasing food insecurity. Along desert margins, huge swaths are being degraded, such that they are converted to desert at a current rate of 12 million hectares per year (Brauch and Spring, 2009, p. 2) according to preliminary estimates.

Climate change is closely linked to land degradation because land degradation causes huge carbon dioxide emission from the soil, while reduced or changing precipitation patterns increase soil vulnerability to land degradation (for example, climate change leads to an increase of torrential rains and flooding in drylands, increasing erosion, which contributes to land degradation). Climate change further poses threats to security by decreasing agricultural productivity in large parts of the tropics and subtropics.

1.2.3 CONFLICT CONSTELLATIONS

The most important conflict constellations arising from the intersection of climate change and land degradation are water scarcity, loss of land, and food insecurity (Schubert et al., 2008; Van Schaik and Dinnissen, 2014; Behrend, 2015).

1.2.3.1 WATER SCARCITY

Land degradation and climate change increase water scarcity in those regions of the world that either already suffer from physical water scarcity or are approaching it, as shown in Figure 1.2.1.[1] This image also shows those regions that suffer from economic water scarcity. Whereas physical water scarcity implies an insufficient availability of water to meet demand, economic water scarcity designates an inability by significant parts of the population to satisfy their need for water, often because they are

---

[1]CAWMA (2007, p. 11), cited in Schubert et al. (2008, p. 81).

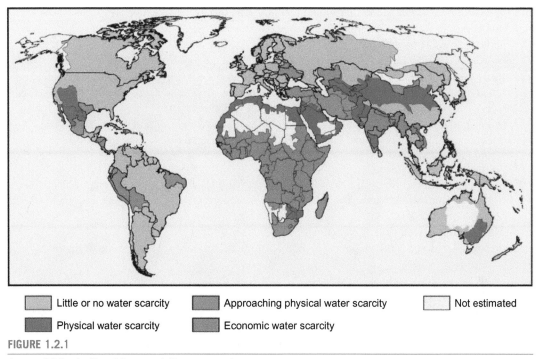

Little or no water scarcity   Approaching physical water scarcity   Not estimated

Physical water scarcity   Economic water scarcity

FIGURE 1.2.1

Areas of "physical" and "economic" water scarcity.

too poor. Economic water scarcity is particularly present in central parts of Africa and South and Southeast Asia.

The decline of freshwater availability in these drylands may lead to conflicts when the inhabitants in downstream riparian zones of a river catchment area no longer have sufficient freshwater once those in upstream riparian zones have satisfied their own needs. While until now, water conflicts have apparently enhanced the cooperation between concerned parties (i.e., between Israel and Palestine or between Egypt and Israel, as demonstrated by cooperation in the context of the Mediterranean Action Plan; Schubert et al., 2008, p. 30), the situation may reverse itself along the most exploited rivers of the world—the Nile, Euphrates, Tigris, and Jordan rivers (WWAP, 2014, Figure 2.4)—once water scarcity threatens the livelihoods of large parts of the population that rely on water from these rivers.

Land degradation increases water scarcity because it reduces the soil's water-holding capacity, thus increasing the runoff after heavy rain showers and thunderstorms. Furthermore, shrinking vegetation cover reduces the evapotranspiration from the ground and, thus, the atmospheric water vapor content. The regions with the greatest vulnerability to land degradation are shown in Figure 1.2.2 (USDA-NRCS, 1998).

Climate change contributes to water scarcity mainly in those regions where precipitation is decreasing. Figure 1.2.3 shows the changes in precipitation projected by the IPCC in its most recent assessment report (IPCC, 2013, Figure SPM 7b). Only regions that are dotted on the figure should be considered because only in these regions are the changes significant. In addition, increasing temperatures will worsen water scarcity because higher temperatures lead to higher levels of evapotranspiration.

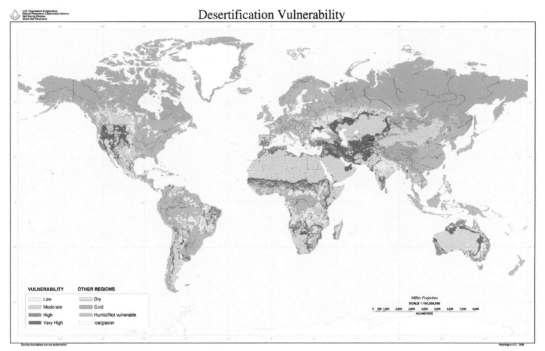

FIGURE 1.2.2

Vulnerability of the soils to land degradation.

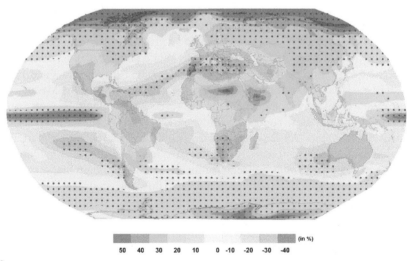

FIGURE 1.2.3

The 39-model mean annual precipitation of the projection for 2081–2100, compared to 1986–2005, calculated for a scenario according to which radiative forcing increases by 8.5 W/m² during the 21st century. The dots indicate regions where the 39-model mean is greater than two standard deviations of natural internal variability in 20-year means and where at least 90% of models agree on the sign of change.

Melting mountain glaciers, which is also caused by climate change, will leave hundreds of millions of people who rely on the glacier-fed rivers of the Himalayas and Andes with water shortages during the dry season in only a few decades (IPCC, 2007, p. 187; IPCC, 2014b, p. 241). The second harvest in the catchment area of rivers like the Indus or Ganges, thus, also will be threatened.

1.2.3.2 LOSS OF LAND

Land degradation and rising sea levels caused by climate change are reducing the agricultural land available worldwide. This threatens the livelihoods of many, especially those in poor, developing countries, as they neither have the means to increase agricultural productivity nor the resources to buy food at the increasing prices demanded by world markets. The annual desertification of 12 million hectares (as discussed in section 1.2.2) corresponds to almost 0.1% of global land area. A further 0.01% (as calculated by Nicholls, 1995, p. 13) will be inundated yearly by projected sea level rise (almost 1 meter for this century; see IPCC, 2013, p. 21). Therefore, land degradation might take about 10 times more agricultural land out of production than sea level rise caused by climate change throughout the 21st century if land degradation will continue to take agricultural land out of production at the same rate as during the last decades, underscoring the significant impact of land degradation on food insecurity.

Climate change also intensifies tropical cyclones (IPCC, 2013, p. 5), causing increasingly high floodwaters; they may already reach up to 4 m (reported from Typhoon Haiyan in the central Philippines, which occurred in November 2013; Reuters, 2013). Tropical cyclones constitute a risk for the population; this is true, in particular, within regions where they hit flat, densely populated

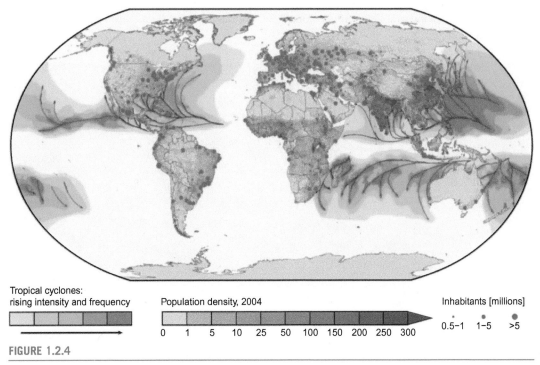

Tropical cyclones:
rising intensity and frequency

Population density, 2004

Inhabitants [millions]

0   1   5   10   25   50   100 150 200 250 300

0.5–1   1–5   >5

FIGURE 1.2.4

Tropical cyclone threat to urban agglomerations.

coastal regions. Figure 1.2.4 illustrates the intensity of tropical cyclones, as well as the population density and the location of urban agglomerations, in order to deduce where the most intensive tropical storms may hit the most densely populated, low-lying coastal regions. As shown, the coasts of Southeast Asia (Southeast China and the Philippines), the Bay of Bengal, and the Caribbean are most vulnerable (Schubert et al., 2008, p. 105).

1.2.3.3 FOOD INSECURITY

Water scarcity and the loss of land both have a huge impact on food insecurity: water scarcity because 70% of consumed water is used for agricultural irrigation (WWAP, 2014, p. 22), and loss of land because it significantly reduces the land available for agricultural purposes. In addition, climate change is reducing agricultural yields in those regions where water for agricultural production is scarce, as this increases the difficulty of irrigation (IPCC, 2014c, p. 510). This is primarily a factor in the tropics and subtropics.

In those regions where food insecurity is already an issue, and also in those where land degradation, climate change, or both already have a negative impact on food security, decreasing supply of local food supply might trigger conflicts. The regions most affected are those where large parts of the population are too poor to buy their food, instead relying on subsistence agriculture. This has been shown to be the case with many examples: from 2007 to 2008, crop failures caused by droughts in grain-exporting countries, especially Australia, led to almost a doubling of the price of internationally traded food. As a result, many relatively rich countries closed their borders to agricultural exports in order to ensure food supply for their own population. This finally led to unrest in 48 countries (Brinkmann and Hendrix, 2010, p. 2). In 2010, another rise in food prices was considered to be one of the triggers of the Arab Spring the revolutions which occurred in many Arabian countries, forced dictators to step down (like Tunisia, Egypt and Libya) and requested democracy.

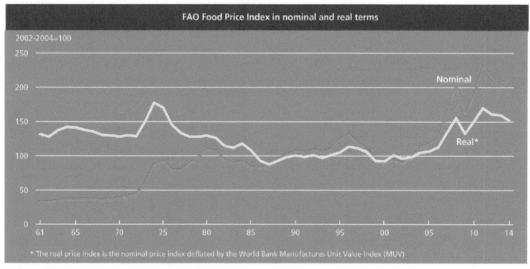

FIGURE 1.2.5

FAO Food Price Index. Note: the real price index is the inflation-adjusted World Bank Manufactures Unit value index (World Bank, 2014), quoted in FAO (2014).

The gradual increase in international food prices started as early as 2001, as can be seen in Figure 1.2.5. The main reasons for this change in prices include the increased demand caused by a growing population, in tandem with rising living standards, the increasing production of biofuels, and low agricultural investments (Brinkmann and Hendrix, 2010, p. 13, Figure 5).

As a conflict constellation, food insecurity might increase in influence considerably, as demand for food is projected to increase by about 50% by 2050 (WWAP, 2014, p. 46), while the availability of further agricultural lands is limited physically—in fact, they cannot be extended by more than 13% in the long run if agriculture is practiced sustainably (Schubert et al., 2009, p. 67).

Land grabs (i.e., foreign investments in agricultural lands) are adding to this conflict constellation, especially in poor, developing countries, because it decreases the food and freshwater resources as land cultivated by local farmers is leased or sold to foreign investors without adequate compensation being made to locals. Land degradation has a huge impact on this phenomenon because of its impact on agricultural productivity: if a relatively rich country cannot feed itself due to land degradation, and food prices on the world markets are too high, there is an incentive to import food cultivated abroad by their own nationals. Climate change also affects land grabbing, as crop failures caused by climate change and variability are a significant driver for the increase in, and volatility of, food prices on the global markets. Food importers tend to prefer cultivating food on purchased or leased agricultural land abroad and import it from there to their home countries once that becomes cheaper than buying the food on the international market. They receive the best offers mostly from poor, developing countries because they are most in need of money. In 2013, land under foreign control, which was already cultivated by the investor, had tripled, showing the rapidly increasing impact of land grabs (Land Matrix, 2014).

Biofuels, which are being deployed on a large scale to reduce transport-related greenhouse gas emissions and, thus, to mitigate climate change, are already the cause of one-third to two-thirds of all land grabs (HLPE, 2013, p. 84). Despite the fact that WBGU expert report, *Future Bioenergy and Sustainable Land Use*, revealed that the use of biofuels will hardly reduce transport-related greenhouse gas emissions (Schubert et al., 2009, p. 188), the hype about biofuels continues.

1.2.4 CONFLICT PATHWAYS

Conflicts may be intensified by the interactions of the above mentioned conflict constellations, caused by land degradation, climate change, or both: if water and food resources in a specific region decrease to the extent that local populations can no longer ensure their livelihoods and are incapable of adapting to the situation but are able to migrate to regions with sufficient freshwater and food, they will do so. However, this is often not possible for the poorest echelons of society, especially in the least developed countries. They can only draw on support that may be provided by their governments or authorities, or on external aid like development assistance or support by the World Food Programme.

In less developed countries, therefore, migration is the most frequent way of adapting to climate change and land degradation, as their situation can be improved, particularly if a family member migrates to a richer country in order to support his or her family through remittances. Often, entire families leave the country; mostly, however, people migrate within their own country or to a neighboring country, as most of them do not have the means to travel far (Schraven, 2012), and as most wish to return home as soon as conditions improve.

People primarily migrate from rural to urban areas, thus intensifying urbanization. The most important migration flux during the next few decades will be oriented toward the large cities on the coastal areas of developing countries. For South and Southeast Asia, it is projected that the number of people living in urban floodplains will 5 to 10 times higher in 2060 compared to 2000. The most important driver for that projected increase is migration. In sub-Saharan Africa, this increase is projected to be even higher (Government of the UK, Government Office for Science, 2011, p. 88).

Migration may increase the conflict potential, particularly if it is oriented from rural areas toward the cities. In urban areas, it is easier for people to join forces, which increases the risk of organized protests and riots. Moreover, migrants often have no other choice than to settle in these cities' present and future floodplains, thus being the first to be affected by storm surges. Furthermore, this migration orientation may also increase the conflict potential if the local population also suffers because of land degradation, climate change, or both.

The negative impacts of land degradation and climate change on the environment are two additional factors leading to migration and, consequently, increase the potential to weaken the governance of states. Most migrants leave their home for many reasons, including such important ones as economic, social, political, and demographic aspects of society (Government of the UK, Government Office for Science, 2011, p. 44 ff.). This may change in the future since the negative impacts of climate change, in particular, will clearly intensify.

The conflict potential within a state may increase as a consequence of land degradation, climate change, or both, and the concerned state can become fragile through an erosion of its institutional and material resources if its governance is excessively strained due to the acceptance of migrants, parts of its population are strongly affected by land degradation and climate change, but they are economically too poor to migrate, or both. In these cases, the state might no longer be able to provide the support required by these population groups for their survival since its institutional and material resources have already been overexploited.

In fragile states, the conflict potential is substantially higher than in stable states because their institutions do not have the resources to control certain regions. Within these regions, therefore, the law cannot be fully enforced. For example, terrorist groups have freedom of movement and can join forces, preventing assistance from being provided to the local population.

1.2.5 HOT SPOTS

This section takes a closer look at those regions whose security is threatened most by land degradation and climate change. Figure 1.2.6 highlights these regional hot spots.

1.2.5.1 LAND DEGRADATION IN THE SAHEL

Since at least the 1930s, the Sahara Desert has been spreading southward into the Sahel by 1–3 km/year (calculated from data in UNEP, 2007, p. 9), due to the overexploitation of land. Since the early 1970s, devastating droughts, which hit the entire Sahel, added to this overexploitation. In recent decades, average precipitation amounts have started increasing again and are projected to continue to increase over the next decades, but with huge variations from year to year (UNEP, 2011, p. 35).

The essential conflict constellation in the Sahel revolves around control of agricultural land between nomads in the north and the sedentary farmers settling further to the south. Due to increasing land degradation, the nomads on the southern border of the Sahara continue to move farther south during the dry season in order to ensure their livelihoods; thus, they use the land claimed by the sedentary

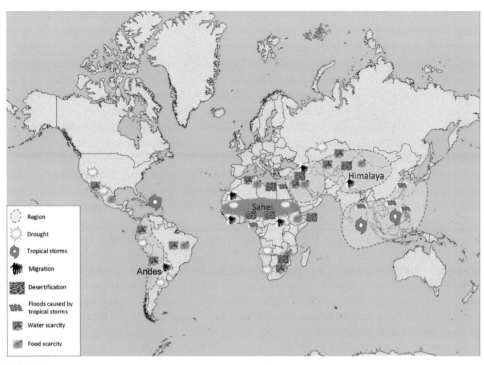

FIGURE 1.2.6

Hot spots of the security-related impacts of climate change and land degradation.

farmers more and more, provoking conflicts. Other important factors that contribute to the development of conflicts in this region include the following:

- The weak governance of almost all the local states (Haken et al., 2013).
- Significant food insecurity: the whole region is still suffering from the repercussions of the drought-induced famines of 2005, 2008, 2010, and 2012. In 2012, a record 18 million people were food insecure, and it was only possible to prevent human disaster thanks to international aid. In early 2014, as many as 20 million people were threatened by food insecurity, despite good harvests (European Commission, 2014, p. 2).
- Land grabs, particularly in Sudan, South Sudan, Mali, and Senegal.
- Migration, as a large influx of migrants is currently moving from the Sahel toward the Gulf of Guinea.

It is highly probable that land degradation will trigger new conflicts in the Sahel in the near future.

1.2.5.2 DROUGHTS IN THE MIDDLE EAST

Large parts of the Middle East are affected by hydrological water scarcity and, recently, precipitation has been decreasing in wide regions, while temperatures have been rising, especially in the Levant region (IPCC, 2013, pp. 194, 203). The entire region is experiencing severe land degradation (see Figure 1.2.2), while agricultural yields are declining. In many areas, the aquifers are also gradually becoming overexploited by irrigation for agriculture (Brown, 2012, p. 60); the large rivers in the region (i.e., the Jordan, Euphrates, and Tigris) are overexploited (Brown, 2012, p. 62).

Recently, the region has been (and continues to be) affected by several conflicts, primarily the wars in Iraq and Syria. The decline in agricultural yields caused by a severe drought that started in 2006 is one of the major triggers for the Syrian civil war (Femia and Werrell, 2013).

The IPCC projects that precipitation will continue to decline over the next few decades (see Figure 1.2.3). This will intensify land degradation, further decreasing both agricultural productivity and yields. Reduced water availability is, therefore, likely to threaten food production significantly. This may fuel the outbreak of additional conflicts in the near future, especially in the region's poorer countries, such as Yemen or Iraq, which currently do not have enough foreign currency to buy or lease agricultural lands in other countries. Moreover, the dams on the upper Euphrates and Tigris rivers in Turkey, as well as the Israeli occupation of Palestine, may very well lead to conflicts over water resources.

1.2.5.3 LAND DEGRADATION AND WATER SCARCITY IN CENTRAL ASIA

Central Asia is characterized by low precipitation, a very high vulnerability to land degradation, and extensive hydrological water scarcity. The hot spot in this region is the Aral Sea, the volume of which has been continuously decreasing due to the immense water consumption of its bordering states. Most of the water is used for cotton cultivation in Kazakhstan, Uzbekistan, and Turkmenistan, where it is drained from the lakes' tributaries.

Since most central Asian states (with the exception of Kazakhstan) are at a relatively low level of development, while their consumption of resources increases continuously due to population growth, land degradation may also intensify conflicts if no suitable countermeasures are taken, primarily with regard to improvements in agriculture and water management.

1.2.5.4 TROPICAL CYCLONES IN SOUTH ASIA AND SOUTHEAST ASIA

Bangladesh, southeast China, and the Philippines are affected, in particular, by tropical cyclones (Schubert et al., 2008, p. 112), which are accompanied by storm surges and strike densely populated coastal areas. Due to a number of factors (including rising sea levels, an increasing number of migrants settling in urban floodplain regions, sinking urban land due to groundwater overexploitation, and increasingly high storm surges from tropical cyclones), such storm surges destroy the resources on which ever more people depend. Typhoon Haiyan in November 2013 is the most recent example of what these countries may expect in the future.

Special attention should be given to Bangladesh, as it is almost at sea level and has one of the highest population densities worldwide. Already, a large flow of migrants has moved from Bangladesh to India, to the extent that India felt it necessary to build a fence to keep migrants out (Kleber and Paskal, 2012, p. 207).

1.2.6 CONCLUSIONS AND RECOMMENDATIONS

Simply stated, the negative impacts of land degradation and climate change are fueled by population growth and rising living standards around the world. These two factors also contribute to the decreasing availability of food and water, particularly in the drylands. Where water is limited (especially when this coincides with insufficient agricultural land), food becomes harder to grow. While the majority of consumed water is earmarked for food cultivation, primarily in irrigated agriculture in the drylands, land degradation and sea level rise are reducing agricultural land (and affecting fresh water supply by, for example, salinating groundwater stores). In poor, developing countries, arable land continues to become scarcer due to land grabs, which are significantly fueled by crop cultivation for biofuels.

Land degradation and climate change are among the greatest challenges of our time: they have already changed environmental conditions in many regions, especially in the four discussed hot spot regions, whose environment-related conflict potential continues to grow.

During the last few decades, land degradation has already reached enormous proportions, while the impacts of climate change have become more noticeable only recently. They will, however, continue to intensify over the decades to come. While it seems to be possible to restore land and to thus mitigate land degradation through sustainable farming and water management, measures aiming at the reduction of greenhouse gas emission will take several decades before affecting the global climate.

The key measures for addressing land degradation and climate change simultaneously are sustainable farming and water management techniques (most notably agroforestry), which is the intentional combination of crops or animals and trees on agricultural parcels and is particularly useful in semiarid regions. Research by World Agroforestry Centre scientists (Garrithy et al., 2010) shows that agroforestry supports the mitigation of climate change, the adaptation to climate change, and the improvement of food security, in addition to mitigating land degradation. Thus, the deployment of agroforestry furnishes a quadruple-win situation.

The trees planted in agroforestry add organic carbon and fertility to the soil through the nutrients that their roots pump up and their leaves deposit. The addition of carbon to the soil is a powerful means for mitigating climate change, as soil can absorb huge amounts of carbon dioxide from the atmosphere and transform it, chemically, into soil carbon. They also buffer fields against droughts and floods and provide cheap and effective erosion control, thus also strongly supporting climate change adaptation. Through these effects, trees can help farmers generate cereal yields of several tons per hectare on plots that, without trees, would not have any significant yields (Sileshi et al., 2012), thereby also improving food security considerably.

Because of these advantages, agroforestry is spreading in the Sahel and East and southern Africa. In Niger, 5 million hectares of faidherbia parklands have boosted yields by about 500,000 tons per year (Garrithy et al., 2010, p. 200). In Zambia, conservation agriculture with trees (CAWT) is spreading among commercial and smallholder farms. And Ethiopian prime minister Meles Zenawi committed his country at the 2011 Durban climate conference to planting 100 million fertilizer trees by 2015. But there are still many obstacles to the spread of agroforestry, including the lack of agroforestry expertise in agriculture ministries, weak rural extension services, inadequate tree seedling supplies, unclear land and tree tenures, relatively scarce funding, and, most of all, widespread farmer ignorance of agroforestry technologies and their potential.

Other sustainable farming and water management techniques that can efficiently mitigate land degradation are holistic grazing management of grasslands, mulching, no-till agriculture, and crop rotations. Further effective measures for mitigating land degradation and climate change are solar stoves for cooking, renewable energy for power generation (IRENA, 2012), and off-grid power generation and the use of minigrids, all of which have also been outlined as measures for improving energy security in section 1.3 of this book. These measures replace the demand for fuel wood because they provide rural communities with alternative sources for cooking and heating, such as the following:

- Solar stoves are a simple technology that can easily be deployed as a substitution for fuel wood in the short run.
- Off-grid and mini-grid production of renewable energy, mainly by photovoltaics and wind energy, can substitute for diesel generators and provide electricity for cooking and heating. These grids require a bit more knowledge to establish, maintain, and service, however. In addition, they require batteries to run during night or when there is no wind.
- The deployment and maintenance of an entire grid for renewable energy require much more funding and technological knowledge.

Due to its high conflict potential, the visibility of land degradation as a conflict amplifier should be vigorously stressed at the UN level. The role and importance of soils should be taken into consideration, regarding the formulation of the UN Sustainable Development Goals, which are scheduled to replace the Millennium Development Goals at a UN conference in September 2015, as well as an important subject of the conferences of Global Soil Week, an event established in 2012 that takes place once a year since then. Soils' role as carbon sinks should also be given high priority in the UN Framework Convention on Climate Change (UNFCCC) negotiations.

Land restoration in poor, developing countries is currently being offered relatively large levels of funding by the Least Developed Countries Fund under the UNFCCC, which funds the measures proposed by the National Adaptation Programs of Action for Least Developed Countries. A huge part of the actions contained in these plans deal with sustainable agriculture and, thus, land restoration. Starting in 2020, industrial countries have committed to provide developing countries with US$100 million per year for measures on climate change, about half of which shall be aimed at adaptation to climate change, which could fund land restoration efforts.

The impacts of land degradation and climate change on human security are increasing quickly and will continue to do so in the decades to come, if no countermeasures are taken immediately. They particularly affect the availability of food and water in drylands, which already experience water scarcity, food insecurity, and even violent conflicts in many places. Therefore, measures for mitigating land degradation and climate change, especially sustainable measures for agriculture and water management in the drylands, should be strongly supported.

REFERENCES

Behrend, H., 2015. Why Europe should care more about environmental degradation triggering insecurity. Global Aff. 1, 67–79.

Brauch, H.G., Spring, U.O., 2009. Securitizing the ground, grounding security. UNCCD Issue Paper 2. Secretariat of UNCCD, Bonn.

Brinkmann, H.-J., Hendrix, C., 2010. World Development Report 2011 Background Paper: Food Insecurity and Conflict: Applying the WDR Framework. World Bank, Washington, DC.

Brown, L., 2012. Full Planet, Empty Plates: The New Geopolitics of Food Scarcity. W. W. Norton & Company, New York.

Caux Initiatives of Change, 2014. Dialogue on land and security. Caux Initiatives of Change http://caux.iofc.org/en/caux-dialogue-land-and-security-0.

Comprehensive Assessment of Water Management in Agriculture (CAWMA), 2007. Water for Food, Water for Life: A Comprehensive Assessment of Water Management in Agriculture. Earthscan, London, and International Water Management Institute, Colombo, Sri Lanka.

European Commission, 2014. ECHO Factsheet: Sahel: Food and Nutrition Crisis. EC Humanitarian Aid and Civil Protection, Brussels.

European Council, 2008. Report on the Implementation of the European Security Strategy: Providing Security in a Changing World. Document S 407/08. European Council, Brussels.

Femia, F., Werrell, C., 2013. Climate change before and after the Arab Awakening: The cases of Syria and Libya. In: Werrell, C.E., Femia, F. (Eds.), The Arabian Spring and Climate Change: A Climate and Security Correlations Series. Centre for American Progress, Washington, DC, pp. 23–32.

Food and Agriculture Organization (FAO), 2014. FAO Food-Price Index. World Food Situation, http://www.fao.org/worldfoodsituation/foodpricesindex/en/ (accessed 17.8.2014).

Garrithy, D.P., Akinnifesi, F.K., Ajayi, O.C., Weldesemayat, S.G., Mowo, J., Kalinganire, A., Larwanou, M., Bayala, J., 2010. Evergreen agriculture: A robust approach to sustainable food security in Africa. Food Secur. 2 (3), 197–214.

Government of the UK, Government Office for Science, 2011. Foresight: Migration and Global Environmental Change. The Government Office for Science, London.

Haken, N., Messner, J.J., Hendry, K., Taft, P., Lawrence, K., Umaña, F., 2013. Failed States Index IX 2013. The Fund for Peace, Washington, DC.

High-Level Panel of Experts on Food Security and Nutrition (HLPE), 2013. Biofuels and food security. Committee on World Food Security, Rome.

Intergovernmental Panel on Climate Change (IPCC), 2007. Climate Change 2007: Impacts, Adaptation and Vulnerability. Contribution of Working Group II to the Fourth Assessment Report of the Intergovernmental Panel on Climate Change. In: Parry, M.L. et al., (Eds.), Cambridge University Press, Cambridge, UK.

Intergovernmental Panel on Climate Change (IPCC), 2013. Climate Change 2013: The Physical Science Basis. Summary for Policymakers. Contribution of Working Group I to the Fifth Assessment Report of the Intergovernmental Panel on Climate Change. In: Stocker, T.F. et al., (Eds.), Cambridge University Press, Cambridge, UK and New York.

Intergovernmental Panel on Climate Change (IPCC), 2014a. Chapter 12: Human security. In: Field, C.B. et al., (Eds.), p. 755–792. Climate Change 2014: Impacts, Adaptation and Vulnerability. Contribution of Working Group II to the Fifth Assessment Report of the Intergovernmental Panel on Climate Change. Cambridge University Press, Cambridge, UK, and New York.

Intergovernmental Panel on Climate Change (IPCC), 2014b. Chapter 3: Freshwater resources. In: Field, C.B. et al., (Eds.), p. 229–269. Climate Change 2014: Impacts, Adaptation and Vulnerability. Contribution of Working Group II to the Fifth Assessment Report of the Intergovernmental Panel on Climate Change. Cambridge University Press, Cambridge, UK, and New York.

Intergovernmental Panel on Climate Change (IPCC), 2014c. Chapter 7: Food security and food production systems. In: Field, C.B. et al., (Eds.), p. 485–533. Cambridge University Press, Cambridge, UK, and New York.

International Renewable Energy Agency (IRENA), 2012. Prospects for the African Power Sector. IRENA Headquarters, Abu Dhabi.

Kleber, C., Paskal, C., 2012. Spielball Erde: Machtkämpfe und Klimawandel. C. Bertelsmann, Munich.

Land Matrix, 2014. Land Matrix Newsletter–January 2014. Land Matrix, published online; http://www.landmatrix.org/en/ (accessed 17.07.2015).

Nicholls, R., 1995. Synthesis of vulnerability analysis studies. In: Beukenkamp, P. et al., (Eds.), Proceedings of World Coast Conference 1993I, Coastal Zone Management Centre, The Hague, pp. 181–218.

North Atlantic Treaty Organization (NATO), 2010. Strategic Concept for the Defence and Security of the Members of the North Atlantic Treaty Organisation. NATO, Lisbon.

Reuters, 2013. "Massive Destruction" as Typhoon Haiyan kills at least 1,200 in Philippines. Reuters, November 9, 2013; http://www.trust.org/item/20131109132543-2fgb9.

Schraven, B., 2012. Policy Perspectives for Environmentally Induced Migration. Network Migration in Europe, http://migrationeducation.de/fileadmin/uploads/SCHRAVEN_01.pdf (accessed 17.07.2015).

Schubert, R., et al., 2008. World in Transition: Climate Change as a Security Risk. WBGU, Berlin, and Earthscan, London.

Schubert, R., et al., 2009. World in Transition: Future Bioenergy and Sustainable Land Use. WBGU, Berlin, and Earthscan, London.

Sileshi, G.W., Debushob, L.K., Akinnifesic, F.K., 2012. Can integration of legume trees increase yield stability in rainfed maize cropping systems in southern Africa? Agr. J. 104 (5), 1392–1398.

United Nations Environment Programme (UNEP), 2007. Sudan: Post-conflict Environmental Assessment. UNEP, Nairobi, Kenya.

United Nations Environment Programme (UNEP), 2011. Livelihood Security. Climate Change, Migration, and Conflict in the Sahel. UNEP, Nairobi, Kenya.

United Nations World Water Assessment Programme (WWAP), 2014. The United Nations World Water Development Report 2014: Water and Energy. UNESCO, Paris.

United States Department of Agriculture—Natural Resource Conservation Service Soils (USDA-NRCS), 1998. Global Desertification Vulnerability Map. Soil Science Division, World Soil Resources, Washington, DC.

Van Schaik, L., Dinnissen, R., 2014. Terra Incognita: Land Degradation as Underestimated Threat Multiplier. Netherlands Institute of International Relations Clingendael, The Hague, the Netherlands.

World Bank, 2014. Manufactures Unit Value Index (MUV). Prospects. http://econ.worldbank.org/WBSITE/EXTERNAL/EXTDEC/EXTDECPROSPECTS/0, contentMDK:20587651~menuPK:5962952~pagePK:64165401~piPK:64165026~theSitePK:476883~isCURL:Y,00.html. (accessed 12.12.2014).

(EM)POWERING PEOPLE: RECONCILING ENERGY SECURITY AND LAND-USE MANAGEMENT IN THE SUDANO-SAHELIAN REGION

1.3

Stela Nenova[1], Hartmut Behrend[2]

Energy and Environmental Policy Consultant[1]
Bundeswehr Geoinformation Centre, German Armed Forces, Euskirchen, Germany[2]

1.3.1 INTRODUCTION

"Those who have no access to modern energy suffer from the most extreme form of energy insecurity" (IEA, 2014a). Energy access is key to promoting inclusive economic growth, social equity, resilience, and environmental sustainability without compromising the livelihoods of current and future generations. The only way to create and ensure fair access to resources and opportunities for all people is to ensure that "the exploitation of resources, the direction of investments, the orientation of technological development, and institutional change are made consistent with future as well as present needs" (Brundtland Report, 1987). Despite considerable progress, nearly 1 billion people nowadays are still living in extreme poverty around the world, without access to energy and water resources, adequate healthcare, or quality education (United Nations, 2015). In the least developed countries, such as countries in the Sudano-Sahelian region, the gap between people who can and cannot access such opportunities is even greater, and they are often the worst hit by food insecurity, economic shocks, and natural disasters due to climate change. Energy access, and thus energy security, are prerequisites for breaking this vicious circle and for fostering equitable and sustainable development.

Furthermore, the demand for food, feed, and biomass for energy production has been growing and will continue to grow in the future, while land resources on a global level remain limited. At the same time, land degradation processes in the form of soil erosion, secondary salinization, or desertification among others, have decreased the productivity of land and its capacity to provide adequately ecosystem goods and services (ELD, 2011) increasing pressure further on both planetary resource boundaries and human livelihoods.

This study explores the links between energy security from the physical, economic, social, technological, and environmental perspectives and the implications of local and global patterns of energy supply and demand on the processes of land-use management and land degradation in the Sudano-Sahelian region. It argues that decoupling the use of energy resources from the use of land resources is crucial for

Land Restoration. http://dx.doi.org/10.1016/B978-0-12-801231-4.00005-7

improving both energy security and land use and management practices in the Sudano-Sahelian region. Breaking the links between demographic and economic growth pressures, reflected in increasing energy demand and consumption, and land resource pressures and impacts will significantly contribute to mitigating and possibly reversing land degradation, while maintaining and improving economic welfare and human well-being. Findings support the argument that improvements in energy security through modern and renewable energy technologies and sustainable land-use management practices, such as agroforestry, can mitigate land degradation and contribute to the improvement of human security by decreasing vulnerability of energy systems, improving energy access to the poor, and supporting human development and well-being. Shifting energy production and use methods and energy demand patterns into more sustainable ones will have a positive impact on local land resources and reduce land degradation. Solutions that value appropriately natural resources and ensure the sustainable management of these resources are key to enhancing energy security, reducing energy poverty, and decoupling energy use from land use. These solutions must have the potential to reconcile an ever-increasing demand for energy with economic development without displacing food production for local communities by biofuel production and without causing overexploitation of land resources. At the same time, halting land degradation processes and promoting land restoration practices through agroforestry in developing countries can have a significant positive impact on the energy security in rural areas and consequently on livelihoods. In addition, practices aimed at restoring land create local employment and sustainable energy access for these communities, thus contributing to human security and well-being.

The study focuses on the Sudano-Sahelian region, which contains many of the least developed countries in Africa. A significant part of the areas of these countries is covered by dry land. The majority of the agricultural land of the countries, however, lies in semiarid regions, and the majority of the populations of these countries still rely on agriculture as the main sector of employment (UNCCD, 2009). In this region, primary energy production and consumption patterns are directly linked to the land and its ecosystem services for the supply of traditional biomass. The Sudano-Sahelian region (see Figure 1.3.1) has been increasingly plagued by land degradation, and this process has lead to a southward extension of the Sahara desert in the last several decades (UNEP, 2007, p. 9), increasingly threatening food, water, and energy access, and ultimately human security. In addition, the region is still suffering from the repercussions of the drought-induced famines of 2005, 2008, 2010, and 2012, and this further intensifies demand pressures on domestic land ecosystem properties, as population and economic growth needs have to be satisfied within the limited natural resources in the region. In early 2014, 20 million people were food insecure, in spite of good harvests in the Sahel region (European Commission, 2014, p. 2). Countries in the Sudano-Sahelian region share several common characteristics related to energy security: a significant part of the population is dependent on traditional biomass as a primary energy source; the region is highly prone to land degradation due to both natural causes and human-induced factors (as discussed in section 1.2 of this chapter); energy production practices are largely unsustainable; demographic growth is very high, while the technological base is still very weak.

Throughout the Sudano-Sahelian region, the primary source of energy is traditional biomass, which is used mostly for cooking, heating, and lighting. Local populations rely heavily on the forests and shrubs in their surroundings to meet their energy demand and for subsistence. In the longer term, fuelwood extraction affects strongly the structure and composition of forests, leading to deforestation, degradation of the surrounding vegetation cover, and eventually land degradation, as degrading soils lose their capacity to provide adequate ecosystem services over time. In the long run, the overextraction of biomass required for satisfying the needs of the local population reduces even further if soils are continuously overused (cf. UNEP, 2007, p. 200). Traditional biomass is the most affordable and accessible local source of energy for people in the Sudano-Sahelian region, as they often have no access to other

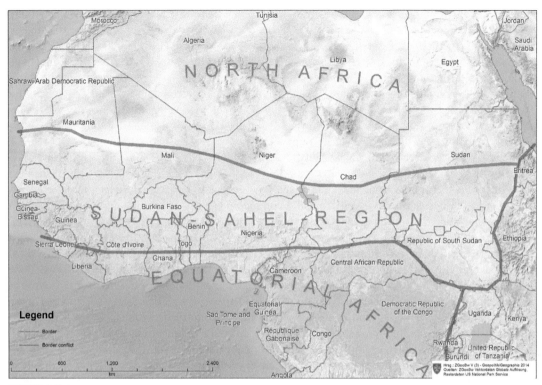

FIGURE 1.3.1

Map of the Sudano-Sahelian region.

energy sources. However, this practice leads to a vicious circle, where the unsustainable use of biomass exacerbates land degradation processes and induces further energy scarcity as biomass resources decrease over time. Ultimately, this process is leading to ever-higher energy insecurity for the very poor and isolated communities in the region. Biomass in itself has a great potential to replace fossil fuels as a source of renewable energy, reducing emissions and improving energy self-reliance and security, but only if the production of biomass avoids any land-use change. However, if biomass production and exploitation practices do not take into account environmental factors such as sustainable use of resources and land, they can lead to land degradation and threaten energy security for local populations, which largely depend upon this land for wood and charcoal to fulfill their basic energy needs.

Pervasive unsustainable energy production and land-use practices in the Sudano-Sahelian region threaten significantly not only current development opportunities for local people, but also future generations' livelihoods. Major drivers for land degradation can be both climate-related extreme events like droughts and heavy precipitation or human-induced factors such as deforestation, fuelwood extraction, and the overexploitation of land by overgrazing or unsustainable agricultural practices. It is estimated that fuelwood extraction has contributed to roughly 7% of total land degradation on a global level (UNEP, 2002, p. 64)[1]. The last assessment of degraded lands shows that during the time period 1981–2002, 24% of the land worldwide was degrading, while at the same time, only 16% was

---

[1]The only data available are from 1996 and were estimated by the Food and Agriculture Organization (FAO).

improving (Bai et al., 2008). Data for the Sudano-Sahelian region (countries in Tables 1.3.1–1.3.4) shows that almost 20% of the population within the countries analyzed in this article was living on degraded land in 2010 (cf. Table 1.2.4). The contribution of the traditional use of biomass to land degradation is largest within the dry savannah on the margins of the Sahara, which are also the most vulnerable to desertification areas (cf. USDA-NRCS, 1998) and around urban areas (cf. UNEP, 2007, p. 200). Land degradation reduces both the agricultural productivity and soils' holding capacity for water, which over time leads to decreasing agricultural production, while demand for it is increasing as population grows. With an annual average population growth rate of almost 3% throughout the entire region (cf. Table 1.3.4), the holding capacity of the soils for water is tremendously important because in most parts of the region, a significant portion of water is consumed by agriculture. In Senegal and Mauritania, it is above 90%, and in Sudan and Mali, it is above 95% (FAO, 2014a). Over time, if the negative trends of biomass extraction are not reversed, land degradation processes directly threaten energy access opportunities and food security for rural communities.

The other dimension of land-use management assessed in this chapter is the unsustainable cultivation of biofuel feedstock as a potential driver for land degradation in the Sudano-Sahelian region, because it has conflicting implications for energy security from a domestic and international point of view. Very often, large-scale foreign investments in land in the region, also known as "land grabbing," are targeted toward the cultivation of land for the production of biofuel crops. The use of biofuels may improve energy security in some parts of the world by decreasing reliance on fossil fuels (for example, in some industrialized, oil-importing countries), while in other parts of the world, the impact of biofuel production on local arable land can potentially cause or even exacerbate land degradation (for example, in some poor, developing countries, such as those in the Sudano-Sahelian region). Governments' needs for revenues are a main driver for renting out or selling agricultural land to foreign investors. In the Sudano-Sahelian region, for example, already 1%– 3% of agricultural land has been subject to transactions with foreign investors for either the sale or the lease of the land for the cultivation of biofuel feedstock (cf. Table 1.3.2). On the other hand, land grabbing seems to contribute to land degradation and internal displacement because farmers who previously cultivated their crops on that land will have to cultivate them somewhere else in the future, very often on land that might be less fertile and thus more vulnerable to land degradation. In addition, unsustainable land management practices on any of the land that has switched hands can exacerbate land degradation in the long run and threaten both energy and food security in the region.

The rest of this study is organized as follows: Section 1.3.2 provides a general theoretical framework for the paradigm shifts of energy security and land degradation; section 1.3.3 takes a closer look at the current energy production and consumption patterns in the Sudano-Sahelian region and discusses links between energy security and land degradation, while section 1.3.4 takes a closer look at future energy scenarios and analyzes the potential impacts of energy security objectives and choices on land resources, taking into account population growth and resource demand, changes in wealth within rural regions, and the future trends for foreign land investments for biofuels of the countries assessed here. Section 1.3.5 looks at two case studies, Sudan and Mali, as examples of the interlinkages of energy security and land degradation, how they can be managed to achieve sustainable outcomes and how pressures on land, energy, and resource scarcity affect human security and conflict prevention. Section 1.3.6 proposes methods for mitigating land degradation and improving energy security at the same time. Finally, section 1.3.7 concludes with some recommendations for addressing current and future challenges.

Table 1.3.1 Energy-Related Indicators

Country	Access to Electricity 2011** (% population)	Primary Energy Use 2011**** kilogram oil equivalent per capita	% Biomass	Households Cooking with Biomass 2010 (% of all) ******
Benin	28	388	56.2	94
Burkina Faso	13			95
Chad	4*			98
Cote d'Ivoire	50	571	77.6	79
Gambia	19			95
Ghana	72	433	57	84
Guinea	19			96
Guinea-Bissau	15			98
Mali	27			98
Mauritania	30*			63
Niger	9			94
Nigeria	48	719	82.2	75
Senegal	57	258	45.8	56
Sierra-Leone	15			98
Sudan*****	29*	460	67.1	93
Togo	27	405	82.1	95
Entire region (average)	**38***	**454**		**78**

*Source: UNDP/WHO (2009) for 2005.
**Source REN21 (2014), p. 21.
***The number was calculated using the population data from 2009.
****Source for total primary energy use from World Bank (2015). Data for the entire region derived from primary energy use data and population data for 2011.
*****in its borders before 2011.
******Source: Ren21 (2014), p. 30.

1.3.2 PARADIGM SHIFTS: ENERGY SECURITY AND LAND DEGRADATION

1.3.2.1 THE PARADIGM OF ENERGY SECURITY

The polysemic nature of the term *energy security* calls for a definition and analysis of energy security that will vary according to the needs of a country (energy importing, producing, or exporting), the various energy sources and carriers, the actors that shape energy markets (state or nonstate ones), the regions, and the types of consumers. Energy security, therefore, will mean different things to developed, energy-exporting, or energy-importing, developing or least-developed countries. Naturally, answers to questions such as who the target beneficiaries of energy security are, what values and principles should be protected, from what threats and risks, at what costs, how, and within what time frame, are important to consider for the assessment of energy security (Cherp and Jewell, 2014) and for the development of adequate mechanisms to improve resilience and lower risks for energy systems.

Table 1.3.2 Indicators for Land Grabbing***

Country	Agricultural Land 2012* (1000 ha)	Foreign Land Deals Concluded (% of agricultural land)	Foreign Land Deals Concluded for Biofuels (% of agricultural land)
Benin	3150	1	1
Burkina Faso	6070	3	0
Chad	4930	0	0
Cote d'Ivoire	7400	0	0
Gambia	445	7	7
Ghana	7400	12	4–9
Guinea	3700	23	20
Guinea-Bissau	550	0	0
Mali	7010	2	1–2
Mauritania	410	0	0
Niger	16,000	0	0
Nigeria	36,700	1	0
Senegal	3420	8	6-7
Sierra-Leone	1900	68	12–52
Sudan **	21,070	9	0–4
Togo	2850	0	0
Entire region	**123,005**	**5**	**1–3**

*Source: FAOSTAT (2015).
**in its current borders because for South Sudan, no data of agricultural area are available.
***Extracted from Landmatrix (2014) on November 6, 2014. Data only include foreign land grabbing where deals have already been concluded with foreign investors. Land grabbing for forest conservation, tourism, wood and fiber, and livestock farming is also not included because it does not change the arable land available. The data for biofuels show huge uncertainty ranges because often the Landmatrix does not disaggregate between land used for harvesting the feedstock of biofuels and land used for other purposes. In addition, it is assumed that where palm oil is harvested on land but the reason for it is not indicated, the crop has been grown as feedstock for biofuels.

Energy security is most often defined as "the reliable and adequate supply of energy at reasonable prices" (Bielecki, 2002), which signifies the presence of uninterrupted supplies of energy for the economy. Energy supply, demand, and price dynamics, especially those related to fossil fuels, are increasingly shaped within a global and interconnected market where geopolitical, financial, weather, supply-chain vulnerabilities, conflicts, and other uncertainties may significantly affect the availability and the pricing of fossil fuel energy commodities. Prices become a product of both national and international dependencies/trends and may be subject to the volatility and risks associated with price fluctuations at the global level. To be reasonable, energy prices should be cost-based and dependent on the supply and demand balance of energy markets (Bielecki, 2002), which means that energy security depends upon the availability of adequate physical supplies to cover demand.

Energy security is further defined "as avoiding the loss of economic welfare that may occur as a result of a change in the price or availability of energy" (Bohi and Toman, 1996, p. 1). The International Energy Agency (IEA, 1995, p. 23) stated that "energy security is simply another way of avoiding market distortions" because "smoothly functioning international energy markets' will deliver "a secure—adequate, affordable and reliable—supply of energy" (IEA, 2002, p. 3) which makes the role of markets the central factor in ensuring energy security. "Energy security always consists of both a physical unavailability component and a price component, (but) the relative importance of these depends on market structure, and in particular the extent to which prices are set competitively or not" (IEA, 2007, p. 32). While energy security as a market product is determined by the efficient functioning of markets through physical supply and price mechanisms, the role of governments is key to shaping supply market dynamics through regulatory policy and contractual prices (and thus energy security) at both the national and global levels. Appropriate policy decisions, strategies, and regulatory frameworks can guarantee the proper functioning of energy markets and mitigate risks related to the unavailability of energy supplies, physical disruptions, or cushion price shocks to protect domestic economies if and when such shocks occur.

A more inclusive definition of energy security in the European Commission's green paper "Towards a European Strategy for the Security of Energy Supply" noted that "energy supply security must be geared to ensuring, for the well-being of its citizens and the proper functioning of the economy, the uninterrupted physical availability of energy products on the market, at a price which is affordable for all consumers (private and industrial), while respecting environmental concerns and looking towards sustainable development" (European Commission, 2000, pp. 1–2). "Security of supply does not seek to maximise energy self-sufficiency or to minimise dependence, but aims to reduce the risks linked to such dependence" through "balancing between and diversifying of the various sources of supply (by product and by geographical region)" (European Commission, 2000, pp. 1–2). While this definition mostly expresses the long-term objective of the European Union (EU) to reduce its own dependence on imported fossil fuels, it can be equally applicable to any country that seeks to balance energy demand and supply with environmental, social, and economic constraints, while mitigating the risks related to internal or external shocks and vulnerabilities of the energy system.

Sovereign nations play a key role in shaping market function, as the institutions and processes in which markets function are continuously reshaped through the interconnections of political and policy mechanisms. Therefore, a purely market-focused definition of energy security is not enough to address energy security aspects such as affordability and energy poverty, sustainability, changes in demand and supply, and various policy-making mechanisms. Energy security is intrinsically linked to both the physical security of supply (network infrastructure, strategic stocks, supplies' diversification), as well as the regulatory and policy consistency in view of the specific challenges that have to be addressed by governmental decisions regarding the energy mix and policies (reducing demand, diversifying supply sources and routes, stimulating investments in clean energy, and ensuring access to energy supplies for all citizens). National governments have a significant role to play in ensuring energy security, both from the perspective of balancing dependence on energy imports or exports and geopolitical developments and in fostering greater reliance on competition in energy markets and in designing adequate regulatory frameworks to sustain these markets (Chester, 2010) and to capitalize on the benefits of these principles for reducing energy poverty.

In addition, energy security can be further characterized as a public good (noncompetitive and nonexcludable); i.e., a good that is not valued adequately by the markets but which can equally benefit all citizens, regardless of whether they pay to have it (Bielecki, 2002). The result of such inadequate pricing may be less optimal production levels of energy security for society. Supply shocks can trigger inflation, gross domestic product (GDP) loss, or high employment, among others (Bielecki, 2002). With the problem of price volatility of commodities such as oil or gas, the security of supply of a country that is highly dependent on the import of these resources can be subject to price shocks due to unexpected global price fluctuations, and consumers may lose from the fact that they cannot always benefit from affordable prices.

Energy security, thus, requires physical infrastructure security, economic security, and supply continuity. While in the short to medium term this mainly translates to reliability of supplies to meet immediate demand without interruption and "the ability of the energy system to react promptly to sudden changes within the supply-demand balance" (IEA, 2014b) in the long term, energy security is inextricably linked to the ability of a country to ensure reliable and adequately priced or affordable supplies of energy for its economy through ensuring availability of energy (through domestic production or imports and adequate storage) to meet consumer demand, and having the necessary adequate physical network for transportation to connect energy supplies to users. In addition, in the long term, energy security can be ensured only through adequate, timely, and ultimately substantial investments to guarantee that energy supply can meet economic and environmental needs, in developing renewable energy technologies to gradually replace the use of fossil fuels, as more sustainable means to provide security for future generations.

A functional approach in analyzing energy security is the commonly used framework of the four "As" of energy security: availability (geological energy resources), accessibility (geopolitical factors), acceptability (social acceptability/public acceptance and environmental sustainability aspects), and affordability (economic/investment costs) (APERC, 2007). While the "4 As" definition echoes to a large extent the traditional basis for energy security in terms of economic costs and physical availability of energy, it does not directly address aspects such as the risks, resilience, and vulnerabilities of energy systems (Cherp and Jewell, 2014), which are especially important in discussing long-term energy security from a cross-cutting perspective.

Therefore, a more comprehensive and systemic approach to energy security is needed in order to evaluate security from a long-term perspective. Cherp and Jewell (2014) define energy security as "the low vulnerability of vital energy systems," where vital energy systems are those energy systems (energy resources, technologies, infrastructure, and uses linked by energy flows) that support critical social and economic functions (Cherp and Jewell, 2014).[2]

---

[2]"Vulnerabilities," within the context of this definition are "combinations of exposure to risks and resilience capacities" where physical or economic risks can be distinguished based on the nature or origins of the risk (Cherp and Jewell, 2014), or where the energy systems can be exposed to either short-term disruptions ("shocks") or long-term "stresses" (Cherp and Jewell, 2014). The "resilience perspective" in this definition of energy security explains the origins of risk due to unpredictable social, economic, and technological factors, where aspects such as spare production capacities, diversification of suppliers, and emergency preparedness have an impact (Cherp and Jewell, 2014). Based on this definition, the "sovereignty perspective" is largely connected to risks associated with "the deliberate actions of foreign actors" (interests, intentions, power), while the "robustness perspective" is associated with the impacts of risks connected to natural or technological phenomena (resource scarcity, aging infrastructure, natural disasters) and their ability to disrupt the energy system.

This study uses the "4As" approach as a framework to evaluate energy security in the short-to-medium term and the Cherp and Jewell approach to discuss further challenges related to long-term energy security in terms of risks to energy systems and resilience. In the long term, additional factors that will influence energy security will be the environmental impact of energy systems, growing demand for energy and energy services, and the depletion of natural resources. Other factors that matter in the long term are related to infrastructure development and upgrade, the effectiveness of policy and regulatory frameworks in place, market developments, financial mechanisms, and investment climate, as well as governance and institutional frameworks (Jewell, 2011).

1.3.2.2 THE PARADIGM OF LAND DEGRADATION

Land degradation is "the reduction or loss of the biological or economic productivity and complexity of rainfed cropland, irrigated cropland, or rangeland, pasture, forest and woodlands resulting from land uses or from a process or combination of processes, including processes arising from human activities and habitation patterns, such as: (i) soil erosion caused by wind and/or water; (ii) deterioration of the physical, chemical, and biological or economic properties of soil; and (iii) long-term loss of natural vegetation" (UNCCD, 2012).

Land degradation, as a process, is "the persistent reduction or loss of land ecosystem services, notably the primary production service" (Quang et al., 2014; MEA, 2005). On the one hand, land is seen as a terrestrial ecosystem, including "soil resources, vegetation, water, landscape setting, climate attributes, and ecological processes" (MEA, 2005), which ensures the functioning of the system. On the other hand, the level of provision of land ecosystem services is the second very important indicator for processes of land degradation. The ecosystem services of the land are directly linked to the benefits that these ecosystems bring to humans in terms of provisioning, regulating, cultural, or supporting services (MEA, 2005). Provisioning services of ecosystems may include products such as food, fiber, fuels (wood, dung, biological materials as sources of energy), genetic resources, biochemicals, fresh water, etc. (MEA, 2005). In this definition, the primary production function of land is of key importance, as it supports the assimilation and accumulation of energy and nutrients by organisms, helps with sequestration of carbon dioxide from the atmosphere and serves as a natural habitat for species (MEA, 2005).

Common indicators for the state of lands include vegetation cover and soil or "net primary productivity (NPP) as a fraction of its potential" (Vlek et al., 2010). Land degradation can be observed when "the potential productivity associated with a land-use system becomes non-sustainable, or when the land within an ecosystem is no longer able to perform its environmental regulatory function of accepting, storing, and recycling water, energy, and nutrients" (Vlek et al., 2010).

While land degradation can occur as a result of natural processes, there is a widespread opinion that it mostly happens as a result of the impact of users' activity on the land and is often a "social problem," which can be prevented if the underlying causes are addressed properly (Vlek et al., 2010). The increasing demand for food, feed, fuels (including biofuels), and fodder linked to an increase in human population and a conversion of land through deforestation, environmental services, irrigation, and pollution, among other issues, contributes significantly to land degradation.

In addition, land degradation is often defined as the "reduction in the capacity of the land to provide ecosystem goods and services over a period of time" (ELD, 2011), which emphasizes the time aspect of this process. This definition is especially useful when the process of land degradation is discussed in the context of the development of resilient and secure energy systems over time, while simultaneously tackling environmental concerns.

While there seems to be an acknowledgment of the negative impacts of land degradation on energy security, food and water security, biodiversity, and ecosystem services, there is no detailed discussion of the interlinkages between land and energy systems, the interactions between the two, and the directions in which the process of land degradation and the challenges of energy security affect human livelihoods, especially in highly vulnerable areas such as the dry lands in the Sudano-Sahelian region.

This study aims to drive this debate forward and to assess how vital energy systems in the Sudano-Sahelian region are shaped by land-use management practices within the context of increasing demand for energy, the competition between energy crops and food crops, environmental sustainability concerns, and climate change.

1.3.3 CURRENT PATTERNS OF ENERGY PRODUCTION AND CONSUMPTION AND THE LINKS BETWEEN ENERGY SECURITY AND LAND DEGRADATION

1.3.3.1 ENERGY INDICATORS

Energy security is crucial for achieving a more equitable and sustainable development in the Sudano-Sahelian region. Current patterns of energy production and consumption in the region indicate that the predominant part of the population is still largely depending on the use of traditional biomass as a cheap and widely available resource over modern forms of energy at the expense of environmental, social, economic, and health concerns.

Throughout the entire Sudano-Sahelian region, energy consumption is dominated by the use of mainly unprocessed biomass (wood, charcoal, and cattle manure) predominantly for household use (cooking, heating, and lighting). The use of energy at the household level as a share of final energy consumption is still much higher than the consumption of energy for industrial or transport purposes (REN21, 2014). A significant part of the population lives in rural areas, where access to modern forms of energy and modern energy services is particularly difficult or not existent at all. Only a very small part of the population, mainly in urban areas, has access to electricity and fuels for transport. Countries in the region share a number of common characteristics related to their energy systems: overreliance of households on traditional biomass and lack of access to electricity and modern energy services (cf. Table 1.3.1):

- **Per capita primary energy consumption**[3]**:** Per capita primary energy consumption varies widely between 0.26 million tons of oil equivalent (mtoe) for Senegal and 0.72 mtoe for Nigeria. For the countries in the Sudano-Sahelian region for which data is available (cf. Table 1.3.1), the average primary energy consumption per capita amounts to 0.46 mtoe, which is equivalent to less than a third of the world average and to about 15% of the Organisation of Economic Cooperation and Development (OECD) average[4]. Precise data, however, is available only for the more developed countries of the Sudano-Sahelian region. The geographic borders of the Sudano-Sahelian region do not coincide with the geographic country borders. Most energy and economic data

---

[3]Only data for Benin, Cote d'Ivoire, Ghana, Nigeria, Senegal, Sudan and Togo are available.
[4]Data were deducted from IEA (2013), p. 42 for data on population and p. 69 for data on total primary energy demand. These data result in a primary energy demand of about 2 mtoe per capita for the world as a whole and 4 mtoe per capita for the OECD countries.

are only available for countries as a whole (cf. Tables 1.3.1–1.3.4)[5]. Except for Senegal, these countries also cover large areas outside the Sudano-Sahelian region that are often wealthier (such as the southern parts of Nigeria, Ghana, Benin, and Togo along the Gulf of Guinea). It is, therefore, more likely that in the wealthier areas, the primary energy consumption is higher compared to consumption per capita in the less wealthy areas of the countries considered. Therefore, the primary energy consumption per capita in the Sudano-Sahelian region itself could even be much lower in reality than the average primary energy consumption estimated previously, which in other words is equivalent to less than one-tenth of the OECD average.

- **Use of traditional biomass for cooking**[6]: In the Sudano-Sahelian region, 78% of the households use traditional biomass for cooking. This statistic may be slightly underrepresenting the situation because in rural areas, the percentage of households relying on biomass could reach up to 98%[7](cf. Table 1.3.1). This ratio varies from country to country between 56% of the households in Senegal and 98% in Chad, Guinea-Bissau, and Mali relying on traditional biomass (mostly wood and charcoal) as a primary source of energy for cooking. In most of the countries in the region, the share of rural population relying on solid fuels for cooking equals or exceeds the share of urban population using such biomass at the household level. While traditional biomass appears to be the most affordable and widely available source of energy, the high dependence on traditional biomass as an energy source is particularly costly from an environmental and health point of view and reflects challenges related to both equitable access to energy and economic feasibility. Cleaner and more efficient options for household use, however, such as electric or gas cook stoves, are not affordable for a large part of the population because electricity or gas may still be too expensive, especially in rural areas, and such modern solutions are not as readily available as traditional biomass. The infrastructure necessary for enabling the use of such equipment is often lacking as well. Incentivizing the use of modern biomass cook stoves has the potential to decrease fuelwood consumption considerably due to these stoves' higher efficiency. In addition, it can bring significant health benefits by reducing carbon monoxide and particulate matter emissions (IEA, 2014a, p. 118).

- **Access to electricity**[8]: In Ghana, more than 70% of the population has access to electricity, while in Senegal, Cote d'Ivoire, and Nigeria, this is true for about half of the population (cf. Table 1.3.1). Yet, the percentage is much lower in the other countries in the region; for example, it is below 10% in Chad and Niger (cf. Table 1.3.1). For the entire region of sub-Saharan Africa, the average national electrification rate stands at 32%, while in rural areas, it is only 7.5% (IEA, 2012, p. 535). These rates reflect national averages, and as such, may mask even wider disparities between access to electricity for populations in urban and rural areas, with rural areas being still largely isolated from national power grid systems and with no access to electricity.

- **Electricity consumption:** Per capita electricity consumption for sub-Saharan Africa, excluding South Africa, stands at 153 kWh/year, which is equivalent to just 6% of the world average

---

[5]Countries in the Sudano-Sahelian region, considered here, include Benin, Burkina Faso, Chad, Côte d'Ivoire, the Gambia, Ghana, Guinea, Guinea-Bissau, Mali, Niger, Nigeria, Senegal, Sierra-Leone, Sudan, South Sudan, and Togo. The data used do not include the Central African Republic and Cameroon because only a small part of these countries are part of the Sudano-Sahelian region.

[6]Data are from different sources and base years because for many countries, the most recent data are not available.

[7]REN21 (2014). Data for 2010. p. 31, Table 4.

[8]Data are from different sources and base years because for many countries, the most recent data are not available.

(IRENA, 2012, p. 2). Even in areas where populations have access to electricity, frequent power outages, especially during peak hours or during drier seasons in areas heavily reliant on hydropower production, challenge the reliability of the energy system and limit electricity access (REN21, 2014). Such system vulnerabilities in many areas create further challenges to industrial competitiveness, reliability of electricity supply, and ultimately energy security, and would often require more expensive backup power generator technologies for load shedding or limiting the negative impact of supply disruptions for households, industries, and other facilities. The energy intensity in the region is a further challenge to be addressed when designing policies to improve energy security. Diesel accounts for 17% of the electricity generation in West Africa (IRENA, 2012, p. 9). Diesel generators are mostly used to overcome power outages. In rural regions, they are also used in a system with power minigrids as a solution to support energy security but an increasing number of microgrid systems are integrating renewable energy sources to provide modern energy services in West Africa (REN21, 2014).

1.3.3.2 LAND GRABBING AND LAND DEGRADATION

Land degradation most often develops as a result of fuelwood extraction, pastoralism, and overgrazing by animals, rainfed irrigation, or unsustainable agricultural practices. While the transactions of land renting and land purchase can be of a purely commercial nature, the purpose of the deals and the land management practices in place can have highly controversial consequences on natural resources such as soil, water, energy, etc. Unsustainable land management practices lead to land degradation and resource depletion, negatively affect energy security and greenhouse gas (GHG) emissions, and threaten local livelihoods and the potential for reaching a more sustainable and inclusive growth for all. Sustainable land management policies and practices are the only instrument to avoid land degradation and to offset it in places where it has spread through land restoration in order to preserve available productive land for both short- and long-term needs.

More and more countries in the Sudano-Sahelian region rent out or sell domestic arable land to wealthier nations for the production of feedstock for either food crops or energy crops, especially biofuels. Currently, already 5% of the agricultural land[9] of the Sudano-Sahelian region is in foreign hands as a result of land grabbing, and 25%–50% of that is earmarked for the cultivation of feedstock for biofuel production (cf. Table 1.3.2). Although currently only Guinea, Sierra Leone, Senegal, and Ghana are leasing or selling major parts of their arable land for the cultivation of biofuel feedstock (Landmatrix, 2014), data from other countries in the region indicates that this tendency of large-scale land transactions is spreading quickly. Foreign land purchase/land grabbing, particularly in Guinea, Sierra Leone, Sudan, and Ghana (cf. Landmatrix, 2014; Table 1.3.2) has resulted in the change of ownership of significant amounts of domestic arable land from local farmers to either larger commercial investors or foreign investors without necessarily resulting in the fair distribution of benefits or compensation for local farmers or owners of the land. Nigeria, Mali, Burkina Faso, and Senegal seem to be increasingly attractive targets for new land deals for harvesting biofuel feedstock. Worldwide, one- to two-thirds of land-grabbing transactions are already earmarked for biofuel feedstock

---

[9]Agricultural land (termed "agricultural area" by FAO) is comprised of permanent cropland and arable land. Arable land itself consists of temporal agricultural crops, temporary meadows, or pasture, but not land that is potentially available for agriculture (FAOSTAT, 2015)

(HLPE, 2013, p. 84). These estimates should be viewed critically, as data uncertainty is very high, and most likely, they underrepresent the real extent of land acquisitions since such transactions are often kept secret to avoid international criticism of land grabbing in view of the controversial impacts that biofuel production can have on land use, food security, and climate mitigation.

Out of the global land area [13,200 million hectares (Mha)], currently about 1800 Mha is used for cultivating various crops (FAOSTAT, 2015), which is equivalent to nearly 14% of the total land area. Biofuel feedstock production is concentrated already on nearly 3% of the global cropland. In the EU, about 2% of the agricultural land is used for growing crops for the production of biofuel[10]. In 2010, one-third of the feedstock for biofuels consumed in the EU was cultivated on land outside the EU[11]. Since then, the interest in purchasing or renting land abroad for cultivating biofuels seems to have increased considerably (cf. Landmatrix, 2014). However, indications of the frequency of land grabbing point to a constantly evolving market for land and land resources on a global level, and it is increasingly more complex to track down, evaluate, and measure the use of land resources and the impact of human activity on land. Land grabbing feeds into the discussion of land degradation when human action on the land that is rented out or purchased results in the overexploitation of natural resources and when the rate of utilization of land resources exceeds the rate of recovery of resources, consequently leading to the decrease or loss of ecosystem services that the land could otherwise provide and to a decrease of the biological or economic productivity of the land.

1.3.3.3 LINKING ENERGY SECURITY TO LAND DEGRADATION IN THE CONTEXT OF THE SUDANO-SAHELIAN REGION

The sustainable development and use of energy resources that "meets the needs of the present without compromising the ability of future generations to meet their own needs" (Brundtland Report 1987) is in the basis of energy security. To qualify as secure, energy systems need to be designed to prevent or mitigate risks, associated with unpredictable social, economic, or technological factors, and to be resilient; i.e., to withstand unpredictable factors adequately (Cherp and Jewell, 2014). Energy security would rely upon the proper combination of adequate energy intensity, spare capacities, diversity of energy sources, and technologies functioning in such a way as not to compromise the functioning of the overall system in the long run (Cherp and Jewell, 2014).

In the majority of the countries of the Sudano-Sahelian region, energy security is directly linked to and cannot be ensured without the deployment of sustainable land-use and land-management practices. The energy systems and the land systems of these countries are strongly intertwined, as the majority of the local population depends directly on the resources and the services of the land to provide fuelwood and charcoal for their basic energy needs.

Availability of Energy Resources

In the Sudano-Sahelian region, the majority of the population, especially in rural areas, has been heavily dependent on traditional biomass as the primary energy source for households (cf. Table 1.3.1), and this overreliance poses a significant challenge to soil sustainability and land

---

[10]http://europa.eu/rapid/press-release_MEMO-12-787_en.htm.
[11]Cf. European Commission (2013a, p. 11): According to that report, 63.9% of all biofuel feedstock consumed in the EU was cultivated on land within the EU, while the remainder of the feedstock was imported from countries outside the EU.

management. The unsustainable extraction and use of fuelwood and charcoal as traditional energy sources can have a significant negative impact on both the primary production properties of the land and on the provisioning services that the land can supply in the long term (i.e., food, fiber, fuels, genetic resources, water, etc.). Degraded land loses its capacity to provide soil conservation, water and nutrients, carbon sequestration, and natural habitat and biomass. In the long run, land degradation becomes detrimental to the supply of biomass, which could otherwise be a sustainable and clean energy solution at both local and national level. Loss of biomass feeds into a negative loop of reduced ground cover, increased evaporation, decrease of water storage, and evapotranspiration, which consequently reduces vegetative cover, productivity of ecosystems, and resource use efficiency (like water, nutrients, and energy) (UNCCD, 2012). The dwindling supply of biomass resources in turn can lead to both deforestation and to more difficulties of poor communities to collect traditional biomass for household needs. Restoring soils and their productivity is key to ensuring energy, water, and food security for these communities for the future, as well as to build resilience and contribute to climate change adaptation (UNCCD, 2012).

The high rate of extraction of fuelwood to satisfy energy demand in the region will directly and negatively affect the physical, chemical, biological, and economic properties of soils if such practices are not counteracted by efforts to restore the lost vegetation cover. As a result, the economic value of degraded land and its potential benefits to support energy security through the supply of bioenergy in the region will decrease in the long term as well.

As traditional biomass is the cheapest, most affordable, and easiest resource to access as an energy solution for the majority of the population, especially in very poor or isolated communities in the Sudano-Sahelian region, the rate of extraction of fuelwood and the energy production methods where biomass is utilized as the primary source of energy, are very important to evaluate in terms of their impact on local vegetation, human health, the environment, and economic development.

Thus, energy security in the Sudano-Sahelian region, evaluated through the lens of physical supply availability, has conflicting implications. Traditional biomass is widely available in many regions as a source of energy, but its unsustainable use has extremely negative impacts on health and the environment and on land and its capacities to provide effectively long-term ecosystem services. Under current practices, the overextraction of fuelwood as a supply source will lead to less availability of this source in the future, thus threatening energy security from a local point of view for rural populations, who are widely dependent on it for meeting their energy demands.

Accessibility of Supplies

Depending on whether a country is an energy importer, energy producer, or energy exporter, the interests at stake will vary with reducing losses, maximizing benefits, exerting influence abroad, or keeping power domestically through resource availability and pricing. While a number of countries within the Sudano-Sahelian region (Ghana, South Sudan, Sudan, and Chad) have proven oil reserves, the majority of these countries' populations do not adequately benefit from the potential profits and economic boosts that national governments could gain from oil exports. Most of the oil that is produced is exported, but the revenues are not invested in developing services and goods to improve socioeconomic development within those states. Only Cote d'Ivoire has small oil reserves and production that is supporting domestic markets since the country has developed its own refinery capacities. Chad is exporting most of its crude oil production, while South Sudan hardly exports any oil currently because of the tensions with Sudan, which controls the majority of the pipeline infrastructure for oil exports.

Ghana has not yet developed a significant oil production capacity to become a major oil exporter for the region. Despite the availability of oil supplies in these countries, oil does not play a major role for energy security in these economies currently due to a number of political and economic factors, and its potential benefits are not exploited to improve local livelihoods in the countries that own crude oil reserves. For the countries in the region that do not have any proven oil reserves, the oil imports reinforce energy dependence of these economies on geopolitical factors, and thus the risks to their energy systems are higher when this dependence is factored in.

At the global level, the growing interlinkages between food, energy, land, water, mineral, and financial markets, create and reinforce yet another type of interdependencies between energy-importing and energy-exporting countries: biofuel trade and land trade. At the global level, industrialized countries' policy objectives of fostering energy security domestically through the use of renewable energy for transport (for example, biofuel) may lead in the longer term to more intense agricultural production both domestically in these countries (such as some EU member states and the United States, among others) and abroad, possibly inducing indirect land-use change and land degradation in other regions of the world. Switching from fossil fuels to biofuels can have controversial costs and benefits when emissions due to land-use change and the competition in trade between biofuel feedstock and food crops are factored in. The debate on the costs and benefits of biofuels is controversial depending on the accounting principles used to estimate the GHG emission reduction potential if indirect land use change is included into the equation.

In addition, the production of biofuels as a process has complex implications for both land management and energy management. When biofuels are produced on existing agricultural land in one region, the overall demand for food and feed crops remains, and may lead to the production of more food and feed in another region to meet this growing demand (European Commission, 2012), and this may have significant implications for land use, ownership, and sustainability. Demand for biofuels is expected to increase in the future, which will put significant upward pressure on natural resources such as land and soils, water and energy use, etc. In the long run, an increase in the demand for biofuels given the constraint of limited global land availability and a higher rate of land degradation will lead to more competition for land resources. The process of land degradation, on the other hand, if not halted or reverted, ultimately will lead to a relative shortage of valuable land (i.e., land that can be used to grow energy crops) in respect to the demand for such crops. Land grabbing, if accompanied by unsustainable land-use management and practices, will create greater challenges for the sustainability of soil capital and ecosystem services at the global level and might perpetuate hunger and poverty in some places. Measures to reduce land degradation and offset the negative impacts of human activity can contribute to both improving the capacities of degraded land and for contributing to energy security through enabling sustainable land management practices to ensure that advanced biofuels can be developed as clean fuel alternatives without harming ecosystems or land capacity to deliver provisioning services. More advanced biofuels, currently in the phase of research and development, have greater potential to contribute positively to the reduction of emissions without compromising land resources as their production will not threaten or displace food crop production and will not lead to deforestation for the production of biofuel feedstock.

In this context, land use itself can become a proxy to measure the capacity and the capability of a state to ensure its own energy security through the sustainable domestic production of biofuel on its own land or on foreign land, to benefit from international trade of biofuels, or, in the case of a number of developing countries in the Sudano-Sahelian region, to rent or sell some of their domestic land to foreign investors in order to support their own government revenues. From a sovereignty perspective, due to these dependencies, the energy systems of the majority of countries in the Sudano-Sahelian region are largely prone to risks associated with foreign actors' interests and economic power, and this is a major challenge to energy security and sustainable development in these countries in the long run. Regardless of the policies and strategies to enhance domestic energy security pursued by any state, unsustainable land use will threaten most of the energy supply security and food security in the poorest regions of the world, as they are most vulnerable to international price shocks and least able to mitigate such shocks.

Acceptability

Biofuels, a form of bioenergy,[12] are regarded as having significant potential to reduce GHG emissions as a clean alternative to fossil fuels and to improve domestic energy security through reducing dependence on fossil fuel imports. A number of industrialized countries are, therefore, promoting the use of biofuels with the objective to mitigate GHG emissions in the transport sector. However, more recent studies show that biofuels' potential for mitigating GHG emissions is small (or even negligible) once the GHG emissions from fertilizers' use (such as nitrous oxide) and emissions due to indirect land-use change are taken into account (Schubert et al., 2009, p. 188). In the case of converting forests or peat land into land for agricultural use, it may be a long time before savings of GHG emissions as a result of increased use of biofuels can compensate for the induced carbon loss due to land-use change for their production (Fonseca et al., 2010). A potential solution could be the cultivation of biofuel crops on land that is not suitable for food crop production, which will contribute to both cleaner fuels when the life cycle is taken into account and to reduced competition between food crops and biofuel crops. In this context, energy security objectives could align with land restoration objectives through simultaneously improving marginal lands, generating cleaner energy sources, and contributing to GHG emission reductions in the long term. Such methods could further foster the development of more resilient and less vulnerable energy systems by limiting risks related to land degradation and competition between food and energy crops.

Affordability

Affordability of energy is a crucial factor for ensuring energy access to local populations and thus for energy security. Affordability refers both to users' ability to access energy sources in terms of affordable pricing and to having adequate infrastructure in place to connect energy sources with consumers. Energy affordability is especially problematic across the majority of countries within the Sudano-Sahelian region where access to electricity (cf. Table 1.3.1) is still an exclusive right of a minority, as opposed to being a universal right for all citizens, both due to lack of energy resources and lack of proper infrastructure to connect energy flows with users and to develop the potential of renewable energy sources.

---

[12]Bioenergy is generated from biomass (for example, agricultural crops, forestry products, agricultural and forestry waste and by-products, manure, microbial matter, and waste from industry or households). It includes different forms of energy, including heat and electricity from burning biomass and biofuels. Conventional biofuels are produced using starch-bearing grains (corn, wheat), sugarcane and sugar beet, oil crops (rape, soybean, oil palm), and in some cases animal fats and used cooking oils. Advanced biofuel technologies are conversion technologies, which are still in the research and development, pilot, or demonstration phase. They include biofuels based on lignocellulosic biomass, such as cellulosic-ethanol, biomass-to-liquid-diesel, and bio-synthetic gas, algae-based biofuels, and the conversion of sugar into diesel-type biofuels using biological or chemical catalysts (IEA, 2015).

Affordability of energy for citizens will be a function of the strategic investments and the policy priorities of a state, resulting in the choice and development of one energy system as opposed to another (in terms of energy mix and physical infrastructure to support the diversity of domestic energy sources). In the Sudano-Sahelian region, poorly developed or missing infrastructure is a major challenge to overcome in order to connect energy generation with users and to ensure access to reliable and affordable electricity, especially for isolated or rural communities. High technical losses and maintenance problems, frequent power outages depending on the season, and rainfall variability and energy demand are posing enormous financial challenges to national governments and perpetuate problems of lack of access to electricity and reliability and affordability of basic energy supplies for their citizens (REN21, 2014).

Infrastructure development and deployment can be costly depending on the source of energy and technologies pursued, but appropriate government regulations, policies, and incentives need to be developed in these countries to prioritize and support affordable, clean, and sustainable energy technologies and to make energy affordable and accessible from both the economic and technical perspectives.

The interconnections between agricultural systems and global energy systems, land-use change, water scarcity, and climate change have become increasingly intertwined and more complex, and they further influence domestic and international prices of goods and energy down the chain. More and more often, "energy and food are converging in a world where energy becomes food and food can become energy" (TCG, 2011). Energy security becomes a function of the rate of land and resource utilization (including water), especially in the context of biofuel production and the interplay between domestic and global biofuels supply and demand, as well as market and regulatory behaviors. On a global scale, agriculture is by far the largest consumer of fresh water (nearly 70% of fresh water withdrawals), while in some developing countries, agriculture accounts for more than 90% of water consumption (Halstead et al., 2014). The food production supply chain accounts for nearly one-third of the world's total energy consumption (UN WWAP, 2014). The rising demand for food and the food production process has put an upward pressure not only on agricultural land expansion (very commonly at the expense of forests), but also to more intensive agriculture on existing land, leading to greater need for energy and water and possibly to resource depletion on existing land (Halstead et al., 2014). Analysis of the impacts of increasing biofuel demand on food markets and food prices indicates a significant contribution to the increase in food prices between 2000 and 2007, albeit not as a main driving factor (Fonseca et al., 2010). Such exogenous shocks can further put constraints on national energy security when the impacts of price volatilities are disproportionately felt across different parts of the population. In the long run, however, a resilient energy system should be able to respond adequately to such shocks and to help mitigate them, as opposed to exacerbating them, while energy security objectives would support land restoration rather than overexploitation of land resources.

1.3.4 VULNERABILITIES, RISKS, AND RESILIENCE OF ENERGY SYSTEMS FROM A LONG TERM PERSPECTIVE

1.3.4.1 THE IEA'S NEW POLICY SCENARIO

The projections for future energy demand used for this study of the Sudano-Sahelian region are based on the New Policy Scenario of the African Energy Outlook (AEO) 2014 of the IEA (2014a). The New Policy Scenario assumes that all policies and measures that have an impact on the energy sector, which already have been agreed upon, as well as those which are currently under consideration and whose implementation seems rather likely, will be implemented. The IEA uses as an indicator of future performance the track record of policy implementation in the past in the respective countries (IEA, 2014a, pp. 70, 73).

According to the New Policy Scenario, the demand for bioenergy in the energy mix in Africa will further increase in absolute terms by 40% by 2040, but its share in the primary energy mix will decrease from 61% in 2012 to 47% by 2040 (cf. IEA, 2014a, p. 77) as a result of the expected increasing penetration of modern, efficient cooking stoves and more advanced fuels. Traditional biomass is expected to continue to be the predominant energy source, which might lead to significantly increasing pressures on land resources and forest stocks. The share of biogas and pellets in the energy mix is projected to increase by 2040 as well.

The foreign investments in land for cultivating feedstock for the production of biofuels might also increase significantly in the Sudano-Sahelian region. According to the New Policy Scenario of the World Energy Outlook (WEO) 2014, the worldwide demand for biofuels will more than triple from 2010 to 2040 (cf. IEA, 2014a, p. 4). Unless the future production of biofuel relies on certain sustainability criteria, however, increasing global demand for biofuel risks exerting additional pressures on ecosystems and governments in the Sudano-Sahelian region to rent out or sell agricultural land for financial profits to foreign investors for cultivating biofuel feedstock.

Furthermore, the electrification rate in sub-Saharan Africa is expected to increase significantly, especially in urban areas. While the total number of people without access to electricity will increase further by 2025, by 2040, this number is projected to decline by 15% compared to 2012. In total, 315 million people in rural areas will gain access to electricity, with around 140 million of these people through minigrids and 80 million through off-grid systems. Most of that electricity will be produced by solar photovoltaics (PVs) and oil, but wind energy, hydropower and geothermal energy will also be used widely (IEA, 2014a, p. 82). The pace at which modern renewable energies will be deployed in that region is very uncertain however (cf. REN21, 2013, pp. 48 ff.). IEA's New Policy Scenario projects an increase of renewable energy demand (other than bioenergy) from 2% across sub-Saharan Africa in 2012 to 9% in 2040 (IEA, 2014a, p. 79).

Hydropower's role in power supply is projected to increase further by 2040 along with other renewables, such as wind, solar, and geothermal, that are expected to grow even faster. Nuclear energy is not expected to play a significant role in the future, accounting to merely 1% of primary energy demand in South Africa by 2040. The energy intensity of the economy in the sub-Saharan region is expected to fall by 3% yearly on average by 2040 to reach levels of 55% lower values compared to the average energy intensity in 2012. Current energy intensity levels in the region are double the world average and triple the OECD average (IEA, 2014a).

The IEA New Policy Scenario shows benefits for the long term in terms of improving energy access, infrastructure, diversifying the energy mix, and higher deployment of renewable energy sources (RES) to meet energy demand and to achieve significant improvements in energy intensity. In terms of risks to the energy system, this scenario addresses important challenges to the security of national energy systems and projects improvements in energy security for citizens based on accessibility, availability, and affordability criteria. However, the acceptability criteria, reflected in the fact that pressures on land resources will not be easily eliminated if traditional biomass continues to be used intensely as a primary energy source in the future, cannot be met as quickly as it should. Appropriate instruments will be necessary to limit the negative impact of energy resource demand on land and ecosystems.

In the context of this scenario, vulnerabilities to the energy system will persist because the resilience capacities of the energy system (and ultimately energy security) can be endangered if land degradation is not adequately addressed. While governments in the region are becoming increasingly aware of the wide implications of unsustainable land use management for land degradation, policies and measures are needed along the whole value chain to reverse the negative impacts of such practices, and government support would be instrumental through the design and implementation of appropriate regulations, policies, and investments, as well as more effective cross-border cooperation. The already significant negative impact of land degradation on food security, water availability, and human security calls for urgent action to mitigate land degradation as soon as possible. Otherwise, the continuous overuse and depletion of land resources by 2040 may have a significantly higher negative impact on energy security in the region than expected. The alternative IEA long-term scenario for Africa, the African Century Case scenario, which assumes higher investments into the energy sector compared to the New Policy Scenario on the grounds of very high expectations of better economic performance of countries in sub-Saharan Africa, does not fully address the abovementioned concerns of land-use and biomass sustainability either.

1.3.4.2 IRENA SCENARIO

While neither the New Policies Scenario of AEO 2014 nor the more progressive African Century Case scenario project an increase in modern renewable energies for sub-Saharan Africa that outpaces the increasing demand for traditional biomass in the medium range, the International Renewable Energies Agency (IRENA) has developed a scenario in which modern renewable energies in Africa could be introduced rapidly, thus reducing demand on land resources and addressing land degradation challenges (IRENA, 2012, p. 13). This renewable scenario examines the impact of policies in Africa to actively promote the transition to a renewable-based electricity system to meet the growing needs of its citizens for electricity, boost economic development, and improve electricity access. More important, this scenario projects that all citizens will have electricity access by 2030 and assumes concerted government action to support the development and deployment of renewable energies and to stimulate significant improvements in the area of energy efficiency.

The projections of IRENA (2012) assume that Africa as a whole will install another 250 GW of electricity-generating capacity from 2010 to 2030 in order to meet the growing demand for energy. That corresponds to almost double the amount of the capacity installed in 2010 (147 GW). By 2030, 80% of the newly installed electricity capacities will be based on renewable energy sources and the share of renewables in the power sector is projected to reach 50%. The scenario assumes further that energy demand will increase faster than population growth and that the urbanization process will support better access to electricity for more people as they migrate to the cities.

The most significant challenges for the IRENA scenario are the availability of sufficient funding and the development of an industrial manufacturing base, as well as installations and service industries, which would be required to support the investments in RES technologies (IRENA, 2012, p. 44). As the deployment of renewable energy technologies requires significant initial investments, securing sufficient investments is key for the success of that scenario.

The IRENA Scenario for West Africa, which covers most of the countries of the Sudano-Sahelian region, projects (IRENA, 2012, p. 48) that 37% of the increase in electricity supply will be met largely by hydropower, followed by wind (16%). By 2030, the whole electricity supply is projected to be based entirely on renewables in many West African countries (IRENA, 2012, p. 49).

IRENA's scenario for the development of the energy mix and the energy system in sub-Saharan Africa and the Sudano-Sahelian region takes into account significant improvements in energy access for citizens and availability, diversified supply, improvements in infrastructure, energy efficiency, and a cleaner energy mix. In this case, the availability, accessibility, and acceptability criteria for the development of a more secure energy system by 2030 could be met through such a scenario. However, the affordability for governments to mobilize the necessary investments to transform the energy system would be a crucial factor for the success of such a scenario to improve energy security. In addition, if proper measures are deployed, this development scenario has significant potential to reduce negative impacts on land potentially faster and more efficiently, as opposed to the IEA scenario discussed previously. To achieve such a scenario, governments will have to play a crucial role in incentivizing significant domestic and international investments to transform and improve energy systems in the region and to meet adequately the projected increase in demand for energy through more sustainable energy sources, technologies, and practices, while adequately valuing resources and natural capital. Such a scenario has a higher value in terms of energy security as it addresses more successfully potential risks while contributing to long-term resilience and lowering the vulnerabilities of both energy and land systems.

1.3.5 CASE STUDIES

This section discusses the case studies of Sudan (including South Sudan after 2011) and Mali to illustrate the interconnections between energy security objectives and land degradation processes in the Sudano-Sahelian region and what measures and instruments can be deployed to decouple energy use from land use and to restore land and simultaneously improve energy security. While energy security and land-use management are in the focus of this analysis, the case studies illustrate that weak energy security in the Sudano-Sahelian region and land degradation have further contributed to the outbreak or deterioration of certain conflicts in the region and that land restoration practices have the potential to improve significantly not only energy security, but also food security, and economic development and to prevent or mitigate violent conflicts.

Sudan and Mali share numerous common characteristics with the other countries in the Sudano-Sahelian region related to socioeconomic development and poverty, land degradation, scarcity of resources, weak governance, and conflict risks, among others.

Sudan[13] has been continuously plagued by violence and the Darfur conflict, termed by UN secretary general Ban Ki-moon as "the first climate change conflict" (UNU, 2009), led to the separation of the state into two parts, Sudan and South Sudan, in 2011. Mali, on the other hand, was largely considered as a rather stable country until violent conflict broke out in the north in early 2012. Pervasive processes of land degradation in the two countries significantly threaten local livelihoods, as they limit the access of poorer rural societies to fertile land for agricultural development and to land for collecting energy sources such as fuelwood or charcoal for basic energy needs. In terms of energy resources, while Sudan and South Sudan have significant domestic oil resources, which they could exploit to foster energy security and economic development, Mali relies on importing oil and oil products to sustain its economy. The two countries have

---

[13]Sudan is considered here with its borders from before July 2011, including South Sudan.

significant renewable energy potential, which has remained largely underexploited up to now but which has the potential to improve socioeconomic development and human well-being in those countries if properly supported and developed by government reforms and mechanisms.

Data was collected based on secondary sources. Research was conducted through the review of relevant literature on land degradation, energy security and conflict mitigation, and legal documents, as well as international statistical data on economic, energy, and human development indicators.

The analysis of these case studies aims to shed more light on the question of what the main factors shaping the energy security landscape in the Sudano-Sahelian region are and how the interplay between energy demand and supply and land resources links local energy systems and land systems in Sudan and Mali within the context of environmental concerns and resource constraints.

Through the cases of Sudan and Mali, the study aims to elaborate on further solutions to enhance the sustainable management of resources (both energy and land) as well as sustainable land-use practices, to foster energy access and security for local communities, and ultimately to prevent or mitigate conflicts in the long run. The findings of this study can be of relevance to other countries in the Sahel region that face similar challenges of energy poverty, land degradation, and food insecurity, and that can benefit from similar practices and solutions to reduce vulnerabilities and build resilient and secure energy and land systems.

1.3.5.1 CASE STUDY 1: SUDAN AND SOUTH SUDAN

Socioeconomic Context

The territories of Sudan and South Sudan, which until July 9, 2011, were within the borders of the state of Sudan, extend from the Sahara Desert in the north to the wet savannah in the south. For the purpose of this study, all statistical data for indicators (including 2011) considered Sudan and South Sudan as one country, Sudan. After 2011, and for the future projections for energy supply and demand and energy security aspects, the two states were analyzed separately[14]. Ethnically and by religion, Sudan (the region of Sudan and South Sudan) is split into Arabs, which are predominantly Muslim in the north (70% of the population) and Animist (25%) and Christians (5%) in the South. The population in the South was marginalized since many people belonging to these groups were enslaved during the colonial era. The most important ethnic groups are Arabs in the north, which are Muslim, followed by Nubians and the Dinkas in the south.

Most people in Sudan make their livelihoods via agriculture and livestock farming. Sedentary farmers in the south cultivate their crops predominantly through rainfed agriculture, while in the north, many nomadic and seminomadic tribes like the Baggara in Darfur (UNEP, 2007, p. 35) practice livestock farming. In some areas in the north, the mechanization of agriculture was introduced during the former British colonial period. Along the Nile River, and especially between the Blue and the White Nile south of Khartoum, most of the population relies on irrigated agriculture.

The Energy Landscape

Sudan is a poor, developing country, but until 2013 it earned a significant portion of its revenues by the export of crude oil. In Sudan, petroleum revenues accounted for more than 50% of total government revenues on average in the period between 2001 and 2010 (IMF, 2012 in IEA, 2014c). After the division of Sudan into two states in July 2011, most of the oil production capacity of the state remained in South

---

[14]The Republic of South Sudan formally seceded from Sudan on July 9, 2011, after an internationally monitored referendum in January 2011, and it was admitted as a new member-state by the UN General Assembly on July 14, 2011.

Sudan. The division of Sudan and South Sudan led to the loss of nearly 75% of Sudan's oil production and nearly 55% of its previous fiscal revenues from oil exports (IMF, 2013). On the other hand, South Sudan, as a landlocked country, still has to rely on Sudan's export infrastructure and seaport for its oil exports (US EIA, 2014). Political uncertainties and disputes over oil revenues continue to challenge the ability of both Sudan and South Sudan to keep up oil production and to rely on stable revenues from oil exports.

The Human Development Index for Sudan ranks the country among the lowest in the world, indicating that despite the significant government revenues from oil exports, their distribution has not been proportionately supporting poorer population and socioeconomic development, and poverty has continued to plague both Sudan and South Sudan. The vast majority of the population (nearly 75%) does not have access to electricity, and 93% of households still rely fully on traditional biomass for cooking purposes. Biomass accounts for more than 67% of primary energy use in Sudan (Table 1.3.1). The rate of undernourishment is one of the highest in the region (24% of the population). This is in stark contrast to expectations if gross national income (GNI) per capita (reflecting government revenues from crude oil exports) in Sudan is compared to that of other countries in the region: in Sudan, GNI is higher than in most of the other countries in the region, except for Nigeria, also an oil-exporting country, and Mauritania (cf. Table 1.3.4). Improving governance will be crucial in order to ensure that resource revenues are distributed equitably to support economic development and improve local livelihoods. Good governance is a prerequisite for attracting investments into adequate infrastructure and new energy technologies to provide affordable and accessible energy to citizens and to improve energy security in both Sudan and South Sudan.

Given the high reliance of households on traditional biomass for energy consumption, vulnerabilities due to the depletion of potential sources of energy for local populations increase further. The extraction of fuelwood, especially within the dry Savannah south of the Sahara Desert and in areas close to urban agglomerations, has been one of the most important causes of land degradation in Sudan. The decrease in vegetation cover as a result of fuelwood extraction, overgrazing, and forest clearing and replacement with agricultural land has led to the degradation of large areas of land and to the depletion of local ecosystem resources. Sudan lost more than 10% of its forest coverage in the relatively brief period between 1990 and 2005 (UNEP, 2007, p. 10). Unsustainable fuelwood extraction practices lead to land degradation, which in turn has had an increasingly negative impact on the productivity of land and on its capacity to be a provider of biomass for energy use for the poorer rural communities. The growing energy demand has led to a higher rate of extraction of fuelwood and thus to increasingly higher rates of deforestation to meet the needs of the growing population, ultimately exacerbating land degradation. This is where the direct link between energy demand, energy supply source, and land degradation is particularly evident. When taking into account that fuelwood, the only energy source available and affordable for the majority of the population, becomes progressively scarcer as land degrades due to human practices, ensuring energy security in the Sudan region becomes an even greater development challenge for the short term to midterm.

Within the long-term projections of the New Policy Scenario, by 2040, Sudan will transform itself from a net exporter of crude oil to a net importer of oil for its refineries (IEA, 2014a, p. 110) while the South Sudan region is projected to face a significant drop in oil exports altogether. As resource-related revenues were a significant source of government revenues for both Sudan before 2011 and South Sudan currently, the risk of macroeconomic instability due to oil-export dependencies will continue to be an important factor within the political, social, and energy development in both countries in the future. While the presence of domestic conventional energy sources may be a significant positive factor in ensuring a country's energy supply availability and energy security, the projected change in oil resource availability for Sudan in contrast to South Sudan can significantly and negatively affect the physical energy supply, energy prices, and overall energy security of Sudan in the longer run. In the context of both Sudan and South Sudan, the overdependence of these economies on revenues from oil exports, thus, is a significant economic and geopolitical risk to the energy security of both states. This indicates that both states' energy systems are still highly vulnerable to external disruptions in energy flows, resource availability and prices, infrastructure inadequacies, and environmental constraints such as land degradation and loss of biomass.

In order to support the development of more resilient energy systems, therefore, both Sudan and South Sudan will need to reduce such dependencies and risks and to enhance their ability to withstand disruptions more effectively through more sustainable land and energy management practices and new technologies. Sudan is estimated to have one of the highest onshore wind energy potentials in Africa (IEA, 2014a) and hydropower capacity is expected to contribute as well to Sudan's energy supply by 2040 (IEA, 2014a)[15]. According to IRENA (2014), Sudan also has one of the highest PV and concentrating solar power (CSP) geographic potentials in Africa, which if developed adequately in the long term, can contribute significantly toward enabling energy access for citizens and fostering less reliance on fossil fuels, better energy security, and increased economic development of the country. Various scenarios point to a significant overall technical potential of renewable energy sources to contribute to the development of a cleaner and more sustainable energy system in Sudan. Especially in the long run, renewable energy technologies can contribute to reducing pressures on land as a source for energy crops, if competition between agricultural land and land designated as having the highest energy potential is avoided. In the discussion of energy potential, it is important to point out that depending on the source of energy and the technology that is chosen to be developed and deployed, significant land surface may need to be available. This, however, does not have to come at the expense of arable land; rather, it may supplement the state's clean energy objectives as part of the development of an integrated sustainable energy system by deploying renewable energy technologies in places where the geographical and technical potential to harness energy is greatest on nonarable land and where the social benefit would be greatest.

While energy security is reflected through the elements of availability, accessibility, affordability, and acceptability, the overall security of the energy system is particularly influenced by external factors, such as weak state governance, extreme environmental pressures (droughts), and land degradation, coupled with high population growth and political instability and conflicts. These factors (albeit very different in their origins and nature) all negatively affect the development of the Sudan's and South Sudan's energy systems and their ability to withstand risks and remain resilient, resulting in the loss of opportunities for improvement in their livelihoods.

Land Degradation: Causes and Consequences

Land degradation in the Sudan region is a serious threat to local communities, as 57% of the work force is engaged in the agricultural sector and depends on fertile land for its survival (FAO, 2005). Signs of land degradation can be observed throughout the last few decades through the patterns of rainfall and vegetation cover, whose monitoring started back in the 1930s. Since the late 1960s, the Sudan region

---

[15]In WEO's scenario, Sudan and South Sudan are evaluated as separate entities, while the IRENA outlook refers only to Sudan.

has been plagued by recurring severe drought, with the most severe instances between 1980 and 1984. Until the early 2000s, land degradation processes lead to a southward progression of the Sahara Desert of 50–200 km, which roughly translates into a rate of 1–3 km per year (UNEP, 2007, p. 9). This process forced the nomads living in the north to move ever farther south during dry seasons (northern winter) in order to support their livelihoods, which created competition and conflict between nomads and the local sedentary farmers. Sudan's population is growing fast, currently at a rate of 2.4% per year (cf. Table 1.3.3) which means that at the same growth rate, the population will likely double within the next 30 years. Population growth is expected to put upward pressure on the demand for energy, water, and natural capital and resources such as land. The growth in livestock numbers has been explosive—from 28.6 million in 1961 to 134.6 million in 2004 (UNEP, 2007, p. 10) and overgrazing of the rangelands has put additional negative pressure on local land resources in a very short period of time.

Both Sudan and South Sudan are characterized by extremely weak state governance systems, which does not facilitate the task of tackling land degradation and simultaneously meeting energy security objectives. The most recent fragile states index that already separates between Sudan and South Sudan classifies Sudan as the fifth-worst-performing country and South Sudan as the worst globally (Haken et al., 2014). Sudan started to lease or sell significant parts of its agricultural land to foreign investors around 2007 (cf. Landmatrix, 2014), despite the fact that many people suffer from hunger. Currently, nearly 10% of Sudan's agricultural land is under foreign control[16] (including both land purchased by foreign investors and rented by foreign actors; cf. Table 1.3.2). The largest agricultural land parcels have been rented or purchased by private or public investors from the United Arab Emirates (UAE), the United States, South Korea, Saudi Arabia, and Egypt. The investments are predominantly made in land that is used for the cultivation of wheat, sugarcane (for biofuel production), cotton, fodder, sunflower, and other grains, according to the limited data available (Landmatrix, 2014). Inadequate rural land tenure regulations are a direct consequence of weak state governance and a major obstacle to sustainable land use, as small-scale farmers have little or no incentive to invest in the land and to protect natural resources (UNEP, 2007).

Land degradation in the Sudan region has contributed to violent conflicts in Sudan and exacerbated conflicts over resource exploitation in the region. On the one hand, violent conflicts lead to population displacement, lack of governance, conflict-related resource exploitation, and underinvestment in sustainable development. On the other hand, competition over land, timber, and land use issues related to agricultural land are all important causes for the instigation and perpetuation of conflict and instability in the Sudan region. The outbreak of the Darfur conflict in 2003 was very much triggered by the competition for access to land, especially between sedentary farmers and nomads. Land degradation in northern Darfur combined with significant population growth resulted in insufficient land resources to satisfy the needs of both sedentary farmers and nomads, leading to the largest number of international migrants and internally displaced persons (more than 5 million people in 2005 or 10% of the entire population of the country (UNEP, 2007, p. 11). In Darfur, the conflict has resulted in the displacement of more than 2.4 million since the outbreak. Sudan imports more than a quarter of its consumption of foodstuff (cf. Table 1.3.4) and receives significant amounts of food aid from the EU and the World Food Programme. On the other hand, the large population displacements as a result of internal conflicts have arguably also contributed to further land degradation processes and energy poverty across Sudan.

---

[16]Data are given for Sudan in its current borders only because there are no data about arable land and permanent cropland for South Sudan available.

Natural resource scarcity, land degradation, and overdependence on biomass, along with political instability and incessant conflicts, have all acted as barriers to the improvement in energy security and to providing basic access to energy for poor communities in the Sudan region. Improving the energy resource availability for the poorer parts of the population through methods such as agroforestry, which can restore land and provide additional fuelwood, has the potential to contribute significantly in the short term to improving local land properties, supporting energy and food security, reducing energy system vulnerabilities, and improving local livelihoods in the short term. The deployment of various modern and renewable energy technologies (solar stoves/modern cooking stoves, minigrids, or off-grid solutions in rural regions) has the potential to significantly reduce fuelwood extraction, mitigate land degradation, and prevent violent conflicts over energy resources in the future. In addition, the health, environmental, and socioeconomic benefits of such solutions outweigh by far the costs that local and national governments will have to invest upfront for the deployment of such solutions in the first place.

1.3.5.2 CASE STUDY 2: MALI[17]

Geographic and Socioeconomic Context

Mali is a landlocked country in West Africa whose territory extends from the Sahara Desert in the north to the Woodland Savannah in the south. Ethnically, it is split into a number of sub-Saharan populations, with the Manding group representing the majority (40%), followed by the Sudanese group (20%), the Voltaic group (17%) (Coulibaly, 2006, p. 5), the Touaregs and Moors in the north (10%), and a number of other small groups. The population is predominantly rural (63.2% in 2011) with nearly 17% nomadic groups (Touaregs, Moors, and Peuhls), supporting their subsistence with pastoralism. However, urbanization has increased quickly in the past decade, from 31% in 2004 to 39% in 2014 (World Bank, 2015).

The Energy Landscape

Mali is a poor, developing country. The electrification rate for 2011 was 27% on average at the national level. The rate for urban population was 54%, while in the rural areas, only 14% of the population had access to electricity (REN21, 2014). With urban populations accounting for only 36.7% of the total population (2011), the challenge of increasing access to electricity for all citizens will remain in the coming years (World Bank, 2015).

The share of the population using solid fuels for cooking in total was 98% in 2010 (compared to 99.8% in 2006), with more than 95% of both urban and rural populations relying on either wood (82.6%), charcoal (14.5%), or other alternative fuels (2.9%) for cooking (REN21, 2014). Moreover, the share of traditional biomass in the total final energy consumption for 2010, including the use of biomass in all sectors in Mali has been 85.4%. Modern biomass[18] accounted for only 1.4% of total final energy consumption in 2010. Hydropower's share in the total final energy consumption amounted to 1.5% (2010). Solar and wind power did not play any role in the energy mix of Mali.

In addition, Mali is highly dependent on oil imports, which puts additional pressure on the country's economic development and the access to reliable energy supplies for both rural and urban populations.

---

[17]Data are from Tables 1.3.1–1.3.4, which include their sources.
[18]The FAO defines this as biomass produced in a sustainable manner from solid wastes and residues from agriculture and forestry.

Oil price volatility can have a strong impact on a country whose fast population growth and poverty additionally challenge the state's ability to manage international fuel price shocks (SREP, 2011). Furthermore, due to the high variability of rainfall, the electricity supply of Mali, 55% of which largely relies on hydropower generation, is highly prone to weather-related risks, and this vulnerability can severely affect the country's electricity supply on a large scale (SREP, 2011). With climate change–related weather extremes, the security of electricity supply from hydropower and biomass production patterns can be easily threatened in the future (SREP, 2011). The current energy system of Mali, therefore, is highly vulnerable to both external supply and economic shocks and to natural phenomena and technological risks, indicating a very low level of robustness and sovereignty of the overall energy system and thus a very weak energy security level.

As alternatives for solid fuels for cooking, liquefied petroleum gas (LPG) has been increasingly popular in some of the West African countries (i.e., Ghana). Although not a renewable source of energy, LPG is much more beneficial in terms of environmental and health impacts compared to the burning of wood and charcoal and can limit deforestation by decreasing local demand for fuelwood. However, the existence of adequate infrastructure for LPG transport is crucial for ensuring access to households, and such infrastructure is still largely underdeveloped in some areas in Mali. Other possibilities for expanding access to cleaner cooking fuels for local populations could be the deployment of more efficient cooking stoves, the electrification of urban areas, and the development of off-grid rural electrification schemes. Due to inadequate grid developments, electricity continues to be relatively expensive and inaccessible for the majority of the population (SREP, 2011). Furthermore, existing regulations to promote rural electrification have not yet been sufficient to promote the involvement of the private sector in order to support grid developments to connect isolated areas and the market deployment of cleaner energy solutions (SREP, 2011). The development of such modern energy forms and services in the future will decrease significantly the pressure on land to supply fuelwood, especially in rural areas, and could contribute enormously to bringing access to energy for the poorer and the isolated communities across Mali and to improving the overall resilience and security of the energy system.

Minigrids and off-grid systems can be cost-effective solutions for increasing the access of rural populations to electricity and for remote areas and can further reduce the negative impact of unreliable central grid services (REN21, 2014), ensuring more reliable energy supply and energy security for these areas. In addition, access to affordable energy supplies and services can have a tremendously positive impact on rural development and social and economic opportunities for the poor and isolated communities.

The creation of the West African Power Pool in 1999, in which Mali participates, has had until now and will continue to have a significant impact on the expansion and development of infrastructure and cross-border connections between countries to support trade of electricity, allowing the country to expand access to energy services, decrease its dependence on fossil fuel imports and improve energy security. Mali's renewable energy potential is estimated to be significant (IRENA, 2014), especially in terms of small-scale hydropower capacity and solar PVs. Harnessing this potential effectively in the future can have tremendous benefits on both energy security and land use. In addition, it can mitigate the reliance on fossil fuels and the use of traditional biomass, while simultaneously stimulating socioeconomic development and reducing poverty and increasing resilience of Mali's energy system.

In the past two decades, Mali has progressed with the development and deployment of more RES technologies, such as increased hydropower generation, solar PV systems for lighting, hybrid minigrid systems, solar thermal systems applications for collective heating, modern cooking stoves in rural areas, and the development of jatropha-based biofuels as more sustainable alternatives to traditional biomass (Mali, 2011). However, these solutions account still for a very small part of the overall energy contribution, and while they represent a transition toward a more resilient and secure energy system, there is a lot more that needs to be achieved on a larger scale. Grid-connected installed capacity of renewable energy sources as of 2014 accounted for a very small share of the electricity production of Mali, with hydropower (300 MW) being the predominant source, while no grid-connected solar, wind, or modern biomass capacity was installed as of 2014. The grid-connected hydropower capacity in 2014 accounted for a share of 65.5% of the total overall capacity of the country, including nonrenewable generation (REN21, 2014, p. 33).

The Malian government has further developed regulatory support mechanisms for renewable energy and a detailed renewable energy policy with ambitious renewable targets to reach up to 25% of installed RES capacity by 2021. The national strategy for the development of biofuels (Stratégie nationale pour le Développement des Biocarburants), adopted in June 2008, has the objective to promote the development and use of biofuels to meet local energy demand and reduce the reliance of Mali on oil imports (Mali, 2011). In addition, Mali has indicated intentions to develop future small-scale hydropower capacities considerably, to be supported by the Mali Investment Plan for scaling up renewable energy (SREP, 2011; REN21, 2014). Fast and efficient implementation of policies and mechanisms supporting these objectives will enable Mali to develop a cleaner, more resilient, and less vulnerable energy system in the longer run, eliminating current pressures on ecosystems and land resources and improving livelihoods.

Land degradation: Causes and Consequences

Most people in Mali support themselves with agriculture, pastoralism, livestock farming, and fishery. The sedentary farmers to the south of the Sahel rely on rainfed agriculture for cultivating their crops or practice livestock farming. Irrigated agriculture is practiced predominantly in the inner Niger delta and has been recently spreading in the south. To date, irrigated agriculture accounts for more than 5% of the agricultural area of Mali (FAO, 2013), and agriculture itself accounts for almost all of the water consumption in Mali (97%, cf. FAO, 2013).

Farmers maintained the traditional land-use system of agroforestry parklands up until the 1970s (Garrithy et al., 2010, pp. 204ff). Demographic, economic, environmental, and social developments since then have put pressure on these traditional land-use systems. Modern Sahelian forest laws and the ways that they are locally enforced have discouraged farmers from optimum parkland management and led to the degradation of many parklands (Garrithy et al., 2010). According to World Bank data, the total forest area of Mali has been continuously decreasing since the 1990s, from 11.5% of the land area of Mali in 1990 to 10.1% in 2012.

Hunger is widespread in the country, and in 2013, the UN estimates that 4.6 million people were food insecure, 450,000 suffered moderate acute malnutrition, and 210,000 suffered severe acute malnutrition (cf. OCHA, 2013, p. 7). Despite the significant number of people suffering from hunger, the Malian government started leasing and selling significant parts of its agricultural land to foreign investors (cf. Landmatrix, 2014). Currently, already 2% of Mali's agricultural land is under foreign control, with the majority of this land serving for growing biofuel feedstock. Data from Landmatrix (2014) indicates a trend of increasing interest from foreign investors in purchasing or renting more land to grow energy crops in Mali. The biggest share of investments in agricultural land in Mali currently comes from the United States, Saudi Arabia, and Libya (Landmatrix, 2014). With only 14% of the total land of Mali being arable, any unsustainable land management practices risk to have a significant negative impact on the poorer rural communities' access to food, energy, and economic opportunities.

Mali has been plagued by land degradation and desertification for many decades, while population growth has remained high throughout that period. Currently, 60% of the Malian population is living on degraded land (cf. Table 1.3.4). Between 1968 and 1974, the country was hit by a series of extreme droughts which led to severe famine and the deaths of many people. Subsequently, many periods of drought resulting in periods of famine have taken place, with the most recent occurring in 2005, 2008, 2010, and 2012. In addition, precipitation variability has increased in past decades, leading to heavy flooding during the rainy season in the south (i.e., 2010 and 2012), both of which were also years of famine because of the droughts in the preceding years.

In addition, land degradation processes, especially in the center of Mali around Mopti (European Commission, 2013a, p. 154) led to more tensions for land resources between the nomads in the north, mostly Touareg, and the sedentary farmers in the south; in addition, they increased pressures on local land systems and land services. These tensions were one of the reasons for the outbreak of the Mali conflict in 2012. The severe drought in 2011 and increasing food prices on the international markets exacerbated food insecurity, while the transboundary movements of weapons by the Touaregs further contributed to the outbreak of conflict the next year (cf. Van-Schaik and Dinnissen, 2014, pp. 51ff). The armed conflict in 2012 consequently led to the additional displacement of more than 400,000 people and caused additional significant food insecurity in the country (Van Schaick and Dinnissen, 2014, p. 51).

Due to degraded land systems and population displacements, Mali currently receives significant amounts of humanitarian aid via the UN Sahel Strategy in order to support its recovery. The vast part of this aid is earmarked for improving food security and nutrition. In total, the humanitarian aid amounted to US$150 million in 2012 and $265 million in 2013. For 2014, $570 million was required (OCHA, 2014, p. 13).

In addition, natural disasters like droughts have put increasing pressure on water resources and soil quality in the region in the past decade. As water is an important part in both agriculture and biofuel production, water shortages and increasing rain variability can significantly endanger both food and energy security in Mali.

Despite the significant RES potential of Mali, renewable energy sources are still underutilized in the overall energy mix of the country, and Mali continues to be heavily dependent on fuelwood for meeting energy demand in rural areas and on the import of oil for energy generation. The collection of fuelwood, however, is contributing to land degradation and has detrimental impacts on human health, air quality, and GHG emissions. Furthermore, land grabs for growing the feedstock for biofuels has already resulted in the loss of a significant portion of arable land for the production of food crops, which in turn further exacerbates food insecurity in Mali and increases the risk pf propagating land degradation and energy poverty.

Measures to Improve Energy Security and Mitigate Land Degradation

However, to meet future challenges and to strengthen the resilience and the robustness of the energy system in Mali, the government would need to introduce further incentives to promote sustainable bioenergy sources and RES development and deployment on a larger scale. Such incentives can be in the form of financial, policy, and regulatory mechanisms, allowing the country to attract both domestic and foreign private investments more readily and to strengthen its reliance on its own sustainable resources in the future. Challenges such as high grid expansion costs, aging infrastructure, and the deployment of modern energy services will require the mobilization of significant financial resources to deploy renewable energy solutions for the poor. Otherwise, severe climate events will continue to cause additional disruptions to Mali's energy system and to limit energy access for citizens, threatening energy security, at least for the short to medium term.

Some incentives that the Malian government has introduced to foster RES development in the past decade include value-added tax reductions for renewable energy projects, reduction of import duties on renewable energy components, as well as the development of financial support from project grants and low interest loans (REN21, 2014). These are all useful mechanisms to support the transition to a cleaner and more secure energy supply for the population in the long run. However, the lack of appropriate fiscal and regulatory framework for renewable energy and biofuels continue to hinder the development of modern energy services for Malian citizens (Mali, 2011). In this context, measures such as increasing agroforestry practices to reverse land degradation processes, the use of more efficient cooking stoves, and the development of minigrids and off-grids for providing electricity to rural areas all have the potential to significantly reduce fuelwood extraction and mitigate land degradation, in addition to potentially preventing violent conflicts over energy, land, water, and food resources in the future. Furthermore, without measures to regulate appropriately land use and to limit land grabbing for biofuel crops' production, it will be very challenging to improve food security and reduce poverty in Mali.

1.3.6 POLICY OPTIONS FOR MITIGATING LAND DEGRADATION AND IMPROVING ENERGY SECURITY

For the governments in the Sudano-Sahelian region, the adverse impacts of land degradation are becoming increasingly evident and alarmingly detrimental to socioeconomic development and energy security. Many of the countries in the region are already considering or implementing policies and measures to mitigate land degradation and to develop and deploy sustainable energy solutions, as illustrated in the cases of Mali and Sudan (Chapter 5).

The measures that have the greatest potential to simultaneously mitigate land degradation and improve energy security are

- Agroforestry practices
- Sustainable cooking technologies
- Off-grid and minigrid solutions
- Renewable energies for power generation

These will be discussed next.

1.3.6.1 AGROFORESTRY PRACTICES

Agroforestry is the most effective and efficient method to mitigate land degradation. At the same time, agroforestry practices can increase energy security considerably in the short run because they can deliver additional fuelwood within a few years, increase agricultural productivity and restore land properties and services (Garrithy et al., 2010, p. 199). The intercropped trees in agroforestry practices sustain green cover on the land throughout the year to maintain vegetative soil cover, bolster nutrient supply through nitrogen fixation and nutrient cycling, generate greater quantities of organic matter in soil surface residues, improve soil structure and water infiltration, and increase direct production of food, fodder, fuel, and fiber (Garrithy et. al, 2010). Agroforestry is becoming more and more popular, especially in Niger and Burkina Faso. In Niger, agroforestry practices were applied on 50,000 ha of land in 2009, which is estimated to have increased food production by 500,000 tons per year compared to traditional agricultural methods without agroforestry. This additional food production can theoretically feed 2.5 million people, provide additional income to the local population from selling fuelwood and timber on the local markets, and increase resilience of those villages that introduced agro-forestry (Reij et al., 2009, p. 2).

Agroforestry practices contribute to land restoration by increasing soils' carbon content, as well as their holding capacity for water and nutrients. At the same time, agroforestry has the potential to improve food security by increasing the agricultural productivity of land by twice or even more, depending on the specific method of agroforestry applied in a particular region (cf. case studies in Bayala et al., 2011). In addition to mitigating land degradation, agroforestry is a very important method for both climate change adaptation and climate change mitigation as it contributes to restoring the land's properties to provide carbon sequestration services in the soil. According to Lal (2010, cited in Garrithy et al., 2010, p. 209), evergreen agriculture, a specific kind of agroforestry, can increase the storage of carbon in the soil tenfold. Agroforestry also mitigates land degradation, not only because of its capacity to improve soils' fertility, but also because the trees planted for providing shade and protection for the agricultural crops add to the physical supply of fuelwood and restore biomass.

Shifting traditional agricultural practices to more sustainable agroforestry practices does not require too much additional knowledge, skills, or resources for the smallholders[19] practicing traditional agriculture. However, the short-term cost of switching to agroforestry practices (i.e., the fact that agricultural productivity is often reduced within the first year of the shift) has to be considered as well. Furthermore, farmers need security for their land tenure rights from their national governments, and guaranteeing land tenure rights to smallholder farmers in the Sudano-Sahelian region still appears to be a significant challenge.

1.3.6.2 SUSTAINABLE COOKING FUELS AND MODERN COOKING TECHNOLOGIES

Sustainable cooking fuels and modern cooking technologies are other key instruments for mitigating land degradation and simultaneously improving energy security throughout the Sudano-Sahelian region. Fuelwood is the primary energy source used for cooking at the household level, and cooking accounts for more than 80% of the energy use of households of sub-Sahara Africa, excluding South Africa (IEA, 2014a, p. 131). This puts a significant burden not only on biomass resource availability in the region, but also on the people collecting the fuelwood, especially women and children. In addition, indoor air pollution resulting from the direct burning during cooking with traditional biomass (wood and charcoal) has a tremendous negative impact on human health and on GHG emissions.

Modern cooking technologies that have the most potential to reduce fuelwood extraction in the short term include gas cookers fueled with kerosene or LPG. This is a feasible solution, especially in coastal countries, because they can easily import kerosene and LPG and then export it to landlocked neighboring countries. Nowadays, such cooking stoves are already widely deployed in some West African countries, where LPG has been further subsidized by national governments: In Senegal, 41% of

---

[19]Smallholders are holders of small tracts of land.

the households use LPG for cooking, and in Cote d'Ivoire and Ghana 14% and 10%, respectively (REN21, 2014, p. 30; IEA, 2014a, p. 136). Gas cookers do not require any fuelwood, but they do contribute to GHG emissions and climate change. Furthermore, as most of the kerosene and LPG resources in the region have to be imported, this may put additional challenges on both governments and consumers once price fluctuations, transport, and infrastructure limitations, as well as physical supply risks to the international markets, are taken into account. Thus, LPG and kerosene-fired cook stoves are not the most cost-effective and efficient solution to avoid overall GHG emissions or to ensure readily available and affordable energy supply for poor households, but they can limit significantly the negative health impacts and contribute to gender equity in energy access.

Governments in the Sudano-Sahelian region should prioritize technology solutions with the highest potential to reduce fuelwood extraction, increase energy efficiency, reduce emissions and indoor pollution, and improve equitable energy access to rural areas (and thus energy security). For example, rocket stoves (GTZ, 2007) and gas-burning stoves, operated through the method of micro-gasification (GIZ, 2011) are comparatively new developments that do not cost much more than traditional stoves and do not require people to change their traditional usage habits. The deployment of such stoves started only very recently, first in Malawi and Uganda, and then also in the Sudano-Sahelian region (cf. GTZ, 2007). While rocket stoves still depend on fuelwood and charcoal as fuel, they are much more efficient than traditional biomass-burning stoves and open fires and have the potential to reduce the demand for fuelwood by more than half. The gas-burning or also microgasifier cooking stoves can be fired with any kind of solid biomass like residues or cattle manure. In the short run, rocket stoves are the best solutions in regions where enough fuelwood is still available, while microgasifier cook stoves are the best solution in poor regions, where fuelwood is already very scarce (GIZ, 2011, p. 1).

Modern electric cooking stoves using renewable energy electricity generation are the most cost-effective, efficient, and equitable option in the long term because they not only considerably increase energy access and security, but further eliminate the problem of GHG emissions and indoor pollution, do not put pressures on the use of agricultural land for fuelwood extraction, and can further mitigate land degradation in the Sudano-Sahelian region. Solar stoves, for example, were promoted widely in the last two decades in regions where tree vegetation and fuelwood were scarce or not available at all, like Tibet and the Alti Plano in the Andes (GTZ, 2007). In other regions, such as the Sudano-Sahelian region, the public acceptance of solar stoves in poor and rural regions was very low because such stoves were still far too expensive, required people to change their cooking habits, and could not provide any heat during the night (GTZ, 2007, p. 19). Furthermore, in many parts of Africa, the establishment of a viable industry to supply spare parts and to ensure the maintenance of such stoves (GTZ, 2007, p. 3) is still a challenge that will take time to overcome.

1.3.6.3 OFF-GRID AND MINIGRID SOLUTIONS

Off-grid and minigrid solutions deploying PVs and wind energy in rural areas for electricity generation can be cheap [roughly US $300 per MWh (IEA 2014a, p. 128)] and efficient solutions for providing electricity to isolated regions that are not connected to any power grid yet and where it might not be economically or technically feasible to connect households to the national grid. Minigrids further offer the possibility to integrate and combine different energy sources (such as hybrid, diesel, or renewable energy power systems). Furthermore, the indicated levelized costs for minigrids and off-grid solutions are less[20] if these are powered by PVs, small hydropower capacities, or wind energy than when they

---

[20]As an average for sub-Saharan Africa and a diesel price of $1 per liter.

solely rely on diesel generators for electricity production (IEA, 2014a, p. 128), although they require batteries in order to continue providing electricity once it is dark or if there is not enough wind. The most important barriers to implementing such solutions at present are the lack of knowledge about their advantages to provide clean and reliable electricity access to rural and isolated areas and the lack of financial capital and resources for deployment and long-term maintenance.

1.3.6.4 RENEWABLE ENERGIES FOR POWER GENERATION[21]

The large-scale deployment of renewable energy technologies for power generation can offer the Sudano-Sahelian region a unique chance to provide people with access to electricity; reduce the demand for traditional biomass use for cooking, heating, and lighting; reduce costly and unsustainable fossil-fuel based power generation; and invest in a sustainable future, both in terms of environmental protection and in economic terms. Furthermore, the deployment of renewable energy technologies provides an opportunity for the region to take advantage of its significant yet underexploited renewable energy potential, to apply the experience from the recent technological progress and cost reductions in large-scale renewable power generation technologies, benefit from the low running costs in the long term, and possibly leapfrog the path taken by industrialized countries by moving directly to renewable-based energy systems.

According to IRENA (2012, p. 16), hydropower has the greatest potential for generating power at low costs, especially large-scale hydropower capacities, in some countries in the Sudano-Sahelian region, such as Mali. Wind energy (cf. section 1.3.4) has significant potential to contribute to the region's energy security by providing abundant clean energy and easier access to it for both rural and urban populations.

IRENA estimates that in the Sudano-Sahelian region, solar energy has the largest technical potential, especially PVs and CSP, but their deployment is still quite costly, especially CSP technologies (IRENA, 2012, p. 21). On the other hand, the costs for solar energy technologies are decreasing quickly, and large-scale deployment across countries where solar energy is particularly abundant might bring costs down and make these technologies competitive in the future.

The major challenges for the deployment of renewable energies in the region are the lack of funding to develop and deploy modern technology solutions and grids, the lack of an industrial manufacturing base, and the lack of installations, maintenance, and service industries that would be required to support these investments and operate the infrastructure in the long run (IRENA, 2012, p. 44). These challenges are further exacerbated by weak governance at the national level and the high political and financial risk for investors in most of the countries throughout the region.

1.3.7 CONCLUSIONS AND RECOMMENDATIONS

The findings of this study illustrate that the lack of energy access and of affordable and clean energy solutions for the majority of the population of the Sudano-Sahelian region undermines energy security in these countries and leads to a significant demand for fuelwood, subsequently contributing considerably to land degradation. Even considerable improvements in energy access and in the use of modern

---

[21]Cf. IRENA (2012, 2014).

energy technologies like those reflected in the New Policy Scenario of the IEA are still likely to lead to an increase in the demand for fuelwood and consequently to further land degradation.

The findings indicate that energy systems in the countries in the region are not adequate to prevent or mitigate supply risks or to withstand unpredictable factors such as climate-related extreme events, energy supply shocks, and price shocks on international markets, or infrastructure challenges. These systems' security is often extremely dependent on both natural phenomena such as droughts or heavy rainfall and on technological factors, such as aging or missing infrastructure links and lack of economic and technical capacity to build, maintain, and develop modern infrastructure and to provide modern energy services. The examples of Sudan, South Sudan, and Mali illustrate that the functioning of their energy systems is easily threatened by such disruptions or stresses, which indicates that they are not robust and resilient enough to withstand such risks and to meet them adequately without threatening local livelihoods. Land degradation as a phenomenon is a direct risk to energy systems and adversely affects energy security in the Sudano-Sahelian region but on the other hand, it is also a consequence of the vulnerabilities and inadequacies of these energy systems. In addition, land degradation processes further challenge the ability of states to meet the objectives of supplying secure, affordable, and clean energy for their citizens and impedes any efforts to resolve fundamental challenges of food insecurity, poverty, and human security. In this context, it is important to understand the links between energy security and land management practices and to translate these into solutions that enhance both energy security through sustainable energy sources and technologies and through ensuring that land resources are adequately protected.

Therefore, appropriate measures are required in order to reverse these trends as soon as possible and to take the pressure off land resources. Some of the proposed technical solutions would require significant upfront investments, such as the widespread penetration of RES, but tailor-made and innovative policy, regulatory, and financial mechanisms have the potential to generate or attract such investments for the deployment of sustainable solutions and for solving numerous environmental, health, and socioeconomic challenges in the long run.

The future development of the bioeconomy will further shift the traditional paradigm of natural capital, land and water resources and will offer a new development path to allow both industrialized and developing countries to meet together growing challenges due to global resource scarcity or depletion.

In light of these development challenges, the study offers four technological recommendations, which should be accompanied by the design and implementation of appropriate policies, regulations and institutional measures to enable governments to successfully implement them and reap the benefits of these solutions.

1.3.7.1 TECHNOLOGICAL RECOMMENDATIONS

1. The development of sustainable agricultural practices, including agroforestry, has the potential to contribute the most to restoring degraded land and ecosystem services, improving soil quality and productivity, and at the same time strengthening energy supply and energy security. The ultimate benefit of agroforestry in the Sudano-Sahelian region would be to mitigate land degradation and support the poorest communities in having access to land and energy sources as a way to ensure their livelihoods in the short term to midterm.
2. The large-scale deployment of renewables-based technological applications, such as minigrids, solar stoves, and off-grid solutions, can further support self-sufficiency for isolated households and rural communities to allow their own energy production. Such solutions are key to enhancing

energy security, reducing energy poverty, and decoupling land degradation due to the overuse of biomass and the extensive cultivation of biofuel feedstock from energy supply and access concerns in the region. Furthermore, they can eliminate the problem of biofuel crops displacing food crops and food production for local communities, which will ultimately help in solving the problem of undernourishment. In addition, such technological applications can create local employment opportunities and deliver sustainable energy for poorer and isolated communities, thus contributing to local development and human security and well-being.

3. The Sudano-Sahelian region has significant geographical and technical potential for the development and deployment of renewable energies. Appropriate infrastructure needs to be built to support renewable energy penetration. Additionally, the development and deployment of infrastructure should be pursued by connecting local technological solutions such as minigrids to the national grid wherever this is economically and technically feasible, and by exploiting the potential of cross-border energy cooperation in the long run. Renewable energies will be very important in bringing energy independence, secure and affordable fuel supplies at the local level, and improving energy security and sustainable resource management for countries in the Sudano-Sahelian region.

4. The possibilities for hydrogen production and use should be further explored in the long run, including hydrogen production from domestic resources (fossil fuels combined with carbon sequestration methods), natural gas, and biomass or waste or using renewable energy sources, such as wind, solar, or hydroelectric power. Hydrogen can be produced at different scales: by large central plants, semicentrally, or in small distributed systems located at or very near the point of use (refueling stations, stationary power sites, etc.), which allows a variety of tailor-made solutions depending on the specific needs of local communities and the energy sources available. Hydrogen and fuel cell applications can have enormous benefits to local communities such as reducing GHG emissions, eliminating fossil fuel pollution, reducing overall dependence on fossil fuel imports, and improving economic and energy independence.

Policy, Institutional, and Regulatory Actions—Recommendations

A stable legal, institutional, and policy framework is a prerequisite for the development of a favorable environment and proper incentives for the commercial deployment of renewable energy technologies to support access to clean, affordable, and secure energy for citizens and for the development of robust, resilient, and secure energy systems in the Sudano-Sahelian region.

Land ownership security provides the incentives required for farmers to cultivate their own land in a sustainable manner and to invest in that land in the longer run as well. If land is just taken away from smallholder farmers without adequate compensation for the purpose of renting this land out to foreign investors, other farmers will also lose the motivation to cultivate their land in a sustainable manner. Therefore, guaranteeing land tenure and property rights is crucial to ensure that land is managed in a sustainable manner and that risks of land degradation, land-use conflicts, or displacement of people, resources, production, and land use are avoided altogether in the future.

The better the governance, the better the administration can guarantee to farmers that their rights are properly protected and that they do not have to fear being forced to leave their land. In this respect, a decentralized administration structure seems to have advantages because it allows the administration to have better access to the farmers in their district, while farmers can gain more confidence in the administration at the same time. Mali, for example, is in the process of establishing a decentralized administration system to support better property rights protection and law implementation at both the local and national levels.

As bioenergy demand is anticipated to increase in the next decades on a global level as a means to reduce fossil fuel dependence, sustainable land-use management will be important to offset pressures on land conversion in some parts of the world to satisfy the demand for biofuel production. Sustainable resource management practices should be developed and implemented in the Sudano-Sahelian region to ensure that energy sources and land resources are not interdependent, as they currently are in very poor or rural communities. National governments need to make sure that private actors follow certain sustainable land-use management practices and supply chains. In addition, the global market for biofuels will add more pressure on different participants, both producers and consumers, to benefit from the international trade of biofuels. Sustainability criteria for biofuels and biofuel supply chains and proper certification schemes could be useful instruments to stimulate the development of advanced biofuels and sustainable land-use management, especially across the most vulnerable regions.

Incentives such as tax credits, national regulations that prioritize clean energy feed-in before fossil fuels, grant schemes, and low-interest loans all have the potential to help poorer people gain access to modern and clean energy solutions, and they can encourage both domestic and foreign investors to develop and promote renewable energy technologies across different types of users and different markets. In addition, public-private partnerships in the field of renewable energy development can provide an additional boost to solutions with great public value and benefits that otherwise would be too expensive for a single private investor to pursue on a larger scale. Such partnerships can further bridge the gap between technology solution development and market deployment and can attract more foreign capital and technology exchange with industrialized countries to support clean and sustainable development objectives in the region.

Schemes like the German Renewable Energy Act, which supports financially renewable energy technologies in order to improve their attractiveness and competitiveness and increase their penetration could be established in the short term. In the longer run, progressively decreasing subsidies will allow these technologies to be self-sustaining and reach competitiveness with other traditional technologies for energy production. On the other hand, fewer subsidies for fossil fuels, or even a carbon tax, could make renewable energies more attractive and eventually more competitive.

The support of the private sector for more expensive or larger-scale energy projects is of fundamental importance because the majority of the countries in the Sudano-Sahelian region do not have the capacity to invest significant domestic capital to finance such projects. So, in case they are not co-funded by bilateral development aid or the international development institutions or banks, the private sector capital is the only possibility to attract or generate funds for such large scale investments.

Support by the international community of capacity-building efforts to inform and train local smallholder farmers and to promote sustainable land-use practices is crucial to the Sudano-Sahelian region due to the significant extent of land degradation and severe resource scarcity challenges. Capacity building is important both at the local and community level and at the governmental level so that governments, nongovernmental organizations (NGOs) and businesses are equally aware of the impact of land-use management practices on the energy balance and security of their respective regions. Therefore, they can be equally well equipped to support the local efforts in improving sustainable land, forest, and resource management through the successful implementation of such practices. In addition, technology transfer and exchange would be very important to enable countries in the region to strengthen their technological base, improve local know-how, tackle poverty and lack of access to education and information, and support the effective implementation of both environmentally friendly technologies and best practices.

REFERENCES

Asia Pacific Energy Research Centre (APERC), 2007. A Quest for Energy Security in the 21st Century: Resources and Constraints. Institute of Energy Economics, Tokyo, Japan.

Bai, Z.G., Dent, D.L., Olsson, L., Schaepman, M.E., 2008. Global Assessment of Land Degradation and Improvement. 1. Identification by remote sensing. GLADA Report 5, Version November 2008 World Soil Information, FAO, Rome.

Bayala, J., Kalinganire, A., Tchoundjeu, Z., Sinclair, F., Garrity, D., 2011. Conservation Agriculture with Trees in the West African Sahel—a review. World Agroforestry Centre, Occasional Paper 14. Nairobi, Kenia.

Bielecki, J., 2002. Energy security: Is the wolf at the door? Q. Rev. Econ. Finance 42 (2), 235–250.

Bohi, D.R., Toman, M.A., 1996. The Economics of Energy Security. Kluwer Academic Publishers, Boston.

Brundtland Report, 1987. Report of the World Commission on Environment and Development: Our Common Future. Transmitted to the General Assembly as an Annex to document A/42/427 - Development and International Co-operation: Environment, http://wwwun-documents.net/our-common-future.pdf. Accessed on 15/12/2014.

Cherp, A., Jewell, J., 2014. The concept of energy security: Beyond the four As. Energy Policy 75 (2014), 415–421.

Chester, L., 2010. Conceptualising energy security and making explicit its polysemic nature. Energ. Pol. 28 (2), 887–895.

Coulibaly, A., 2006. Country Pasture/Forage Resource Profile Mali. FAO, Rome. http://www.fao.org/ag/agp/AGPC/doc/Counprof/PDF%20files/Mali-English.pdf.

Deutsche Gesellschaft für Technische Zusammenarbeit (GTZ), 2007. Here Comes the Sun. Options for Using Solar Cookers in Developing Countries. Eschborn, Germany. http://cedesol.org/docs/english/gtz-en-here-comes-the-sun-2007.pdf.

Deutsche Gesellschaft für Internationale Zusammenarbeit (GIZ), 2011. Micro-gasification: Cooking with Gas from Biomass. An Introduction to the Concept and the Applications of Wood-Gas Burning Technologies for Cooking. Eschborn, Germany. https://energypedia.info/index.php?title=File:Micro_Gasification_Cooking_with_gas_from_biomass.pdf&page=1 (accessed 27.11.2014).

Economics of Land Degradation (ELD), 2011. The Economics of Desertification, Land Degradation and Drought. Toward an Integrated Global Assessment. ZEF—Discussion Papers on Development No. 150. Bonn, Germany.

Energy Information Administraion of the United States of America (US EIA), 2014. Country Analysis Brief: Sudan and South Sudan. Update of September 3, 2014. http://www.eia.gov/countries/cab.cfm?fips=su.

European Commission, 2014. ECHO Factsheet: Sahel: Food and Nutrition Crisis. EC Humanitarian Aid and Civil Protection, Brussels.

European Commission, March 27, 2013a. Report from the Commission to the European Parliament, the Council, the European Economic and Social Committee, and the Commission of the Regions. Renewable Energy Progress Rep. Com, Brussels.

European Commission, 2013b. Soil Atlas of Africa. European Commission, Joint Research Centre, Luxembourg.

European Commission, 2012. Indirect Land-use Change. Press release of October 17, 2012. http://europa.eu/rapid/press-release_MEMO-12-787_en.htm.

European Commission, November 29, 2000. Green Paper: Towards a European Strategy for the Security of Energy Supply.

Fonseca, M.B., et al., 2010. Impacts of the EU Biofuel Target on Agricultural Markets and Land Use: A Comparative Modelling Assessment. Institute for Prospective Technological Studies. Joint Research Institute of the European Commission Reference Report, Seville, Spain.

Food and Agriculture Organisation of the United Nations (FAO), 2014a. AQUASTAT. Irrigation Water Requirement and Water Withdrawal by Country. http://www.fao.org/nr/water/aquastat/water_use_agr/index.stm.

Food and Agriculture Organisation of the United Nations (FAO), 2014b. The State of Food Insecurity in the World. Strengthening the enabling environment for food security and nutrition. FAO, International Fund for Agricultural Development, World Food Programme. FAO, Rome, Italy.

Food and Agriculture Organisation of the United Nations (FAO), 2013: AQUASTAT.FAO country profile of Mali. http://www.fao.org/nr/water/aquastat/countries_regions/mli/index.stm (accessed 17.12.2014).

Food and Agriculture Organisation of the United Nations (FAO), 2005. AQUASTAT. FAO country profile of Sudan. http://www.fao.org/nr/water/aquastat/countries_regions/SDN/index.stm (accessed 13.12.2014).

Garrithy, D.P., et al., 2010. Evergreen agriculture: A robust approach to sustainable food security in Africa. In: Food Security. The Science, Sociology, and Economics of Food Production and Access to Food2Springer 2, 197–214. http://dx.doi.org/10.1007/s12571-010-0070-7.

Haken, N., et al., 2014. Fragile States Index 2014. The Fund for Peace. Washington, DC. http://library.fundforpeace.org/library/cfsir1423-fragilestatesindex2014-06d.pdf.

Halstead, M., Kober, T., van der Zwaan, B., 2014. Understanding the Energy-Water Nexus. Energy research Centre of the Netherlands (ECN). Petten, the Netherlands.

High-Level Panel of Experts on Food Security and Nutrition, Committee on World Food Security, Food and Agriculture Organisation of the United Nations (FAO) (HLPE), 2013. Biofuels and Food Security. A Report by the High-Level Panel of Experts on Food Security and Nutrition of the Committee on World Food Security.

International Energy Agency (IEA), 2015. Website of IEA on bio-energy. http://www.iea.org/topics/renewables/subtopics/bioenergy/.

International Energy Agency (IEA), 2014a. African Energy Outlook. A Focus on Energy Prospects in Sub-Sahara Africa. World Energy Outlook Special Report, International Energy Agency, Paris.

International Energy Agency (IEA), 2014b. What is energy security? http://www.iea.org/topics/energysecurity/subtopics/whatisenergysecurity/ (accessed 07.01.2015).

International Energy Agency (IEA), 2014c. World Energy Outlook 2014. IEA. Paris, France.

International Energy Agency (IEA), 2013. World Energy Outlook 2013. IEA. Paris, France.

International Energy Agency (IEA), 2012. World Energy Outlook 2012. IEA. Paris, France.

International Energy Agency (IEA), 2007. Energy Security and Climate Policy: Assessing Interactions. OECD/IEA, Paris (March).

International Energy Agency (IEA), 2002. Energy Security. OECD/IEA, Paris.

International Energy Agency (IEA), 1995. The IEA Natural Gas Security Study. OECD/IEA, Paris.

International Monetary Fund (IMF), 2013. Sudan 2013 Article IV Consultation. IMF Country Report No. 13/317. October 2013, IMF, Washington, DC.

International Monetary Fund (IMF), 2012. Fiscal Regimes for Extractive Industries: Design and Implementation. IMF, Washington, DC.

International Renewable Energies Agency (IRENA), 2014. Estimating the Renewable Energy Potential in Africa. IRENA Headquarters, Abu Dhabi, United Arabic Emirates.

International Renewable Energies Agency (IRENA), 2012. Prospects for the African Power Sector. IRENA Headquarters, Abu Dhabi.

Jewell, J., 2011. The IEA Model of Short-Term Energy Security (MOSES): Primary Energy Sources and Secondary Fuels. Working Paper, IEA, Paris.

Landmatrix, 2014. Website of "International Land Coalition", "CIRAD (French organisation for agricultural development)", "Ub", "Leibnitz Institut für Globale und Regionale Studien", and "Gesellschaft für Internationale Zusammenarbeit". http://www.landmatrix.org/en/.

Mali, 2011. Republic of Mali, Ministry of Energy and Water Resources, National Department of Energy. Renewable Energy Mali: Achievements, Challenges, and Opportunities. Bamako, Mali.

Millennium Ecosystem Assessment (MEA), 2005. Ecosystems and Human Well-being: Synthesis. Island Press, Washington, DC.

OCHA, 2013. Sahel Regional Strategy. Mid-year review 2013. https://docs.unocha.org/sites/dms/CAP/MYR_2013_Sahel_Regional_Strategy.pdf (accessed 17.07.2015).

OCHA, 2014. 2014–2016 Strategic Response Plan Sahel Region. http://www.fao.org/fileadmin/user_upload/newsroom/docs/Regional%20Sahel%20SRP%20Final.pdf (accessed 12.12.2014).

Quang, B.L., Nkonya, E., Mirzabaev, A., 2014. Biomass Productivity-Based Mapping of Global Land Degradation Hotspots, ZEF—Discussion Papers on Development Policy No. 193, Center for Development Research, Bonn, Germany.

Reij, Ch., Tappan, G, Smale, M., 2009. Agroenvironmental Transformation in the Sahel: Another Kind of "Green Revolution." International Food Policy Research Institute, Discussion Paper 00914, Addis Ababa, Ethiopia.

REN21, 2013. Renewables Global Futures Report 2013. Paris.

Renewable Energy Policy Network for the 21st Century (REN21), 2014. ECOWAS Renewable Energies and Energy Efficiency Status Report 2014. Paris.

Scaling-Up Renewable Energy Program for Low Income Countries (SREP), 2011. Republic of Mali Ministry of Energy and Water, National Directorate of Energy, SREP—Mali Investment Plan: Scaling Up Renewable Energy, Vol. 1: Investment Plan Annexes. Bamako, Mali. http://www.afdb.org/fileadmin/uploads/afdb/Documents/Project-and-Operations/SREP-Mali_IP_Volume1_EN_21Sept%20(2).pdf. Accessed 09/01/2015.

Schubert, R., et al., 2009. World in Transition: Future Bioenergy and Sustainable Land Use. Flagship Report WBGU 2007. Berlin.

Statistics of the Food and Agriculture Organisation of the United Nations (FAOSTAT), 2015. http://faostat3.fao.org/home/E (accessed 03.02.2015).

Terrestrial Carbon Group Project (TCG), 2011. Innovative Approaches to Land in the Climate Change Solution. Policy Brief.

United Nations Convention to Combat Desertification (UNCCD), 2012. Policy Brief. Zero Net Land Degradation. May 2012. Bonn, Germany. http://www.unccd.int/Lists/SiteDocumentLibrary/Rio+20/UNCCD_PolicyBrief_ZeroNetLandDegradation.pdf (accessed 09.11.2014).

United Nations Convention to Combat Desertification (UNCCD), 2009. African Dryland Commodity Atlas. Secretariat of the United Nations Convention to Combat Desertification and the Common Fund for Commodities. Bonn, Germany.

United Nations Environmental Programme (UNEP), 2007. Sudan: Post-conflict Environmental Assessment. UNEP, Nairobi. http://postconflict.unep.ch/publications/UNEP_Sudan.pdf (accessed 06.12.2014).

United Nations Environment Programme. (UNEP), 2002. Global Environment Outlook 3. UNEP, Nairobi, Kenya.

United States Department of Agriculture—Natural Resource Conservation Service Soils (USDA–NRCS), 1998. Global Desertification Vulnerability Map. Soil Science Division, World Soil Resources, Washington, DC.

United Nations, 2015. The Millennium Development Goals Report 2015. New York, USA.

United Nations Convention to Combat Desertification (UNCCD), 2012. Policy Brief. Zero Net Land Degradation. May 2012. Bonn, Germany. http://www.unccd.int/Lists/SiteDocumentLibrary/Rio+20/UNCCD_PolicyBrief_ZeroNetLandDegradation.pdf.

United Nations Development Programme (UNDP), 2014. Human Development Report 2014. Reducing Vulnerabilities and Building Resilience. UNDP, New York, Sustaining Human Progress.

United Nations Development Programme (UNDP)/World Health Organisation (WHO), 2009. The Energy Access Situation in Developing Countries. A Review Focusing on the Least Developed Countries and Sub-Saharan Africa, UNEP, New York.

United Nations Office for the Coordination of Humanitarian Affairs (OCHA), 2013. Sahel Regional Strategy. Midterm Review 2013. https://docs.unocha.org/sites/dms/CAP/MYR_2013_Sahel_Regional_Strategy.pdf.

United Nations University (UNU), 2009. Does Climate Change Cause Conflict? Our World. Brought to You by the United Nations. http://ourworld.unu.edu/en/does-climate-change-cause-conflict. (accessed 07.12.2014)

United Nations World Water Assessment Programme (UN WWAP), 2014. The United Nations World Water Development Report 2014: Water and Energy. UNESCO, Paris.

Van Schaik, L., Dinnissen, R., 2014. Terra Incognita: Land Degradation as Underestimated Threat Multiplier. Clingendael ReportNetherlands Institute of International Relations, The Hague, the Netherlands.

Vlek, P.L.G., Le, Q.B., Tamene, L., 2010. Assessment of land degradation, its possible causes, and threat to food security in sub-Saharan Africa. In: Lal, R., Stewart, B.A. (Eds.), Advances in Soil Science: Food Security and Soil Quality. Taylor and Francis Group, LLC, pp. 57–87 1. Food security. 2. Soils—Quality. I. Lal, R. II. Stewart, B. A. (Bobby Alton), 1932– III. Series: Advances in soil science (Boca Raton, Fla.).

World Bank, 2015: Online Data Bank. http://data.worldbank.org/indicator#topic-19.

APPENDIX DEVELOPMENT INDICATORS

Table 1.3.3 Development Indicators(1)

Country	Human Development Index**	Fragile States Index***	Population Growth****	Fertility Rate****
	2013	Rank 2014	% per Year (2000–2012)	Birth per Woman (2000–2015)
Benin	165	74	3.1	4.9
Burkina Faso	181	39	2.9	5.7
Chad	184	6	3.4	6.3
Cote d'Ivoire	171	14	1.7	4.9
Gambia	172	59	3.1	5.8
Ghana	138	118	2.5	3.9
Guinea	179	12	2.2	5
Guinea-Bissau	177	16	2.2	5
Mali	176	36	3.1	6.9
Mauritania	161	28	2.8	4.7
Niger	187	19	3.7	7.6
Nigeria	152	17	2.6	6
Senegal	163	62	2.8	5
Sierra-Leone	183	35	3.1	4.8
Sudan*	166	5	2.4	4.5
Togo	166	41	2.6	4.7
Entire region			**2.7**	

*In its borders before 2011, South Sudan was ranking first in the Fragile States Index 2014
**Source: UNDP (2014).
***Source: Haken et al. (2014).
****Source: World Bank (2015).

Table 1.3.4 Development Indicators (2)

Country	Undernourished People**	Cereal Import Dependency Ratio***	Population Living on Degraded Land****	Gross National Product per Capita*****
	% of Total Population 2012–2014	2009–2011	% of Total Population 2010	In Purchase Power Parities, 2012
Benin	9.7	36.2	1.6	1570
Burkina Faso	20.7	10.4	73.2	1510
Chad	34.8	9.5	45.4	1320
Cote d'Ivoire	14.7	61.3	1.3	1960
Gambia	6	43.8	17.9	510
Ghana	<5	26.4	1.4	1940
Guinea	18.1	14.3	0.8	980
Guinea-Bissau	17.7	31.4	1	550
Mali	<5	4.7	59.5	1160
Mauritania	6.5	74	23.8	2520
Niger	11.3	7.9	25	650
Nigeria	6.4	21.7	11.5	2420
Senegal	16.7	51.2	16.2	1920
Sierra-Leone	25.5	19.7	N/A	1360
Sudan*	24.3******	26.6	39.9	2030
Togo	15.3	15.8	5.1	920
Entire region			18.4	

*In its borders before 2011.
**Source: FAO (2014b). Note: most recent data from the Sahel Strategy are differing considerably especially in Mali and Niger at rates above 20% of the population in 2014.
***FAOSTAT (2015).
****Source: UNDP (2014).
*****World Bank (2015).
******Data for the average from 2008–2010.

ENABLING GOVERNANCE FOR SUSTAINABLE LAND MANAGEMENT

1.4

Jonathan Davies

International Union for Conservation of Nature

1.4.1 INTRODUCTION

Living on the planet Earth, we often forget that human welfare depends on land—on earth itself—and yet land is becoming a scare commodity. A total of 12 million ha of land, where 20 million tons of grain could have been grown, disappear every year due to human activities (UNCCD, 2011). Land is declining in terms of actual, physical area as it is exhausted, abandoned, or put to other uses. It is also declining in productivity as soil fertility is mined and water resources are overexploited (Bruinsma, 2003). This precipitous decline in land has led to major increases in land prices in some countries, which is driving speculation and accumulation of land in the hands of a few, with inevitable consequences for the poorest people and the poorest countries: that is, those with the lowest purchasing power and the weakest land rights. The outcome has been a rise in the phenomenon of "land grabbing": since the 2008 food crisis, between 15 and 20 million ha of farmland in developing countries has changed hands (von Braun and Meinzen-Dick, 2009). Around the world, some 50 million people may be displaced in the 10 years up to 2022 as a result of desertification (UNCCD, n.d.).

However, land degradation not only affects the poor, but it also extends to all segments of global society. Land degradation, for example, contributes to climate change by releasing carbon into the atmosphere and reducing the capacity of land to absorb carbon. More than 2700 gigatons (Gt) of carbon are stored in soil worldwide, which is significantly more than the combined total of the atmosphere (780 Gt) and biomass (575 Gt) (Lal, 2008). In addition, vast quantities of atmospheric carbon can be sequestered through sustainable land management practices, such as in the 5 billion ha of rangelands worldwide that store up to 30% of the world's soil carbon; improved rangeland management has the biophysical potential to sequester 1300–2000 megatons (Mt) of carbon dioxide equivalent (MtCO2e) worldwide by 2030 (Wilkes and Tennigkeit, 2008).

Land degradation contributes to a decline in other ecosystem functions besides carbon cycles. It can have a major impact on hydrological cycles, reducing infiltration and increasing runoff, which contribute to cycles of flood and drought. Vegetation cover and soil organisms play vital roles in water

Land Restoration. http://dx.doi.org/10.1016/B978-0-12-801231-4.00006-9

infiltration and, therefore, in maintaining soil moisture and recharging aquifers. As a consequence of land degradation, groundwater resources, and especially shallow unconfined aquifers, can be seriously affected (FAO, 1993).

Land degradation also contributes, both directly and indirectly, to the loss of biodiversity. In some cases, this is obvious, such as when habitat is destroyed by clearing forests to create agricultural land. Agricultural land may still be sustainably managed, though the implications for biodiversity can be far-reaching. Converting rangelands to cropland can lead to a major decline in termite populations, whose winged "alates" provide vital nourishment for migratory birds; we remain poorly informed of the role of land use changes and desertification in the loss of trans-Saharan migratory birds (Davies et al., 2012). Land degradation has very serious economic consequences as well. Globally, the cost of deforestation and land degradation has been estimated at up to €1.5–3.4 trillion or 3.3%–7.5% of the global gross domestic product (GDP) in 2008 (TEEB, 2008). We take land for granted, and only now are we discovering that it is a finite resource that is nearing its limits.

1.4.2 LAND DEGRADATION AND CONFLICT

A number of reports have been published suggesting links between conflict and the use and scarcity of environmental resources in general, and more specifically between conflict and land degradation. UNEP (2009) reports that at least 18 violent conflicts since 1990 have been fueled by the exploitation of natural resources, and in the previous 60 years, as much as 40% of intrastate conflicts were linked to natural resources. This includes some of the "usual suspects," such as minerals and oil, but also timber and land. For example, conflicts in Darfur and the Middle East are reportedly related to the control of fertile land and water. Such conflicts are projected to rise as the global population grows and as resources become comparatively scarcer, particularly where compromised by climate change. Food scarcity and rising food prices were identified as significant triggers of the recent conflicts in the Middle East and North Africa and may increasingly result in social unrest globally (Lagi et al., 2011).

Land degradation may contribute to conflict at both the local and international levels. If land degradation contributes to overall land scarcity, it may be a factor in conflicts where one group acquires land over which another group claims ownership. At an international level, land degradation may lead to dust storms or siltation of rivers, which can lead to disagreements between nation-states. Most of the world's largest rivers, for example, are shared by more than one country, and the water supplies of millions of people depend on international cooperation between governments.

Sudan's conflict in Darfur, which according to multiple media sources led to over 200,000 deaths and displaced 2.5 million people, has been portrayed by some as a conflict over land, evidenced by the scorched-earth tactics of militias. Although links have been made to land degradation (UNEP, 2007), the situation is more complex in reality. Bantekas (2010) finds that the confluence of a number of environmental factors contributed to the conflict, including a significant decline in rainfall, an increase in the human population, and land mismanagement that led to soil depletion and deforestation. Interestingly, flooding is also cited as a factor alongside drought, indicating an overall breakdown of fundamental ecosystem services as a result of land degradation. Yet, other factors—including ethnic rivalries and government biases—are considered to be of greater significance than the environmental ones. Some authors warn of the risk of conflict discussions being distorted by pre-ordained narratives, when in reality, the situation is too complex to empirically attribute to environmental factors (Forsyth and Schomerus, 2013; Kevane and Gray, 2008).

Attributing conflicts to natural resource scarcity, and particularly to land degradation, is overly simplistic. Conflicts usually have multiple overlapping factors, such as ethnic and religious differences, historical grievances, the legacy of colonialism, and economic adversity. Conflicts related to land have arisen where different groups claim rights over the same resource, such as in the Sahel, where pastoral and crop farming communities have historically shared rights of access or use (Cotula, 2006). Conflicts have arisen as a result of the uncertainty created by the legacy of nationalization of land, through changing political ideologies, or through the legacy of colonialism, creating uncertainty over legal arrangements and complex, overlapping claims (IUCN, 2011). Although there is a link between land degradation and conflict—whether it is land degradation as a result of conflict and poorly managed resource competition, or vice versa—we should not be too eager to blame the victims.

Furthermore, competition is not conflict. Competition over natural resources is to be expected, and in most cases, it does not escalate into conflict. Many communities have strong and long-standing traditions of resource sharing, and these are at the heart of governance of the commons (Ostrom, 1990). Where conflicts arise over local resource use, the pertinent question is, "What happened to these local arrangements?" Often, the answer lies in changing power relations between different groups, both internally (for example, as a result of population growth and changing social fabric) and externally (for example, the emergent power of government and nongovernmental actors) (Herrera et al., 2014; IUCN, 2011; Toulmin and Quan, 2000).

1.4.3 GOVERNANCE: A COMMON DENOMINATOR[1]

In many cases, both land degradation and conflict are the outcomes of governance weaknesses. This may be particularly the case in drylands, which are remote, poorly served by such basic services as security, and have weak legal recognition of land rights. In many cases, customary governance arrangements have been eroded in recent years, with the state emerging as a powerful actor in determining resource rights while often lacking the capacity to take responsibility for sustainable resource use. The confluence of weak property rights and poor enforcement is an important emerging factor in conflicts today (Glazer and Konrad, 2003).

In some cases, it may be difficult to differentiate between conflict and competition over resources; indeed, the two may lie on a continuum. It is clear, however, that where there is competition over resources, there is a need to enforce rules and regulations. There is a growing understanding of the role of governance, including the rule of law, in sustainable natural resource management, particularly for communally managed resources (including air, water, and the majority of land and forests); this has contributed to a growth in initiatives to strengthen or restore governance systems. Since governance is embedded in culture and behavior as much as in legal statutes, strengthening it can be a complex task. To simplify this, governance can be broken down into three main elements (IUCN, 2000):

- Rules, norms, institutions and processes that determine how power and responsibilities are exercised
- Decision making
- Citizens' participation

---

[1]Many of the examples given here are taken from Herrera et al. (2014), which includes case studies from Botswana, Cameroon, Jordan, Kenya, Lebanon, Mongolia, Morocco, Spain, the United States, and West Africa.

Governance can be defined as the expression of the relationship between citizens, leaders, and public institutions to manage land and other resources (DFID, 2006). In terms of the government's role, there are three main factors to consider (Kaufmann et al., 1999):

- Government's capacity to accomplish goals
- Responsiveness of public policies and institutions to the needs of citizens
- Accountability

From the perspective of citizens, the factors under consideration are their participation in decision making and how society addresses compliance with the established laws and accounts for the government's actions. The sustainable use of natural resources cannot be achieved without fair access and control of natural resources. Improving governance, therefore, starts with empowering communities, recognizing their rights and responsibilities, and ensuring equity and economic development (IUCN, 2000). The following sections of this part highlight a number of examples that illustrate some of the key principles in strengthening local governance over resources (Herrera et al., 2014).

1.4.3.1 RESTORING THE INFLUENCE OF TRADITIONAL LEADERSHIP

Governance in many societies is vested in customary institutions; thus, the key to strengthening governance is often to revive and reform those institutions. This is a delicate task that is achieved through extensive negotiations with communities, a concerted effort to empower those communities, and thorough adherence to the principles of Free, Prior, and Informed Consent (FPIC). For example, in Morocco, work to strengthen the traditional practice of protecting communal rangeland areas in what are known as *Aghdals* found that acknowledging tribal rights was the cornerstone of success. Specifically, tribes established criteria to reorganize rural communes for communal resource management (Boutaleb and Firmian, 2014). Similarly, strengthening communal rangeland management in Kenya was made possible by closely involving tribal institutions in the development of new management systems and through legal recognition of the customary institutions of the Boran community (Roba, 2014).

1.4.3.2 RECOVERING ANCESTRAL KNOWLEDGE

In many communities, local and indigenous knowledge is critical to sustainable land management, as it is local institutions and agreements that enable such knowledge to be deployed effectively. Local knowledge has enabled land users in Lebanon and Jordan, for instance, to restore degraded land and to put in place management regimes for sustainability, which has improved both agricultural productivity and conservation of biological diversity (Sattout, 2014; Haddad, 2014).

1.4.3.3 UPDATING TRADITIONAL SYSTEMS

Whilst traditional governance and knowledge systems offer many possibilities for sustainable land management, they need assistance in many cases to keep up to date with the ever-changing context, including changes in climate, the economy, and national politics. As experiences in East Morocco (Boutaleb and Firmian, 2014) and Kenya (Roba, 2014) illustrate, effective governance requires not only the revival, but also the evolution of local institutions. Such evolutions must come from within the society, often posing particular challenges for marginalized groups within the society, such as women. Updating traditional systems is a sensitive process requiring skillful facilitation and extensive effort at empowerment.

1.4.3.4 EMPOWERING WOMEN THROUGH GOVERNANCE

Caution must be taken in strengthening customary institutions so that historical inequities are not reinforced. There are particular concerns about safeguarding and promoting the role of women in natural resource management, which must be considered by overall efforts to strengthen governance. Through dialogue with communities, it is possible to allow women to raise their voices and contribute to their empowerment to give them a greater say in the way natural resources are governed. In Jordan, this has been achieved by emphasizing women's rights and responsibilities and by giving explicit attention to securing their land rights through customary and statutory law (Haddad, 2014).

1.4.3.5 ENHANCING SOCIAL FABRIC AND GRASSROOTS ORGANIZATIONS

Grassroots organizations often play a central role in strengthening local resource governance. Supportive social networks, such as those provided by community-based organizations, cooperatives, and associations, are important for empowerment, accountability, and capacity, and they are often the link between government and communities. To be effective, therefore, these networks must be locally and culturally acceptable and usually must be deeply embedded in the community. Interventions that effectively strengthen governance usually pay close attention to the attributes of good governance that exist already (Herrera, 2014).

1.4.3.6 LEGAL RECOGNITION OF LOCAL RULES AND REGULATIONS

Legal recognition of local rules and regulations has been shown to be an effective means of strengthening governance in a variety of countries, including Kenya (Roba, 2014) and Spain (Herrera, 2014). In some cases, including Morocco (Dominguez, 2014), Lebanon (Sattout, 2014), and Cameroon (Moritz et al., 2014), it has been demonstrated that a proper translation of customary rules into the legal framework is important. Legal recognition cannot be proposed as an absolute requirement for stronger governance, but it clearly adds a level of legitimacy, reflecting the importance of the rule of law in many governance frameworks.

On the other hand, experiences suggest that it is easier to gain government support for the sensitive process of strengthening governance if work is conducted within the framework of existing laws. Fortunately, most countries offer legal avenues for protecting both communal and private land, and often the challenge is one of less implementation of existing laws, which is often the result of low capacity and awareness, sometimes combined with limited political will and civic awareness. However, addressing these shortcomings can establish a productive working environment for building relationships between communities and government as the basis for stronger natural resource governance (Herrera et al., 2014).

1.4.3.7 PRESERVING NATURAL INFRASTRUCTURE

Local land users often value natural infrastructure that is critical to their overall landscape management but is poorly recognized or protected. Pastoralists, for example, often maintain natural water sources, salt pans, and access corridors which are critical to their livelihoods, but which also attract competition from other users. The pastoralist infrastructure in the Logone floodplain of Cameroon, for instance, is largely hidden, which makes protecting it a challenge. In this instance, government is responding by developing regulations to protect livestock corridors and, to date, some 150 km of transhumance corridors have been protected (Moritz et al., 2014). A similar process is more advanced in Spain, where the Vías Pecuarias Act of 1995, with support from several regional laws and council plans, now safeguards up to 125,000 km of livestock corridors (known as *Cañadas*) covering 400,000 ha.

1.4.4 OVERALL LESSONS FOR IMPROVED GOVERNANCE AND CONFLICT MANAGEMENT

A common thread running through the governance work cited here is the emphasis on participation and empowerment. This reflects the delicacy of the task, the potential risks associated with interfering in customary institutions, and the centrality of enabling local people to make informed decisions that can have profound consequences. Participatory tools can improve people's capacity in management and planning, but their participatory skills are often lacking, which is a major barrier to scaling up good governance. The sensitivity of the work may explain why more efforts have not been made in the past to strengthen governance and natural resource rights: as experience is gained, such concerns should become less of a deterrent. Some countries have adopted policies of participation—for example, Botswana (Buckham-Walsh and Mutambirwa, 2014) and Jordan (Haddad, 2014)—which create further space and legitimacy for strengthening governance.

The Jordanian case shows the importance of using a participatory approach to strengthen governance, supported by local accomplishments. The work focuses on four goals: building capacity, empowering people, opening access to decision making, and engaging underprivileged groups. Capacity building in this case means the following: (i) developing a focused vision that clearly identifies problems and long-term solutions; (ii) gathering reliable and thorough information; (iii) acknowledging political choices from various proposed solutions; (iv) assessing and reducing risks both in the short and long term; and (v) advocating for local and underprivileged people. Empowering local groups, including women, provides a foundation for success, allowing local communities to develop the management system of their choice. Five factors have been highlighted as contributing to empowerment (Haddad, 2014):

- Authentic local participation
- Networks and communication mechanisms between participants and stakeholders
- Shared strategies with participants
- Early achievements from pilot projects
- A gender-sensitive approach that enhances the voices and capacities of women for participation and promotes their role within the community governance structures

Governance is, to a large extent, about relationships between individuals, but it is also about relationships between citizens and their government. Rule of law is often cited as essential for good governance, and experience shows that to strengthen governance at a local level, it is important to work within legal frameworks and support government in implementing its laws and policies. One of the critical factors in strengthening governance has been developing a mutual understanding—between communities, government, and development partners—of the government's role in local resource governance. In most cases, the role is relatively light, following the principle of subsidiarity in which responsibility is devolved to the lowest practical level, which in many cases means the level of land users and local communities. However, as governance is strengthened at the local level, challenges may become apparent at higher levels, demonstrating a further role for government (for example, in the responsible governance of water resources on the landscape scale). Decentralization policies have been widely helpful in enabling this devolution of responsibility in a number of countries.

Security has emerged as another factor in improved governance, both as a contributor and a beneficiary. The experience in Kenya, for example, was made possible after significant increases in security that led to a period of stability and growth, as well as an improvement in relationships between the community and government, and indeed an overall improvement in the sense of Kenyan citizenship (Roba, 2014). At the same time, stronger governance can contribute to security through better dialogue and negotiations and greater reliance on the rule of law.

Tenure security has also both contributed to and benefited from improved governance. Where legal avenues have not initially been followed, improved governance, participation, and empowerment nevertheless have strengthened land rights and responsibilities among land users through informal processes and through improved organization to defend claims. In other cases, the application of land laws has been the starting point for enabling governance, particularly where land is managed communally, and therefore misappropriation of land is easier. However, it is essential to recognize that land rights lie on a continuum (from simple rights to use a resource to the full right of ownership and alienation) and to understand where rights need to be strengthened along that continuum (Schlager and Ostrom, 1992). Critical to sustainable management are the rights of management and of exclusion (which sometimes is periodic or seasonal). The right of alienation—the right to buy and sell land as property—does not appear to be necessary for sustainable land management. This is an important lesson that greatly influences the acceptability of governance in the eyes of some governments.

The benefits of improved governance are multiple and varied: from improvements in rights and recognition to stronger economies and environmental conservation. The importance of better relationships between citizens and government, and between citizens, has already been discussed in this part of the chapter. Securing natural resource governance can be seen as an entry point to broader rights-based development and equity. Benefits can also be observed in local and national economies as improved governance leads to more sustainable management and overall greater productivity from land and natural resources. In Jordan, for example, improvements in rangeland management led to an increase in overall biomass production, which supported larger numbers of livestock as well as the harvesting of a number of marketable medicinal and cosmetic plant products (Haddad, 2014).

Governance has also been shown to benefit biodiversity, often including the return of locally extinct fauna and flora. These in turn offer further economic opportunities to local communities. Community-based conservation based on improved local governance is increasingly being accepted as an important component of conservation strategies, providing local habitats in de facto community-conserved areas, as well as connectivity and dispersal areas for more traditional and exclusionary conservation practices. Furthermore, as a result of sustainable land management, governance contributes to improvements in ecosystem functions, such as safeguarding hydrological cycles that not only protect against flood, but also store water in the natural infrastructure, thereby reducing the risk of drought. Sustainable land management improves soil formation and protects soil from desertification, thereby both sequestering and protecting soil organic carbon and mitigating climate change. Although the cost of land degradation, as outlined earlier, is dramatic, the benefits of sustainable land management can be equally impressive, and through appropriate governance, they can be achieved at relatively low cost (Herrera et al., 2014).

1.4.5 CONCLUSION

While it is important not to overattribute conflict to land degradation, the two are nevertheless connected. Improvements in natural resource management and ecosystem health can contribute to peace building directly (through improved relations and cooperation) and indirectly (through economic development and employment). Moreover, since conflict and land degradation share a common denominator— shortcomings in governance—they can be simultaneously addressed through a common approach. The UN's "Report of the High-Level Panel of Eminent Persons on the Post-2015 Development Agenda" (UN, 2013) notes that governance was not reflected in the Millennium Development Goals and should be integral to the post-2015 sustainable development agenda, where it needs to be seen as a core element of well-being rather than an option. At the time of writing, this recommendation is being considered as part of the proposal for the Sustainable Development Goals (Open Working Group, 2014). Natural resource governance also needs specific attention as a subset of governance, where issues of land tenure and resource security can be better targeted as the foundation of sustainability.

Land degradation also needs urgent global attention to ensure that we stop exhausting this vital resource. The options before us are limited: either we convert the last remaining natural refuges for food production, or we radically change the way we use land. Improvements in both governance and sustainable land management must take into account changing social dynamics, including urbanization, population growth, changing market patterns, and the emergence of female-headed households, as well as environmental trends, such as climate change and a growing demand for resources. Greater attention is needed to both sustainable management and restoration of natural resources.

According to a recent study, restoring 150 million ha of degraded landscapes around the world would inject over US$80 billion annually into the world economy, net of costs, in terms of food security, jobs, and other direct benefits (Ferwerda, 2013). At the same time, maintaining and expanding sustainable land management practices offer some of the lowest-cost options for moving toward a land degradation neutral world.

REFERENCES

Bantekas, I., 2010. Environmental Security in Africa. In: Abass, A. (Ed.), Protecting Human Security in Africa. Oxford University Press, Oxford, pp. 43–63.

Boutaleb, A., Firmian, I., 2014. Community Governance of Natural Resources and Rangelands: The Case of the Eastern Highlands of Morocco. In: Herrera, P.M. et al. (Eds.), The Governance of Rangelands: Collective Action for Sustainable Pastoralism. Routledge, London, pp. 94–107.

Bruinsma, J., 2003. World Agriculture: Towards 2015/2030. An FAO Perspective. FAO and Earthscan. http://www.fao.org/fileadmin/user_upload/esag/docs/y4252e.pdf.

Buckham-Walsh, L., Mutambirwa, C.C., 2014. Strengthening Communal Rangelands Management in Botswana: Legal and Policy Opportunities and Constraints. In: Herrera, P.M. et al. (Eds.), The Governance of Rangelands: Collective Action for Sustainable Pastoralism. Routledge, London, pp. 214–235.

Cotula, L., 2006. Land and Water Rights in the Sahel Tenure: Challenges of Improving Access to Water for Agriculture. Issue Paper No. 139. International Institute for Environment and Development, London. http://pubs.iied.org/pdfs/12526IIED.pdf.

Davies, J., et al. 2012. Conserving Dryland Biodiversity. IUCN, United Nations Convention to Combat Desertification, UNEP-WCMC, IUCN World Commission on Protected Areas (WCPA), IUCN Commission on Ecosystem Management (CEM), and the IUCN Commission on Environmental, Economic and Social Policy (CEESP). Published by IUCN, Gland https://www.iucn.org/about/union/commissions/cem/cem_resources/other_cem_publications_and_papers/?uPubsID=4715.

Department for International Development (DFID), 2006. Eliminating World Poverty: Making Governance Work for the Poor. DFID, London.

Dominguez, P., 2014. Current Situation and Future Perspectives for the Governance of Agro-Pastoral Resources in the Ait Ikis Transhumants of the High Atlas (Morocco). In: Herrera, P.M. et al. (Eds.), The Governance of Rangelands: Collective Action for Sustainable Pastoralism. Routledge, London, pp. 126–144.

Ferwerda, W., 2013. Nature Resilience: Organising Ecological Restoration by Partners in Business for Next Generations. Rotterdam School of Management–Erasmus University. https://portals.iucn.org/library/efiles/documents/2012-068.pdf. (accessed 15.01.15.).

Food and Agriculture Organization (FAO), 1993. Land Degradation in Arid, Semiarid, and Dry Subhumid Areas: Rainfed and Irrigated Lands, Rangelands, and Woodlands. http://www.fao.org/docrep/X5308E/X5308E00.htm.

Forsyth, T., Schomerus, M., 2013. Climate Change and Conflict: A Systematic Evidence Review. In: JSRP Paper 8 http://www.lse.ac.uk/internationalDevelopment/research/JSRP/downloads/JSRP8-ForsythSchomerus.pdf. (accessed 15.01.15.).

Glazer, A., Konrad, K.A., 2003. Conflict and Governance. Springer, Berlin.

Haddad, F., 2014. Rangeland Resource Governance—Jordan Case. In: Herrera, P.M. et al. (Eds.), The Governance of Rangelands: Collective Action for Sustainable Pastoralism. Routledge, London, pp. 45–61.

Herrera, P.M., 2014. Searching for Extensive Livestock Governance in the Northwestern Inland of Spain: Achievements from Two Case Studies in Castilla Y León. In: Herrera, P.M. et al. (Eds.), The Governance of Rangelands: Collective Action for Sustainable Pastoralism. Routledge, London, pp. 191–213.

Herrera, P.M., Davies, J., Manzano Baena, P., 2014. The Governance of Rangelands: Collective Action for Sustainable Pastoralism. Routledge, London.

IUCN, February 2000. Policy on Social Equity in Conservation and Sustainable Use of Natural Resources. Adopted by IUCN Council Meeting.

IUCN, 2011. The Land We Graze: A Synthesis of Case Studies about How Pastoralists' Organizations Defend Their Land Rights. IUCN ESARO office, Nairobi, Kenya, viii+48 pp.

Kaufmann, D., Kraay, A., Zoido-Lobaton, P., 1999. Governance Matters. In: Policy Research Working Paper. World Bank, Washington, DC.

Kevane, M., Gray, L., 2008. Darfur: rainfall and conflict. Environ. Res. Lett. 3, 1–10.

Lagi, M., Bertrand, K.Z., Bar-Yam, Y., 2011. The Food Crises and Political Instability in North Africa and the Middle East. arXiv:1108.2455v1 [physics.soc-ph] http://necsi.edu/research/social/food_crises.pdf. (accessed 13.08.15.).

Lal, R., 2008. Sequestration of atmospheric CO_2 in global carbon pools. Energ. Environ. Sci. 1 (1), 86–100. doi:10.1039/b809492f.

Moritz, M., Catherine, L.B., Drent, A.K., Kari, S., Mouhaman, A., Scholte, P., 2014. Rangeland Governance in an Open System: Protecting Transhumance Corridors in the Far North Province of Cameroon. In: Herrera, P.M. et al. (Eds.), The Governance of Rangelands: Collective Action for Sustainable Pastoralism. Routledge.

Open Working Group, 2014. Open Working Group Proposal for Sustainable Development Goals. Open Working Group of the United Nations General Assembly on Sustainable Development Goals. Document A/68/970 http://undocs.org/A/68/970.

Ostrom, E., 1990. Governing the Commons: The Evolution of Institutions for Collective Action. Cambridge University Press, Cambridge, UK.

Roba, G., 2014. Strengthening Communal Governance of Rangeland in Northern Kenya. In: Herrera, P.M. et al. (Eds.), The Governance of Rangelands: Collective Action for Sustainable Pastoralism. Routledge, London, pp. 181–190.

Sattout, E., 2014. Rangelands Management in Lebanon: Cases from Northern Lebanon and Bekaa. In: Herrera, P.M. et al. (Eds.), The Governance of Rangelands: Collective Action for Sustainable Pastoralism. Routledge, London, pp. 145–155.

Schlager, E., Ostrom, E., 1992. Property-rights regimes and natural resources: a conceptual analysis. Land Econ. 68, 249–262.

The Economics of Ecosystems and Biodiversity (TEEB), 2008. An Interim Report: European Communities. http://www.teebweb.org/publication/the-economics-of-ecosystems-and-biodiversity-an-interim-report/.

Toulmin, C., Quan, J., 2000. Evolving Land Rights, Policy, and Tenure in Africa. International Institute for Environment and Development (IIED), London.

United Nations (UN), 2013. A New Global Partnership: Eradicate Poverty and Transform Economies Through Sustainable Development. Report of the High-Level Panel of Eminent Persons on the Post-2015 Development Agenda. http://www.un.org/sg/management/pdf/HLP_P2015_Report.pdf.

United Nations Convention to Combat Desertification (UNCCD), 18 November 2011. Land and Soil in the Context of a Green Economy for Sustainable Development, Food Security and Poverty Eradication. In: Submission of the UNCCD Secretariat to the Preparatory Process for the Rio+20 Conference.

United Nations Convention to Combat Desertification (UNCCD), n.d. Desertification Land Degradation and Drought: Some Global Facts and Figures. http://www.unccd.int/Lists/SiteDocumentLibrary/WDCD/DLDD%20Facts.pdf.

United Nations Environment Programme (UNEP), 2007. Sudan Post-Conflict Environmental Assessment. United Nations Environment Program UNEP, Nairobi.

United Nations Environment Programme (UNEP), 2009. From Conflict to Peacebuilding: The Role of Natural Resources and the Environment. United Nations Environment Programme. http://www.iisd.org/pdf/2009/conflict_peacebuilding.pdf.

von Braun, J., Meinzen-Dick, R., 2009. Land Grabbing by Foreign Investors in Developing Countries. International Food Policy Research Institute (IFPRI), Washington D.C. http://www.ifpri.org/publication/land-grabbing-foreign-investors-developing-countries.

Wilkes, A., Tennigkeit, T., 2008. Carbon Finance in Rangelands: An Assessment of Potential in Communal Rangelands. IUCN World Initiative for Sustainable Pastoralism, Nairobi, Kenya. http://cmsdata.iucn.org/downloads/microsoft_word___carbon_finance_english.pdf.

CONCEPTS AND METHODOLOGIES FOR RESTORATION AND MAINTENANCE

TENETS OF SOIL AND LANDSCAPE RESTORATION

2.1

Rattan Lal

Carbon Management and Sequestration Center, Ohio State University, Columbus, OH, USA

2.1.1 INTRODUCTION

Soil, the essence of all terrestrial life, is a nonrenewable resource over the time frame of one to two human generations (i.e., 25–50 years). With a rapid increase in human and livestock populations over the 20th century, natural resources are under great stress. Soil degradation adversely affects efforts at environmental conservation and sustainable development. Thus, protecting soil resources and restoring degraded lands are of prime importance.

Some key determinants of soil quality are the concentration and pool of soil organic carbon (SOC), along with their spatial (Jacinthe and Lal, 2006) and temporal variability. Thus, a strong depletion of the SOC concentration/pool sets in motion soil degradation processes. A decline in SOC concentration below the threshold level (ranging from 1.5%–2% in the root zone) (Aune and Lal, 1997; Loveland and Webb, 2003) leads to a reduction in soil structure, as denoted by decreased aggregates and aggregate strength, an increase in soil erodibility to water- and wind-driven processes, an increased susceptibility to crusting and compaction, a decrease in water infiltration capacity with an attendant increase in the risk of runoff and erosion, and a reduction in plant-available water capacity. These degradation trends reduce the efficient use of nutrients and water, adversely affect plant growth, reduce net primary productivity, decrease economic profitability, and jeopardize long-term sustainability.

This chapter aims to describe the importance of soil quality and its key characteristics, as well as to discuss the principles of soil restoration and the strategies that may be employed to do so in diverse ecoregions. The following section lays out the science behind the soil organic matter, its dynamics, processes governed by its quality and quantity, and the principles and strategies of soil and landscape restoration.

2.1.2 SOIL EROSION AND ORGANIC CARBON DYNAMICS

Changes in land use and management influence the SOC pool, and thus soil quality and productivity. For example, converting natural ecosystems to agricultural uses decreases the SOC pool by reducing biomass carbon input and increasing biomass losses. Land use change further exacerbates losses of the SOC pool by increasing the rate of oxidation of SOC by microbial activity through changes in soil moisture and temperature regimes and accelerating soil erosion by water and wind. The vulnerability to soil erosion is increased because of the soil disturbance and the decrease in protective vegetation

Land Restoration. http://dx.doi.org/10.1016/B978-0-12-801231-4.00002-1

cover. In addition to removing the nutrient-rich top soil, soil erosion carries away large amounts of SOC because it is lighter than the mineral fraction (low bulk density) and is concentrated in proximity to the soil surface where erosional processes are most active. There are numerous mechanisms governing the dynamics (loss or gain at a given landscape position) of the SOC pool by accelerated erosion (Table 2.1.1). Predominant among these are the breakdown of structural units (aggregates) and exposure of the SOC to microbial processes, leading to an increase in the decomposition rate. The erosion-induced losses of the SOC pool and the attendant emissions of greenhouse gases (GHGs), such as carbon dioxide (CO_2), methane (CH_4), and nitrous oxide (N_2O), are important factors accelerating the human-induced effects on climate change.

Table 2.1.1 Mechanisms Affecting Soil Carbon Dynamics under Erosional Processes

	Mechanisms/Processes	Reference
1	Microbial processes: differential microbial utilization of SOC between eroding and depositional sites	Dungait et al. (2013)
2	Aggregate breakdown	Bremenfeld et al. (2013)
3	Stabilization against microbial decay in depositional sites	Wang et al. (2013)
4	Change in the ratio of microbial biomass carbon (MBC) to total downslope SOC	Nie et al. (2013a, 2013b)
5	Erosion-induced soil degradation may increase CO_2 emissions	Mchunu and Chaplot (2012)
6	Interrill process	Kuhn et al. (2012)
7	Rate of dynamic replacement of eroded soil carbon	Nadeu et al. (2012)
8	Soil aggregation and stabilization of SOC as affected by erosion/deposition	Wang et al. (2014)
9	Soil organic carbon redistribution and the emission of CO_2	Wang et al. (2014)
10	Geomorphic and pedogenic processes along eroding hill slope	Wiaux et al. (2014)
11	Erosion-induced changes in soil biochemical and microbiological properties	Park et al. (2014)
12	Erosional impacts on decomposition and humification	Haring et al. (2013)
13	Soil redistribution of organic carbon	Zhang and Li (2013)
14	Nutrient dynamics under erosional processes	Nie et al. (2013a; b)
15	Concentration of particulate and mineral-associated organic matter	Cheng et al. (2010)
16	Role of geomorphology	Hancock et al. (2010)
17	Erosion can be a source or sink depending on the magnitude and type of land use change	Boix-Fayos et al. (2009)
18	Spatial intensity of water erosion	Yadav and Malanson (2009)
19	Loss of more labile SOC fraction (POC)	Martinez-Mena et al. (2008)
20	Downslope movement of SOC to alluvial fans, which alters the mosaic of SOC metabolism and storage	Smith et al. (2007)
21	Formation of new SOC at eroding sites and burial of eroded SOC	Van Oost et al. (2005)
22	CO_2 emission through transport of carbon in sediments, carbon burial, and recovery	Page et al. (2004)

Interrill processes (e.g., detachment and transport) that stem from raindrop splashes on soil and result in erosion, which is dependent on drop size and intensity of rainfall (see, for example, http://extension.psu.edu/agronomy-guide/cm/sec1/sec11d), also affect SOC dynamics. The redistribution of SOC-laden sediments in eroding landscapes, as well as the subsequent deposition elsewhere, can increase SOC in low-lying sites where sediments accumulate. For the landscape on the whole, however, erosion decreases the soil/ecosystem carbon pool, nutrients, and clay/silt reserves. Over and above the adverse effects of such processes on soil quality and ecosystem functions and services, eroding landscapes are also a source of GHG emissions, especially CO_2, CH_4, and N_2O. The impact of water and wind breaks down structural aggregates and exposes the SOC hitherto protected against microbial processes. In addition, changes in soil temperature and moisture regimes also influence the emission of CO_2, CH_4, and N_2O. Thus, implementing strategies to control soil erosion are important to sustaining soil quality, agronomic productivity, and reducing emissions of GHG from soils of agroecosystems.

Accelerated soil erosion also exacerbates the risk of desertification, particularly land/soil degradation in arid and semiarid climates through the loss of water by surface runoff, increased evaporation (due to the loss of vegetation cover and increase in soil temperature), and lower water productivity. Thus, restoration of degraded landscapes is important for enhancing agronomic/biomass productivity, as well as improving the quality and quantity of renewable water resources, mitigating climate change, and enhancing resilience against extreme events.

2.1.3 STRATEGIES OF SOIL AND LANDSCAPE RESTORATION

Basic strategies for restoring degraded soils and landscapes include the following (Figure 2.1.1):

- Conserving biocomplexity and integrating multiple functions to create sustainable landscapes (Figure 2.1.2)
- Maximizing soil water (green water) storage in order to conserve and manage water resources, and minimizing erosion losses through no-till farming and cover cropping (Figure 2.1.3 through 2.1.5)
- Anticipating and managing future tradeoffs to improve ecosystem functions and services (Reed et al., 2013)
- Building upon traditional knowledge, strengthening communication between scientists and agricultural stakeholders, and using a participatory approach (Figure 2.1.1) to promoting inclusive ecological restoration (Dixon and Carr, 1999).

A participatory approach, involving land managers and farmers to identify and implement recommended management practices, is useful to create a more sustained and inclusive effort in restoration and conservation, making it a good strategy to use even in developed economies. As such, Berglund et al. (2013) demonstrated that it is important to cultivate communication to implement participatory approaches even in developed economies, such as Iceland.

In this context, the 10 tenets of soil/landscape restoration and the corresponding practices are outlined in Table 2.1.2. These include reducing losses of water, nutrients, and other materials (e.g., SOC, clay, and silt) from the landscape through erosion or leaching; improving soil structure; creating positive carbon and nutrient budgets; strengthening biogeochemical cycles by enhancing biocomplexity; managing favorable soil reaction and salt balance; and creating disease-suppressive soils by improving soil quality and its functional capacity.

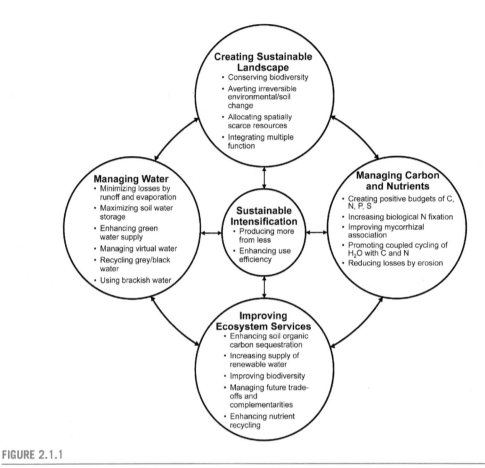

FIGURE 2.1.1

Strategies of intensification for sustainable development and ecological restoration.

FIGURE 2.1.2

Female farmers in sub-Saharan Africa, recycling nutrients by applying manure and household wastes. They grow a range of crops (including maize, yam, cassava, sorghum, and beans) together to enhance biocomplexity.

FIGURE 2.1.3

Direct seeding of rice with no-till farming in Cambodia.

FIGURE 2.1.4

The presence of crop residue mulch is essential to conserving water, recycling nutrients, moderating soil temperature, and reducing risks of soil erosion.

FIGURE 2.1.5

Incorporating a leguminous cover crop in the rotation cycle improves benefits of no-till farming by increasing soil fertility and providing surface cover.

Table 2.1.2 10 Tenets of Soil/Land Restoration

	Principle	Best Management Practices
1	Conserve water	Mulch farming, conservation agriculture, vegetative hedges
2	Control erosion	Continuous soil cover, no-till farming, cover cropping
3	Improve soil structure	Minimize disturbances, mulch cover, composting, cover cropping
4	Recycle nutrients	Deep-rooted plants, residue management, agro-forestry, mixed farming
5	Enhance soil biodiversity	Conservation agriculture, mulch farming, no-till, cover cropping
6	Create a positive carbon budget	No-till, mulch farming, biomass-carbon input, controlled grazing
7	Replace nutrients harvested (N, P, K, Zn, Mo)	Integrated nutrient management, manuring, biological nitrogen fixation (BNF), mycorrhizae, balanced fertilizer use
8	Manage soil pH	Liming, slow release fertilizer
9	Maintain a favorable salt balance	Good internal drainage, leaching of salt, good quality irrigation water
10	Create disease-suppressive soil	Improve soil biodiversity, vermiculture

2.1.4 IMPLEMENTATION OF ECOLOGICAL RESTORATION

While technological options to implement these tenets are well known, their site-specific adaptability needs to be tested and validated, as outlined in Table 2.1.3. The goal is to enhance ecological complexity and mimic natural ecosystems. Locally grown food based on ecoregion-specific plants is an important example of using traditional knowledge (Figure 2.1.6). Similarly, developing viable and clean sources of household energy can reduce the rate of deforestation in regions that rely on wood and charcoal (Figure 2.1.7). Watershed management and water harvesting approaches at the household and community levels can also increase access to clean water while improving health and alleviating poverty (Figure 2.1.8).

Table 2.1.3 Some Examples of Landscape Restoration

Concept	Application	Reference
Enhancing ecological complexity	Mediterranean, Southern Mexico, Degraded drylands	Roose et al. (2011) Dalle et al. (2011) Westley et al. (2010)
Adopting problem-solving approach	Site-specific condition	McAlpine et al. (2010)
Developing and implementing ecological networks	Multiactor planning	Opdam et al. (2006)
Managing and minimizing risks	Risk analysis	Wurzel (2010)
Enhancing environmental sustainability	Landscape restoration	Masnavi (2013)

FIGURE 2.1.6

Locally grown food, based on plants adapted to the ecoregions, is an important strategy to achieve food security.

FIGURE 2.1.7

Firewood as a source of household fuel is a major cause of deforestation in sub-Saharan Africa.

FIGURE 2.1.8

Lack of access to clean water remains a major issue in developing countries.

Thus, ecological restoration requires a holistic approach (Zhang, 2001). In order to ensure that activities are complementary rather than negating each other's effects, every aspect of the ecosystem must be taken into account. While the basic principles may be known, effectiveness of these techniques depend on site-specific implementation, the variants of which are discussed next.

2.1.4.1 IMPLEMENTATION AT THE LANDSCAPE LEVEL

Implementation of restorative techniques at the landscape level is critical to enhancing the effectiveness of any policy implementation for sustainable land use. Di Pietro's (2001) work studying villages in France indicated that agricultural practices designed in accordance with landscape (e.g., location within the geography and valley level) ensure long-term sustainability of resources and landscape productivity, rather than short-term economic gains when practices are implemented only at the field level. Therefore, it is recommended that the best management practices outlined in Table 2.1.2 are implemented at the landscape level rather than the field level.

2.1.4.2 HARMONIZING THE ECOLOGICAL EFFECTS WITH CURRENT AND FUTURE SOCIAL DEMOGRAPHIC CHANGES

While ecological restoration is important (Zhang, 2001), its consequences are complex and difficult to predict when employed in rapidly changing economies, such as in China (Ma et al., 2013). Thus, it is essential to harmonize the ecological effects with due consideration to present and future social/demographic changes, as the goal is to benefit both nature and the human society.

In addition to addressing the short-term and ever-growing needs of human society, the long-term effects on nature must be addressed. For example, large-scale afforestation in China's arid and semiarid regions has led to several unforeseen consequences. Cao et al. (2011) reported that the use of inappropriate species and an overemphasis on tree and shrub planting has compromised the environmental good. Excessive reliance on afforestation has led to the deterioration of soil ecosystems, decreased vegetation cover, and exacerbated water shortages because climatic, pedological, hydrological, and landscape factors were ignored (Cao et al., 2011).

Similarly, indiscriminate use of inputs to rapidly close yield gaps (Foley et al., 2011) can adversely affect biocomplexity. Phalan et al. (2014) observed that closing the yield gaps to attainable levels to meet the projected demands for food and feed by 2050 could preserve a land area equivalent to that of the Indian subcontinent, but it also would reduce the biodiversity of existing croplands. For instance, significant shift of floristic composition from monocotyledonous weeds in nonreclaimed sodic land (i.e., high in sodium) to dicotyledonous weeds in the reclaimed land in Uttar Pradesh, India, has been shown to occur even within 10 years (Srivastava et al., 2011). Thus, habitat protection is a critical goal, meaning that weed control must be achieved by eradicating nothing while decreasing the competition (Dixon, 2003). Similar principles apply to pest control; the goal is not to oversupply the natural predators, but rather to create a balance between predator and parasitic populations.

Furthermore, the heritage and aesthetic value of traditional landscapes should be preserved by specific land management plans with due consideration for the emerging issues of climate change and variability. Based on case studies done in the Denver basin in Flanders, Belgium, Dupont and Van Eetvelde (2013) observed that some plans lead to deterioration of the traditional character of the land and threaten the natural-scientific, historical, and aesthetic values of the landscape. Thus, the quality and heritage value of the landscape must be preserved, and climate change vulnerability and adaptation maps must be considered in any landscape management plans (Dupont and Van Eetvelde, 2013).

2.1.4.3 BUILDING UPON TRADITIONAL KNOWLEDGE

While today's problems largely cannot be solved by yesterday's technologies, it is important to build upon traditional knowledge and combine its concepts with the modern principles. In lowland Namaqualand, South Africa, Botha et al. (2008) observed that using the cumulative restorative knowledge of users of those lands provides important practical insights for restoring degraded lowlands. Furthermore, the long-term stewardship of the landscape embedded in traditional knowledge must be the motivation driving restoration activities. Diemont and Martin (2009) emphasize the importance of indigenous groups that have designed and managed the agroecosystems of the Lacandon Maya of the Chiapas rain forest in Mexico. Technologies used by ancient Mayan cultures may offer strategies for regional restoration and conservation initiatives in Mesoamerica (Mexico and Central America). Creating woodland islets, to reconcile ecological restoration with conservation and agricultural land use, is another example of linking traditional knowledge with current innovations to address the needs of modern society (Benayas et al., 2008).

2.1.4.4 RISK ASSESSMENT AND MANAGEMENT

Risks to the environment and human health are associated with all land use and management styles, because all processes and systems in nature are interconnected (Commoner, 1972). It is possible that significant harm can actually be caused by activities designed to prevent damage. Thus, an objective risk assessment is needed for all interventions. To be effective, a multidisciplinary or interdisciplinary approach is essential and may involve regular discussions and joint activities among ecologists, biologists, conservationists, ecological engineers, hydrologists, geologists, medical specialists, and economists. In addition, a transdisciplinary approach ensures saving of time, energy, and costs involved (Burger, 2008) through continuous interaction among participants and stakeholders. Ecologically, emergy [it is a definition of available energy (exergy) that is consumed in direct and indirect transformations needed to make a product or service] may be used, rather than energy, though its rigor and utility is contested on thermodynamic grounds (Wei et al., 2008).

2.1.4.5 MULTIPLE BENEFITS OF LANDSCAPE RESTORATION

An effective strategy of landscape restoration must produce several co-benefits. For example, the eroded landscape in the Loess Plateau of China can be converted to Miscanthus planting for biofuel production, while restoring the environment by reducing erosion and mitigating climate change by sequestering carbon in biomass and soil (Liu and Sang, 2013; Guzman and Lal, 2014). Restoration of mine lands by establishing biofuel plantations is one example of multiple benefits (Guzman et al., 2014; Ussiri and Lal, 2005).

2.1.4.6 ECOLOGICAL ENGINEERING

The strategy of ecological engineering involves multifunctional mechanisms of landscape restoration through environmentally creative technology or ecological engineering technology. It refers to sustainable land use balanced with ecosystem stability, as a parallel to an intensive cultivation for agronomic production (Schuller et al., 2000). Wakatsuki and Masunaga (2005) describe the usefulness of the Sawah technique for managing wetlands in sub-Saharan Africa. The Sawah system is designed for efficient irrigation and drainage of local topography, which can be adjusted on 10 million hectares (Mha) of wetlands along the Niger River. This system can increase the mean rice paddy yield from 1.3–1.7 tons per hectare (t/ha) to 4–5 t/ha and include vegetables and upland crop production (Wakatsuki and Masunga, 2005; Tanko, 2013).

By conducting a landscape ecology experiment on a 1000-ha area in Germany, Schuller and colleagues (2000) demonstrated that small, unused patches and linear structures connecting them are integral to nature conservation. Creating and connecting a range of smaller ecosystems within a larger landscape is important to preserving biocomplexity, cycling nutrients, and improving the productivity of adjacent cultivated landscape. A similar strategy of creating woodland islets was used by Benayas et al. (2008) to reconcile ecological restoration, conservation, and agronomic land use. In southern Peru, Whaley et al. (2010) restored a highly degraded dry forest by building upon vegetation relics in conjunction with building cultural capacity and environmental engagement using an ecosystem approach. Environmental engagement efforts included community involvement in developing sustainable forest products, festivals, school programs, publications aimed at the local community, and engaging all stakeholders (e.g., owners, agribusiness, government, and nongovernmental organizations/NGOs).

All strategies outlined here emphasize the importance of protecting landscape functions and ecosystem services both for humans and nature during restoration activities, including habitat conservation for all wildlife, not just one or two species selected for political or economic reasons (e.g., pandas or bald eagles).

2.1.5 ESTABLISHING VEGETATION COVER

The first step in soil and landscape restoration—and an important one at that—is establishing an effective vegetation cover comprising a mixture of native and adaptable species. While introduced species can provide rapid cover, the risks of their invasive attributes must be objectively assessed. There are numerous examples of invasive species that are difficult to eradicate, such as leucaena and lantana in India, white clover or Dutch clover in Iceland, kudzu in the southeastern United States, and eupatorium in West Africa.

On the other hand, revegetating with salt-tolerant or holomorphic (salt-tolerant) species can help to restore saline and sodic soils. Even the partial restoration of salt-affected land through revegetation with halophytes (salt-tolerant plants) can set in motion the reclamation process. Such deep-rooted halophytes lower the water table through transpiration but also lead to salt accumulation in the root zone. Barrett-Lennard (2002) observed that replacing deep-rooted perennial native vegetation with shallow-rooted annual crops in southern Australia resulted in a rising water table and exacerbated secondary salinization. Therefore, it is widely recognized that restoring salinized landscapes necessitates reintegrating perennial plants (e.g., trees, shrubs, and fodders) into the farming system.

Restoration of coal mining landscapes has become an important area of focus with the pressure to reduce coal combustion, since coal combustion is a major contributor to the total anthropogenic emissions of 35 Gt CO_2/yr (Victor et al., 2014). Restoration of these lands is critical to ecosystem functions and services. Forest restoration is widely practiced in reclaiming disused coal mines in the Appalachian region of the eastern United States. It is estimated that more than 6000 km^2 in Appalachia have been disturbed by coal mining since 1990. Zipper et al. (2011) reported that the productivity of these drastically disturbed lands can be restored to support forest vegetation by improving the soil's physical and chemical quality. Sequestration of SOC is an important aspect to soil quality restoration (Shrestha and Lal, 2006, 2007, 2008; Ussiri and Lal, 2005; Jacinthe and Lal, 2006; Liu and Lal, 2013, 2014; Shukla et al., 2004, 2005).

Similar to these disturbed mined soils, those degraded by erosion can also be restored by afforestation. In some cases, fire management is critical to establishing the forest cover. Understanding local perceptions of degradation and attitudes to fire management are important to restoring forest vegetation cover. Similarly, forest remnants on indigenous land can provide the ecological and experiential inspiration to restore forest cover on adjacent degraded pastoral lands, thus accelerating the rate of landscape restoration. In this way, the preservation of forest remnants can strategically contribute to the restoration process of degraded landscapes.

Similar to the restoration of forest lands, establishing vegetative cover is also critical to restoring grasslands and reversing desertification. In desert areas in China, Yan et al. (2013) studied the effects of different land use intensities on restoring sandy landscapes. Sand dune stabilization involves promoting natural recovery of vegetation suited to sandy conditions (i.e., native species) by establishing natural vegetation enclosures (see also section 2.2 in this chapter). Soil moisture conservation is the critical factor under desert conditions. Establishing forest shrubs (*Atriplex* sp) has been effective at restoring soil chemical quality in semiarid areas of Morocco (Zucca et al., 2011). While improving the vegetation cover, fodder shrubs also have an important economic value. If soil moisture and nutrient reserves are adequate, seed variability is a strong factor for restoring species-rich grasslands on ex-arable lands, as plant species on degraded landscapes have limited dispersal abilities (Kardol et al., 2008).

Deep-rooted perennial shrubs play an important role in soil-water dynamics and in making water available to shallow-rooted plants. Hydraulic redistribution of water via roots from moist to dry soils, which occurs by a process known as *hydraulic lift*, is an important mechanism by which plants cope with drought. Hydraulic-lifted water released into adjacent soil can support the growth of adjoining plants capable of lifting (Liste and White, 2008). This mechanism should be considered in restoring vegetation cover in dry and desert ecosystems.

2.1.6 WATER MANAGEMENT

Drought stress is a serious constraint in landscape restoration in both arid and semiarid regions. There are several types of drought: (i) *meteorological*, caused by a long-term deficit of rainfall; (ii) *hydrological*, caused by a deficit in river flow; (iii) *pedological*, attributed to a shortage of soil-water storage (green water) in the profile or soil solum (surface and subsurface layers that have the same history of formation as soils); (iv) *agronomic*, caused by a shortage of plant-available water to maintain the physiological needs of evapotranspiration during the growing season; and (v) *sociological*, caused by competing uses to meet social/recreational needs. Landscape restoration can be strongly affected by all types, but especially by pedological and agronomic droughts, which adversely influence seedling establishment and crop stand development.

Drought can also be a symptom of poor management and planning decisions at the local and/or regional levels (Manakou et al., 2011). Poor management involves the unregulated utilization of water resources, which leads to inappropriate consumption, inefficient water use based on wasteful practices, and poverty. These problems are confounded by rapid population growth, industrialization, and urbanization, among other factors. Thus, rational management of water within the hydrological basin is important to landscape restoration and management. The Indus Basin is an example of the problems exacerbated by sharing water among four countries: Pakistan, India, Nepal, and Bangladesh. While management of short water cycles can prevent irreversible water loss from the landscape/watershed (Ripl and Eiseltova, 2009), long-term planning is essential to the sustainable use of finite resources.

Sustainable water resources management practices include managing both supply and demand. The increase in poorly managed demand is causing the rapid depletion of groundwater in Ogallala aquifer in the southwestern United States, in the North China Plains, and in the Indo-Gangetic Plains. Groundwater extraction to support rice-wheat systems in a semiarid/arid environment of northwestern India and eastern Pakistan caused rapid groundwater depletion (Laghari et al., 2012). To meet future demand, India plans a vast canal network: a 15,000-km-long network of canals and tunnels to move 174 km^3 of water each year from surplus areas to deficit regions within India (Bagla, 2014). However, the environmental impacts of such a network need to be carefully considered, including invasive species, risks of inundation and salinization, and increase in incidence of malaria and other diseases.

In addition to the quantity of water available for supplemental irrigation, public health issues are important to consider with regard to the use of reclaimed water for landscape irrigation in such areas as for golf courses and sports areas (Mujeriego et al., 1996). Integrated sustainable water and landscape management are also essential to the cities of the future, as this is where 70% of the population will reside by 2050. Novotny and Brown (2007) addressed sustainable water and landscape management for future cities with specific reference to the following: (i) urban water sustainability (reliable water supply and wastewater and storm water management) based on the involvement of beneficiary communities; (ii) the impact of extreme events; (iii) mass balance (i.e., total flow through all channels of water in and out of the land in question) and pollution of water in urban centers (e.g., by incorporating flow through or from contaminated sources); (iv) hydrological and pollution stresses; (v) identifying solutions at water and landscape levels; and (vi) use of future urban hydrological and ecological systems (Mulenga, 2008).

Ecosystem-specific attributes can also cause issues specific to certain landscapes and ecoregions. For example, stream and riparian restoration involves specific issues characteristic of this ecoregion with regard to biodiversity and eutrophication or the specific vegetation to be used in afforestation. In the United States, riparian ecosystems have been widely degraded by urbanization and other land uses, thus requiring ecological restoration to protect the land through which streams flow. Any future conservation of riparian areas must consider conserving biodiversity in streams and including restored stream functions. Some important measures are (i) implementing a joint planning of stream restoration and protection of riparian open land; (ii) incorporating the monitoring and measurement of stream restoration in the initial objectives; and (iii) including conservation management plans for protected open land, which can use stream restoration actions, some of which are briefly discussed in the following section.

2.1.7 LANDSCAPE RESTORATION AND ECOSYSTEM SERVICES

Landscape and soil degradation involve a reduction in ecosystem functions and services. Thus, decisions about sustainable landscape management must consider the restoration of essential ecosystem services (Forouzangohar et al., 2014). It is pertinent to identify the hot spots of important ecosystem services within a landscape and to prioritize these areas. Important among these services, with numerous co-benefits, are soil/ecosystem carbon sequestration (Lal et al., 2013), renewable water supply, and biodiversity. There is a close interaction between ecosystem services and biodiversity, which are mutually beneficial or even symbiotic (Schneiders et al., 2012). Conversion of natural lands to agroecosystems reduces biodiversity with adverse effects on the latter. Therefore, restoring biodiversity within agricultural landscapes is important to improving ecosystem services while sustaining agronomic productivity. Land sharing, rather than land separation can enhance farmed environments, and involves biodiversity-based agricultural practices (Benayas and Bullock, 2012). Afforestation of agroecosystems to enhance biodiversity has the co-benefits of generating another income stream for carbon credits (Perring et al., 2012; Schneiders et al., 2012). Thus, the economic analysis of landscape restoration must consider all co-benefits and tradeoffs.

The flow of ecosystem services and their contribution to the well-being of humans and nature must be included in any economic assessment (Costanza et al., 1997, 2014; Bateman et al., 2011; Zhang et al., 2013). Generating carbon credits is a by-product of one of the most important ecosystem services: carbon sequestration provided through landscape/soil restoration (Lal, 2004; Lal et al., 2004; Curran et al., 2012). It is also an example of the multiple outcomes possible through landscape restoration (Perring et al., 2012). However, enhancing one ecosystem service can potentially suppress another. For example, advancing food security by intensification and monoculture can decrease biodiversity. Similarly, removing crop residues for biofuel production can reduce SOC concentration and the SOC pool. Therefore, anticipating and managing future tradeoffs and complementarities between ecosystem services is an important guiding principle of landscape restoration (Reed et al., 2013).

2.1.8 CONCLUSIONS

Many of the global issues of the 21st century (climate change, food and nutritional security, water quality, eutrophication, and dwindling biodiversity) can be addressed through soil and landscape restoration and by implementing practices based on sound ecological principles. Rather than being anthropocentric in our approach, when humanity's relation with nature is driven by its growing and insatiable demands (e.g., resources for economic development, recreation, and comfort, rather than solely for survival), it is important to realize that humans, like other biota, are a part of nature. In fact, humans belong to nature, not the other way around. The latter misconception—namely, that humans "own" land and natural resources with the solemn right to conquer and dominate it at will—has been responsible for widespread problems regarding resource exploitation, with attendant land/ecosystem degradation and desertification. Adopting a nature-centric (or biocentric) egalitarian approach would lead to greater habitat preservation, enhanced biocomplexity, and restored ecosystem functions and services.

Soil and landscape restoration using the ecological principles discussed in this chapter involves combining both approaches: meeting human needs while sustaining ecosystem functions and services. Human survival depends on learning to live in harmony with nature. Therefore, long-term sustainability must be made a higher priority than the immediate demands for short-term gains based on myopic approaches.

REFERENCES

Aune, J., Lal, R., 1997. Agricultural productivity in the tropics and critical limits of properties of Oxisols, Ultisols, and Alfisols. Tropical Agric. (Trinidad) 74, 96–103.

Bagla, P., 2014. India plans the grandest of canal networks. Science 345 (6193), 128.

Barrett-Lennard, E.G., 2002. Restoration of saline land through revegetation. Agri. Water Manage. 53 (1–3), 213–226.

Bateman, I.J., Mace, G.M., Fezzi, C., Atkinson, G., Turner, K., 2011. Economic analysis for ecosystem service assessments. Environ. Res. Econ. 48 (2), 177–218.

Benayas, J.M.R., Bullock, J.M., 2012. Restoration of biodiversity and ecosystem services on agricultural land. Ecosys. 15 (6), 883–899.

Benayas, J.M.R., Bullock, J.M., Newton, A.C., 2008. Creating woodland islets to reconcile ecological restoration, conservation, and agricultural land use. Front. Ecol. Environ. 6 (6), 329–336.

Berglund, B., Hallgren, L., Aradottir, A.L., 2013. Cultivating communication participatory approaches in land restoration in Iceland. Ecol. Soc. 18 (2), 35.

Boix-Fayos, C., Vente, J., Albaladejo, J., Martinez-Mena, M., 2009. Soil carbon erosion and stock as affected by land use changes at the catchment scale in Mediterranean ecosystems. Agri. Ecosys. Environ. 133 (1–2), 75–85.

Botha, M.S., Carrick, P.J., Allsopp, N., 2008. Capturing lessons from land-users to aid the development of ecological restoration guidelines for lowland Namaqualand. Biolog. Conserv. 141 (4), 885–895.

Bremenfeld, S., Fiener, P., Govers, G., 2013. Effects of interrill erosion, soil crusting, and soil aggregate breakdown on in situ CO_2 effluxes. Catena 104, 14–20.

Burger, J., 2008. Environmental management—Integrating ecological evaluation, remediation, restoration, natural resource damage assessment, and long-term stewardship on contaminated lands. Sci. Total Environ. 400 (1–3), 6–19.

Cao, S.X., Chen, L., Shankman, D., Wang, C.M., Wang, X.B., Zhang, H., 2011. Excessive reliance on afforestation in China's arid and semi-arid regions: Lessons in ecological restoration. Earth-Sci. Rev. 104 (4), 240–245.

Cheng, S.L., Fang, H.J., Zhu, T.H., Zheng, J.J., Yang, X.M., Zhang, X.P., Yu, G.R., 2010. Effects of soil erosion and deposition on soil organic carbon dynamics at a sloping field in Black Soil region, northeast China. Soil Sci. Plant Nut. 56 (4), 521–529.

Commoner, B., 1972. The Closing Circle Nature, Man and Technology. Bantam Books, New York.

Costanza, R., et al., 1997. The value of the world's ecosystem services and natural capital. Nature 387, 253–260.

Costanza, R., et al., 2014. Changes in global values of ecosystem services. Global Environ. Change 26, 152–158.

Curran, P., Smedley, D., Thompson, P., Knight, A.T., 2012. Mapping restoration opportunity for collaborating with land managers in a carbon credit-funded restoration program in the Makana Municipality, Eastern Cape, South Africa. Rest. Ecol. 20, 56–64.

Dalle, S.P., Pulido, M.T., de Blois, S., 2011. Balancing shifting cultivation and forest conservation lessons from a "sustainable landscape" in southeastern Mexico. Eco. Appl. 21 (5), 1557–1572.

Di Pietro, F., 2001. Assessing ecologically sustainable agricultural land-use in the central Pyrenees at the field and landscape level. Agr. Eco. Environ. 86 (1), 93–103.

Diemont, S., Martin, J.F., 2009. Lacandon Maya ecosystem management sustainable design for subsistence and environmental restoration. Eco. Appl. 19 (1), 254–266.

Dixon, R.M., 2003. Weeds eradicating nothing but doing something. The Imprinting Foundation. Published online www.imprinting.org/scientific-conservation/Land%20IMPRINT%20SPECIFS%20SUSTAINABLE%20AGRICULTURE%20AND%20EC%20RESTOR%20CORNBELT.htm.

Dixon, R.M., Carr, A.B., 1999. Land imprinting specifications for ecological restoration and sustainable agriculture. In: Proceedings of Conference 31—International Erosion Control Association (IECA), Palm Springs, CA, pp. 21–25. February.

Dungait, J., Ghee, C., Rowan, J.S., McKenzie, B.M., Hawes, C., Dixon, E.R., Paterson, E., Hopkins, D.W., 2013. Microbial responses to the erosional redistribution of soil organic carbon in arable fields. Soil Biol. Biochem. 60, 195–201.

Dupont, L., van Eetvelde, V., 2013. Assessing the potential impacts of climate change on traditional landscapes and their heritage values on the local level: Case studies in the Dender basin in Flanders, Belgium. Land Use Policy 35, 179–191.

Foley, J.A., et al., 2011. Solutions for a cultivated planet. Nature 478 (7369), 337–342.

Forouzangohar, M., Crossman, N.D., Macewan, R.J., Wallace, D.D., Bennett, L.T., 2014. Ecosystem services in agricultural landscapes A spatially explicit approach to support sustainable soil management. Sci. World J. 2014: 483298. http://dx.doi.org/10.1155/2014/483298.

Guzman, J.G., Lal, R., 2014. Miscanthus and switchgrass feedstock potential for bioenergy and carbon sequestration on minesoils. Biofuels 5 (3), 313–329.

Guzman, J.G., Lal, R., Byrd, S., Apfelbaum, S.I., Thompsom, R.L., 2014. Carbon life cycle assessment for prairie as a crop on reclaimed mineland. Land Degradation Development (Land Degrad. Develop). http://dx.doi.org/10.1002/ldr.2291.

Hancock, G.R., Murphy, D., Evans, K.G., 2010. Hillslope and catchment scale soil organic carbon concentration An assessment of the role of geomorphology and soil erosion in an undisturbed environment. Geoderma 155 (1–2), 36–45.

Haring, V., Fischer, H., Cadisch, G., Stahr, K., 2013. Implication of erosion on the assessment of decomposition and humification of soil organic carbon after land use change in tropical agricultural systems. Soil Biol. Biochem. 65, 158–167.

Jacinthe, P.A., Lal, R., 2006. Spatial variability of soil properties and trace gas fluxes in reclaimed mine land of southeastern Ohio. Geoderma 136 (3–4), 598–608.

Kardol, P., Van Der Wal, A., Bezemer, T.M., De Boer, W., Duyts, H., Holtkamp, R., Van Der Putten, W.H., 2008. Restoration of species-rich grasslands on ex-arable land: Seed addition outweighs soil fertility reduction. Biol. Conserv. 141 (9), 2208–2217.

Kuhn, N.J., Armstrong, E.K., Ling, A.C., Connolly, K.L., Heckrath, G., 2012. Interrill erosion of carbon and phosphorus from conventionally and organically farmed Devon silt soils. Catena 91 (4), 94–103.

Laghari, A.N., Vanham, D., Rauch, W., 2012. The Indus basin in the framework of current and future water resources management. Hydrol. Earth Sys. Sci. 16 (4), 1063–1083.

Lal, R., 2004. Soil carbon sequestration impacts on global climate change and food security. Science 304 (5677), 1623–1627.

Lal, R., Griffin, M., Apt, J., Lave, L., Morgan, M.G., 2004. Ecology—Managing soil carbon. Science 304 (5669), 393.

Lal, R., Lorenz, K., Hüttl, R.F., Schneider, B.U., Von Braun, J., 2013. Ecosystem Services and Carbon Sequestration in the Biosphere. Springer, Dordrecht, Netherlands.

Liste, H.H., White, J.C., 2008. Plant hydraulic lift of soil water—Implications for crop production and land restoration. Plant and Soil 313 (1–2), 1–17.

Liu, R., Lal, R., 2013. A laboratory study on amending mine soil quality. Water, Air and Soil Pollution 224 (1679), 1–17.

Liu, W., Sang, T., 2013. Potential productivity of the Miscanthus energy crop in the Loess Plateau of China under climate change. Environ. Res. Lett. 8 (4). 044003. http://dx.doi.org/10.1088/1748-9326/8/4/044003.

Loveland, P., Webb, J., 2003. Is there a critical level of organic matter in the agricultural soils of temperate regions: a review. Soil Tillage res. 70, 1–8.

Ma, H., Lv, Y., Li, H.X., 2013. Complexity of ecological restoration in China. Ecol. Eng. 52, 75–78.

Manakou, V., Tsiakis, P., Tsiakis, T., Kungolos, A., 2011. Management of the hydrological basin of Lake Koronia using mathematical programming. In: Kungolos, A., Karagiannidis, A., Aravossis, K., Samaras, P., Schramm, K.W. (Eds.), Proceedings of the Third International Conference on Environmental Management,

Engineering, Planning, and Economics (CEMEPE 2011) & SECOTOX Conference. Grafima Publishers, Thessaloniki, Greece, pp. 165–170.

Martinez-Mena, A., Lopez, J., Almagro, A., Boix-Fayos, C., Albaladejo, J., 2008. Effect of water erosion and cultivation on the soil carbon stock in a semiarid area of southeast Spain. Soil Till. Res. 99 (1), 119–129.

Masnavi, M.R., 2013. Environmental sustainability and ecological complexity developing an integrated approach to analyse the environment and landscape potentials to promote sustainable development. Intl. J. Environ. Res. 7 (4), 995–1006.

McAlpine, C.A., et al., 2010. Can a problem-solving approach strengthen landscape ecology's contribution to sustainable landscape planning? Landscape Ecol. 25 (8), 1155–1168.

Mchunu, C., Chaplot, V., 2012. Land degradation impact on soil carbon losses through water erosion and CO_2 emissions. Geoderma 177, 72–79.

Mujeriego, R., Sala, L., Carbo, M., Turet, J., 1996. Agronomic and public health assessment of reclaimed water quality for landscape irrigation. Water Sci. Technol. 33 (10–11), 335–344.

Mulenga, M., 2008. Book review: Cities of the Future Towards Integrated Sustainable Water and Landscape Management. In: Novotny, V., Brown, P. (Eds.), Environ. Urban. 20, 275–276.

Nadeu, E., Berhe, A.A., De Vente, J., Boix-Fayos, C., 2012. Erosion, deposition, and replacement of soil organic carbon in Mediterranean catchments: A geomorphological, isotopic, and land use change approach. Biogeosci. 9 (3), 1099–1111.

Nie, X.J., Zhang, J.H., Su, Z.A., 2013a. Dynamics of soil organic carbon and microbial biomass carbon in relation to water erosion and tillage erosion. PLoS One 8 (5), e64059.

Nie, X.J., Zhao, T.Q., Qiao, X.N., 2013b. Impacts of soil erosion on organic carbon and nutrient dynamics in an alpine grassland soil. Soil Sci. Plant Nut. 59 (4), 660–668.

Novotny, V., Brown, P., 2007. Cities of the Future: Towards Integrated Sustainable Water and Landscape Management. International Water Association, IWA Publishing, London.

Odum, E.C., 2004. Emergy analysis of shrimp mariculture in Equador: A review. Ecological Modelling 178, 239–240.

Opdam, P., Steingrover, E., Van Rooij, S., 2006. Ecological networks: A spatial concept for multi-actor planning of sustainable landscapes. Landscape Urban Plan. 75 (3–4), 322–332.

Page, M., Trustrum, N., Brackley, H., Baisden, T., 2004. Erosion-related soil carbon fluxes in a pastoral steepland catchment, New Zealand. Agr. Eco. Environ. 103 (3), 561–579.

Park, J.H., Meusburger, K., Jang, I., Kang, H., Alewell, C., 2014. Erosion-induced changes in soil biogeochemical and microbiological properties in Swiss Alpine grasslands. Soil Biol. Biochem. 69, 382–392.

Perring, M.P., et al., 2012. The Ridgefield Multiple Ecosystem Services experiment: Can restoration of former agricultural land achieve multiple outcomes? Agr. Eco. Environ. 163, 14–27.

Phalan, B., Green, R., Balmford, A., 2014. Closing yield gaps perils and possibilities for biodiversity conservation. Philos. Trans. Royal Soc. B—Biol. Sci. 369 (1639), 1471–2970.

Reed, M.S., et al., 2013. Anticipating and managing future trade-offs and complementarities between ecosystem services. Ecol. Soc. 18 (1), 5.

Ripl, W., Eiseltova, M., 2009. Sustainable land management by restoration of short water cycles and prevention of irreversible matter losses from topsoils. Plant Soil Environ. 55 (9), 404–410.

Roose, E., Bellefontaine, R., Visser, M., 2011. Six rules for the rapid restoration of degraded lands: Synthesis of 17 case studies in tropical and Mediterranean climates. Secheresse 22, 86–96.

Schneiders, A., Van Daele, T., Van Landuyt, W., Van Reeth, W., 2012. Biodiversity and ecosystem services: Complementary approaches for ecosystem management? Ecol. Ind. 21, 123–133.

Schuller, D., et al., 2000. Sustainable land use in an agriculturally misused landscape in northwest Germany through ecotechnical restoration by a "Patch-Network-Concept." Ecol. Eng. 16 (1), 99–117.

Shrestha, R.K., Lal, R., 2006. Ecosystem carbon budgeting and soil carbon sequestration in reclaimed mine soil. Environ. Intl. 32 (6), 781–796.

Shrestha, R.K., Lal, R., 2007. Soil carbon and nitrogen in 28-year-old land uses in reclaimed coal mine soils of Ohio. J. Environ. Quality 36 (6), 1775–1783.

Shrestha, R.K., Lal, R., 2008. Land use impacts on physical properties of 28-year-old reclaimed mine soils in Ohio. Plant and Soil 306 (1–2), 249–260.

Shukla, M.K., Lal, R., Ebinger, M.H., 2004. Soil quality indicators for reclaimed minesoils in southeastern Ohio. Soil Sci. 169 (2), 133–142.

Shukla, M.K., Lal, R., Ebinger, M.H., 2005. Physical and chemical properties of a minespoil eight years after reclamation in Northeastern Ohio. Soil Sci. Soc. Amer. J. 69 (4), 1288–1297.

Smith, S.V., et al., 2007. Soil erosion and significance for carbon fluxes in a mountainous Mediterranean-climate watershed. Ecol. Appl. 17 (5), 1379–1387.

Srivastava, P.K., Baleshwar, B.S.K., Singh, N., Tripathi, R.S., 2011. Long-term changes in the floristic composition and soil characteristics of reclaimed sodic land during eco-restoration. J. Plant Nut. Soil Sci. 174 (1), 93–102.

Tanko, A.L., 2013. Agriculture, livelihoods and fadama restoration in northern Nigeria. In: Wood, A., Dixon, A., Mccartney, M. (Eds.), Wetland Management and Sustainable Livelihoods in Africa. Routledge, New York, pp. 205–228.

U.N. 2014. World Urbanization Prospects. Department of Economic and Social Affairs, Population Division (ST/ESA/SER.A/352), 27 pp.

Ussiri, D., Lal, R., 2005. Carbon sequestration in reclaimed minesoils. Crit. Rev. Plant Sci. 24 (3), 151–165.

Van Oost, K., Govers, G., Quine, T.A., Heckrath, G., Olesen, J.E., De Gryze, S., Merckx, R., 2005. Landscape-scale modeling of carbon cycling under the impact of soil redistribution: The role of tillage erosion. Global Biogeochem. Cycles 19 (4). GB4030, http://dx.doi.org/10.1029/2005GB002541.

Victor, D.G., Gerlagh, R., Baiocchi, G., 2014. Getting serious about categorizing countries. Science 345 (6192), 34–36.

Wakatsuki, T., Masunaga, T., 2005. Ecological engineering for sustainable food production and the restoration of degraded watersheds in tropics of low pH soils: Focus on West Africa. Soil Sci. Plant Nut. 51 (5), 629–636.

Wang, X., Cammeraat, L.H., Wang, Z., Zhou, J., Govers, G., Kalbitz, K., 2013. Stability of organic matter in soils of the Belgian Loess Belt upon erosion and deposition. Eur. J. Soil Sci. 64 (2), 219–228.

Wang, X., Cammeraat, H., Erik, L., Romeijn, P., Kalbitz, K., 2014. Soil organic carbon redistribution by water erosion—The role of CO_2 emissions for the carbon budget. PLoS One 9 (1). e96299.

Wei, J.B., Xiao, D.N., Zeng, H., 2008. Sustainable development of an agricultural system under ecological restoration based on energy analysis: A case study in northeastern China. Intl. J. Sustain. Dev. World Ecol. 15 (2), 103–112.

Westley, F., Holmgren, M., Scheffer, M., 2010. From scientific speculation to effective adaptive management: A case study of the role of social marketing in promoting novel restoration strategies for degraded dry lands. Ecol. Soc. 15 (3), 6.

Whaley, O.Q., Beresford-Jones, D.G., Milliken, W., Orellana, A., Smyk, A., Leguía, J., 2010. An ecosystem approach to restoration and sustainable management of dry forest in southern Peru. Kew Bull. 65 (4), 613–641.

Wiaux, F., Cornelis, J.T., Cao, W., Vanclooster, M., Van Oost, K., 2014. Combined effect of geomorphic and pedogenic processes on the distribution of soil organic carbon quality along an eroding hillslope on loess soil. Geoderma 216, 36–47.

Wurzel, K.A., 2010. Risk Assessment in land restoration. In: Brown, K., Hall, W.L., Snook, M., Garvin, K. (Eds.), Sustainable Land Development and Restoration. Elsevier, Philadelphia, pp. 345–362.

Yadav, V., Malanson, G.P., 2009. Modeling impacts of erosion and deposition on soil organic carbon in the Big Creek Basin of southern Illinois. Geomorphol. 106 (3–4), 304–314.

Yan, W., Zhao, H.L., Zhao, X.Y., 2013. Effects of land use intensity on the restoration capacity of sandy land vegetation and soil moisture in fenced sandy land in desert area. Contemp. Prob. Ecol. 6 (1), 128–136.

Zhang, X.S., 2001. Ecological restoration and sustainable agricultural paradigm of mountain-oasis-ecotone-desert system in the north of the Tianshan Mountains. Acta Botanica Sinica 43, 1294–1299.

Zhang, J.H., Li, F.C., 2013. Soil redistribution and organic carbon accumulation under long-term (29 years) upslope tillage systems. Soil Use Manage. 29 (3), 365–373.

Zhang, J.J., Fu, M.C., Zeng, H., Geng, Y.H., Hassani, F.P., 2013. Variations in ecosystem service values and local economy in response to land use: A case study of Wu'an. China. Land Deg. Devel. 24, 236–249.

Zipper, C.E., Burger, J.A., Mcgrath, J.M., Rodrigue, J.A., Holtzman, G.I., 2011. Forest restoration potentials of coal-mined lands in the eastern United States. J. Environ. Quality 40, 1567–1577.

Zucca, C., Julitta, F., Previtali, F., 2011. Land restoration by fodder shrubs in a semi-arid agro-pastoral area of Morocco: Effects on soils. Catena 87, 306–312.

STABILIZATION OF SAND DUNES: DO ECOLOGY AND PUBLIC PERCEPTION GO HAND IN HAND?

2.2

Nir Becker[1], Meira Segev[2], Pua Bar (Kutiel)[2]

Department of Economics and Management, Tel-Hai College, Upper Galilee Israel[1]

Department of Geography and Environmental Development, Ben-Gurion University, Beer-Sheva Israel[2]

2.2.1 INTRODUCTION

Land use change is a pressing conservation issue. Understanding the causes and consequences of these changes requires spatial and temporal models (Foley et al., 2005). Additionally, researchers are interested in explaining the causal relationships between individual choices and the outcomes of land use change. Therefore, there is a need to integrate social and natural sciences in order to establish ecosocial economic models (Spurgeoon, 1998; Irwin and Geoghegan, 2001).

Some argue that ecosystems and their services cannot or should not be valued in monetary terms (e.g., Spash et al., 2005). However, it can also be argued that every decision involves value judgments and that providing an estimate of the value of the contribution made by ecosystems allows this to be done more rationally than by assigning zero value, which is often inferred when no value is given. After all, "we cannot avoid the valuation issue because as long as we are forced to make choices, we are doing valuation" (Costanza and Folke, 1997, p. 50). Furthermore, avoiding explicit valuation of ecosystem services is not an option in the face of rapid global environmental changes and ecosystem degradation and loss, particularly since ecosystems and their services may be forgotten when no value is given.

Coastal areas in general, and sand dunes in particular, are considered to be areas in which human impact and the level of development are among the highest (Drees, 1997; Holdgate, 1993; Van der Meulen and Salman, 1996).

Approximately 70% of Israel's population resides in a 500 km^2 area of its 190-km-long Mediterranean coastline. Consequently, coastal dunes that have previously been mobile because of traditional grazing and cutting are now either built-up areas (approximately two-thirds of the dune area) or are stabilizing due to the encroachment of local shrubs and invaded by such species as *Acacia saligna* (Tsoar and Blumberg, 2002; Kutiel et al., 2000, 2004; Kutiel, 2001; Levin and Ben-Dor, 2004).

Land Restoration. http://dx.doi.org/10.1016/B978-0-12-801231-4.00011-2

Despite the limited area of the coastal sand dunes, their ecological and nature conservation importance is significant (Koniak et al., 2009; Kutiel, 2001). The xeric (i.e., very dry) conditions of the sand, together with the continuous connection to the dunes of the Negev and northern Sinai desert, makes them an isolated area within the Mediterranean region (Kutiel, 2001; Kutiel et al., 2004). Furthermore, 26% of Israel's endemic plant species are concentrated in this region. This is the highest rate of endemism in one habitat. Likewise, 22 arthropod species occur in Israel only along the coastal sand dunes [Bar (Kutiel), 2013]. These species represent diverse habitats, and some of them are of biogeographic importance.

Mobile coastal dunes are also hydrologically important because they are the main sink for water supply that supports aquifers and the ecosystem as a whole. About 45% of the annual rainfall is lost in stabilized dunes as a result of evapotranspiration and hydrophobic processes (unpublished data). In addition, more than a third of Israel's irrigation water is wastewater that is filtered in the coastal dunes as part of a tertiary treatment (Silberman et al., 1992). Finally, mobile sand dunes are also a tourist attraction and are preferred by visitors (e.g., Drees, 1997).

In order to decide whether stabilized or mobile dunes should be preferred, we need to estimate the costs and benefits of maintaining each dune type. This section of the chapter tries to estimate the benefits of the two dune types by employing two approaches: (i) a contingent valuation method (CVM) study of the preferred landscape management and (ii) by attaching values from an ecological value index (EVI). Besides valuing a nonmarket good (landscape), we can use the two approaches to split the value into social and ecological motives. This is important especially for understanding whether these estimates are biased by using only one dimension (either social or ecological) of the conservation process.

2.2.2 STUDY SITE

The Nizzanim Long-Term Ecosystem Research (LTER) nature reserve is located in the southwestern part of the Israeli Mediterranean coast (see Figure 2.2.1). The nature reserve covers 20 km^2 and is the only large nature reserve along the Mediterranean coast of Israel that contains dunes with various levels of stabilization (Kutiel, 2001). About 60% of the dunes are semistabilized, 20% are stabilized, and only 20% are mobile (Rubinstein et al., 2013). Since 2005, the Israel Nature and Parks Authority has invested significantly in preventing further stabilization of the semistabilized dunes and, in the short term, to transition some of the stabilized dunes back to mobile dunes in order to expand and encourage the reestablishment of plant and animal species that thrive in sandy soils (psammophilic) (Kutiel et al., 2000).

2.2.3 METHODS

2.2.3.1 DICHOTOMOUS CHOICE

In this CVM study, the dichotomous choice (DC) method is used to investigate people's willingness to pay (WTP). The DC method was first used by Bishop and Heberlein (1979), while Hanemann (1984) developed the conceptual and theoretical arguments. The key feature of this method is that individuals are asked whether they would pay a suggested price in a hypothetical market situation. The possible answers are usually "Yes" or "No." WTP is calculated based on the probability of saying "yes."

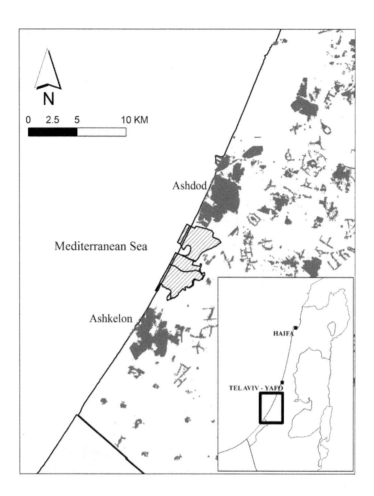

FIGURE 2.2.1

Map of Nizzanim LTER nature reserve.

2.3.3.2 SURVEY DESCRIPTION AND ADMINISTRATION

We sampled 300 respondents in our survey. All respondents were over 18 years old and, if possible, we tried to contact heads of households. The surveys were personal interviews, primarily because of the proven effectiveness of this approach, especially in similar cases, when a relatively complex set of environmental goods is entailed. The final version of the questionnaire was composed as follows.

The first section introduced the issue to the respondents and contained one page explaining the situation of the sand dunes along the Israeli coastline. This included the ecological importance of sand dunes compared with human activities and the resulting decrease in preserved area. It also included an explanation of dune stabilization processes and the consequences of this process. In contrast, we explained the motivation to develop infrastructure projects in a crowded country such as Israel.

The second section was the WTP section. It explained that, in order to base public policy on educated decisions, the government is eager to know the value of each type of land use to the public. To keep the area in its natural form, funding would be provided through a closed environmental fund in a single payment. A special "cheap talk" paragraph was added just before the payment question, which was intended to remind the respondents about their budget limitations, in particular, and the hypothetical nature of the survey. Respondents were offered four initial payment bids to keep the land in its natural state: $50, $100, $200, and $400. Depending on their answer, follow-up bids were presented: a slightly higher price point if they answered positively to the initial bid, or a slightly lower one if they answered negatively. The different formats were A (50, 75, 25), B (100, 150, 75), C (200, 300, 150), and D (400, 600, 300).

The third section elicited motivations and helped spot protest bids. Motivations included use (actual visits), bequest (not visiting, but want to keep the land for future generations), option (to visit in the future), and existence (do not intend to visit nor have any bequest motives, but think that it is important to keep nature in its natural form). Meanwhile, protest bids were detected by a zero WTP, followed by circling the option that it is not their duty to finance such preservation plans. A total of 5 respondents were classified as protests and were excluded from the survey; they were replaced with other respondents to keep the 300 responses.

The last section explored respondents' sociodemographic characteristics. These included gender, age, origin (place of birth), number of children, membership in green environmental organizations, education, income, visit frequency, general knowledge of nature conservation, and specific knowledge about sand dune stabilization processes. Respondents were first asked to circle their preferred landscape of the two types. Their payment statement is thus related to the type of dune structure they prefer.

2.2.3.3 ECOLOGICAL MEASURE

The process for stabilizing sand dunes is followed by changes in plant composition, from psammophilic species to more generalistic and opportunistic species. In order to estimate the specific value that is lost due to stabilization, we index each species by its uniqueness and importance in nature conservation. We use the ecological value index (EVI), developed by Cohen and Bar (Kutiel) (2005). Positive numbers (1–8) were given to important species (pssamophilic, protected, endemic, rare, very rare, and endangered), while negative numbers (−1−−4) were given to nonendemic, opportunistic, and invasive species.

Since changes in the plant assemblages occur during the stabilization process, the ecological value of individual species and of the ecosystem as a whole is expected to change as well, reflecting the total ecological value of the specific dune type.

2.2.4 RESULTS

2.2.4.1 DESCRIPTIVE STATISTICS

About 48.67% of respondents preferred active landscape management (remobilization of semistabilized dunes by removing the shrubs), while the rest preferred passive management (see Table 2.2.1). The mean respondent visited such areas 0.98 times a year, and only 7% belonged to "green" (i.e., environmental) organizations. Women constituted 57% of respondents and the mean levels of education and income were 3.03/5 and 2.57/4, respectively. The mean number of children

Table 2.2.1 Descriptive statistics

Variable	Mean	SD	CI (95%)
Gender (1 = female)	0.567	0.065	±0.129
Age (5 decedaes)	2.667	0.169	±0.339
Origin (1 = Native)	0.767	0.055	±0.110
Children	1.400	0.192	±0.384
Green (member = 1)	0.067	0.032	±0.065
Education (1–5)	3.033	0.119	±0.238
Income (1–4)	2.567	0.110	±0.220
Visits (yearly frequency)	0.980	0.359	±0.717
Knowledge 1 (1–3)	1.600	0.104	±0.208
Knowledge 2 (1–3)	1.483	0.096	±0.192
Landscape preference (1 = mobile dunes preference)	0.487	0.500	±0.001

N=300

living in the household at the time of the survey was 1.4 per household (about 1 child fewer than the national average, but consistent with the age profile of visitors). Israeli natives accounted for 77% of respondents, while 23% were immigrants. Finally, on a three-point scale, respondents had knowledge levels about sand dunes and stabilization processes of 1.6 and 1.5, respectively.

The percentage of positive responses (to both the original and follow bids) for the lowest bid is always higher than that for the highest bid (Table 2.2.2). This confirms the theory that support for a measure drops as taxes rise.

Use versus nonuse value: We calculated the specific values by identifying which motive the respondent circled. For each value component, the respondent had a 5-unit Likert scale to express his or her agreement with the statement. This enabled us to allocate a relative amount that could be associated with a specific value. We summed the specific values of the entire sample population and divided the result by 300 to get an average WTP value per respondent for each value type. The relative shares were 0.21, 0.26, 0.24, and 0.29 for use, option, bequest, and existence values, respectively. Thus, existence value was the biggest one, and the difference between total nonuse values and the use values given by respondents was 3.7.

Table 2.2.2 WTP for the first and second questions

Version type	YY	YN	NY	NN	TOTAL
A (50, 75, 25)	33	27	9	20	89
B (100, 150, 75)	24	25	8	25	82
C (200, 300, 150)	7	13	16	42	78
D (400, 600, 300)	3	8	6	34	51

2.2.4.2 ECONOMETRIC ESTIMATION

It appears that not all coefficients are significant. Those that are significant and positively correlated with WTP are education, income, knowledge about sand dune stabilization (knowledge 2), and active landscape management (dummy variable where 1 = prefers active management and 0 = prefers passive management). Meanwhile, those that were significant and negatively correlated with WTP were age and bid level. Insignificant variables were gender, origin, number of children, membership in green organizations, visits to such sites, and general knowledge about environmental issues.

2.2.4.3 ESTIMATION OF THE BENEFITS OF THE VARIOUS SAND DUNE TYPES

Public decision differs according to the criterion used. If public decision is accepted by a majority, a passive management regime should be chosen. However, if efficiency dictates the results, active management prevails, as the benefit per household is significantly larger compared to the relative number of households that voted for a specific scenario.

Choosing a specific land use does not mean that the households that preferred the other type will not enjoy the landscape as well. We took both uses partially into account by adding only a fraction of their benefit from their desired landscape. This was done by comparing the ratio between the first and second choices and scaling down the benefit by the same ratio (Table 2.2.4).

2.2.4.4 ECOLOGICAL VALUE MEASURED IN MONETARY TERMS

The total ecological indexes for stabilized and nonstabilized sand dunes were 2.7 and 1.95, respectively. This was mainly due to the disappearance of psammophile species during the stabilization process.

The relative economic value of each species under either active management or without management is calculated by multiplying its relative ecological contribution by the total value per household for that specific sand dune type. If we subtract the endemic species, which disappear with stabilization, we reduce the value of the stabilized dune to 67% of its previous value. The landscape is valued only by 58% of its previous value if active management is not used (Table 2.2.3).

2.2.5 DISCUSSION

Unexpectedly, membership in green organizations is not a significant variable, possibly because it is not more widespread; visit frequency is also insignificant, possibly because of the low share (21%) of use-value out of the total value (use + bequest + option + existence). General knowledge is also insignificant, but the value of general knowledge may be lost due to the inclusion of landscape management.

The total value of the sand dunes was found to be 189 M NIS (new Israeli shekels) for passive management and NIS 298 million for active management (Table 2.2.4). A 5% amortization brings the value to NIS 9.45 and 14.9 million annually, respectively, with 21% derived from use values and 79% from nonuse values.

The annual cost of active management lies between NIS 245,000 and NIS 305,000 (personal communication with Nature and Parks Authority officials). These costs include vegetation cover removal

Table 2.2.3 Regression estimates for landscape valuation

Variable	Logit		Probit	
	Value	**S.E.**	**Value**	**S.E.**
Constant	0.84	0.60	0.45	0.36
Gender	−0.103	0.186	−0.061	0.113
Age*	−0.174	0.098	−0.107	0.059
Origin	−0.219	0.215	−0.131	−0.130
Children	0.016	0.102	0.012	0.061
Green	0.169	0.279	0.091	0.168
Education**	0.235	0.104	0.143	0.063
Income***	0.307	0.111	0.189	0.067
Visits	0.031	0.030	0.161	0.181
Knowledge 1	0.215	0.140	0.131	0.085
Knowledge 2***	0.381	0.128	0.237	0.077
Active Landscape management***	0.893	0.264	0.507	0.156
Bid***	−0.008	0.001	−0.005	0.001
	LL = −341.84 LR χ^2 = 86.43 Prob > χ^2 = 0.0000 Pseudo R^2 = 0.1067		LL = −362.462 LR χ^2 = 85.19 Prob > χ^2 = 0.0000 Pseudo R^2 = 0.1052	

*90% significant
**95% significant
***99% significant.

Table 2.2.4 Benefit estimation

Value	Landscape management	Logit	Probit	Arithmetic mean
Value per household (IS)	Passive	127.0	164.7	145.85
	Active	238.6	266.1	252.35
National value (M. IS)	Passive	188.8	205.9	197.35
	Active	298.2	389.5	343.85

(amortized capital cost and variable expenses), continued monitoring, and the labor costs of the inspector, biologist, and other indirect professional personnel. Taking the average (NIS 275,000) and comparing it to the annual loss of value due to dune (NIS 7.325 million), we find a benefit-cost ratio of 26.6 for active management, which clearly justifies such an investment both on ecological and social grounds.

From an ecological point of view, using the EVI or any other similar index (Rubinstein et al., 2013) should be a necessary method to estimate the efficiency of targeting specific species by specific programs if there is a budget limitation.

2.2.6 SUMMARY AND CONCLUSIONS

A landscape's value varies depending on whether esthetic criteria (texture, colors, etc.), social criteria (e.g., recreation), or cultural criteria are included. These can be characterized by some subjective value for individuals (Briggs and France, 1980), while landscape valuation can be based on objective assessment of ecological components (Blackmore et al., 2014). In this study, we used two approaches that are based on one common denominator: CVM evaluates subjective valuation, and EVI evaluates objective ecological value.

The results show that if active management does not take place and stabilization continues, sand dunes will lose 33% of their ecological value due to the loss of endemic and psammophlic species, which are characteristic of mobile and semistabilized dunes. However, the public attaches another value reduction of 9%, probably in part because people attach ecological importance to mobile dunes and are willing to pay more than the ecological losses.

From a social point of view, we find that the yearly value of the sand dunes in the Nizzanim LTER nature reserve may increase from NIS 9.45 to 14.9 million if active management is used to destabilize the sand dunes. This NIS 5.45 million increase annually is composed of both use and nonuse values (21% and 79%, respectively). Based on the total value of the two sand dune types, the annual CVM loss due to passive management is 42%, while the EVI loss due to passive management is only 33% of the total value of the sand dunes. Therefore, contrary to intuitive belief, nonexperts may judge the full consequence of the loss due to nonoptimal landscape use policy. Of course, the final verdict should be based on a cost-benefit analysis (CBA). As shown, the benefits heavily outweigh the costs; thus, policy makers can be certain that tax money will be spent wisely.

Finally, planning and valuation should incorporate one into the other. We do not think that one method outweighed the other; rather, both value assessments should be presented to decision makers. Their task could be made much easier if a CBA based on the two methods has a clear message one way or the other, or indeed attracts attention to opposing viewpoints and values. The final verdict would be far more likely to harness public support and ecological success, if tax money is spent wisely on effectively valued land management decisions.

REFERENCES

Bar (Kutiel), P., 2013. Restoration of coastal sand dunes for conservation of biodiversity—The Israeli experience. In: Martinez, M.L., Gallego-Fernández, J.B., Hesp, P.A. (Eds.), Springer Series on Environmental Management, New York, pp. 173–186.

Bishop, R., Heberlein, A., 1979. Measuring values of extra market goods: Are indirect measures biased? Amer. J. Agri. Econ. 61, 926–930.

Blackmore, L., Doole, G., Scilizzi, S., 2014. Practitioner versus participant perspectives on conservation tenders. Biodivers. Conserv. 23, 2033–2052.

Briggs, D.J., France, J., 1980. Landscape evaluation: A comparative study. J. Environ. Manage. 10, 263–275.

Cohen, O., Bar (Kutiel), P., 2005. Effect of invasive alien plant—Acacia saligna—on natural vegetation of coastal sand ecosystems. J. Forests, Wood., Environ. 7, 11–17 (in Hebrew, with an English abstract).

Costanza, R., Folke, C., 1997. Valuing ecosystem services with efficiency, fairness and sustainability as goals. In: Daily, G. (Ed.), Nature's Services: Societal Dependence on Natural Ecosystems. Island Press, Washington, DC, pp. 49–70.

Drees, J.M. (Ed.), November 1995. Coastal dunes, recreation, and planning. In: Proc. European Union for Coastal Conservation, Leiden, the Netherlands.

Foley, J.A., et al., 2005. Global consequences of land use. Science 309 (5734), 570–574.

Hanemann, W.M., 1984. Welfare evaluations in contingent valuation experiments with discrete responses. Amer. J. Agri. Econ. 66, 332–341.

Holdgate, M.W., 1993. The sustainable use of tourism—A key conservation issue. Ambio 22, 481–482.

Irwin, E.G., Geoghegan, J., 2001. Theory, data, methods: Developing spatially explicit economic models of land use change. Agri. Ecosys. Environ. 85, 7–23.

Koniak, G., Noy-Meir, E., Perevolotsky, A., 2009. Estimating multiple benefits from vegetation in Mediterranean ecosystems. Biodivers. Conserv. 18, 3483–3501.

Kutiel, P., 2001. Conservation and management of the Mediterranean coastal sand dunes in Israel. J. Coastal Conserv. 7, 183–192.

Kutiel, P., Cohen, O., Shoshany, M., Shub, M., 2004. Vegetation establishment on the southern Israeli coastal sand dunes between the years 1965–1999. Landscape Urban Plan. 67, 141–156.

Kutiel, P., Peled, Y., Geffen, E., 2000. The effect of removing shrub cover on annual plants and small mammals in a coastal sand dune ecosystem. Biolog. Conserv. 94, 235–242.

Levin, N., Ben-Dor, E., 2004. Monitoring sand dune stabilization along the coastal dunes of Ashdod—Nizanim, Israel, 1945–1999. J. Arid Environ. 58, 335–355.

Rubinstein, Y., Groner, E., Yizhaq, H., Svoray, T., Bar (Kutiel), P., 2013. An eco-spatial index for evaluating stabilization state of sand dunes. Aeolian Res. 9, 75–87.

Silberman, J., Gerlowski, D.A., Williams, N.A., 1992. Estimating existence value for users and nonusers of New Jersey Beaches. Land Econ. 68, 225–236.

Spash, C., Stagl, S., Getzner, M., 2005. Exploring Alternatives for Environmental Valuation. Routledge, Oxon, UK.

Spurgeoon, J., 1998. The socio-economic costs and benefits of coastal habitat rehabilitation and creation. Marine Poll. Bull. 37, 373–382.

Tsoar, H., Blumberg, D., 2002. Formation of parabolic dunes from barchan and transverse dunes along Israel's Mediterranean coast. Earth Surf. Proc. Landforms 27, 1147–1161.

Van der Meulen, F., Salman, A.H.P.M., 1996. Management of Mediterranean coastal dunes. Ocean Coast. Manage. 30, 177–195.

TRUST BUILDING AND MOBILE PASTORALISM IN AFRICA[1]

2.3

Alan Channer

Initatives for Lands, Lives, and Peace, Geneva, Switzerland

2.3.1 BACKGROUND: MOBILE PASTORALISM AND GRASSLANDS

"Our animals are everything to us. We drink their milk, eat their meat, use their skins, and exchange them. When our animals perish, we perish."

This remark by a Tuareg elder holds true for the many pastoralist communities that range—or that used to range—across Africa (Gwin, 2011).

Pastoralism is a complex and dynamic livelihood system that has evolved over millennia to maximize production in an environment that is both fragile and harsh (Smith, 1992). It has been defined as "a neolithic adaptation to aridity; in biotic terms, a method of extracting protein from otherwise unpalatable cellulose of grasses and shrubs through the secondary use of the products of domestic ruminants" (Bonte and Galaty, 1991, p. 10).

Grassland constitutes 40% of the global land surface area and approximately 50% of the land surface of Africa (Millennium Ecosystem Assessment, 2005). (The category of "grassland" includes savannah: that is, grassland interspersed with trees or shrubs.) The livelihoods of 100–200 million mobile pastoralists worldwide are entirely dependent upon this land (IUCN, 2008).

To ensure timely, sufficient, and high quality of pasture for their livestock, mobility is a highly adapted response by pastoralists to use available resources; both in time (accessing a resource that is differentially available across the seasons) and in space (accessing a resource which is extensively dispersed) (International Institute for Environment and Development and SOS Sahel International UK, 2010). It has been described as "the most viable form of production and land use in most of the world's fragile dry lands" (IUCN, 2008).

As livestock (cattle, camels, goats, and sheep) can live up to 15–50 years of age, well-maintained herds also become stores of wealth and measures of prosperity (Broch-Due, 1999). Much like modern banking systems, livestock can then also be used as loans, insurance, collateral for debts, or debt repayment (Salzman, 2004). Moreover, unlike other forms of capital such as land, water, and oil, livestock is self-reproducing; therefore, pastoralist capital has the potential to expand "naturally"

---

[1]The author thanks Meera Shah for significant contributions to this chapter from her postgraduate research on pastoralism in Kenya.

(Salzman, 2004). Conversely, livestock is vulnerable to droughts, disease, and theft, and therefore requires a significant investment in security.

Pastoral strategies such as mobility, security, and distribution in livestock herding in turn require, and produce, concomitant relationships and forms of social and cultural norms and systems (Smith, 1992). For instance, among the Turkana in northern Kenya, livestock raiding is often employed as means not only to overcome poverty and hunger, but also for sociocultural purposes such as dowry payment, as political retaliation, for opportunistic wealth amassing, and for herd restocking (Opiyo et al., 2012). As such, violent conflicts, as well as ties of exchange and cooperation through self-organization and governance mechanisms, are age-old markers of relations between different pastoralist groups, and between pastoralists and their neighboring communities (Schlee, 2011). Intricate and widely imposed systems of rules, laws and guidelines developed through centuries of finessing ensure pastoralists live a healthy, peaceful and sustainable lifestyle (Okello et al., 2014).

However, these conflicts have recently been exacerbated by the impact of changing climate. Between the early 1970s and mid-1990s, rainfall across the Sahel decreased by an average of 25% (Hulme et al., 2001). Diminishing rainfall and increasing temperatures have contributed to a decrease in the number of trees across the Sahel (Gonzalez et al., 2012). Severe droughts during this period contributed to the deaths of hundreds of thousands of people.

The genders are not affected evenly. Kratli and Swift (1999) highlight the effect of insecurity on women in pastoralist communities, as the fear of abduction and rape leads to girls marrying at younger ages and families to lower bride wealth. Similarly, Hamne (2014) notes a high incidence of traumatic experience among women in Baringo, Kenya, including forced marriages at young ages, violent abuse, and rape. A cycle of insecurity is perpetuated by school closures and the desertion of government and development agencies.

2.3.2 A CONTEXT OF MISTRUST

Despite an apparent advantage in terms of sustainable land management, mobile pastoralists have been consistently neglected, and often actively discriminated against, all over the world. The gulf between many pastoralist communities and African governments is exacerbated by the geographic remoteness and the harsh terrain that these pastoralists inhabit.

The African colonial period brought the privatization of land, ranching, irrigation schemes, extractive industries, wildlife conservation, new settlements, and the creation of reserves for pastoralists, as well as new national boundaries (African Union, 2010; Schlee, 2011). Long-established patterns of transhumance were obliterated or disrupted, resulting in two overarching outcomes:

- Restricted movement and the reduction in land available for grazing made pastoralists and their livestock more vulnerable to starvation during dry seasons and drought. Consequently, pastoralists either found themselves at the heart of conflicts over land and water to which they had been denied access, or were forced to rely on food aid to support their production methods.
- This fed into the second outcome of the colonial practices: the construction of a narrative of pastoralism as an irrational and structurally flawed way of life, and pastoralists as penurious, backward, undeveloped, unwilling to modernize, and prone to conflict. In this way, pastoralist communities were not only geographically remote but were actively marginalized, both culturally and

politically (Kratli and Swift, 1999); the provision of education and health care to nomadic pastoralists was seen by colonial and postcolonial government administrators as virtually impossible.

The seeds of mistrust between government and pastoralist communities were sown, and the ability of many African nations to harness the productivity of vast tracts of grassland was compromised thereafter. Unable to compete with the forces of colonialism, traditional governance systems and mechanisms were also severely undermined. In the case of Somalia, a profound disconnect between the colonial and centralized state system and decentralized Somali pastoral culture was a root cause of the country's turmoil (Farah et al., 2002). Many of the early attempts to improve the livelihoods of mobile pastoralists by governments and international development agencies failed.

2.3.3 FAILED INTERVENTIONS AND AN INADEQUATE THEORETICAL FRAMEWORK

The upheaval in pastoralist livelihoods continued after independence for many countries as Western aid agencies called for improved livestock production through private ranching, and market integration (Fratkin, 2001, p. 6). For instance, the "Freedom from Thirst" government campaign in Sudan, funded by Western donors, brought boreholes to the west of the country in the 1960s. The availability of water, for people and for livestock, was intended to improve the quality of life in the region.

While nomadic pastoralists previously gauged the length of time they spent in these areas by the availability of seasonal water supplies, the introduction of borehole points saw larger and larger numbers of livestock staying for increasingly longer periods at the same location. As a result, grazing and tree-felling for firewood around the borehole points intensified, leading to soil erosion and overall environmental degradation. So, when drought struck western Sudan in the 1970s, the carrying capacity of the land was already at a tipping point, with degradation in many areas. Livestock died or were sold off at rock-bottom prices, herd owners went bankrupt, and some communities never recovered (Peters, 1996).

The "tragedy of the commons" theory (Hardin, 1968) also played a key role in appropriating environmental degradation in these regions to communal (and thus poorly managed) pastures. Since any gains from growing individual herd sizes would belong to the owner of the herd, but the losses arising from consequences of large herd sizes (such as overgrazing) would be shared by all, Hardin suggested that personal best interest would drive herdsmen to keep growing their herds until "(t)he herds exceeded the natural 'carrying capacity' of their environment, soil was compacted and eroded, and 'weedy' plants, unfit for cattle consumption, replaced good plants" (Hardin, 1968, p. 1244). Therefore, privatizing the commons would internalize the impacts, thus transferring management responsibilities to individual herdsmen. With this conceptual framework informing policy makers, mobile pastoralists were seen as the villain when it came to land degradation.

However, Cossins (1978) points out that the communal use of land is not the "free for all" that many think it is: "Communal grazing areas are generally divided into specific grazing territories, each with its own rights of access and set water points and various controls are maintained in inter-area movements." Cossins contends that "communal systems based on pastoral use of land are no more destructive of resources and produce more food per unit area than any other animal-based arid land use system. For example, the Borana of southern Ethiopia produce more protein per hectare per year than

Australian ranches in similar rainfall zones" (Cossins, 1978). He argues that "the real limitation to communal land tenure systems is that few developers believe in them because they have imbibed too fully the Western idea that production increases and resource conservation can only be effectively achieved under private ownership" (Cossins, 1978).

African governments have attempted to engage with mobile pastoralists to varying degrees. In Nigeria, although there have been several interventions geared toward pastoralist education, sedentarization, livestock health, and productivity, they have had little impact (Okello et al., 2014).

Iro (1995) corroborates this finding with respect to the Fulani (the largest grouping of pastoralists in Nigeria): "Every attempt to bring Nigeria's Fulani into the fold of so-called progressive society has failed, leaving the Fulani at the mercy of the weather and faulty government actions that impoverish rather than promote the welfare of the pastoral producers."

The challenge of integrating mobile pastoralists into the modern nation state is by no means confined to Africa. In Norway, for example, the Saami people have long campaigned for *siida*, their own institution regarding land rights, organization, and daily management of reindeer. This was enshrined in law by the Norwegian Parliament in 2007 (Ulvevadet, 2012).

In the seminal academic textbook *Agricultural Development in the Third World,* Carl Eicher (1990) wrote:

> In livestock production, the technical and social science research base for herders is seriously inadequate, especially for dry areas. For example, the International Livestock Centre for Africa (ILCA) recently decided to abandon its 10 year research programme on nomadic and semi-nomadic herding in dry Africa because of the lack of progress in developing improved technology for herders in these areas and the belief that research should be focused on the more favorable areas where there is a potential for a bigger pay off.

Evidently, the general sociopolitical and socioeconomic context across Africa has been inimical to sustainable rangeland management by mobile pastoralists. With productive grasslands diminishing, human populations rising, and development pressures on land use increasing, conflict over access to pastures and water has become more frequent. The increased availability of automatic weapons has only served to make these more deadly.

In order to break this vicious cycle, it is vital to understand the links between conflict and land degradation and to pinpoint how technical and trust-building interventions can be made in tandem. The rest of this section focuses on three examples—from central Darfur in Sudan, Kaduna State in Nigeria, and Baringo County in Kenya—that illustrate these dynamics.

2.3.4 DARFUR, SUDAN—A NEED FOR GOOD GOVERNANCE

Darfur is a vast region straddling desert and savanna. While there are many overlapping causes of conflict in Darfur (de Waal, 2007), changes in land use, leading to degradation, is undoubtedly one of them (UNEP, 2014).

A United Nations Environmental Programme (UNEP) report uses an example from Um Chaloota, in central Darfur, to illustrate one reason why tensions between pastoralists and farmers are increasing. Here, as in other more agriculturally productive parts of Darfur, the population quadrupled between

1973 and 2000. As a result, rainfed agriculture has replaced large tracts of rangelands and forest that were previously used as pasture for livestock, as well as corridors for grazing and access to watering points. Moreover, pastoralists now have to maintain relationships with a larger number of people to ensure safe and continued access to herding and grazing routes. This situation becomes even more delicate when livestock has to move through cultivated land in order to reach old pasture and watering points, creating the potential for conflict between the groups, especially if any animals stray (UNEP, 2014). Thus, in order to reduce the risk of violent conflict, the report concludes that a new system of governance is needed in Darfur, involving elements such as "the physical demarcation of the livestock routes and clarification and protection of legal rights" (UNEP, 2014, p. 11).

The Um Chaloota example is one of many that led the UNEP to draw an overall conclusion that "a holistic and long-term approach is needed to support Darfur's emergence from chronic cycles of violence" and that "relationships amongst communities, government, civil society and the private sector need strengthening as part of rebuilding good governance for natural resources" (p. 48).

2.3.5 KADUNA STATE, NIGERIA: ETHNORELIGIOUS CONFLICT AND SOCIOECONOMIC INCLUSIVITY

Conflict between mobile pastoralists and sedentary farmers is also prevalent in northern Nigeria. As in Darfur, Kaduna State has experienced multiple overlapping conflicts, of which land degradation is one of many causal factors. The lack of investment in agriculture by the Nigerian government is an overarching dimension. Several commentators are scathing about Nigerian government policy over recent decades. For example, Busch (2014a) maintains that "the whole edifice of what passes for governance in Nigeria is grounded upon the production of oil and the theft of its revenues. There is very little productive industry or agriculture in the North [to]… sustain its populace in food and jobs without a regular and hefty payment of money to the Northern states by the Federal Government from revenues derived from the oil industry."

Government agricultural extension services are weak. In a recent study conducted near Kano, Nigeria, Maiangwa et al. (2007) found that while 45% of farmers had access to information on crop production, none had access to information on methods for soil conservation. Yet, various climatic and socioeconomic factors are causing land degradation in northern Nigeria and pushing the Sahara further south into the country. Changes in land use, including agricultural expansion and intensification with reduced fallow periods, as well as wood extraction for fuel and timber, are making the soil susceptible to erosion. Similarly, lower and more unpredictable rainfall patterns have all been linked to land degradation in Nigeria. Consequently, farmers and pastoralists are increasingly encroaching into each other's resource boundaries (such as land, pasture, crop-residue, or water), leading to conflict between them (Macaulay, 2014).

Furthermore, conflict between sedentary farmers and mobile pastoralists in the middle belt of Nigeria often has an ethnoreligious dimension: farmers tend to be Christian and hail from indigenous groups, such as Tiv and Baggi, while pastoralists are often Muslim and of Hausa/Fulani ethnicity (Busch, 2014b). As in Darfur, rising populations and land degradation are putting more pressure on relationships between herders and farmers, as well as between herders from different communities.

There are other, more sinister factors at work. For instance, mobile Fulani herdsmen are a significant source of cross-border arms trafficking for a group of violent insurgents called Boko Haram (Omitola, 2014). The Fulani have also been targeted by Boko Haram as part of its quest for greater influence and power (McGregor, 2014). At the same time, disaffected Fulani are increasingly joining gangs involved in armed robbery and cattle rustling, not only in Nigeria, but also in Niger, Chad, Cameroon, Mali, and Senegal. In February 2012, 23,000 Fulani herders were reported to have entered Cameroon from Nigeria's eastern state of Taraba, following deadly clashes with farming communities (Busch, 2014b). In 2013 and 2014, thousands fled their homes in Kaduna State, following raids by cattle rustlers that killed dozens of people.

A pervasive cycle of chaos and violence is ongoing. Okello et al. (2014, p. 2) note the "growing uptake of arms by pastoralist youths that have become available as a result of various sequelae to the west's 'War on Terror' and the Arab Spring." There is now a danger that the regreening successes in neighboring Niger (see, e.g., Cameron, 2011) could be undermined by the side effects of massive insecurity in northern Nigeria. Having vandalized power lines in Nigeria, tree cutting and demand for firewood in the states controlled by Boko Haram has increased and the destruction has spread across borders (Hayden, 2015). Therefore, Okello et al. (2014, p. 12) conclude that the "existing Fulani self-governance system must be well understood if ongoing approaches towards increased political inclusion, improved service provision and conflict resolution are to be successful."

Toward this end, an attempt is currently underway to resolve tensions between herders and farmers in the Sanga local government area, where about 50,000 people were temporarily displaced in June 2014, after militia groups said to be of Fulani origin went on a rampage (Wuye, 2014). Pastor Dr. James Wuye and Imam Dr. Muhammad Ashafa, two internationally acclaimed peacemakers, are mediating between the Fulani and the chiefdoms of the Ayu, Bodobodo, and Numana ethnic groups. Wuye and Ashafa aim to have the conflicting parties hold a meeting at which forgiveness will be extended and a Peace Affirmation will be signed, in a similar process to that which ended ethno-religious conflicts in Yelwa Shendam, in Plateau State (Channer, 2006).

Kratli and Swift (2001) stress the importance of ritual healing in pastoral societies and note that it has often been neglected by international agencies, as has working with Islamic faith leaders. Part of Wuye and Ashafa's methodology is to link peace affirmations to a subsequent joint social action (Wuye and Ashafa, 2011). In this case, the chosen joint action for land restoration, which celebrates the mutual ties that herders and farmers used to have, could be critical both for long-term peace and for sustainable land management. The two are inextricably linked, and together they can improve the livelihoods of millions.

Peace-building and sustainable land management can even mitigate terrorism. As Mazrui (2012, p. 4) explains, "[t]here is need to restore a sense of self-worth… for disadvantaged young people in Nigeria. This would be the best antidote to political and religious extremism in the unfolding decades of Nigeria's history."

2.3.6 BARINGO COUNTY, KENYA: AN EXAMPLE OF GOOD PRACTICE

Baringo County in Kenya presents one of the most striking examples of a positive interaction between land restoration and trust building. Soil erosion has become severe in the Baringo lowlands as overgrazing of diminishing communal reserves has accelerated, and the combination of little or no vegetative cover and high-intensity storms has led to flash flooding and heavy runoff. This has been exacerbated by deforestation in the foothills, leading to silt buildup in Lake Baringo (Rehabilitation of Arid Environments Charitable Trust, n.d.).

This degradation of local ecosystems is forcing local pastoralist communities to share fewer pasture and water resources, frequently concentrated in small areas around seasonal rivers and riverbeds. While there has always been an element of conflict between Baringo's main ethnic groups, the Tugen, Ilchamus, and Pokot, particularly since cattle rustling is associated with rites of passage and warriorhood, the three communities' elders used to be effective at negotiating reparations and access to dry-season grazing and water points (Rehabilitation of Arid Environments Charitable Trust, n.d.). However, according to the founder of the Rehabilitation of Arid Environments Charitable Trust (RAE Trust), Murray Roberts, the forced sharing of resources has resulted in more encounters and incidences of rustling, sometimes growing into violent conflicts (Roberts, 2014). Roberts (2014) suggests that there has been an overall breakdown of traditional social values during his lifetime: "When I was a child you could leave anything in the bush, come back two weeks later and it would still be there. Now someone will take it." However, in recent years, cattle rustling has become a scourge in Baringo and neighboring counties, with fatalities and interethnic bitterness replacing mere competitiveness.

The RAE Trust, which has contributed to the reclamation of over 4000 acres of pasture and received a Special Mention for the UNCCD Land for Life Award in 2013, has introduced an effective model of land restoration through improved grassland management. One of the key practices that the RAE Trust has introduced is the safeguarding of surplus grass during the rainy season, for storage and use as fodder during the dry season. This is akin to growing grass in temperate countries for use as hay or silage. The method works when communities collectively fence off land, plough, prepare the soil to maximize water harvesting, sow grass seed, and then prevent all grazing until the grass is mature. The resulting grass can be harvested and stored in fodder banks, cut and sold as thatch, or used to fatten livestock. The grass seed can also be harvested and sold. Importantly, the practice also prevents soil degradation during the rainy season by reducing water erosion, and during the dry season by reducing the distances covered by pastoralists in search for pasture.

However, while the land restoration activities were successful, some of the RAE Trust's projects (for example, in Kiserian and Rugus), have broken down due to interethnic conflict. It is evident that in order to restore long-term peaceful coexistence and stability in the region, land restoration will have to be paired with trust-building activities.

To this end, the Initiatives for Land, Lives, and Peace (ILLP)—working through Initiatives of Change (IofC) Kenya—have introduced a distinctive intervention to rebuild trust among these communities. At a meeting of community leaders, Joseph Karanja of IofC Kenya presented the film *An African Answer*, depicting an effective, wholly African approach to conflict resolution between rival ethnic groups. This film draws extensively from the experiences of Imam Muhammad Ashafa and Pastor Wuye, who decided to lay down their arms and work together to end communal violence plaguing Nigeria in the 1990s. Featuring this film encourages communities to foster healing and reconciliation and to rebuild peace and prosperity.

This innovative combination of peace-building and land restoration has been acclaimed as "a milestone in the constructive management of the conflict" in Baringo by UNCCD (2014). As Baringo County's deputy governor, Mathew Tuitoek, clearly stated: "Without peace, we cannot have development in this county" (UNCCD, 2014).

Jospeh Kwopin's experience is pertinent in this regard: (UNCCD, 2014):

Joseph Kwopin from the Pokot community was once a victim of one of the many cattle raids. 42 of his cows were stolen—despite the AK47 he carried with him on his long journeys to find grazing lands. Finally, as a last resort, he turned his attention to degraded land and sowed indigenous grass. Only when its root causes are addressed, can the conflict be solved.

Today, the grass fattens livestock on more than 30 acres of pasture. In addition, his business includes a fodder bank and grass seed, which he sells to other farmers. Joseph Kwopin has demonstrated that with the right techniques, it is possible to maintain pastoralism in Baringo. "It is the people factor that is often the most critical for sustainability," says Elizabeth Meyerhoff-Roberts, a co-director of the Rehabilitation of Arid Environments Trust.

Since the meeting, Kwopin has welcomed community leaders from the Ilchamus and Tugen to learn from his pasture management. A peace committee, catalyzed by IofC Kenya, has resolved to hold an interethnic healing ritual to end mutual grievances and help pave the way to ending deadly cattle raids. The committee also wishes to see the reinstatement of cattle auctions, in which all ethnic groups feel free to participate.

In another example, IofC Kenya signed a memorandum of understanding with the UK-based charity Excellent Development (at the time of writing of this section), to introduce sand dams to Baringo. Ilchamus, Tugen, and Pokot community leaders have formed one community-building organization, so that the introduction of sand dams and newly available dry-season water, benefits them all, and thereby helps consolidate peace.

By fostering partnerships between organizations that promote specific technologies for land restoration on the one hand and trust building on the other, ILLP can help catalyze sustainable land management in environmentally degraded, conflict afflicted regions.

2.3.7 TRUST BUILDING SUCCESSES

Today, the importance of linking trust building with technical interventions in land restoration is being increasingly recognized. Improved pasture management can help build peace, as can holistic management techniques. Recent findings indicate that the holistic management of grassland which mimics the patterns in which diverse herds of wild herbivores graze is also an effective means of reversing land degradation, and even combating climate change (see, e.g., Savory, 2009). For example, the Laikipia Wildlife Forum (LWF) encourages Maasai and Samburu pastoralists to practice both "planned grazing," in which the amount of grass that grows in the rainy season is maximized, and "bunched grazing," where herds move close together, breaking up the ground and fostering water and nutrient flow, while at the same time adding manure and implanting seed (Laikipia Wildlife Forum, 2013).

The benefits of tourism are also felt in wildlife conservancies, where livestock and wildlife live side by side and where pastoralist communities benefit from tourist revenues. This happens, for example, at the Sanctuary at Ol Lentille for the Laikipiak Maasai. But these interventions can also only be applied when there is already peace.

In regions where communities are affected by land degradation and conflict, better dialogue and mutual cooperation are needed among all the stakeholders—state actors and pastoralist community leaders, pastoralists and sedentary farmers, leaders of different pastoralist ethnic groups, and elders and the underemployed, disaffected youth within pastoralist communities—in order to reestablish peace and restore land.

Many are optimistic about the future. In a radio interview broadcast in several African countries, Swift argued that "pastoralists not only have a future in Africa, but that their future will be rather more successful than many others, as climate change makes all drylands more risky with more extreme droughts and floods." Indeed, he continues, "Pastoralists, because of their mobility, are uniquely equipped to respond to this" (Massarenti, 2012).

Communication technology is also helping. For example, many pastoralists now have cell phones and use them in day-to-day business to get information on pasture conditions at the next destination or the price of animals at a distant market. New possibilities for distance learning, adapting interactive curricula already available on the internet, could be applied for communities on the move (Massarenti, 2012).

Mobile pastoralism is likely to evolve increasingly toward agropastoralism, where pastoralists spend more time in one place, with improved pasture management, climate-smart crop cultivation, and other livelihood enterprises. As population increases and climatic conditions become more unpredictable, implementing conflict-mitigating and trust-building strategies will be key. Whatever happens to Africa's pastoralist communities over the next decades will be a determining factor for the future of grassland across the continent.

REFERENCES

African Union, 2010. Policy Framework for Pastoralism in Africa: Securing, Protecting, and Improving the Lives, Livelihoods and Rights of Pastoralist Communities. Department of Rural Economy and Agriculture, Addis Ababa, Ethiopia.

Bonte, P., Galaty, J., 1991. Introduction. In: Galaty, J.G., Bonte, P. (Eds.), Herders, Warriors, and Traders: Pastoralism in Africa. Westview Press, Boulder, p. 10.

Broch-Due, V., 1999. Remembered cattle, forgotten people: The morality of exchange and the exclusion of the Turkana poor. In: Anderson, D., Broch-Due, V. (Eds.), The Poor Are Not Us: Poverty and Pastoralism in Eastern Africa. James Curry, Oxford; Ohio University Press, Athens, OH.

Busch, G.K., 2014a. What is Boko Haram and whence did it arise? Pambazuka News. Available at: http://allafrica.com/stories/201405201229.html (accessed 28.2.2015).

Busch, G. K., 2014b. The Fulani and Kanuri Main Actors in Boko Haram. (accessed 23.02.15.)., from, http://www.elombah.com/index.php/special-reports/23405-the-fulani-and-kanuri-main-actors-in-boko-haram.

Cameron, E., 2011. From vulnerability to resilience: Farmer Managed Natural Regeneration (FMNR) in Niger. Climate and Development Knowledge Network and World Resources Institute. (accessed 23.02.15.). http://www.wri.org/sites/default/files/pdf/inside_stories_niger.pdf.

Channer, A., 2006. Imam and the Pastor [film]. FLTFilms, London.

Cossins, N., 1978. Resource conservation and productivity improvement under communal land tenure. In: Joss, Lynch, Williams, O.B (Eds.), Rangelands: a resource under siege. Proceedings 2nd International Rangeland Congress. Adelaide, Australia, pp. 119–121.

De Waal, A., 2007. Darfur and the failure of the responsibility to protect. Intl. Affairs 83 (6), 1039–1054.

Eicher, C.K., 1990. Africa's food battles. In: Eicher, C.K., Staatz, J.M. (Eds.), Agricultural Development in the Third World. John Hopkins University Press, Baltimore and London, pp. 516–517.

Farah, I., Hussein, A., Lind, J., 2002. Deegan, politics and war in Somalia. In: Lind, J., Sturman, K. (Eds.), Scarcity and Surfeit: The Ecology of Africa's Conflicts. Institute for Security Studies, Pretoria, South Africa, pp. 320–356.

Fratkin, E., 2001. East African pastoralism in transition: Maasai, Boran, and Rendille cases. Afr. Stud. Rev. 44 (3), 1–25.

Gonzalez, P., Tucker, C.J., Sy, H., 2012. Tree density and species decline in the African Sahel attributable to climate. J. Arid Environ. 78, 55–64.

Gwin, P., 2011. Insaisissables seigneurs du Sahara. Natl. Geog, 3–19 October.

Hardin, G., 1968. The tragedy of the commons. Science 162, 1243–1248.

Hayden, S., 2015. Boko Haram seizes military base on Nigeria's border with Chad. Voice News. Available at: https://news.vice.com/article/boko-haram-seizes-military-base-on-nigerias-border-withchad (accessed 28.02.15.).

Hulme, M., Doherty, R., Ngara, T., New, M., Lister, D., 2001. African climate change: 1900–2100. Clim. Res. 17, 145–168.

Hamne, G., 2014 Personal communication [e-mail].

International Institute for Environment and Development and SOS Sahel International UK, 2010. Modern and mobile. The future of livestock production in Africa's drylands. IIED and SOS Sahel UK, London.

International Union for the Conservation of Nature (IUCN), 2008. Pastoralism (World Initiative for Sustainable Pastoralism (WISP)) [website] (accessed 07.02.15), at: http://www.iucn.org/wisp/pastoralist_portal/pastoralism/.

Iro, I., 1995. From nomadism to sedentarism: An analysis of development constraints and public policy issues in the socioeconomic transformation of the pastoral Fulani of Nigeria. University Graduate School in African Studies Department, Howard University, Wasington, DC.

Kratli, S., Swift, J., 1999. Understanding and Managing Pastoral Conflict in Kenya: A Literature Review. Department for International Development (DFID), London.

Laikipia Wildlife Forum (LWF), 2013. Rangeland rehabilitation. Laikipia. http://www.laikipia.org/programmes-top/rangeland-rehabilitation (accessed 28.02.15.).

Macaulay, B.M., 2014. Land degradation in northern Nigeria: The impacts and implications of human-related and climatic factors. Afr. J. Environ. Sci. Technol. 8 (5), 267–273.

Maiangwa, M.G., Ogungbile, A.O., Olukosi, J.O., Atala, T.K., 2007. Land Degradation: Theory and Evidence from the North-West Zone of Nigeria. J. Appl. Sci. 7 (6), 785–795.

Massarenti, J., 2012. Interview: jeremy swift, "Pastoralism has a future in Africa". Afronline. http://www.afronline.org/?p=24112 (accessed 28.02.15.)

Mazrui, A.A., 2012. Nigeria: From Shari'a movement to "Boko Haram." Global Experts Team. United Nations Alliance of Civilizations, New York.

McGregor, A., 2014. Alleged connection between Boko Haram and Nigeria's Fulani herdsmen could spark a Nigerian civil war. Terrorism Mon. XII (10).

Millennium Ecosystem Assessment, 2005. Ecosystems and Human Well-being: Synthesis. Island Press, World Resources Institute, Washington, DC.

Okello, A.L., Majekodunmi, A.O., Malala, A., Welburn, S.C., Smith, J., 2014. Identifying motivators for statepastoralist dialogue: exploring the relationships between livestock services, self-organisation, and conflict in Nigeria's pastoralist Fulani. Pastoralism 4 (12).

Omitola, B., 2014. Between Boko Haram and Fulani herdsmen: Organised crime and insecurity in Nigeria. Department of Political Sciences, Osun State University. From (accessed 23.02.15.), http://www.issafrica.org/uploads/5th-Crime-Conf-2014/X002-Bolaji-Omitola.pdf.

Opiyo, F., Wasonga, O., Schilling, J., Mureithi, S., 2012. Resource-based conflicts in drought-prone Northwestern Kenya: The drivers and mitigation mechanisms. Wudpecker J. Agr. Res. 1 (11), 442–453. Available at http://www.wudpeckerresearchjournals.org.

Peters, C., 1996. Sudan: A Nation in the Balance. Oxfam Professional, Oxfam, UK.

Rehabilitation of Arid Environments Charitable Trust, n.d. Dryland problems and area description. [website] (accessed 23.02.15.), from, http://www.raetrust.org/background.htm.

Roberts, M., 2014. Personal communication. [e-mail].

Salzman, P., 2004. Pastoralists: Equality, Hierarchy, and the State. Westview Press, Boulder, CO.

Savory, A., 2009. Foreword. In: Howell, J. (Ed.), For the Love of Land: Global Case Studies of Grazing in Nature's Image. Book Surge Publishing, Charleston, South Carolina.

Schlee, G., 2011. Territorializing ethnicity: The imposition of a model of statehood on pastoralists in northern Kenya and southern Ethiopia (English). Ethnic Rac. Stud. 36 (5), 857–874.

Smith, A.B., 1992. Pastoralism in Africa: Origins and Development Ecology. Hurst, London.

Ulvevadet, B., 2012. The governance of Sami reindeer husbandry in Norway. University of Tromsø, Tromsø, Norway.

United Nations Convention to Combat Desertification (UNCCD), 2014. Initiative for land, lives, and peace in baringo county. Kenya, Land, Lives, Peace. UNCCD News 6 (1), 13–17. Available at: http://landlivespeace.org/2014/03/28/initiative-for-land-lives-and-peace-in-baringo-county-kenya/ (accessed 28.02.15).

United Nations Environmental Programme (UNEP), 2014. Relationships and Resources: Environmental governance for peacebuilding and resilient livelihoods in Sudan. UNEP, Nairobi.

Wuye, J., 2014. Personal communication. [e-mail].

Wuye, J.M., Ashafa, M.N., 2011. A Resource Guide for Grass-roots Practitioners. In: For the Love of Tomorrow Films/Initiatives of Change, London.

LAND DEGRADATION FROM MILITARY TOXICS: PUBLIC HEALTH CONSIDERATIONS AND POSSIBLE SOLUTION PATHS

2.4

Adam Koniuszewski

Green Cross International, Geneva, Switzerland

2.4.1 MILITARY ACTIVITIES

While still overlooked as a major cause of ecological damage, military activities (during wars, conflicts, and in peacetime) cause widespread and long-lasting environmental degradation. In fact, nature has always been a "victim" of war: Attila burned the landscape behind him in his retreat; the nuclear bombings of Hiroshima and Nagasaki killed hundreds of thousands of people, mutated the DNA of future generations, and contaminated vast areas of land; Agent Orange caused widespread and long-lasting toxic pollution in Vietnam, Cambodia, and Laos. It is, therefore, increasingly accepted that military activities must be regulated and monitored to prevent unnecessary environmental impacts. This process began during the Rio Conference on Environment and Development in 1992, with the inclusion of environmental norms for the military as part of Agenda 21, including the need for cleanup and restoration of the damage caused by military activities (see the subsection entitled "Case Study: Land Contamination in Kuwait After the 1990–1991 Iraqi Invasion," later in this section to illustrate this point).

While military equipment and weapons are meant for use on the battlefield in conflict, most weapons will never be used in combat. In fact, most weapons will only be used in training, and most ammunition is disposed of due to the expiration of their shelf life and from obsolescence.

Training sites, including shooting ranges, pose significant pollution problems as well as an ammunition disposal headache (see the subsection entitled "Case Study: Land Contamination at Shooting Ranges," later in this section, for more details). Environmentally safe and sound alternative methods must be developed and implemented.

Most weapons contain explosives and propellants. In the United States until the mid-1990s, the common practice for the disposal of obsolete conventional weapons—including landmines, grenades, and artillery shells—were the open-burn/open-detonate methods. *Open burning* means draining explosive powder from munitions and setting it on fire in large open pans. *Open detonation* involves the packaging of munitions in barrels, and then typically burying them under 2 m of earth and detonating them. These practices released toxins into the soil, water, and air, causing widespread, long-term pollution. They are still used when closed burn/closed detonate methods, in which the operation takes place in a contained structure, are not possible for safety and security reasons. These sites then require extensive cleaning and remediation. Advocacy for closed-burn and closed-detonation practices started in the mid-1990s. These methods have been implemented in the United States for the remediation of conventional and chemical weapons. Closed-detonation systems and technologies contain the bulk of effluents and collect them in solid and liquid form for further treatment or for sealed storage in toxic landfill sites. Polluting emissions from closed-detonation facilities are scrubbed before being released into the atmosphere.

2.4.2 CHEMICAL WEAPONS

Chemical weapons were first used during World War I. They are banned by the 1997 Chemical Weapons Convention (CWC), which not only requires their complete destruction, but also mandates that environmental and public health be fully protected during the process of disarmament and disposal. The CWC specifically outlaws such disposal methods as open-pit burning, land burial, and sea-dumping. Permitted destruction technologies include incineration and hydrolysis with treatment of the remaining effluents to destroy artillery projectiles, mortars, air bombs, rockets, rocket warheads, spray tanks, and the chemical weapon agents stored in bulk or in munitions. The safe and verified destruction of close to 62,000 metric tons of chemical weapons, stockpiled mostly in the United States and in Russia, has cost over $30 billion so far. This achievement, representing 85% of the declared chemical weapons arsenals worldwide, earned the Organization for the Prohibition of Chemical Weapons (OPCW) the 2013 Nobel Peace Prize.

It is estimated that the destruction of the remaining 11,000 metric tons of chemical weapons (approximately 15% of the total) stockpiled in the United States, Russia, and elsewhere will take up to 10 years to complete. The recent concerns of European citizens over the planned destruction of the 600 metric tons of Syrian chemical weapons, including sarin and mustard gas, have forced the international community to proceed with the neutralization of the weapons offshore, on board the US vessel *Cape Ray*.

Nonstockpiled weapons also pose a great risk to both land and oceans. In the United Sates alone, there are 249 known and suspected sites containing chemical warfare material, some in large quantities that pose great challenges in terms of operational complexity and potential remediation costs. Many chemical weapons have also been dumped at sea. Despite the lack of a comprehensive sea-dumped chemical weapons database, records indicate that 160,000 tons may have been dumped in Russian seas; a large portion of the 303,000 tons left in Germany and the UK after World War II was sea-dumped; and the US military dumped chemical weapon agents on at least 74 occasions between 1918 and 1970. These also represent a threat to human health and the environment. Studies on this topic are still at the early stage.

2.4.3 NUCLEAR CONTAMINATION

Since the beginning of the atomic energy era and the construction of the first nuclear reactor in Chicago in 1942, there have been numerous soil radiation events that resulted from human activities. These include the testing of nuclear weapons, their military use in the final stage of World War II, controlled releases of radioactive wastes from the peaceful uses of atomic energy, and nuclear accidents, spills, and leaks, including those of Windscale (United Kingdom) and Three Mile Island (United States), the crash of a U.S. military airplane carrying thermonuclear bombs containing plutonium (in Spain), and accidents at Chernobyl (the then-U.S.S.R.) and, most recently, at the Fukushima plant (Japan).

Such instances can result in various types of contamination. In severe cases, as in the situation that occurred near the Chernobyl site, the physical removal of topsoil from some areas within the critical 30-km zone for disposal at special sites was required. At worst, topsoil (and floral cover) was so heavily contaminated by longer-lived radionuclides that reclamation for agricultural use was rendered neither economic nor acceptable in the context of local community or public health.

Leakage from storage facilities poses particular threats as well. In the Hanford Nuclear Reservation (in southwestern Washington State in the United States), which operated between 1943 and 1987 to produce plutonium for military purposes, 110,000 tons of nuclear fuel was processed to produce 73 tons of nuclear weapons and reactor fuel. The process generated millions of gallons of highly radioactive waste and other forms of hazardous waste. In 1988, the remediation of the site began. By 2013, the cleanup efforts were estimated to have cost around $40 billion, with another $115 billion required.

2.4.4 DEPLETED URANIUM

Most depleted uranium comes as a byproduct of the production of enriched uranium for use in nuclear power reactors and in the manufacture of nuclear weapons. It is used in a number of peaceful applications, including as ballast in aircraft, for radiation shields in medical equipment, and as containers for the transport of radioactive materials. Due to its high density, depleted uranium also has military applications as defensive armor plate on tanks and as armor-penetrating bullets and projectiles because of its ability to self-sharpen as it penetrates its target and its propensity to ignite on impact.

The pyrophoric nature of depleted uranium used in ordnance and armoring is particularly significant for its potential environmental impacts. When a depleted uranium projectile hits a hard target, it burns and oxidizes, bursting into radioactive dust and aerosolized microparticles. The resulting dispersion of depleted uranium dust is scattered depending on its chemical and physical form (chemical species, particle dimensions, solubility, etc.) and can travel by wind and leaching, depending on local soil characteristics and meteorological and climatic conditions. This can result in widespread environmental contamination and human intake by inhalation or ingestion, or if it enters the body through open wounds. The harmful effects of depleted uranium can be caused by both its radioactivity and chemical toxicity, but the respective mechanisms have not yet been studied in detail, and existing studies often contradict each other. There is a widely shared view among experts that further research is needed to assess the health risks and environmental impacts of the use of arms and ammunition containing depleted uranium.

2.4.4.1 CONFIRMED USE OF DEPLETED URANIUM WEAPONS

The United States and United Kingdom confirmed using depleted uranium weapons during the 1990–1991 Gulf War against Iraq, firing 320 tons and less then 1 ton of such munitions, respectively. It was also reported that depleted uranium friendly-fire and accidental fire incidents contaminated 31 US combat vehicles (namely, 16 Abrams tanks and 15 Bradley armored vehicles) during the conflict.[1]

FIGURE 2.4.1

Radioactive battlefields in Kuwait and Iraq.

American forces then confirmed using depleted uranium weapons during the Balkan wars of the 1990s (in Bosnia-Herzegovina, Serbia, and Kosovo), as well as in the invasion of Afghanistan since 2001 and in Iraq from 2003 onward. United Nations Environment Programme (UNEP) estimates of depleted uranium munitions used during the Iraq conflict range between 1000 and 2000 metric tons, mostly by the US military. The UK Ministry of Defense reported using less than 1.9 tons of these weapons in 2003. This is the first conflict where it was alleged that the US military fired depleted uranium weapons on civilian targets, which prompted US congressman Jim McDermott to call for the public release of the entire depleted uranium firing coordinates.

Depleted uranium ammunition was also used in trials and trainings, such as those held in Madison, Indiana, where ammunition tests between 1941 and 1995 left more than 70 tons of depleted uranium shell fragments. These included leftovers from the tests of radioactive armor piercing shells conducted between 1984 and 1994.

2.4.4.2 ENVIRONMENTAL AND HEALTH CONSIDERATIONS

Concerns about the environmental and health impacts of the use of depleted uranium weapons have led to studies and reports by various institutions, including the World Health Organization (WHO), UNEP, and more recently by the Pax Christi interchurch peace group in the Netherlands. The WHO reported on the implications of depleted uranium, stating that only their military use is likely to have any significant environmental impact and that measurements of sites where such weapons were used indicated localized contamination (and in some cases, potential risks for the food chain) that should be monitored. The WHO recommends cleanup zones where depleted uranium contamination levels are deemed unacceptable by qualified experts. Following its investigation of claims by Iraqi doctors of increased rates of cancer and birth defects, the WHO Summary Document of Prevalence of Congenital Birth Defects investigation in Iraq concluded in September 2013 that there is "no clear evidence to suggest an unusually high rate of congenital birth defects in Iraq."

From 1999 to 2003, UNEP conducted environmental assessments and measurements on depleted uranium targeted sites in Kosovo, Serbia and Montenegro, and Bosnia and Herzegovina. More recently, based on the findings of the UNEP Desk Study on the Environment in Iraq, UNEP initiated a depleted uranium project in Iraq that was focused on capacity building at the Radiation Protection Center of the Iraqi Ministry of Environment. The 2010 UNEP report to the UN indicated that major scientific uncertainty persisted regarding the long-term environmental impacts of depleted uranium, particularly with respect to long-term groundwater contamination. Therefore, it called for the adoption of a precautionary approach, including the cleanup and decontamination of affected areas and measures to raise awareness about the potential risks to the civilian population. The UNEP report is consistent with the findings of Pax Christi, whose 2014 study on the impact of depleted uranium use during the Iraq wars was funded by the Norwegian Ministry of Foreign Affairs, and that identified up to 365 contaminated sites in Iraq that require cleanup. It also stated that toxic waste is being spread by scrap metal dealers and those who worked for them (including children), and that the claims of Iraqi doctors about adverse health consequences should be taken seriously.

2.4.4.3 UN RESOLUTION ON DEPLETED URANIUM

Beginning in 2008, the UN General Assembly has been adopting resolutions on depleted uranium. The last one, Resolution 69/57, was approved on December 2, 2014, supported by 150 votes, with 4 against (France, Israel, United Kingdom, and United States), and 27 abstentions. It calls for additional studies on the potential long-term effects on humans and the environment, recognizes the UNEP conclusions that major scientific uncertainties persist, particularly for long-term groundwater contamination, and calls for a precautionary approach to be adopted. It invited member states that used such weapons to provide affected states, upon request, with detailed information about the location where the weapons were used to help assess the resulting impacts. It also encouraged member states to provide assistance to affected states in identifying and managing contaminated sites and material.

2.4.5 CASE STUDY: LANDMINES AND OTHER REMNANTS OF WAR*

As of October 2014, 56 states and four other areas still have an identified threat from antipersonnel mines. A further six States Parties to the Anti-personnel Mine Ban Convention had either suspected or residual mine contamination. Today, massive antipersonnel mine contamination (defined as more than 100 km^2) is believed to exist only in Afghanistan, Bosnia and Herzegovina, Cambodia, Turkey, and very probably also in Iraq. Heavy antipersonnel mine contamination (more than 20 km^2 and up to 100 km^2) is believed to exist in several states: Angola, Azerbaijan, Croatia, Thailand, and Zimbabwe. The situation in Lao PDR, Myanmar, and Vietnam is not known, but may also be heavy (Landmine Monitor 2014).

Explosive hazards, including landmines and cluster munitions, cause harm to civilians during armed conflict and can do so long after the conflict has ended through the direct consequences of explosive blasts, the soil pollution caused by landmine toxics, and local communities being deprived of access to land and natural resources.

FIGURE 2.4.2

Environmental impact chain of contamination from remnants of conflict.

Source: GICHD

This was highlighted in UNEP's assessment of the cluster bomb airstrikes in Lebanon in 2006, though it is also seen more generally. Valuable pasture can become inaccessible, potentially leading to overgrazing of those areas that are accessible and subsequent habitat degradation. Land scarcity resulting from contamination has the potential to generate new socioeconomic dynamics and set new cycles of poverty and environmental degradation in motion. Faced with growing livelihood pressures, local populations are likely to resort to unsustainable practices and intensify the exploitation of the diminished areas available to meet short-term needs.

This finding is corroborated by the phenomenon of deforestation; this generally accelerates as an indirect consequence of contamination. Where arable land has been mined, the long-term consequences of selling forest and fruit trees give way to immediate pressures to simply survive. Deforestation can in turn affect marshlands and water tables, which has an impact on fish and other wildlife. Thus, remnants of conflict can set in motion series of events leading to environmental harm in the form of soil degradation or deforestation. This can affect entire species populations by degrading habitats

---

*The landmine section is based on the Geneva International Center for Humanitarian Demining (GICHD) paper titled "'Do No Harm' and Mine Action: Protecting the Environment While Removing the Remnants of Conflict," presented at the seminar "Peacebuilding and Environmental Damage in Contemporary Jus Post Bellum: Clarifying Norms, Principles, and Practices," organised by Leiden University in June 2014. The entire article is downloadable from http://www.gichd.org/fileadmin/GICHD-resources/rec-documents/Do-no-harm-and-mine-action-Leiden-University-Jun2014.pdf.

and altering food chains. Disruption of soil structure further exacerbates the erosion problem and leads to increased sediment load in the drainage system.

The terrestrial environment suffers when the remnants of conflict explode. Detonating munitions degrade land through topsoil damage or erosion, with sustained impacts on moisture availability, soil structure, vulnerability to water flows, erodibility, and productivity. As witnessed in Vietnam, soil productivity dramatically decreases if land is contaminated.

Whenever ammunition is used, there is always contamination. Various chemical products are released when ammunition enters the environment. The contents of the ammunition leak out when it breaks up on impact. Colored smoke residue or ashes are released. Leaks of explosive substances into aquifers or their evaporation into the air can be immediate and substantial environmental risks. Research has shown that, in some heavily used military training areas, munitions-related chemicals, such as explosives and perchlorate, can enter soil and groundwater. Furthermore, transpiration of trinitrotoluene (TNT) through the roots and stems of plants causes higher concentrations of this chemical in the leaves, making them dangerous to grazing animals. After a prolonged period, the consequences of the corrosion of ammunition fragments and the release of various alloying elements, such as iron, manganese, chromium, zinc, and copper, start to appear. In agricultural regions, toxic elements can easily penetrate the soil, seep into the water table, and pass into the human food chain.

However, the nature and extent of the environmental effects of remnants of conflict, and especially of their toxic substances, remain incompletely documented. There is still a considerable need for further study on the impact of such contamination on the environment and public health.

The removal of landmines and other remnants of conflict, and the destruction of these substances, is necessary to address the negative consequences of contamination from remnants of conflict and to restore livelihoods. However, some of the methods used may have unintended environmental impacts. When demining manually, only locations where metal detectors have indicated metal contamination will be subject to manual digging. Fertile topsoil has to be removed and soil and root systems disturbed, and lower vegetation (bushes, etc.) may have to be cut in order to get access to a suspected or confirmed contaminated area. Erosion may result from this process. Manual clearance remains the preferred tool, especially in areas with dense vegetation where a primary environmental concern is to conserve as much vegetation as possible. Nonetheless, manual clearance is time-consuming and exhausting; consequently, mechanical systems are often used to speed up this process.

Machines have considerable potential for increasing efficiency, but they cause a greater impact on the soil and the ecosystem. A variety of mechanical systems are used (tiller systems, flails, or converted plant machinery) to process soil in the search for remnants of conflict. Inevitably, this will disturb and possibly damage the soil. Soil might often be moved to another location, where it will be distributed evenly over a large, flat surface and subsequently checked for explosive items or evidence of such. When using flails and tillers, the soil will pass through those systems, even though it will remain in the same location after being processed. The consequences of such practices could take the form of various types of erosion, deforestation, changes to soil composition, reduced soil fertility, and ground pollution.

Mechanical systems remove or destroy vegetative cover, which in turn can lead to increased water runoff and wind erosion. Tillage increases wind erosion rates by dehydrating the soil and breaking it into smaller particles that can be picked up by the wind. Deforestation is closely linked to erosion and mechanical demining. The removal of trees means the removal of litter that plays a crucial role in infiltration, protecting soil from erosion and raindrop impacts. Litter also provides organic matter that is important to the stability of soil structure. Deforestation can allow the wind to cut long, open channels as it travels over the ground at higher speeds, and agricultural land that has its topsoil blown away by the wind may be destroyed. Less fertile soils are naturally associated with losses in agricultural production, thereby putting food security at risk. Furthermore, the resilience of communities can be affected. Indeed, the environment and disasters are inherently linked, and environmental damage such as deforestation or degradation of land may reduce the nature's defense capacity against hazards and, in turn, can even aggravate the impact of disasters. Mindful of this fact, and in order not to undermine the positive impact of mine clearance, mine action organizations should ensure that their operations do not increase the long-term vulnerability of affected communities and that, by mitigating environmental damage, they can effectively contribute to disaster risk reduction, and thereby to sustainable development.

Soil degradation occurs when changes in its depth or its physical or chemical properties reduce its quality. During mechanical demining, the organic layer, as well as surface soil, will generally be processed, and the physical or chemical properties and the structure of the soil might be changed or damaged. This can again affect soil fertility, rooting potential, and water-holding capacity.

FIGURE 2.4.3

What not to do in topsoil clearance.

Source: GICHD.

Not only can mechanical mine clearance result in soil erosion and lead to other environmental damage, but there is also a risk of chemical pollution to soil and water. Contamination might be caused by detonations of explosive items in the ground or by leaking hydraulic fluids and fuel, which can occur when refueling demining machines. When hydraulic fluids enter the environment through spills and leaks from machines or storage areas and waste sites, severe environmental damage can result. Mine action organizations and humanitarian actors must consider the possible negative impacts of their operations to ensure that they "do no harm," do not lead to longer-term vulnerability, and reduce natural disturbance to an acceptable level.

2.4.6 CASE STUDY: LAND CONTAMINATION AT SHOOTING RANGES

Contamination by lead, antimony, copper, arsenic, zinc, and other toxic substances is a significant problem at shooting ranges around the world and a major source of environmental contamination. Lead is usually considered the highest-priority concern because of its severe toxic effects and the vulnerability of children—there is no safe level of lead exposure. Typical health concerns include brain and nervous system damage, reduced IQ, learning disabilities, and reproductive problems, even at very low levels of exposure. Nevertheless, other components of bullets that are present in small amounts also accumulate rapidly in high concentrations in the soil.

Lead typically accumulates in the ground at shooting ranges over many years. In the United States, lead concentrations of 233,142 mg/kg were recorded at one of the 12,000 military small arms and recreational shooting ranges. This represents 580 times what the EPA considers to be the safe limit of 400 mg/kg. For reference, a single 12-gauge shotgun shell contains enough lead (28 g) to contaminate the daily water consumption of 1.9 million people, the size of the population of Houston, Texas. Annually, hunting and shooting result in lead deposits of some 500 tons in Switzerland, 800 tons in Denmark, and up to 60,000 tons in the United Sates.

Aside from mining and waste management, outdoor firing ranges put more lead into the environment than any other major industrial activity in the United States. And while the US military has been pursuing massive cleanup efforts of at least 700 firing ranges, commercial shooting activities remain largely unregulated and uncontrolled.

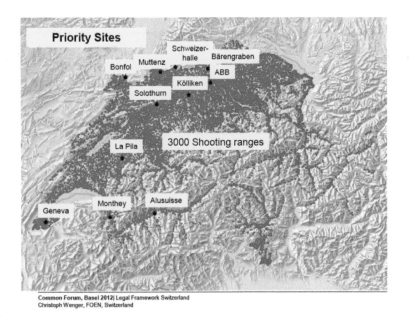

Common Forum, Basel 2012| Legal Framework Switzerland
Christoph Wenger, FOEN, Switzerland

FIGURE 2.4.4

4000 contaminated sites in Switzerland (2012).

Switzerland identified 4000 contaminated sites in the country that require remediation. The cost of remediating lead contamination at 3000 shooting ranges is estimated at €500 million (Wenger, 2012). At the Muchea Air Weapons Range in Australia, a project to remediate soil contaminated with lead, copper, and antinomy successfully removed 3.5 tons of lead from 3000 tons of contaminated soil through a combination of mechanical sieving using gold mining equipment and a chemical process that reduced leachable lead by 99%.

Looking forward, the most promising pollution prevention strategy for firing ranges is the development and mandatory use of the so-called green bullet, which is a slug made from environmentally nontoxic tungsten and tin.

2.4.7 CASE STUDY: LAND CONTAMINATION IN KUWAIT AFTER THE 1990–1991 IRAQI INVASION

The first Gulf War, which followed the 1990 Iraqi invasion, caused massive damage to the fragile desert ecosystems of Kuwait. The burning and gushing of oil from the hundreds of wells led to an unprecedented disaster that severely contaminated large areas of the country. The environmental assessment conducted seven years later clearly showed that while the marine environment recovered over time, the greatest damage had been done to the land.

War debris was strewn over in large areas, threatening contamination from rust and the fuel and solvents remaining in vehicles. Many locations were still full of recovered unexploded munitions that had to be disposed of through controlled detonation. This process continues to this day, with serious concerns of residual contaminants from the demining efforts and subsequent detonations.

A total of 788 wells were sabotaged or set afire during the conflict, causing one of the worst oil pollution events on record. It took experts from 18 nations nine months to put out the so-called Fires of Kuwait. These accounted by themselves for half of the total fires in the history of the petroleum industry. Some 2–3 million barrels of crude oil, burned and unburned, were emitted daily for 300 days. About 60 million barrels of oil seeped into the desert, contaminating an estimated 40 million tons of soil, while another 24 million barrels spilled to form 246 lakes covering 49 km^2 of desert. Smoke and soot contaminated another 953 km^2.

In the areas where firefighting occurred, the salinization of the soil from the saltwater used to extinguish the fires has affected the ability of native plants to compete with salt-tolerant vegetation. Coastal plants, therefore, have tended to replace native species. These new plants are not favored by sheep, which affects the value of these areas for grazing.

Severe impacts also resulted from massive troop movements. During the war, some 3500 tanks, 2500 armed personnel vehicles, and 375,000 military fortifications were brought through Kuwait. Fortifications, bunkers, trenches, and weapons pits were installed, and 1.6 million landmines and around 130,000 tons of ordnance were set, removed, and detonated. In all, up to 50% of the surface of Kuwait suffered from soil compacting, explosions, and the impact of oil mist and soot. Large areas of the desert topsoil were destroyed, as were many of the native plant species.

Seven years after the invasion, the less contaminated areas showed signs of recovery. But the situation in the oil-lake areas was worsening. Contaminated soil had to be cleaned, but in large areas, the oil was seeping deeper and deeper into the sand, making the problem increasingly difficult to solve. The eroding soil and particles were also becoming airborne, creating large-scale chronic health problems. Contaminated soils presented threats to the nearby groundwater sources. Oil trenches stretching for 120 km in the south were covered with sand. A few inches under the surface, black, oiled sand retains a strong petroleum odor. All along these trenches, most of the soil does not support vegetation. In certain areas, especially along the western border of Kuwait with Saudi Arabia, there is concern that these trenches may directly affect the natural runoff and flow of shallow groundwater (2–3 m underground) by acting as a less permeable barrier. The other concern is that they may also introduce hydrocarbon contamination to water that comes in contact with the trenches.

In 2014, a quarter of a century after the conflict, the United Nations Compensation Commission (UNCC) recognized a number of claims for large-scale environmental remediation and restoration to Kuwait, amounting to approximately $3 billion. This represents only a small portion of the overall environmental damage, estimated at $40 billion in the postconflict environmental assessment conducted by the Green Cross International in 1998.

The cleanup is to be a team effort between the Kuwait Oil Company and the United Nations (UN), with the bulk of the financing going toward the remediation of the oil lakes ($2.3 billion) and terrestrial resource restoration (some $643 million). The funds will allow the detection and removal of unexploded ordnance (UXO), the construction of landfills, the removal and disposal of the most contaminated soils (estimated at 26 million m^3), and for bioremediation and revegetation. The awards also called for the disposal of UXO through open burning/open detonation.

The bids for the work, which is expected to take up to 25 years to complete, were launched in October 2013. A total of 17 companies have been prequalified to participate in the remediation process. Cost effectiveness and scalability are important factors in the selection process.

2.4.8 SOIL REMEDIATION

When a site is suspected of being contaminated, the first step is usually to conduct an assessment that would include sampling and a chemical analysis. This would help determine the appropriate course of action, including the remediation strategy to adopt.

Traditionally, contaminated soils were excavated and disposed of in landfills (a procedure known as "dig and dump"), or alternatively, they can be treated using chemical methods or through incineration. Such an approach is expensive. Increasingly, on-site remediation (in situ) is used through a range of methods that are effective in addressing the contamination problems and are proving to be significantly more cost-effective.

Bioremediation is a treatment that uses natural microorganisms to break contaminants into less toxic compounds, using the petrol pollutants as a source of carbon and energy to degrade the petroleum into carbon dioxide and water.

Since 1986, Kuwait has gained significant experience in researching bioremediation. In 1994, a joint research program between the Kuwait Institute for Scientific Research (KISR) and Japan Petroleum Energy Center was initiated on a 1-ha area covered by an oil lake. Some 3000 m^3 of sludge was extracted from the site and transported to a containment facility. Pollution in this area was observed at levels as deep as 270 cm. The oil concentration in the sludge samples ranged from 20%–60%. The project was meant to lay the foundation for the restoration of all of Kuwait's oil-contaminated land. One of its main objectives was to demonstrate its biological technologies to remediate and rehabilitate lakebeds under Kuwaiti conditions. Three bioremediation techniques were tested: land farming, windrow composting piles, and static bioventing piles.

The 12-month results obtained from the field scale experimentation showed excellent oil biodegradation results for up to 84% of sludge. Such bioremediation is labor and water intensive and may not be cost effective for the entirety of the 49-km^2 area of oil lakes and sludge. Other research has shown that passive (i.e., natural) bioremediation can help to remediate large areas if they are not contaminated with more than 1 cm of deposited soot and unburned droplets.

Phytoremediation entails the use of various plant species and their associated microorganisms to remove, degrade, and metabolize various contaminants, including oil-related pollution, heavy metals, and persistent organic pollutants. Through phytoremediation, it is possible to speed up and reduce the level of petrol pollutants further by stimulating oil degradation and increasing the microbial population and activity in contaminated soils. Highly tolerant and effective plants must be screened to fight oil pollution, depending on the types of pollutant and the local conditions. Alfafa vegetation was found to be particularly effective in Kuwait.

The University of Kuwait found that several plant species can survive in contaminated soil and that their roots remain healthy and free of oil because oil-degrading microbes are recruited to these micro-habitats. Analysis showed that land farming reduces light soil contamination by some 80% in 6 months and heavy soil contamination by 80% within 12 months.

Advantages of phytoremediation include cost-effectiveness, the wide range of native plants available that are adapted to local conditions, and the ability to grow on site from seeds, which makes them easy to store and use.

Mycoremediation involves using fungi to degrade contaminants including persistent and highly toxic pollutants like TNT and the nerve gases O-ethyl S-diisopropylaminomethyl methylphosphonothiolate (VX) and sarin. Mushrooms were found to be effective and cost efficient in remediating soils contaminated with oil, chemicals, heavy metals, and radiation.

Vermiremediation is another method of remediation that uses earthworms to absorb and remove heavy metal and pesticide contamination from soil. Earthworms disperse the toxins and help reduce the overall concentration of contaminants. The positive results of vermiremediation following tests in the state of Gujarat, India, have resulted in great interest for the use of earthworms because they are economical, use local resources, and avoid reliance on limited landfill space.

ACKNOWLEDGMENTS

The author wishes to thank Ambassador S. Batsanov (Russian Federation, retired), who is currently the director of the Geneva Office of Pugwash Conferences on Science and World Affairs; and Dr. Paul Walker, director of the Green Cross International Environmental Security and Sustainability Programme.

REFERENCES

Brown, P., 2003. Gulf Troops Face Tests for Cancer, The Guardian. April 25, 2003.

CRC Care, September 6, 2013. Lead Contamination in Shooting Range Soils. http://www.crccare.com/case-study/lead-contamination-in-shooting-range-soils.

Green Cross International, An Environmental Assessment of Kuwait—Seven Years After the Gulf War, December 1998.

Nafeez, A., 2013. How the World Health Organization covered up Iraq's nuclear nightmare. The Guardian.

National Research Council, 2012. Redstone Arsenal: A Case Study, in: Remediation of Buried Chemical Warfare Materiel. National Academies Press, Washington, DC. pp. 66–74.

Ong, C., Chapman, T., Zilinskas, R., Brodsky, B., Newman, J., 2009. Chemical weapon munitions dumped at sea: An interactive map. James Martin Center for Nonproliferation Studies. August 6. http://cns.miis.edu/stories/090806_cw_dumping.htm

Peterson, S., 1999. The Gulf War Battlefield: Still "Hot" with Depleted Uranium. Middle East Research and Information Project, Washington DC.

Scott, R.I., 2001. Lead Contamination in Soil at Outdoor Firing Ranges. Princeton. November 24. http://www.princeton.edu/~rmizzo/firingrange.htm.

United Nations, 69th session, 57th resolution. "Effects of the use of armaments and ammunitions containing depleted uranium". December 2, 2014.

United Nations Environmental Programme (UNEP), 2003. Depleted Uranium in Bosnia and Herzegovina: Post-Conflict Environmental Assessment.

Wenger, C., 2012. FOEN, Switzerland.

World Health Organization (WHO), 2001. Depleted Uranium: Sources, Exposure and Health Effects—Full Report.

Zwijnenburg, W., 2013. In a State of Uncertainty, Impact and Implications of the Use of Depleted Uranium in Iraq. Pax Christi.

FLOOD AND DROUGHT PREVENTION AND DISASTER MITIGATION: COMBATING LAND DEGRADATION WITH AN INTEGRATED NATURAL SYSTEMS STRATEGY

2.5

Rhamis Kent

Permaculture Research Institute, Australia

2.5.1 INTRODUCTION

Desertification is the process by which fertile land becomes a desert, typically as a result of drought, deforestation, climate change, or inappropriate agriculture. Although there are diverse views on the complex relationship between climate change and anthropogenic causal factors of desertification, climate change must be recognized as largely driven by environmental changes resulting from human activities. The conventional modern paradigm often adopted for increasing agricultural production is to rely more heavily on genetically modified food technologies, allocating larger tracts of land to monocropping, dedicating scarce water resources to irrigation, and increasing the use of synthetic fertilizers, pesticides, herbicides, and fungicides to boost productivity.

A variety of fieldwork and research studies have reported and documented the root causes of land degradation historically. Works such as *Collapse: How Societies Choose to Fail or Succeed* (Diamond, 2005), *Topsoil and Civilization* (Dale and Carter, 1975), *Dirt: The Erosion of Civilizations* (Montgomery, 2007), and *Conquest of the Land Through Seven Thousand Years* (Lowdermilk, 1975) have highlighted virtually the same dynamics and factors that ultimately lead to the inevitable collapse of civilizations. These factors have resulted in the failure of ecosystems' capacity to produce food and essential ecological functions. According to Diamond's historical survey, the most dominant and frequently repeated factors are the following:

- Deforestation and habitat destruction
- Soil problems (loss of soil fertility, increased salinity, and erosion)
- Water management problems

Land Restoration. http://dx.doi.org/10.1016/B978-0-12-801231-4.00014-8

The current paradigm for generating agricultural output breaks the ecosystem's natural cycle of production and ultimately ends in failure and collapse: developing large tracts of agricultural land focused on the monocultural production of annual crops results in widespread deforestation and habitat destruction; using chemicals and monocropping destroys biodiversity and causes soil degradation, including erosion, loss of soil fertility, and eventual loss of topsoil; and diverting more water resources to irrigation drains precious groundwater reserves and ultimately disrupts the hydrological cycle.

According to the United Nations (UN) Food and Agriculture Organization (FAO, 2002), too much or too little water has always been the natural curse of agriculture. Despite greatly improved knowledge of weather systems, the use of meteorological satellites, and advanced computer simulation of climate, farmers today are more exposed to climate extremes than ever. Vulnerability has increased for a number of reasons other than climate change:

- Population densities have increased.
- Marginal land is increasingly used to grow inappropriate crops, leading to potential soil erosion and flash floods.
- Deforestation has denuded steep land of its protective vegetative cover.
- Powerful machinery strips the land of its vegetation in a fraction of the time compared to traditional methods.
- Economic pressures on farmers to increase productivity with intensive industrial farming requiring high chemical input have led to unstable and unsustainable farming practices.

It will prove impossible to maximize agricultural production from limited water resources unless the factors that accentuate the effects of natural disasters can be corrected.

The cumulative effect of these activities has reached epic proportions. The global society must shift the current production paradigm from one that destroys ecological systems to one that reconstructs them. Practices must become regenerative (or sustainable, at the very least) and have a positive effect on both the environment and the local community.

2.5.2 SOIL EROSION: CAUSES AND CONSEQUENCES

The soil, which forms the upper part of the Earth's "skin," is a living environment. The topsoil layer, typically 4 in. or less in desert climates, includes billions of microorganisms per cubic foot, as well as plant roots, fungi, worms, and insects. One part of this living tissue grows into living organisms, such as plants, mushrooms, and small animals; the other part helps break down dead organic material into components that serve as nutrients for the regeneration of new life. In order to function optimally as a spawning bed for life, soil needs sun, air, water, and plant residue (Zeedyk and Jansens, 2009, p. 1).

Soil erosion occurs when there is insufficient cover to protect the soil's surface from raindrop impact or the shear stress of flowing water. Erosion worsens with increasing slope angle, slope length, and fragility of the soil. These weakened soil conditions then increase the impact of raindrop splash, wind, and storm water runoff. Soil loss (erosion) in the form of sheet flow, rills (small erosional rivulets), and gullies will follow. Eventually, the water table drops as a result of the draining of the soil. When water rapidly runs through clay soils, mineral compounds (such as salts) are leached out (deflocculation). The clay loses its structure and is blown or washed away. Too much air in the clay hampers plant growth and increases underground water drainage, causing tunnel erosion (piping).

Eventually, the soil collapses, which is the beginning of gully erosion. Gullies occur when rills converge in a concentrated flow of surface runoff. As the soil surface steepens, the velocity of the surface flow increases, and the energy of erosive forces increases exponentially. Runoff exerts an abrasive force on the soil. Where the grade steepens, or where the soil hardness changes abruptly, runoff will scour more and create a head cut (Zeedyk and Jansens, 2009, p. 3).

Head cuts travel upstream, disturbing more soil and gutting entire hillsides and pastures. They increase rapid runoff as a result of increased drainage patterns. The result is what is called "badlands": a landscape with a multitude of gullies and head cuts, flat areas consisting of rock and gravel, devoid of vegetation. The regeneration capacity of soils in badlands is minimal due to poor soil structure, very low water-holding capacity, lack of seeds, and absence of microbial life (Zeedyk and Jansens, 2009, p. 4).

2.5.3 RESTORING LANDSCAPE FUNCTION THROUGH SOIL FORMATION AND WATER HARVESTING

As the soil plays a crucial role in the regeneration of life on Earth, it is of utmost importance that the soil structure, composed of mineral particles, decomposing organic matter (humus), and microorganisms, is of optimal quality to help regenerate life (Zeedyk and Jansens, 2009, p. 1). One of the most important factors in the soil's structure is its capacity to absorb water and hold it in its pores like a sponge, a capacity that degraded soils have lost. Indeed, in many arid regions, soils have been seriously degraded by the impact of uncontrolled land uses, such as unmanaged grazing, mining, construction, pollution, and excavation, which have, in many cases, led to a hardening or removal of the top layer of the soil. As a result, rain or snowmelt cannot penetrate the soil, causing precipitation to run over the land surface instead.

To bring life back to degraded land, the soil must be made to retain more water. One way to encourage this process is to direct water to sites where infiltration occurs or is enhanced. Once water is slowed down or retained, it is given more time to soak into the soil. Thus, the soil's crusty structure softens, allowing more water to soak in and cling to soil particles. In addition, it enlivens microorganisms, such as mycelia (fungi), that help transport water from the pores in the soil to plant roots. The roots "wake up" and begin to absorb more water and strengthen their aboveground parts. Moreover, if the moisture is retained long enough and is replenished effectively, dormant seeds in the soil may germinate. The renewed and reinvigorated plant life intercepts and slows down precipitation as it runs off along plant leaves and stems to soak into the soil. Plants also slow air movement and cast shade over the ground, reducing evaporation and keeping more water in the soil. Plant stems and roots hold the soil together; dead plant matter adds to the organic components of the soil and stimulates the proliferation of microorganisms; and plant roots support other microorganisms. Thus, plant life develops soil structure and biological activity (Zeedyk and Jansens, 2009, p. 1).

This biological process also helps retain and catch soil particles, thus reducing soil loss due to erosive forces and reducing the pollution of waterways with sediment. The result of "sponge" restoration is the resurgence of native plant growth and the reduction of unchecked runoff and erosion.

With this, the native ecosystem receives an indispensable boost to its capacity to begin a new cycle of plant succession. Typically, plant diversity will increase, which bolsters the resilience of the plant community to sudden impacts (e.g., fire, pests, flooding, or drought). In addition, plant communities create habitat for an increasing number of animals. Eventually, the landscape may become productive again for managed human use, such as gardens, farms, or pastures (Zeedyk and Jansens, 2009, p. 2).

2.5.3.1 GENERAL SOIL RESTORATION TECHNIQUES

Soils can be healed through water-harvesting and soil-improvement techniques. In dry landscapes, effective rainfall for plant growth is scarce and often further limited due to unintended water losses. Therefore, it is important to harvest as much precipitation as possible and make it infiltrate the soil. Water that is stored in the soil evaporates slowly and flows gradually downhill to the main watercourses and wetlands. This provides valuable water flow to springs and seeps, which maintain riparian habitat. Water harvesting on slopes can be best achieved by placing barriers on contours, technically forming a small terrace. Regionally appropriate, low-cost harvesting techniques include the following (Zeedyk and Jansens, 2009, p. 4):

- Structures that retain or divert storm water runoff, such as rolling dips, diversion drains, swales (which are large water-harvesting soaking channels) and berms, and microcatchments, are designed to hold the water back, and water should not flow over them, as they are high enough to retain or divert the water flow.
- Structures that slow the flow of water to give it more time to infiltrate, such as one-rock dams, rock lines on contour, straw wattles, and straw bale dams, are designed to be overtopped by water flows and are therefore rather low.
- A protective layer of mulch on top of the soil enriches the soil and protects it against wind erosion and evaporation, while also providing a more cohesive structure to the soil.

The basic concept is to slow, spread, and soak floodwater to facilitate retention and storage within the landscape. Starting at the top of the catchment areas and working down, this will mitigate the damaging effects of rapid flow and large volumes of water. In turn, this will enable the establishment of a perennial, tree-crop-based agroecological system that creates significant increases in soil fertility, organic matter formation, and landscape stability. Ultimately, this moderates the effects of large flood events, in addition to addressing concerns related to local food and water security.

This system in its most stable, mature form converts the destructive events of major floods to creative events that increase soil water storage and continuously recharge the aquifers with fresh rainwater. Surface flows of water in acute flooding events are reduced in both volume and speed with pacified flows ideal for agricultural irrigation. This model of water management and subsequent reestablishment of vegetation can be greatly expanded over time. Other critical features of this system are its antievaporative functional behavior due to shade and windbreak creation, minimizing the loss of water through excessive overexposure to the wind and sun in arid climates. This is, by a wide margin, the primary source of water loss within these climatic contexts.

The following illustrations provide an example of how landscape can be designed and arranged to minimize evaporation caused by excessive exposure to both wind and sun:

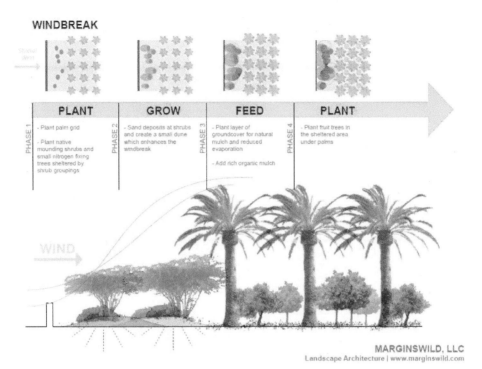

WINDBREAK

PHASE 1	PLANT	PHASE 2	GROW	PHASE 3	FEED	PHASE 4	PLANT

PHASE 1 — PLANT
- Plant palm grid
- Plant native mounding shrubs and small nitrogen fixing trees sheltered by shrub groupings

PHASE 2 — GROW
- Sand deposits at shrubs and create a small dune which enhances the windbreak

PHASE 3 — FEED
- Plant layer of groundcover for natural mulch and reduced evaporation
- Add rich organic mulch

PHASE 4 — PLANT
- Plant fruit trees in the sheltered area under palms

WIND

MARGINSWILD, LLC
Landscape Architecture | www.marginswild.com

FIGURE 2.5.1

Source: Jennifer Cooper/Marginswild, LLC.

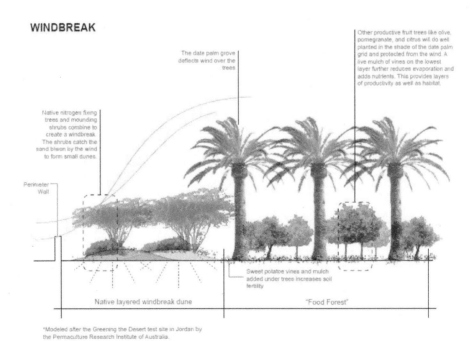

WINDBREAK

The date palm grove deflects wind over the trees

Other productive fruit trees like olive, pomegranate, and citrus will do well planted in the shade of the date palm grid and protected from the wind. A live mulch of vines on the lowest layer further reduces evaporation and adds nutrients. This provides layers of productivity as well as habitat.

Native nitrogen fixing trees and mounding shrubs combine to create a windbreak. The shrubs catch the sand blown by the wind to form small dunes.

Perimeter Wall

Native layered windbreak dune

Sweet potatoe vines and mulch added under trees increases soil fertility

"Food Forest"

*Modeled after the Greening the Desert test site in Jordan by the Permaculture Research Institute of Australia.

MARGINSWILD, LLC
Landscape Architecture | www.marginswild.com

FIGURE 2.5.2

Source: Jennifer Cooper/Marginswild, LLC.

"FOOD FOREST"

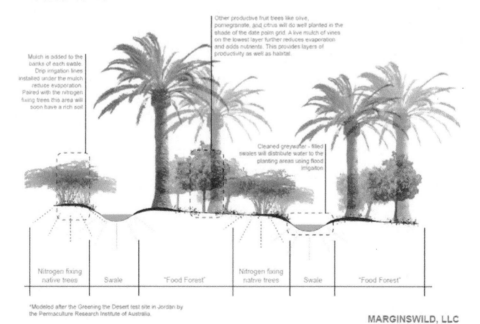

PLANT	FEED	GROW	HARVEST!
PHASE 1 - Plant palm grid - Construct greywater system	**PHASE 2** - Plant young nitrogen fixing trees adjacent to greywater - Plant layer of groundcover for natural mulch and reduced evaporation - Add rich organic mulch	**PHASE 3** - Plant fruit trees in the enriched soil in the shade of the palms	**PHASE 4** - Harvest dates and diverse fruits from Food Forest! - Providing habitat for many species - Self-feeding system has been created

MARGINSWILD, LLC
Landscape Architecture | www.marginswild.com

FIGURE 2.5.3

Source: Jennifer Cooper/Marginswild, LLC.

"FOOD FOREST"

Other productive fruit trees like olive, pomegranate, and citrus will do well planted in the shade of the date palm grid. A live mulch of vines on the lowest layer further reduces evaporation and adds nutrients. This provides layers of productivity as well as habitat.

Mulch is added to the banks of each swale. Drip irrigation lines installed under the mulch reduce evaporation. Paired with the nitrogen fixing trees this area will soon have a rich soil.

Cleaned greywater - filled swales will distribute water to the planting areas using flood irrigation

Nitrogen fixing native trees	Swale	"Food Forest"	Nitrogen fixing native trees	Swale	"Food Forest"

*Modeled after the Greening the Desert test site in Jordan by the Permaculture Research Institute of Australia.

MARGINSWILD, LLC
Landscape Architecture | www.marginswild.com

FIGURE 2.5.4

Source: Jennifer Cooper/Marginswild, LLC.

The following water-catchment features must be established at higher elevations within the wadis (dry river valleys):

- Hand-built gabions beginning at the top of the watersheds serve as an initial means to reduce the speed and force of the water flow concentrated in the wadis, which effectively diminishes soil erosion and incisions caused by floodwaters.
- Gabions combine with swales, progressively becoming larger in size and scale as they are positioned in lower elevations, toward the floor of the wadis, where the work is increasingly performed with earthmoving equipment.
- The floodplains themselves are cleared of *Prosopis* spp. (a leguminous species), in conjunction with additional earthworks features to better manage excess water draining from the wadis.
- The removed *Prosopis* are processed and turned into products, such as mulch, compost, and "bio-char," which can be used to improve organic matter content, hydrological function, fertility, and production potential in local landscapes.

Thus, on the top of each plateau, contour banks, approximately 0.5 m tall, made of uncompressed earth and stone positioned slightly off-contour gradually direct runoff water toward each wadi top head cut. The steep slopes at higher elevations are not accessible to machinery. At the top of each wadi, a succession of hand-built gabions is built to a height of 1–1.5 m as silt traps. As the wadi narrows and the slopes become shallower, the largest volume of silt for minimum rock-infrastructure is trapped.

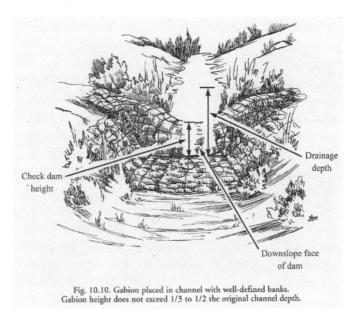

Check dam
height
Drainage
depth
Downslope face
of dam
Fig. 10.10. Gabion placed in channel with well-defined banks.
Gabion height does not exceed 1/3 to 1/2 the original channel depth.

FIGURE 2.5.5

Source: Lancaster (2008: 234).

FIGURE 2.5.6

Source: Neal Spackman, Al Baydha Project.

FIGURE 2.5.7

Source: Neal Spackman, Al Baydha Project.

FIGURE 2.5.8

Source: Neal Spackman, Al Baydha Project.

Each gabion has a level spillway to release water during large rain events to minimize the possibility of these features being overwhelmed. Rock-formed splash aprons are used as an erosion prevention measure for falling water below the spillway. These features are repeated for use in the steep slopes found in higher elevations within the wadi wherever the landscape profile requires it.

FIGURE 2.5.9

Source: Neal Spackman, Al Baydha Project.

FIGURE 2.5.10

Source: Neal Spackman, Al Baydha Project.

FIGURE 2.5.11

Source: Neal Spackman, Al Baydha Project

The lower slopes are accessible to machinery, as they have a gentler gradient and a narrow, flat base within the wadi profile. In this section, the wadi narrows and becomes increasingly shallow, trapping the largest volume of silt for a minimal rock-infrastructure built across the base of the wadi. These features can be formed using bulldozers and large excavators with a rock-grabbing attachment. Gabions on these slopes are comparatively taller and larger rock structures of 1.5–3 m with larger spillways and slash aprons. These measures are repeated throughout the wadi, wherever the landscape profile allows (that is, where the wadi remains relatively narrow and less than 50 m wide).

At the base of the wadi, where the floodplains become wider than 50 m with an incised channel, gabions can be built to the height of the floodplain at the flatter and narrower locations. Swales are positioned across the wadi, connecting to the silt fields upslope of the gabion. These continuous water-harvesting features placed on the contour have a soft earth mound on the lower downslope side. They are placed higher in the landscape, relative to the rock wall, with a trench that is lower than the surface of silt field.

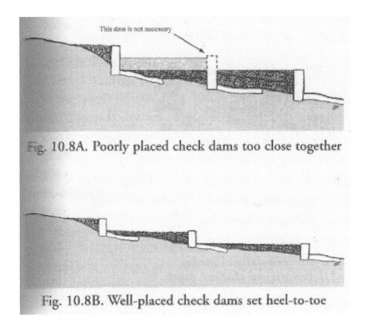

FIGURE 2.5.12

Source: Lancaster (2008: 233).

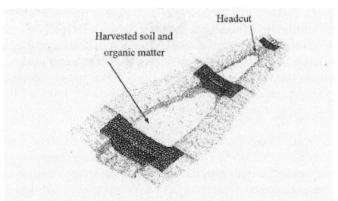

Fig. 10.9. Series of check dams, showing positioning
relative to headcut erosion

FIGURE 2.5.13

Source: Lancaster (2008: 233).

Use of terraces in steep slopes is a highly effective traditional technique in Yemen that can be of great use within the context of this overall strategy.

Fig. 3.6. Narrow terraces on a steep slope

FIGURE 2.5.14

Source: Lancaster (2008, p. 92).

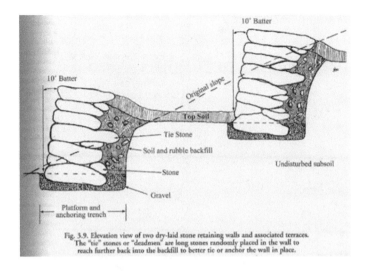

Fig. 3.9. Elevation view of two dry-laid stone retaining walls and associated terraces. The "tie" stones or "deadmen" are long stones randomly placed in the wall to reach further back into the backfill to better tie or anchor the wall in place.

FIGURE 2.5.15

Source: Lancaster (2008, p. 95).

These terraces slow, spread, and soak water into the floodplain before the main channel overflows the primary gabion and continue to do so while water flows in this channel. This creates ideal conditions to establish new tree plantings, initially relying on drip or subsurface irrigation that later will grow with minimal inputs of water and fertilizer (biological, not chemical). Floodwaters become a benefit to the system by providing water for the hydration of the landscape:

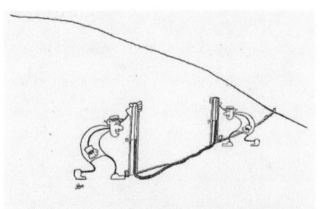

Fig. 2.4A. A contour line identified with a bunyip water level and marked with stakes and a line scratched into the soil for the construction of a contour berm

FIGURE 2.5.16

Source: Lancaster (2008, p. 68).

FIGURE 2.5.17

Source: Neal Spackman, Al Baydha Project.

FIGURE 2.5.18

Source: Neal Spackman, Al Baydha Project.

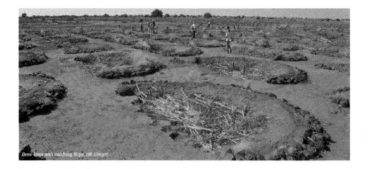

FIGURE 2.5.19

Source: WOCAT (2013, p. 109).

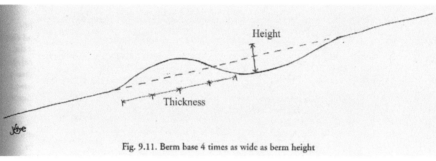

Height
Thickness
Fig. 9.11. Berm base 4 times as wide as berm height

FIGURE 2.5.20

Source: Lancaster (2008, p. 215).

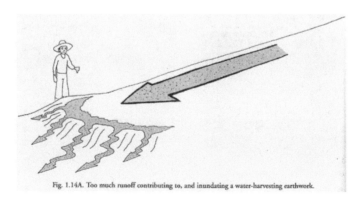

Fig. 1.14A. Too much runoff contributing to, and inundating a water-harvesting earthwork.

FIGURE 2.5.21

Source: Lancaster (2008, p. 52).

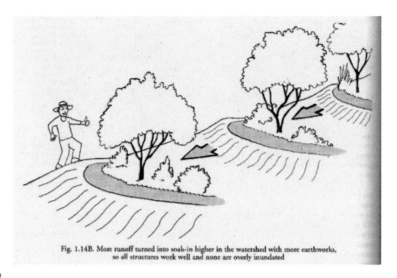

Fig. 1.14B. More runoff turned into soak-in higher in the watershed with more earthworks, so all structures work well and none are overly inundated

FIGURE 2.5.22

Source: Lancaster (2008, p. 52).

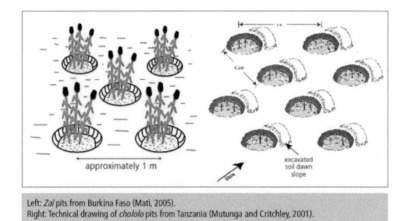

approximately 1 m

excavated soil dawn slope

Left: *Zaï* pits from Burkina Faso (Mati, 2005).
Right: Technical drawing of *chololo* pits from Tanzania (Mutunga and Critchley, 2001).

FIGURE 2.5.23

Source: WOCAT (2013, p. 111).

Swales are to be planted with a diverse mixture of drought-hardy, pioneer leguminous (nitrogen-fixing) trees and productive perennial fruit trees. Between each installation of flood-control infrastructure, the wadi's flow-control channel needs to be widened and chamfered with a maximum slope angle of approximately 30 degrees. These slopes should be planted with a stabilizing perennial cover crop for erosion control.

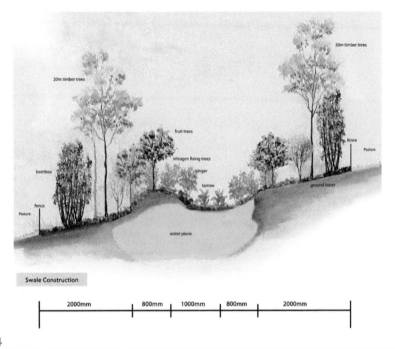

FIGURE 2.5.24

Source: Huggins (2011).

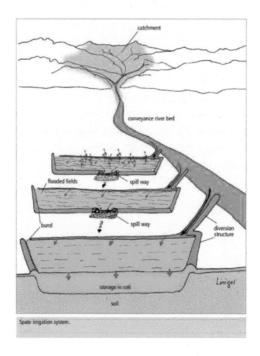

FIGURE 2.5.25

Source: WOCAT (2013: 10).

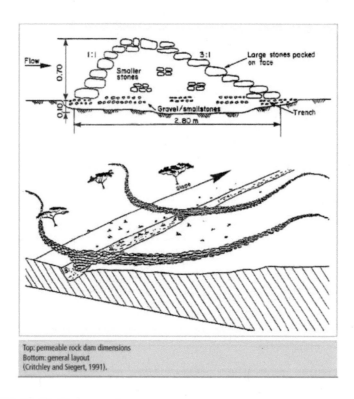

Top: permeable rock dam dimensions
Bottom: general layout
(Critchley and Siegert, 1991).

FIGURE 2.5.26

Source: WOCAT (2013, p. 38).

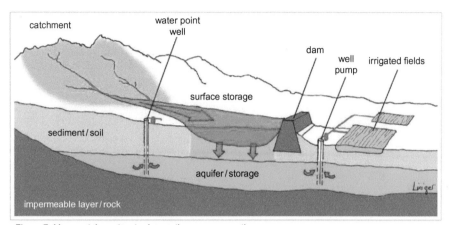

Figure 7: Macrocatchment water harvesting: a cross-section.

FIGURE 2.5.27

Source: WOCAT (2013, p. 11).

Similar work is underway in Saudi Arabia's Mecca Governorate, for the Al Baydha Project:

FIGURE 2.5.28

Source: Neal Spackman, Al Baydha Project.

FIGURE 2.5.29

Source: Neal Spackman, Al Baydha Project.

FIGURE 2.5.30

Source: Neal Spackman, Al Baydha Project.

FIGURE 2.5.31

Source: Neal Spackman, Al Baydha Project.

FIGURE 2.5.32

Source: Neal Spackman, Al Baydha Project.

FIGURE 2.5.33

Source: Neal Spackman, Al Baydha Project.

FIGURE 2.5.34

Source: Neal Spackman, Al Baydha Project.

2.5.4 PROJECT IMPLEMENTATION

The Australia-based nongovernmental organization (NGO) Permaculture Research Institute (PRI) specializes in repairing degraded landscapes in the world's most challenging environments. There has been a particular focus on creating vehicles for education, skills development, and capacity building to provide food and water security, as well as landscape stabilization. This proactively enables those most affected by ecological degradation, which results in floods, droughts, famine, and social unrest, to address the root causes of these problems.

PRI's work in Yemen's Hadramout Valley serves as an example of how it conducts project work and training. According to the United Nations Office for the Coordination of Humanitarian Affairs (UNOCHA), floods and heavy rains that affected eastern Yemen (particularly Wadi Hadramout valley and the coastal areas) from October 24–25, 2008, resulted in one of the most serious natural disasters in Yemen in the last decades. Flash floods and surging waters killed at least 73 people and forced an additional 20,000 to 25,000 people into displacement. At least 3264 predominantly mud-brick houses have been totally destroyed or damaged beyond repair, while hundreds of others are uninhabitable. … In addition to houses, several health facilities and an estimated 166 schools were damaged or destroyed. The flooding and consequences, such as loss of livelihoods, … as surging water caused extensive damage to the local agriculture and honey production, washing away crops, palm trees and soil from the fields (UNOCHA, 2008, p. 1).

This disaster was sudden and severe, affecting one-third of the country. Eastern Yemen's landscape is dominated by rugged mountains and wadis. The area's main valley, Wadi Hadramout, is densely populated and accounted for 70% of the damaged area, where the flood surges reached up to 6 m in some locations.

Environmental degradation has particularly exacerbated the devastation caused by flooding, as have poverty and marginalization, which often cause the poor to live in unsuitable and exposed conditions. The key to effectively breaking this cycle of destruction is to reverse land degradation by restoring landscape function.

According to the International Centre for Agricultural Research in the Dry Areas (ICARDA, 2012, p. ix):

The most effective means to combat land degradation in Yemen have proven to be measures to prevent water erosion, controlling pasture grazing, promoting good farming practices, encouraging fodder production, rehabilitation of degraded lands by using local species … and declaration of protected areas. In Yemen's experience with land resource conservation, the use of conservation-based farming practices and soil and water conservation techniques have been effective in controlling land desertification.

Conservation farming has been effective (though not of widespread use) for reducing water losses from agricultural fields, and for decreasing soil erodibility while increasing its filterability and water holding capacity. Good farming practices to keep soil fertile have been zero tillage, crop selection and rotation, timeliness of cultural practices, manuring, and mulching. The use of indigenous technical knowledge has also been a valuable tool, but remains a highly underestimated resource used for land conservation. Policies oriented to initiating participation and rural community partnerships for land conservation have also been positive.

Education of the wider community is a vital factor in combating local desertification. Strategies must be implemented locally, and regular monitoring and communication is required to ensure that the community fully understands the course of action, that the results are proved, and that the community is experiencing the benefits.

In this project, the local government ministry, its staff, and local communities will participate in three action areas: training, demonstration sites, and monitoring and establishing an historic database. The "regenerative" aspect of the project's focus is to once again make useful those areas that are currently barren and ecologically dysfunctional, as a result of long-term degradation caused by mismanagement.

2.5.4.1 TRAINING

Education and training at the ground level are vital for the success of this project. The ultimate goal is to create local trainers who can build regional capacity and initiate work within local communities. Programs should include the following topics:

- Theory and principles of permaculture and agroecology
- Soil rehabilitation and erosion control
- Recycling and waste management
- Food production
- Water harvesting and management and earthworks
- Ecological pest control
- Drought-proofing
- Ecologically friendly house placement and design
- Dryland strategies
- Livestock management
- Aquaculture and aquaponics
- Catastrophe preparedness and prevention (i.e., flood and drought mitigation)
- Windbreaks and fire control
- Composting, soil biology, and natural fertilizer
- Urban landscape design
- Holistic management
- Keyline design
- Food forests and oasis agriculture
- Renewable and energy efficient technology

2.5.4.2 DEMONSTRATION SITES

The design and approach of these demonstration sites will utilize the findings and experiences gathered from other case studies. For example, the Iraqi "agricultural oases" model cited later in this section provides an excellent basis upon which to build. Similar examples can be referenced from both Yemen and Oman.

Specific project details are proposed as follows:

- The most critical factor to consider is access to water. It must be utilized to its best capacity. For example, buried, gravity-fed, drip-irrigation systems create large wicking beds. Provisions should be made for the creation of water storage on elevated stands that have groundwater periodically pumped up to them, and then slowly distributed through subsurface drip-irrigation lines buried throughout the landscape.

- Mulch material is supplied and distributed over these areas using hydromulchers and hydroseeders. Over time, the biomass accumulated through revegetation of the demonstration site will eventually be the primary source of all the organic matter required.
- These fields should be further prepared using livestock to drop manure and urine, boosting fertility and organic matter content. Livestock may also be strategically managed and utilized in the scarification, inoculation, and distribution of seeds (for example, multifunctional, hardy leguminous pioneer species).
- The use of solar-powered, mobile electric fencing to manage the grazing areas is critical. The utilization of holistic management or mob grazing is an ideal feature of the integrated system.
- Chippers, mulchers, and shredders should be made available to process the organic matter to be used on site.
- The planting of palm and fruit trees among the legumes is key to the establishment of a tree-based cropping system within the proposed production system. With this orchestrated progression and succession model, an effective tree canopy and windbreak can be established relatively quickly, minimizing excessive evaporation and desiccation caused by the sun and wind and setting the stage for other food crops (perennial and annual varieties) to be grown. Additionally, more livestock can be introduced to the system with the improved management of water. If this arrangement is implemented over a large enough area, a more favorable microclimate will be generated within the region, which helps to restore the proper functioning of the hydrological cycle.

The demonstration sites are to be integrated with the training courses. The initial cohort will ideally comprise agricultural engineers and personnel from agricultural and environmental ministries. Each cohort will be allotted a demonstration site to design, develop, monitor, and maintain. PRI has already initiated this training, with courses being taught in Wadi Hadramout since December 2011.

Most of the additional costs would take the form of nonrecurring capital equipment, such as hydroseeders and hydromulchers; chippers, shredders, and mulchers (arboreal equipment and tools); agricultural implements (i.e., subsoilers and Yeomans Plows); tractors; rubber tracks for tractors and trucks; earthmoving equipment (i.e., excavators and bulldozers); trucks or transport equipment; compost (compost turners), compost tea (compost tea brewers), biofertilizer, and micronutrient preparation equipment; and mobile solar-powered fencing to manage livestock.

Each site should have staff to establish these systems and monitor progress in the initial stages. Staff would include site managers, administrative office staff, and field operators (including the possible use of interns on site).

2.5.4.3 MONITORING AND ESTABLISHING A HISTORIC DATABASE

The magnitude and impacts of desertification vary greatly from place to place and change over time. This variability is driven by the degree of aridity combined with the pressure that humans put on ecosystem resources. However, there are wide gaps in our understanding and observation of desertification processes and the underlying factors. A better delineation of desertification and degraded land areas through large-scale monitoring would assist in the planning of rehabilitation efforts and enable cost-effective actions in affected areas.

The monitoring and analysis work of this component is critical and will be handled by other information systems, specifically, geographic information systems (GISs) and remote sensing tools. GIS are designed to capture, store, display, and analyze data related to a variety of Earth surface phenomena. The data produced from the GIS and remote-sensing components will be utilized in the reporting and to provide recommendations and guidelines to prioritize future rehabilitation efforts.

A wide variety of tools are also currently available, allowing the creation of geographically referenced, three-dimensional (3D) orthographic maps constructed from drone imagery and GIS software. This allows for improved accuracy and better planning of any and all intended work on the ground.

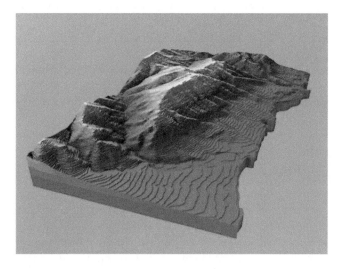

FIGURE 2.5.35

Source: Georgi Pavlov/HUMA.

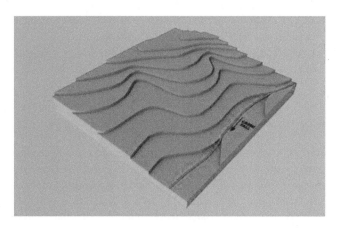

FIGURE 2.5.36

Source: Georgi Pavlov/HUMA.

2.5.4.4 PROGRAM OUTLINE

Objectives: Capacity Building Through Mutual Knowledge Sharing and Training Workshops

- Initial Agricultural Experiments

Initial Agricultural/Agroecological Experiments

- Water features
- Tree Crops
- Implement Planned Grazing

Broad-acre Permaculture Approaches

- Agroforestry, Silvo-Pastoralism
- Watershed Restoration

Holistic Management Workshops

- Goal Setting
- Decision Making
- Grazing Planning
- Land Planning
- Financial Planning
- Biological Monitoring
- Policy Design and Analysis

The proposal summarized here builds upon the greatest social resources of the Middle East–North Africa (MENA) region's inhabitants: namely, the deep love for the land that has great significance and value in the rich history, cultural identity, and traditions of its people. Both the old and the new adaptations of life and culture within these regions can be built upon through an introduction to, and training in, several recent proven approaches from applied sustainability science, which include agroecology, permaculture, holistic management, and keyline design. Each of these holistic, practical approaches has much to offer the specific social and ecological situation regarding how best to proceed, given recent developments within much of the MENA region.

As such, the main goal of the project is to enhance and maintain the highly valuable ecosystem services provided by the region's landscapes through a process of knowledge sharing and capacity building for long-term local stewardship. Developing stewardship capacity, leadership, and skills must be approached in the social realm: by building further on already existing institutions, by "training the trainers" in land design and management sciences, by expanding the diversity of livelihood strategies in line with stewardship principles, and by practicing the proposed implementation itself.

The interventions proposed here will particularly focus on restoring proper hydrological dynamics and function in landscapes. While increasing local capacity for applying integrated land use design and management, the aforementioned regenerative interventions will impart substantial beneficial impacts on food security, livelihood diversification, and regional adaptive capacity. In addition, this framing will help to develop greater adaptive capacity by increasing communication, responsibility, and accountability among the various social and governance institutions that already exist within the region.

REFERENCES

Al-Rashed, M.F., Sherif, M.M., 2000. Water resources in the GCC countries: an overview, water Resources management 14. Kluwer Academic Publishers, pp. 59–75. (Online, accessed 27.01.15.), http://www.ce.utexas.edu/prof/mckinney/ce397/Topics/Gulf/Al-Rashed_2000.pdf.

Altieri, M.A., 2000. Agroecology: principles and strategies. The Overstory, 95. (Online, accessed 30.11.14.). http://www.agroforester.com/overstory/overstory95.html.

Altieri, M.A., 2009. Livestock raising, an important element of the production system of oases, monthly review. Agroecology, Small Farms, and Food Sovereignty. 61 (03). http://monthlyreview.org/2009/07/01/agroecology-small-farms-and-food-sovereignty/ (July-August) (Online, accessed 30.11.14.).

Dale, T., Gill Carter, V., 1975. Topsoil and Civilization, revised ed. University of Oklahoma Press, Norman, OK.

Dewar, J.A., 2007. Perennial polyculture farming: seeds of another agricultural revolution? RAND Corporation. http://www.rand.org/content/dam/rand/pubs/occasional_papers/2007/RAND_OP179.pdf (Online, accessed 30.11.14.).

Diamond, J., 2005. Collapse: How Societies Choose to Fail or Succeed. Penguin Group, New York.

Dolle, V., 1991. Livestock raising, an important element of the production system of oases. In: Desert Development, Part 1: Desert Agriculture, Ecology and Biology: Advances in Desert and Arid Land Technology and Development, Bishay, A., Dregne, H. (Eds.), vol. 5, p. 541.

Food and Agriculture Organization (FAO), 2002. Crops and Drops: Making the Best Use of Water for Agriculture. FAO, Rome.

Geno, L., Geno, B., 2001. Polyculture production principles, benefits and risks of multiple cropping land management systems for Australia. Rural Industries Research and Development Corporation (RIRDC Publication No 01/34, RIRDC Project No AGC-3A), Canberra. https://rirdc.infoservices.com.au/items/01-034 (Online, accessed 30.11.14.).

Huggins, N., 2011. Permaculture design for horses, people & habitat. The Permaculture Research Institute. (Online, accessed 27.01.15.). http://permaculturenews.org/2011/02/16/permaculture-design-for-horses-people-habitat/.

International Center for Agricultural Research in the Dry Areas (ICARDA), (2012,ix). Combating land degradation in yemen – A national report. (Online, accessed 30.11.14.). http://www.icarda.org/wli/pdfs/OASIS_Country_Report4_Land_Degradation_in_Yemen.pdf.

Lancaster, B. 2008, Rainwater Harvesting for Drylands and Beyond: Volume 2 Water-Harvesting Earthworks, Rainwater Press. HYPERLINK "http://permaculturenews.org/author/craig%20mackintosh%20pri%20editor/" \t "_blank" Mackintosh, C. 6 August 2010, "Letters from Jordan – On Consultation at Jordan's Largest Farm, and Contemplating Transition", (Online, accessed 30 November 2014). URL: http://permaculturenews.org/2010/08/06/letters-from-jordan-on-consultation-at-jordans-largest-farm-and-contemplating-transition/

Lowdermilk, W.C., 1975. Conquest of the Land Through Seven Thousand Years. US Department of Agriculture Soil Conservation Service, Washington, DC.

Mackintosh, C., 2010. Letters from Jordan – on consultation at Jordan's largest farm, and contemplating transition. http://permaculturenews.org/2010/08/06/letters-from-jordan-on-consultation-at-jordans-largest-farm-and-contemplating-transition/ (Online, accessed 30.11.14.).

Mollison, B., 1988. Permaculture: a designers' manual, Second ed. Tagari Publications.

Montgomery, D.R., 2007. Dirt: The Erosion of Civilizations, first ed. University of California Press, Oakland, CA.

Nuberg, I.K., Evans, D.G., Senanayake, R., 1994. Future of forest gardens in the uvan uplands of sri lanka. Environ. Manage. 18 (6), 797–81.

Pretty, J., 2009. Can ecological agriculture feed nine billion people. Mon. Rev. 61 (06), November (Online, accessed 30.11.14.), http://monthlyreview.org/2009/11/01/can-ecological-agriculture-feed-nine-billion-people/.

Rosenzweig, C., Hillel, D., 2008. Climate change and the global harvest. Oxford University Press, New York.

United Nations Millennium Ecosystem Assessment, World Resources Institute 2005. Ecosystems and Human Well-Being Synthesis. (Online, accessed 30.11.14.). http://www.millenniumassessment.org/documents/doc.

United nations office for the coordination of humanitarian affairs' report on their Yemen flood response plan of 2008. Yemen Floods Response Plan Consolidated Appeal Process. UNOCHA 2008, p. 1, Geneva, Switzerland.

UNFAO's (United nations food and agriculture organization) Corporate Document Repository reference "Crops and Drops: Making the Best Use of Water for Agriculture". ftp://ftp.fao.org/docrep/fao/005/y3918e/y3918e00.pdf (Online, accessed 30.11.14.).

World Overview of Conservation Approaches and Technologies (WOCAT), 2013. Water harvesting, guidelines to good practice. Geographica Bernensia, Bern, p.10, https://www.wocat.net/fileadmin/user_upload/documents/Books/WaterHarvesting_lowresolution.pdf (Online, accessed 30.11.14.).

"Ecosystems and Human Well-Being Synthesis", United Nations Millennium Ecosystem Assessment, World Resources Institute 2005, (Online, accessed 30 November 2014). URL: http://www.millenniumassessment.org/documents/document.356.aspx.pdf

World Bank. 2003. China - Loess Pla.

Zeedyk, B., Jansens, J.-W., 2009. An introduction to erosion control, third ed. A Joint Publication from Earth Works Institute The Quivira Coalition and Zeedyk Ecological Consulting. http://quiviracoalition.org/images/pdfs/1902-An_Introduction_to_Erosion_Control.pdf (Online, accessed 30.11.14.).

ENVIRONMENTAL SECURITY, LAND RESTORATION, AND THE MILITARY: A CASE STUDY OF THE ECOLOGICAL TASK FORCES IN INDIA

2.6

Dhanasree Jayaram

Manipal University, Manipal, Karnataka, India

2.6.1 INTRODUCTION

Land restoration is a well-researched issue across the globe, looked at in the sciences, social sciences, and policy studies. Globally, land has been used, misused, and abused to fulfill human requirements—mostly for food, water, and energy—sometimes beyond the restorative capacity of the land. From incorrect cultivation practices to mining to deforestation to overgrazing, many human activities have set the Earth on the road to an unsustainable future, potentially endangering human lives. In India, too, land degradation is a serious issue that state authorities and nonstate agencies alike have failed to curb. A 2010 study by the Indian Council of Agricultural Research and the Department of Space revealed that 120.4 million hectares (Mha) of land in the country is degraded or wasteland (Trivedi, 2010). In the state of Uttar Pradesh, land degraded by water erosion alone constitutes 54% of the area. In Madhya Pradesh, it is 44%, in Karnataka 41%, in Jharkhand 40%, and other states are not far behind (Trivedi, 2010). The scale and severity of the issue has driven the newly elected national government in 2014 to announce that it would launch a new program to make the country "land degradation neutral" by 2030. This initiative is not a mere addendum to its main objective of poverty eradication but rather an indispensable element, according to the government (PTI, 2014b).

Land restoration is being pursued actively in many countries, but on a global level, its urgency is yet to be fully understood by the policy-making community. There have been attempts to raise the urgency of the issue of land degradation by emphasizing its importance for security. Yet in India and many other countries, the security establishment does not perceive environmental change and land degradation as an existing security threat, although concerns exist regarding their probable effects in the future. In India, steps have been taken by the National Security Council (NSC) and the National Security Advisory Board (NSAB) to expand the notion of security beyond the traditional notions, but the relative lack of expertise and literature (from an Indian perspective) in fields such as environmental security has fettered the whole process of realization of a national security policy that encompasses nontraditional

Land Restoration. http://dx.doi.org/10.1016/B978-0-12-801231-4.00015-X

163

security threats (environmental security being one among many). The inability of the security-related actors (state and nonstate) to sustain the discussion on the environmental aspects of security has resulted in them being relegated to issues of secondary importance nationally.

The Indian decision makers' general propensity to adopt a viewpoint that is independent of the Western line of thinking on security issues has also led to a blinkered understanding of the problems posed by environmental change and land degradation to India's security (Patil, 2014). For instance, the Western discourse on environmental security is dominated mainly by resource scarcity, environment-conflict nexus (intrastate, interethnic, and interstate rivalries), and now environmental risk assessment, while the Indian thought process on the environment encompasses issues such as social justice (distribution of resources, poverty alleviation), development (for better adaptation to environmental problems), and socioeconomic viability/sustainability. This has essentially resulted in the Indian establishment's reluctance to "securitize" environment, in line with the Western reference point, on both the theoretical and policy level.

Security implications are one aspect; the other factor, which has hardly received attention, is the role of the military in environmental security and land restoration. Military dimensions of environmental security, as well as the environmental dimensions of the military, are seldom discussed in academic or policy circles. Although environmental security as a concept has found its place in the security discourse at both normative and empirical levels, the role of the military in this debate has been largely restricted to the domain of the environment-conflict thesis—that is, to the military's duty of maintaining internal peace in the event of violent conflict caused by environmental stress. The military's goal is no longer just to win wars but to deter them, which requires identifying all possible causes, including environmental ones. At the same time, the military until recently has only been associated with their traditional responsibility of upholding territorial integrity and, in recent times, more nontraditional tasks such as international peacekeeping and disaster relief. Although the military is recognized more as a political force, its economic and social functions (such as postconflict reconstruction and protection of sea lanes of communication/maritime trading routes) have also begun to be acknowledged. However, environmental protection, conservation, and restoration are yet to be mainlined within military strategy or policy, despite the fact that the armed forces in India have the personnel [1,325,000 active forces and 1,155,000 reserves (India as a Great Power, 2013), in addition to a huge pool of former participants in the armed forces] and resources to undertake such activities on a greater scale.

In this context, an attempt is made in this section to define environmental security from the point of view of land restoration and to analyze the close links between environmental security, land degradation, and the military. It looks at the ways in which land degradation affects human and national security. Investigating the land restoration angle of this theme, involvement of the Indian armed forces in land restoration in the country is reflected on, and to substantiate this point further, a case study of the reclaimed Bhatti mines in the national capital of India (Delhi) is presented here. Based on field visits, this study elucidates a firsthand account of the constructive engagement of the military in environmental security initiatives in India.

2.6.2 LAND DEGRADATION AS PART OF THE ENVIRONMENTAL SECURITY SPECTRUM

Before delving into the military dimensions of environmental security, it is important to cast light on how land restoration matters to environmental security and why the framework of environmental security itself matters to this discussion. Since the military is considered the primary security apparatus of a nation-state, its role in land restoration as part of the larger environmental security policy is an important part of the analytical framework. Furthermore, securitization of the environment raises the urgency of environmental issues and calls for a more concerted approach to the problem, in which every single constituent of the state, including the military, is a stakeholder. Although the military is usually referred to as an instrument of last resort, even in the aftermath of natural disasters, it invariably is the first responder, especially in the event of national emergencies (caused by calamities such as earthquakes, floods, cyclones, and so on). Land restoration is indeed an issue on which the civil and military agencies need to work together on a war footing (that is, with greater urgency and effectiveness). As a provider of security, the military could be perceived as not only an instrument of force, but also as a provider of human and environmental security to citizens.

Regarding land restoration's relevance to environmental security, the conceptual debate on environmental security has still not reached a universal consensus, even though it has been identified by many states as a critical factor in their security calculus. Environmental security could be (and indeed has been) defined in many different ways. One of the most oft-quoted and general definitions states, "Environmental security is a state of the target group, either individual, collective, or national, being systematically protected from environmental risks caused by inappropriate ecological process due to ignorance, accident, mismanagement or design" (Landholm, 1998). However, the range of definitions given by states and supranational actors, such as the United Nations (UN), is extensive, and more important, the working definitions in different contexts vary. It is not only the intangible perceptions and tangible implications of environmental security that differ from context to context, but also the responses of actors that are contingent on factors such as the vulnerability of the region and its capacity to adapt, among others. At the same time, the fundamental principles on which these definitions have been based are the same, as represented in Figure 2.6.1.

The food-energy-water security nexus (an integral part of the environmental security debate), which encompasses the fundamental threats posed by a set of wide-ranging and interconnected problems, provides a straightforward example of how land degradation indeed could be human civilization's greatest threat. With the pressure on land resources mounting day by day due to an increasing population making greater demands for food, water, and energy, coupled with the limited amount of land resources, land degradation has more or less become inevitable. Correspondingly, the food, water, and energy services that are supported by land also become degraded and depleted—reducing productivity and thereby affecting human lives directly. The poor (marginal farmers and denotified,[1] nomadic, and seminomadic tribes) remain the most affected, as they usually disproportionately inhabit degraded land. Due to the scarcity of food, water, and energy, they are forced to move to marginal land (forests, steep slopes, and so on) for cultivation, causing further degradation (UNEP, 1999). Therefore, the problem of land degradation forms a vicious circle. Globally, the connections between the food-energy-water nexus and sociopolitical instability have also been established (World Economic Forum, 2011). Take for instance, a country like India, where agriculture accounts for almost 90% of water consumption, which is more water than used per unit by the industry that uses only 6% of the total available freshwater. (Brooks, 2007; FICCI, 2011). Conflicts arise between agricultural and industrial

---

[1]In 1871, the British colonial state passed the Criminal Tribes Act and classified (notified) some nomadic and seminomadic tribes as being "criminal" under it. Members of these communities were seen to be "addicted to the systematic commission of non-bailable offences" and once a tribe was notified, all its members were obliged to register with the local magistrate, or else they could be indicted for committing a crime under the Indian Penal Code. After independence, the new Indian government repealed the act in 1952 and denotified the tribes.

FIGURE 2.6.1

Three components of environmental security.

Based on information extracted from Belluck et al. (2004).

users and urban and rural users, as well as different states that share common water resources. In the Palakkad district of Kerala in southern India, the largest Coca-Cola plant was forced to shut down due to prolonged agitation spearheaded by farmers (and other locals) in the region who accused the plant of "leaving farms parched and land poisoned" (Brown, 2003; Mathew, 2010). It could also be argued that the "resource curse" is as big a potential trigger for violence and conflict as resource scarcity, as has been seen in the Middle East and North Africa, but this is more in the context of oil and minerals. Yet another reasoning for the environment-conflict nexus could be explained through the lens of political ecology, which emphasizes that it is the (mostly unequal or disproportionate) distribution of resources that sparks conflict between the haves and the have-nots.

Land degradation has been a factor in many cases of social and political instability in India, although it cannot be pinpointed as the only cause. Environmental degradation caused by illegal logging and mining in regions inhabited by denotified nomadic and seminomadic tribes and poor farmers has in many cases pitted these so-called powerless communities against state authorities. The $12 billion Pohang Iron and Steel Company (POSCO) steel plant, to be installed in Odisha (Orissa), has met unyielding resistance from the tribal and rural communities for nearly eight years. Concerns regarding the plant arise from the imminent displacement of locals and the loss of livelihoods (which are central to human security). These concerns are embedded within the human-environmental systems of the region, as the livelihoods of these communities are sustained by the forest and coastal/estuarine ecosystems and other resources that could be endangered by the steel plant. On many occasions, these protests have turned violent, with clashes between locals and the state police (Khalid and Kumar, 2014).

Similarly, the Maoist/Naxalite insurgency, which has been labeled as "the single biggest internal security challenge ever faced by our country" by former prime minister Manmohan Singh (Kennedy, 2014), is also linked closely to natural resource management and land degradation. The Western Ghats, along India's southwest coast (an ecological hot spot), has been at the center of controversy due to reports released by expert panels set up by the government that propose zoning in the region, thereby restricting development projects in certain areas (Shrivastava, 2014; Mazoomdar, 2013). Although the movement itself has very little to do with environmental degradation ideologically, politically, or organizationally, it is possible that environmental management issues have a role in fueling the conflict by driving displaced tribes into the movement (Gowda, 2014; Prebble, 2013; Team Mangalorean, 2009).

The critics of this view argue that Naxalism was originally an outcome of agrarian unrest in Naxalbari, and therefore, to say its root cause is environmental degradation would be incorrect. In this argument, Naxal/Maoist insurgents are exploiting conditions created by high levels of poverty and the lack of economic development (aggravated by themselves through their resistance to developmental projects in the affected areas) to make inroads into vulnerable regions. Moreover, they have been accused of using mining and illegal logging as a financial resource base (mainly through extortion; Mauskar, 2014). Whether environmental degradation leads to growth in Naxal/Maoist recruitment or whether it is abused by the insurgents to increase their own power, the fact of the matter is that environmental degradation is very much part of the problem; hence, it cannot be overlooked. These examples show how environmental degradation could pose human, environmental, as well as national security challenges.

Even though environmental security has been defined largely as the impact of environmental degradation on security (and sometimes vice versa) as shown by the abovementioned instances, it is highly difficult to define environmental security within a single framework, as it operates differently in different contexts involving different actors and agencies. In this discussion, the working definition should incontrovertibly place the issue of land degradation and the military at the center, keeping in mind the three components of environmental security elucidated earlier. One of the definitions discussed by the Millennium Project of the UN very systematically links "restoration of the environment damaged by military actions, and amelioration of resource scarcities, environmental degradation, and biological threats that could lead to social disorder and conflict" to environmental security (Landholm, 1998). This definition could prove to be a stepping stone toward building and analyzing linkages between land restoration, environmental security, and the military. Although the definition reflects more on the military aspect, it must be remembered that any act of restoring destroyed or degraded lands would help maintain environmental security, not only in the specific location but also elsewhere, as the natural environment has no borders. It also reinforces the requirement for the military to be an active stakeholder in environmental protection and restoration. However, the military need not necessarily always operate in military environments when it comes to land restoration; they could also partner with civil agencies in nonmilitary environments.

These two parallel roles can be explained using specific examples. In India, the Siachen glacier, the world's highest battleground and the site of deployments by Indian and Pakistani soldiers, is reportedly melting at an accelerated pace. It was found to have receded by nearly 800 m during the course of 23 years, and another data set suggests that the glacier had shrunk by nearly 10 km in 35 years. According to environmental experts, the heavy military presence in the region (in addition to global warming) has accelerated the melting. Furthermore, it has been found that the military presence has resulted not only in the deterioration of the ecosystems but also in the pollution of the Indus River, which flows through both countries. The main cause for this ecological deterioration is the humongous amount of waste generated by the militaries of both countries that consists mostly of metal and plastics, "which simply merge with the glacier as permanent pollutants, leaching toxins like cobalt, cadmium, and chromium into the ice." A report of the International Union for the Conservation of Nature (IUCN) reveals that on the Indian side alone, on average, 900 kg of human waste is dumped every day (Agence France-Presse(AFP), 2012; Qamrain, 2013). To prevent this degradation, the Indian army is carrying out the Clean Siachen-Green Siachen campaign "to plant trees on Siachen to maintain ecological balance and save the glaciers" on the Indian side of the border (PTI, 2013). This initiative stems largely from the need for the Indian army to prevent serious environmental degradation that could hamper its operations in the Siachen and render its presence in the area unviable in the long run due to accumulation of human waste. Another reason for the initiative was criticism over the negative impacts of the military's presence in the area. This is an example of the armed forces involved in the restoration of land damaged by military deployment. The Indian army has been engaging in land restoration activities in nonmilitary settings as well, which are elaborated upon later in this section.

2.6.3 MILITARY DIMENSIONS OF ENVIRONMENTAL SECURITY: INDIAN AND GLOBAL PERSPECTIVES

The connections between environmental security, land restoration, and the military at the definitional and conceptual levels have been ascertained using several examples in India. At this stage, these linkages ought to be fortified further by reflecting on the existing perspectives on the issue in India and around the world. The increasing role of the armed forces in tackling environmental security issues through programs such as Humanitarian Assistance and Disaster Relief (HADR) in the event of environmental disasters has elevated the level of understanding and analysis in this domain to some extent, yet operationalization of military dimensions within the environmental security discourse is far from being realized. There have been very few attempts to use theory for linking the military with environmental security, except in the case of environmental degradation caused by armed conflict, despite the fact that there have been an untold number of cases of involvement of the military in environmental issues other than the aforementioned context. In 1993, the UN released a report titled *Potential Uses of Military-Related Resources for Protection of the Environment*. At the time this report came out, the Cold War had come to an end, and the scope of security had broadened to embrace nonmilitary security threats and economic issues had begun to predominate over military ones. With environmental deterioration becoming a grave concern across the globe, and with several nations, including the United States (US), China, Russia, and Germany, diverting military resources to the civilian sector, the focus on the military as a stakeholder in the exercise of environmental protection and conservation began to gain momentum (UNEP, 1993). In Brazil, the role of the armed forces in safeguarding the environment, including through preventing accidents such as oil spills and wildfires and through protecting endangered species, is constitutionally guaranteed despite the military dictatorship that ruled the country from 1964 to 1985. The armed forces are known to provide logistical support to various environmental agencies and to institute environmental educational programs for both military personnel and civilian populations residing around military units (UNEP, 1993).

There are many reasons that states think along the lines of integrating the military with environmental policy. First and foremost, the armed forces tend to be among the largest landholders in many countries, and the territories they occupy in some cases are environmentally fragile and vulnerable to degradation and disasters of various types, as is the case in India (Apte, 2014). Thus, it is imperative for them to preserve the environment in the areas that are directly under their jurisdiction, the positive results of which would spill over into other parts of the particular country as well. Second, although environmental security is an international issue while the military is national, there is clearly a need to integrate all national actions (in which the military should be an equal partner) to tackle various environmental concerns at the international level. Since the problems and symptoms are international in nature, the solutions and approaches should also be transnational. Third, one of the central motives for involving the military in environmental activities is to reduce the cost of environmental protection and restoration through "redeployment, reorientation, and retraining" (UNEP, 1993). The military has personnel (sheer numbers), technological capabilities (including monitoring ability), logistical and administrative infrastructure, and command, control, communications, and intelligence systems, which could be extensively put into use during peacetime.

Yet another reason for this development is that the military is considered more or less self-sufficient and capable of showing leadership. It has been contended that the military, "as a large institution, is well suited to lead a whole of government approach to sustainability" (Hartman et al., 2012). A whole-government approach entails the involvement of departments and agencies—both military and civilian—cutting across portfolio boundaries and amalgamating their skills, expertise, knowledge, and capacities to build an integrated response to specific issues and to work toward a common goal. This is expected to ease and improve decision making, policy implementation, and administration. Military officials from the US have made the case for the need for the military to work toward achieving sustainability at three levels—strategic (to identify and secure a resource base aimed at fulfilling national and international military/security objectives); operational (energy efficiency and reduction in operational and total life cycle costs); and tactical (reduction in environmental footprint, as well as risks to fighters). In addition to these military imperatives, actions by the armed forces directed toward environmental protection and conservation are expected to promote "sustainable communities" and a better quality of life, not only among armed personnel but also in the surrounding areas (Hartman et al., 2012).

In India, the establishment is currently not very interested in enhancing the role of the military in environmental security. Traditionally, the military is trained to fight wars. It is commonly considered, particularly in the Indian milieu, that the military ethos might not gel with environmental ethics or policy because their working knowledge and practice is generally restricted to black or white areas at the operational level, while environmental issues are full of gray areas. An oft-quoted instance is the reluctance of the Indian military to intervene in internal security theaters such as the Maoist conundrum that, as already pointed out, has been considered one of the biggest internal security threats presently facing the country (Mauskar, 2014). The military's role is constitutionally restricted, with total control of the armed forces (army, navy, and air force) in the hands of the center and operations limited to "aid to civil authorities" outside the purview of the cantonments and border areas.[2] During the British era,

---

[2]Under the Seventh Schedule, Article 246 of the Indian Constitution, List I (the Union List) includes the following: 2A. Deployment of any armed force of the Union or any other force subject to the control of the Union or any contingent or unit thereof in any State in aid of the civil power; powers, jurisdiction, privileges and liabilities of the members of such forces while on such deployment; and 3. Delimitation of cantonment areas, local self-government in such areas, the constitution and powers within such areas of cantonment authorities and the regulation of house accommodation (including the control of rents) in such areas. The Union List contained in the Part XI of the Indian Constitution consists of 97 items on which the centre has exclusive authority to act and this list includes the naval, military, and air forces; and any other armed forces of the Union.

the military was well represented in all the major government boards and committees, including the Defence Committee of the cabinet. This system that India inherited after independence slowly faded. The colonial army was powerful, but since independence, the military's role has been reduced to providing assistance to civilian agencies—a fundamental pillar of India's democratic system and a deliberate and much-thought-out strategy. From the time of India's first prime minister, Jawaharlal Nehru, measures have been adopted to subdue the role of the armed forces, a policy that was continued by Nehru's successors. Nehru strongly believed that "professional competence, not political initiatives was the first requirement of the Indian Army" (Pardesi, 2011).

The other rationale for not involving the military in environmental activities is the fact that it is regarded as a sacrosanct entity that should not compromise territorial integrity or security in exchange for environmental protection (Kamath, 2014). After all, its primary duty is to protect the country from any form of external aggression and intrusion, and in the process, they invariably pollute the environment and degrade natural resources in conflict zones. In fact, military forces are considered one of the biggest polluters and exploiters of the environment in the world (Rosenburg, 1995). Although the environment will always be given secondary importance during conflict situations, efforts are being made globally to reduce pressures on the environment and natural resources in post-conflict situations, as seen in the efforts by the United Nations Environment Programme (UNEP) to reduce the environmental footprint of 120,000 peacekeepers spread throughout the world through operational efficiency and sustainable service delivery. The UNEP's strategy covers a wide gamut of issues, such as water, energy, waste, and wildlife, among others (UNEP, 2012).

2.6.4 THE ROLE OF THE MILITARY IN LAND RESTORATION IN INDIA

Even though there has been a general reluctance to use military experience in environmental activities on a larger scale, there are instances in which the government has taken proactive steps to engage the military in environmental activities, particularly in mine reclamation, watershed management, and afforestation. In the early 1980s, severe environmental degradation in the Himalayas, mainly in the Shivaliks (the foothills of the Himalayas), led environmentalists in the region to raise concerns. Norman Borlaug, who is famously known as "the father of the Green Revolution," was alarmed by the degree of deforestation on the Shivalik hills, and he urged Indira Gandhi, then–prime minister of India, not to depend on civilian agencies alone to restore the area that had been ravaged by limestone mining. It was his proposal to involve the armed forces in land restoration in the Shivaliks due to the urgency of the problem. Gandhi accepted this proposition, but since serving military personnel cannot be withdrawn from their regular duties, young ex-servicemen from the particular region were chosen to carry out the task instead. In this way, the world's reportedly first ecological unit of the Territorial Army, the Ecological Task Force (ETF), was formed to employ afforestation to reclaim the mining areas in the Mussoorie hills (Mohan, 2005). The primary objective of bringing in the armed forces was to instill discipline and dedication into the whole exercise. This thought was shared by armed forces personnel as well, according to whom they could "execute specific ecology-related projects with a military-like work culture and commitment" (Mohan, 2005).

Currently, eight ETF battalions consisting of several units of the Indian army have been deployed in various parts of India.[3] Table 2.6.1 provides an account of these units and their assigned tasks.

---

[3]An ETF unit (in a particular location) comprises a headquarters and one or more companies under an officer, junior commissioned officers (three or more), and soldiers of other ranks (at least 100, depending on the number of companies).

Table 2.6.1 Ecological Task Forces in India

ETF Unit	Regimental Affiliation	State	Year of Raising	Composition	Tasks
127 Infantry Battalion	Garhwal Rifles	Uttarakhand	1982	Two-company ETF until 2008; two additional companies added by the government of Uttarakhand in April 2008	Soil conservation and afforestation in the Mohand region of Shivalik ranges, particularly in Shahjahanpur Range, near Saharanpur; and restoration of Aglar watershed north of Mussoorie hills
128 Infantry Battalion	Rajputana Rifles	Rajasthan	1983	Two-company ETF	Afforestation in the Pugal area on the left bank of the Indira Gandhi Canal; development of Amarpurna lake and the bird sanctuary; and greening of Mohangarh in Jaisalmer
129 Infantry Battalion	Jammu and Kashmir Light Infantry	Jammu and Kashmir	1988	One-company ETF with affiliated Forest Department personnel	Restoration of catchments and watershed in Samba; and afforestation and soil conservation on Bahu and Jindra mountain ranges
130 Infantry Battalion	Kumaon Regiment	Uttarakhand	1994	Two-company ETF until 2008; two additional companies added by the government of Uttarakhand in April 2008	Restoration of forests in Kumaon region of Pithoragarh district
132 Infantry Battalion	Rajput Regiment	Delhi	2000	Two-company ETF; second company added in 2005	Conversion of barren land of Bhatti mines into thick forest and parks; compensatory plantation in Ujawa near Najafgarh, NCT of Delhi
133 Infantry Battalion	Dogra Regiment	Himachal Pradesh	2006	One-company ETF	Ecological restoration of Sutlej basin
134 and 135 Infantry Battalions	Assam Regiment	Assam	2007	Three-company ETF	Afforestation

Based on information from Gautam (2009b) and Mohan (2005).

As such, the twin agendas of raising "ecobattalions" have been ecological restoration and rehabilitation of ex-military. On the one hand, the military is more used to operating in difficult terrain and extreme weather conditions than civilian agencies and, on the other, short-service commission officers who are still young, energetic, and enthusiastic but have retired from the armed forces could be motivated to reenter public life in a different role. The initiative is supported by the Ministry of Environment, Forests, and Climate Change (previously called the Ministry of Environment and Forests), the Ministry of Defence (MoD), and particular state governments. The ETFs are financed by the MoEFCC, the state governments, or both, while the personnel come from the MoD. The Forest Department is known to impart training to the uniformed military, especially with regard to the choice of saplings and planting techniques. In this respect, the Forest Department also benefits from the fact that a large proportion of the soldiers hail from rural backgrounds and have a certain amount of awareness about agriculture and forestry. Therefore, a significant proportion of the soldiers who join the ETF come in with a basic knowledge of the environment around them, making the Forest Department's task easier. Besides, military personnel are known to attain greater awareness of different ecological regions, including biodiversity hot spots such as the Himalayas in the north and Western Ghats in the south, through their postings all over India (Gautam, 2009b).

ETFs have been credited with many successes across the country. The 127th Battalion, set up in Dehradun in 1982, managed to reclaim a mining area of nearly 2500 ha through massive afforestation and sound watershed management, in addition to the construction of soil conservation structures. ETF 127 has tackled not only recurring landslides and difficult terrain (sometimes at a height of above nearly 2500 metres), but also human interventions in the form of grazing, fires, and damage to fencing by the villagers. The "Green Warriors," as they have begun to be known as, restored those parts of Mussoorie that had seen negligible vegetation. Subsequently, the 128th Battalion was deployed in the Thar Desert of Rajasthan, where their efforts led to sand dune stabilization. Here too, the main reason behind using the services of the ETFs was the region's demanding terrain, which required enormous effort to prolong and sustain the survival rate of the saplings. The armed forces are accustomed to harsh and vulnerable terrain—mountains, deserts, rivers, jungles, and so on—by virtue of their training. Their successes in Mussoorie and Jaisalmer prompted the central government and many state governments to extend their presence in other ecologically degrading or degraded parts of the country like Jammu and Kashmir, Pithoragarh, Delhi, Himachal Pradesh, and Assam (National Afforestation & Eco-Development Board, 2011). Gradually, they also began to engage with civilians for environmental awareness campaigns and the promotion of tourism. The original plan had been to withdraw the ecological battalions once the assigned task was fulfilled (in three to five years), but until now, all of them have been redeployed to sustain the effects of the accomplishments of the ETFs in those regions and in most cases, more land has been allocated and more battalions have been added.

2.6.5 BHATTI MINES IN THE CAPITAL: A CASE STUDY

The national capital, Delhi, is now among the most polluted cities of the world. A World Health Organization study released in 2014 stated that Delhi has the highest concentration of atmospheric particulate matter (PM) of 2.5 μ (less than 2.5 μ and classified as "carcinogenic") in the world (PTI, 2014a). The concentration of this PM has increased from 168 μg/m^3 in January 2011 to 183 μg/m^3 in January 2014 in Delhi (Mazoomdar, 2014). The permissible level is 50 μg/m^3 (24-h mean)

(Road to Urban Future, 2014). This is despite the fact that the Delhi government (as per the 1998 directive of the Supreme Court of India) was credited with curbing the menace of air pollution significantly by introducing alternative fuel—compressed natural gas (CNG)—to the public transport system between 2002 and 2008. In addition to this decisive step, the establishment took an active interest in creating parks and tree plantations to clean up the air of the capital. As far as the CNG revolution is concerned, when new cars (specifically diesel ones) were added to the already congested roads of Delhi, the gains made by the introduction of CNG were lost (Tankha, 2014). However, efforts aimed at afforestation continued to create "lungs" within the National Capital Territory (NCT) of Delhi, as degraded lands and wastelands were targeted for tree plantation and ecological restoration. The wrecked Bhatti mining area, located on the Southern Ridge of the Aravalli Ranges on the Delhi-Haryana border, is one such area that was earmarked by the government of Delhi for reclamation, conservation, and protection. In this mining zone, the ETFs have been collaborating with the Delhi government's Department of Environment (Forest Department).

The Bhatti mining area was densely forested until the start of the 20th century. However, illegal and unregulated mining in the Badarpur region near Bhatti destroyed the whole ecological terrain of nearly 15 square kms. More than 250 mines, some of them 300 m in diameter and 50 m in depth, were created to extract mineralized quartzite for construction purposes. This area was officially given to the ETFs for ecoregeneration in 2001 by the Delhi state government after civilian agencies, nongovernmental organizations (NGOs), and the Forest Department could not achieve much success in restoring it. Initially, they were allocated approximately 2000 ha of land in 2001, and thereafter, the allotted land was raised by more than 4000 ha. ETF 132, deployed in the Bhatti area, restored the whole area by planting 1.3 million trees, creating 70 water bodies, biofencing area into subplots and constructing tracks for water tankers. In fact, five mining pits have been transformed into permanent water bodies, mainly meant for wildlife and avifauna. ETF 132 overcame the low survival rates of the saplings by scattering seeds and ensuring that even in the rocky areas, *dhak* and *amaltas* grew naturally through pollination. In the region, there are currently 46 varieties of indigenous plants, including *sheesham*, *peepul*, and *siris*. In many instances, the survival rate of the plants has been estimated to be in the range of 70%–87% (Gautam, 2009b). Moreover, the Bhatti forest is also home to more than 100 species of birds (including migratory ones), such as sandpipers, crested pied cuckoos, Eurasian golden orioles, and painted sandgrouse. Many other species of butterflies, reptiles, and mammals have also found refuge in this area (Prakash, 2011). In short, the area has now become a biodiversity hot spot.

Until 2004, the region had not received any rainfall at all since illegal mining in the area began. After ETF 132 took over the area and embarked on the land restoration process, the large-scale afforestation carried out by the ETF ensured that moisture was put back into the air through the process of evapotranspiration of plants (the sum of evaporation from the land surface plus transpiration from plants), causing rainfall. In 2005, the region witnessed rainfall on 12 days, and in 2013, the area witnessed rainfall on 72 days. ETF 132 has also been instrumental in preventing human encroachment, illegal mining, and other prohibited interventions in the area. Furthermore, in 2009, it claimed carbon credits through the Clean Development Mechanism (CDM) of the Kyoto Protocol and secured funding from Germany—the first military unit in the world to do so (Gautam, 2009a). This brought to light the efficacy of the afforested area in carbon sequestration and storage. ETF 132's success prompted the Delhi government to add a portion of the neighboring Asola Wildlife Sanctuary to the lands under ETF 132's care. As of November 2014, 1094 acres of this sanctuary are being afforested by the ETFs, and the remaining 3206 acres are expected to be handed over to the ETFs in the coming years (Lalchandani, 2011). ETF 132 is credited with several other valuable accomplishments, such as:

- *Decreased encroachment in the free wildlife sanctuary and stoppage of illegal mining in the area*— Within four years of taking over the Bhatti mining area, ETF 132 ensured that illegal mining had stopped completely in the whole area. Their success in preventing encroachment by villagers in this area has prompted the Forest Department to rope in their resources to stop fresh encroachment in the adjoining Asola Bhatti Wildlife Sanctuary as well, as already stated earlier (Singh, 2014).
- *Improvement of water bodies and check dams and influence on water availability*—A total of 36 of the 200 mining pits in the Bhatti area had been converted to water bodies by 2011. Thereafter, in 2011, 2012, and 2013, 11 ponds were transformed into larger water bodies and five check dams were constructed in areas where natural waterfalls or natural/artificial depressions were present. This has resulted in a rise in the water tables in the area (Lalchandani, 2011; Prakash, 2011; Singh, 2014).
- *Potential source of medicinal resources*—Besides afforestation, the ETF has been credited with the planting of species of trees native to Aravalli, like *Butea monosperma*, *Salvadora*, *Prosopis cineraria*, *Acacia nilotica*, and *Anogeissus pendula*, among others, which are also known for their traditional medicinal values (Prakash, 2011). For instance, *Butea monosperma* is used to cure diabetes, ulcers, piles, eye-related diseases, diarrhea, urinary infections, and various other diseases (Sindhia, 2010). *Acacia nilotica* helps check bleeding gums, mouth ulcers, and genitourinary disorders (La-Medicca).

2.6.6 CONCLUSION

The positive example of the ETFs in Bhatti could be emulated elsewhere in India. Despite these success stories, there is still much resistance to greater involvement by ETFs in land restoration activities. The commanding officer of ETF 132 pointed out that such an initiative of the MoEFCC and the MoD ought to be expanded so that ex-military members who are no longer being trained for battlefield operations could be made a part of environmental restoration programs. While the establishment recognizes the constructive implications of involving ETFs, there is always reluctance on the part of the MoEFCC to fund such projects due to their cost, in spite of the fact that other (civil) agencies might not be in a position to achieve what the ETF could accomplish in merely 10 years in a completely ravaged area like Bhatti. Certain state governments are enthusiastically fully funding such projects without depending on funds from the MoEFCC, as seen in the case of ETF 133 in Himachal Pradesh (Mohan, 2005). In another case, in response to regular forest fires on Mount Japfu, Nagaland, the state's forest department officials have also begun to contemplate raising an ETF with equal funding from both the state and central governments to improve the degenerating soil profile in the area (Nagaland Forest Fire, 2014).

There have been incongruous views about the role of the ETFs in some states. For example, when the army proposed an ETF to deal with environmental degradation caused by mining projects like that of POSCO in 2014, the Orissa state government felt that instead of conservation or tree plantation, ETFs should be entrusted with the task of protecting forest officials from attacks by Maoist insurgents. Here too, the issue of funding is yet to be worked out, with the army suggesting that the task force could be financed jointly by the center (70%), the state (25%), and the corporate sector in the district (5%; Mohanty, 2014). The level of coordination between ETFs and the Forest Department is also still too low. The Forest Department should ideally take responsibility for imparting the technical know-how of land restoration, afforestation, and so on to those ETFs that are not currently trained in such activities. Initially, when Indira Gandhi sanctioned the ETF proposal, she had plans to raise ETF units in every state to check the alarming rate of ecodegradation occurring in the country, but this has not materialized yet. Under the present central government led by Narendra Modi, ETFs are expected to be expanded to other ecologically degraded areas as well as to activities such as Ganga rejuvenation. The MoD has proposed to raise 40 units, consisting of 40,000 personnel, who would be assigned the undertaking of restoring the damaged ecosystem along the river (Kulkarni, 2014).

It is a fact that despite it not being part of their mandate, the armed forces of India are carrying out environmental activities such as the protection of catchment areas, wildlife, and forests; working in water rejuvenation; and performing other duties like those discussed so far in this section. The case study of the Bhatti area is a clear illustration of the role of the military in environmental security, but this is a very small step in comparison to the larger role that it could take. This is where the question of whether the military is ready to take on this responsibility becomes significant. On the question of whether the armed forces are ready to adapt to the changing requirements of global and national security (with equal importance given to their human and environmental dimensions), it is inevitable that the whole process would have a psychological impact on them. Besides training in battlefield operations, they would have to be trained in environmental security management, which is currently under way only on a small scale. Ideas such as reuse, repackaging, recovery, recycling are already being instilled in the minds of soldiers in the country, but further work needs to be done on the monitoring and evaluation policies currently in place. The already-practiced bottom-up approach to environmental security issues within the military needs to be complemented with institutionalized and nationwide, top-down policy decisions (the ETF being a mere stepping stone) as well. Instead of training only the lower cadre, there is a need to train the senior leaders in the armed forces (army, air force, and navy chiefs) in order to create, sustain, and nurture environmental leadership in the armed forces. As a case in point, the Integrated Defence Staff (IDS), which consists of officers and personnel from the three services; the Ministry of External Affairs/Indian Foreign Service; Defence Finance/Defence Accounts Department; Department of Defence (Ministry of Defence); and the Department of Defence Research and Development (Ministry of Defence; IDS Headquarters), could institute a cell on environmental security so that there is a single point of contact for the three services for integrated response to environmental security issues. Like any other organ of the government, the military needs to show leadership in terms of environmental responsibility and accountability.

India, being a massive country with a large amount of resources and the second-largest population in the world, has tended to tread the path of irreversible land degradation. To prevent this, every segment of the population has to undertake environmentally sustainable measures to curb this menace that clearly poses social, political, and economic security challenges to the nation. The military is the largest landholder in the country; hence, it is necessary to institutionalize its role in environmental policies, including land restoration in both military areas (as a rule rather than a mere necessity) and nonmilitary areas (through civil-military coordination). In every state of India, there is a need to raise ETFs in different locations affected by severe land degradation. This becomes more essential in light of the inability of civil agencies to carry out demanding environmental tasks in ecologically challenging terrain on their own. The lack of coordination between civil and military sectors in the country is leading to the delay or abandonment of critical issues, including environmental issues. If the two come together on one platform, funds for land restoration projects to be executed by the ETFs could be raised more easily and environmental issues such as land degradation could be tackled more effectively. Instead of being wrapped in a veil of secrecy, the military should be integrated with society (except in cases of serious national security threats) through environmental activities, especially as physical wars have become rarer and peacetime security needs have increased. The military should reflect society and not separate itself from it.

Bhatti mines in 2001.

Bhatti mining area in 2014.

A mining pit turned green by ETF 132.

Accomplishments of 132 Delhi eco-warriors.

Management of avifauna and wildlife by ETF 132.

Value additions attributed to ETF 132.

ETF 132 personnel addressing trainees of the Forest Department.

REFERENCES

Agence France-Presse (AFP), 2012. Siachen Standoff Taking Environmental Toll. The Hindu.

Apte, D., 2014. Chief Operating Officer and Principal Scientist, Armed Forces Cell. Bombay Natural History Society (BNHS). ed.

Belluck, D.A., Hull, R.N., Benjamin, S.L., Alcorn, J., Linkov, I., 2004. Environmental Security, Critical Infrastructure, and Risk Assessment: Definitions and Current Trends. In: Morel, B., Linkov, I. (Eds.), Environmental Security and Environmental Management: The Role of Risk Assessment. The NATO Programme for Security through Science and Springer, Berlin, pp. 3–16.

Brooks, N., 2007. Imminent Water Crisis in India. Arlington Institute, WVA.

Brown, P., 2003. Coca-Cola in India Accused of Leaving Farms Parched and Land Poisoned. The Guardian. Accessed October 16, 2014, from http://www.theguardian.com/environment/2003/jul/25/water.india.

Federation of Indian Chambers of Commerce and Industry (FICCI), 2011. Water Use in Indian Industry Survey. FICCI, New Delhi.

Gautam, P.K., 2009a. The Indian Military and the Environment. Institute for Defence Studies and Analyses, New Delhi.

Gautam, P.K., 2009b. An overview of Ecological Task Forces (ETF) and ecological institutions of the Indian Army. J. United Ser. Inst. India 139, 267–271.

Gowda, A., 2014. Maoists Pose Threat to Tigers in Karnataka's Nagarahole. India Today. Accessed October 16, 2014, from http://indiatoday.intoday.in/story/maoists-pose-threat-to-tigers-in-karnatakas-nagarahole/1/408864.html.

Hartman, J., Butts, K., Bankus, B., Carney, S., 2012. Introduction. In: Hartman, J., Butts, K., Bankus, B., Carney, S. (Eds.), Sustainability and National Security. Centre for Strategic Leadership, US Army War College, Carlisle, pp. vii–xiii.

IDS Headquarters. About IDS. Accessed October 16, 2014, from http://ids.nic.in/aboutids.htm.

India as a Great Power: Know your Own Strength, The Economist. Accessed October 16, 2014, from http://www.economist.com/news/briefing/21574458-india-poised-become-one-four-largest-military-powers-world-end.

Kamath, P.G., 2014. Retired Lt. General of the Indian Army. ed.

Kennedy, J., 2014. Gangsters or Gandhians? Political sociology of the Maoist insurgency in India. India Rev. 13, 212–234.

Khalid, S., Kumar, S., 2014. Posco Steel Project Faces Steely Opposition. Al Jazeera Accessed October 16, 2014, from http://www.aljazeera.com/indepth/features/2014/01/posco-steel-project-faces-steely-opposition-2014122536522239.html.

Kulkarni, P., 2014. Army Plans to Raise 40 Eco-Battalions to Rejuvenate Ganga. Indian Express. Accessed October 16, 2014, from http://indianexpress.com/article/india/india-others/army-plans-to-raise-40-eco-battalions-to-rejuvenate-ganga/.

Lalchandani, N., 2011. Nurture, Nature Pull Bhatti Mines Out of the Pits. The Times of India. Accessed October 16, 2014, from http://timesofindia.indiatimes.com/city/delhi/Nurturenature-pull-Bhatti-Mines-out-of-the-pits/articleshow/9792225.cms.

La-Medicca. Acacia Nilotica. Accessed October 16, 2014, from http://www.la-medicca.com/raw-herbs-acacia-nilotica.html.

Landholm, M., Glenn, J.C., Gordon, T.J., Perelet, R., 1998. Defining Environmental Security: Implications for the U.S. Army. Army Environmental Policy Institute, Georgia.

Mathew, R., 2010. Coca-Cola Liable to Pay Damages Worth Rs. 216.26 Crore. The Hindu.

Mauskar, J.M., 2014. Distinguished Fellow, Observer Research Foundation (New Delhi), and former Joint Secretary. Ministry of Environment and Forests (MoEF). ed.

Mazoomdar, J., 2013. The Real Gap Between Two Western Ghats. Tehelka. Accessed October 16, 2014, from http://www.tehelka.com/2013/05/the-real-gap-between-two-western-ghats/.

Mazoomdar, J., 2014. Why Delhi Is Losing Its Clean Air War. BBC. Accessed October 16, 2014, from http://www.bbc.com/news/world-asia-india-26012671.

Mohan, V., 2005. Battling for Green Cover. The Tribune. Accessed October 16, 2014, from http://www.tribuneindia.com/2005/20051008/saturday/main1.htm.

Mohanty, D., 2014. Army Proposes Eco Task Force, Orissa Differs on Role, Army Proposes Eco Task Force, Orissa Differs on Role. The Indian Express. Accessed October 16, 2014, from http://indianexpress.com/article/india/india-others/army-proposes-eco-task-force-orissa-differs-on-role/.

Nagaland Forest Fire 'Almost' Under Control, Assam Tribune. Accessed October 16, 2014, from http://www.assamtribune.com/scripts/detailsnew.asp?id=feb0514/oth07

National Afforestation and Eco-Development Board, 2011. Evaluation Report on Eco-Task Forces (2009–10). Ministry of Environment and Forests. G.o.I, New Delhi.

Pardesi, M.S., 2011. Instability in Tibet and the Sino-Indian Strategic Rivalry: Do Domestic Politics Matter? In: Ganguly, S., Thompson, W.R. (Eds.), Asian Rivalries: Conflict, Escalation, and Limitations on Two-Level Games. Stanford University Press, Stanford, CA, pp. 79–117.

Patil, S., 2014. Associate Fellow, National Security, Ethnic Conflict and Terrorism studies, Gateway House; and former Assistant Director at the National Security Council Secretariat in Prime Minister's Office. ed.

Prakash, S., 2011. Role of Eco Task Force in Restoration of Degraded Ridge Ecosystem "The Bhatti Mines" Delhi. Indian Wildlife Club. Accessed October 16, 2014, from http://www.indianwildlifeclub.com/ezine/view/details.aspx?aid=728.

Prebble, M., 2013. Renewable Resource Shocks and Conflict in India's Maoist Belt. New Security Beat. Accessed October 16, 2014, from http://www.newsecuritybeat.org/2013/02/renewable-resource-shocks-conflict-indias-maoist-belt/.

Press Trust of India (PTI), 2013. "Green Siachen" Photo Gets National Award. Hindustan Times. Accessed October 16, 2014, from http://www.hindustantimes.com/india-news/green-siachen-photo-gets-national-award/article1-1013349.aspx.

Press Trust of India (PTI), 2014a. Delhi Most Polluted City in the World: WHO. The Indian Express. Accessed October 16, 2014, from http://indianexpress.com/article/cities/delhi/delhi-most-polluted-city-in-the-world-who/.

Press Trust of India (PTI), 2014b. India to be "Land Degradation Neutral" by 2030: Prakash Javadekar. The Economic Times. Accessed October 16, 2014, from http://articles.economictimes.indiatimes.com/2014-06-17/news/50651100_1_combat-desertification-environment-ministry-india-land.

Qamrain, N.U., 2013. Siachen Glacier Receding. The Sunday Guardian. Accessed October 16, 2014, from http://www.sunday-guardian.com/news/siachen-glacier-receding.

Road to Urban Future, The Hindu. Accessed October 16, 2014, from http://www.thehindu.com/opinion/editorial/road-to-urban-future/article5815821.ece.

Rosenburg, D.G., 1995. Peace, Health, and the Environment: Challenging a Conspiracy of Silence. Peace Magazine 11, 24.

Shrivastava, K.M., 2014. Western Ghats: Moily Agrees to Reduce Area of Eco-Senstitive Zone in Kerala, Down to Earth. Accessed October 16, 2014, from http://www.downtoearth.org.in/news/western-ghats-moily-agrees-to-reduce-area-of-ecosenstitive-zone-in-kerala-43680.

Sindhia, V.R., Bairwa, R., 2010. Plant review: Butea Monosperma. Intl. J. Pharm. Clin. Res. 2, 90–94.

Singh, D., 2014. Task Force to Help Forest Department to Restore Asola Bhatti Sanctuary. Hindustan Times, New Delhi, p. 8.

Tankha, M., 2014. Diesel Cars Have Negated CNG Gains, Say Experts. The Hindu. Accessed October 16, 2014, from http://www.thehindu.com/news/cities/Delhi/diesel-cars-have-negated-cng-gains-say-experts/article5657697.ece.

Team Mangalorean, 2009. Naxalite Infestation a Social Problem. Mangalorean. Accessed October 16, 2014, from http://www.mangalorean.com/news.php?newstype=local&newsid=118477.

Trivedi, T.P., Sharma, R.P., Bharti, V.K., Shastri, A., 2010. Degraded and Wastelands of India: Status and Spatial Distribution. Indian Council of Agricultural Research and National Academy of Agricultural Sciences, New Delhi.

United Nations Environment Programme (UNEP), 1993. Potential Uses of Military-Related Resources for Protection of the Environment. Office for Disarmament Affairs, New York.

United Nations Environment Programme (UNEP), 1999. The State of the Environment, in: Global Environment Outlook 2000. UNEP, Nairobi. Accessed October 16, 2014, from http://www.grida.no/publications/other/geo2000/?src=/geo2000/.

United Nations Environment Programme (UNEP), 2012. Greening the Blue Helmets: Environment, Natural Resources, and UN Peacekeeping Operations. UNEP, Nairobi.

World Economic Forum, 2011. Water Security: The Water-Food-Energy-Climate Nexus. Island Press, Washington, DC.

RELEASING THE UNDERGROUND FOREST

CASE STUDIES AND PRECONDITIONS FOR HUMAN MOVEMENTS THAT RESTORE LAND WITH THE FARMER-MANAGED NATURAL REGENERATION (FMNR) METHOD

2.7

Julia Birch[1], Peter Weston[2], Tony Rinaudo[3], Rob Francis[3]

Independent Consultant, Melbourne, Australia[1] World Vision International, Solomon Islands[2]
World Vision Australia, Melbourne, Australia[3]

2.7.1 INTRODUCTION

In many developing countries around the world, farmlands have been degraded and are unable to produce regular crops or provide pasture for livestock. These degraded lands have lost virtually all of the natural vegetation that binds the earth together and provides the organic matter necessary to maintain soil fertility.

Subsistence farmers account up to 70% of the total population of many developing countries' regions. For instance, in Sahelian Africa, this accounts for 40–50 million people. Due to the increasing variability of climate conditions and the higher frequency and severity of drought, a greater number of these people will face hardships and food insecurity more frequently.

However, vast tracts of seemingly treeless land, ranging from deserts to farmlands to degraded forests, conceal an "underground forest" with the potential to rapidly restore a healthy cover of trees. These forests are being released through the practice of Farmer-Managed Natural Regeneration (FMNR), which is allowing the rapid restoration of degraded landscapes at low cost and at scale.

Underground forests are composed of living tree stumps, tree roots, and tree seeds in the soil with the capacity to grow quickly under the right conditions. Roots and tree stumps, for example, often have access to deep soil moisture and nutrients and contain stored energy that can propel rapid growth. It is easy for a casual observer not to notice, or even to discount, the presence of seemingly insignificant but

Land Restoration. http://dx.doi.org/10.1016/B978-0-12-801231-4.00016-1

FIGURE 2.7.1

A satisfied Nigerien farmer manages trees through FMNR in his millet field.

FIGURE 2.7.2

Invisible to us, but not to be underestimated, there is almost as much biomass of roots in the ground, as there is of trunk and branches above the ground.

FIGURE 2.7.3

The stump and root systems of most tree species continue to live and produce new shoots after felling. Regeneration of such shoots can be rapid because of their access to soil moisture and nutrients, and having stored energy to draw on in the root system.

telltale shoots sprouting from stumps, roots, or seeds. The term *underground forest* describes the fact that the forest is present, but largely beneath the earth, unseen and unappreciated.

FMNR is an innovative land regeneration system that primarily involves the protection and regular pruning of the shrubby regrowth sprouting from tree stumps, roots, and self-sown seeds. FMNR began in 1983, as an experiment with a few willing farmers in the Niger Republic, West Africa. After years of very limited success with conventional approaches to reforestation, this alternative practice to restore rural landscapes was explored and developed with pioneering farmers.

FMNR was initially met with some skepticism. Very few farmers kept trees on their land at that time; if they did, they would be going against the accepted norms of society, which is very hard to do in an African rural context. Additionally, fuel wood and building timber was so scarce at the time that any tree that stood would almost certainly be stolen. Consequently, farmers were not motivated to leave trees on their land. Beliefs also played a role; for instance, people believed that trees depressed crop yields and that they took many years to grow and bring benefits. These were great concerns in a food-insecure region, where needy farmers looked for a quick return.

Despite these shaky beginnings, within a window of just 20 years, this simple, inexpensive, and rapid form of land regeneration had been widely adopted across 50% of the nation's farmlands. FMNR helps individuals and communities achieve economic sustainability at the household level through the restoration of their farmlands and in some cases, such as in Ghana (1) and Ethiopia, even in the surrounding bushland. This method has doubled crop yields and family incomes, provided timber for building, cooking, and heating, restored degraded soils, and helped communities become more resilient and adaptive to climate change. FMNR reduces reliance on food relief. For example, in 2005, Eric Toumieux, then director of World Vision Senegal, wrote, "[W]hen a deadly combination of locusts and drought struck Niger, farmers in 36 villages in the Aguie Department of Niger Republic [where FMNR had been adopted] overcame the tragedy by selling firewood and non-timber tree products. As a result, there is no need for any food distribution in this community, unlike what is happening elsewhere in Niger."

FMNR as such was developed, popularized, and extensively promoted by the Serving in Mission (SIM) organization in Niger in the 1980s; however, it is simply a form of ancient tree management practice termed *coppicing* when managing regrowth from tree stumps or *pollarding* when managing regrowth from tree trunks. Examples of various forms of FMNR can be found around the world, both in the past and today. It is not unusual to find remote farmers with no contact with the outside world discovering and utilizing FMNR by themselves. Various organizations, including World Vision, the World Agroforestry Centre, the Africa Regreening Initiative, and the World Resources Institute, have strongly promoted FMNR in recent years on a global scale to policy makers and donors through direct meetings, peer-reviewed articles, media (including films), and the Internet.[1] FMNR has also been promoted at the field level through establishing national regreening committees and the implementation of projects to scale up its adoption. For example, World Vision has implemented FMNR projects in over 18 countries in Africa, Asia, and the Pacific, and national regreening committees have been established in Ethiopia and Uganda. As a result, the technique is gaining popularity at an accelerating rate, year after year. Through awareness workshops, exchange visits, demonstration plots, conferences, the Web, written materials, video clips, and presentations, FMNR is being adopted within and beyond World Vision project boundaries around the world.

This section of the chapter documents the spread of FMNR and some of the more dramatic impacts of this practice. It also highlights some of the main preconditions that have made the rapid spread of FMNR possible.

2.7.2 FMNR: BIRTH AND SPREAD OF A MOVEMENT, NIGER REPUBLIC

FMNR experimentation and promotion began with around 10 farmers in the Maradi Department of the Republic of Niger in 1983 by SIM International[2] project staff, led by Tony Rinaudo and Pasteur Cherif Yacouba from the Evangelical Church of Niger.[3] During the severe famine of 1984 (2), a food for work program introduced some 70,000 people to FMNR and its practice on about 12,500 ha of degraded farmland. From 1985–1999, FMNR continued to be promoted locally and nationally as exchange visits and training days were organized for various nongovernmental organizations (NGOs), government

---

[1]For example, see World Resources Report (2008). "Turning Back the Desert: How Farmers Have Transformed Niger's Landscape and Livelihoods," pp. 142–157; http://pdf.wri.org/world_resources_2008_roots_of_resilience_chapter3.pdf: Inside Stories on Climate-Compatible Development: Niger (2013); http://www.wri.org/publication; The Quiet Revolution: How Niger's farmers are regreening the croplands of the Sahel, ICRAF Trees for Change No. 12 Nairobi; World Agroforestry Centre (ICRAF). http://worldagroforestry.org/content/2013-quiet-revolution-how-niger%E2%80%99s-farmers-are-regreening-croplands-sahel-icraf-trees-change; Africa Regreening Initiative updates: http://www.africa-regreening.blogspot.com/.

[2]SIM International is a Christian interdenominational mission agency. The innovation was pioneered and managed by Tony Rinaudo, now Principal Natural Recourses Advisor, who leads World Vision Australia's FMNR program.

[3]The French name for this is Eglise Evangélique de la République du Niger.

FIGURE 2.7.4 AND FIGURE 2.7.5

Before FMNR in the Maradi region of Niger; Years 2–3 of FMNR in the Mardi region of Niger.

foresters, Peace Corps volunteers, and farmer and civil society groups. Additionally, SIM project staff and farmers were sent to numerous locations across Niger to provide training.

By 2004, FMNR was being practiced on over 5 million ha, which accounts for about half of Niger's total farmland (3). This averages out at a staggering rate of 250,000 ha per year over a 20-year period.

The Niger experience of FMNR prompted geographer and sustainable land management specialist Chris Reij[4] to comment, "this is probably the largest positive environmental transformation in the Sahel and perhaps all of Africa" (4). Meanwhile, in Mali, studies headed up by Chris Reij have shown that almost 0.5 million ha have been regenerated on the Seno Plains, perhaps primarily in response to positive changes in forestry law in 1994. In addition, around 6 million ha of old and aging agroforestry

---

[4]Chris Reij, Senior Fellow, World Resources Institute, Washington, DC. Quoted by Mark Hertsgaard, 2009 "Regreening Africa." The Nation. http://www.thenation.com/doc/20091207/hertsgaard.

Left: Map of Niger Republic, West Africa. FMNR experimentation and promotion began in Maradi Department, south central Niger in 1983. Visitors wanting to learn were received from other parts of the country, and Maradi farmers and project staff were sent to other regions to teach FMNR.

parkland can be found in southern Mali. The emergence of some 300,000 ha of new agroforestry parklands in Yatenga and Zondoma provinces in Burkina Faso coincided with an upsurge in the adoption of water harvesting techniques.[5]

The positive impacts of FMNR on livelihoods in Niger can be summarized as follows:

- The number of trees in many villages in Niger is now 10–20 times greater than the baseline value 20 years ago. In the 100 villages where the first SIM project took place, 88% of farmers practice FMNR in their fields, adding an estimated 1.25 million trees each year (5).
- Gross income (Maradi Region) has grown by $17–21 million due to FMNR (6), which translates to around $1000 per household each year (7). Extrapolating this added income from FMNR to the entire 5 million ha implies an aggregate income of $900 million/year (8), benefiting approximately 900,000 households (with 4.5 million people). This is especially significant since, according to the latest United Nations (UN) Human Development Index score, Niger is the poorest nation in the world.[6]

---

[5]Reij, C. Pers. Comm. April, 2014.
[6]http://hdr.undp.org/sites/default/files/hdr14-report-en-1.pdf.

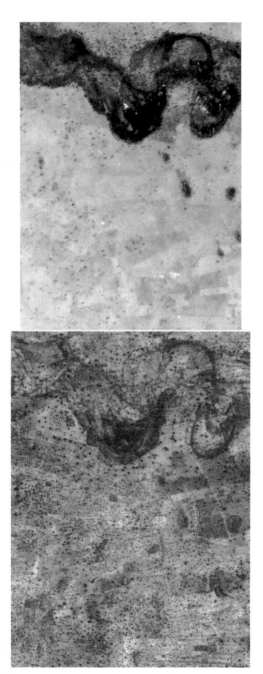

FIGURE 2.7.6 AND FIGURE 2.7.7

Above: US Geological Survey (USGS) satellite photos taken in 1975 (left) and 2003 (right) show greatly increased tree cover in Niger. Trees show up as black dots.

- Due to the implementation of FMNR, farmers in Niger are producing an estimated additional 500,000 tons of cereals per year.[7] This additional production covers the requirements of 2.5 million people out of a total population of about 15 million in 2009. FMNR also has an indirect impact on food security through tree products, which farmers can harvest and sell in local markets. Moreover, despite a near-doubling of the population since 1980, Niger has been able to maintain per-capita production of millet and sorghum, which make up more than 90% of the typical villager's diet. Per capita production remained at approximately 285 kg between 1980 and 2006 (10).
- Despite severe famine as a consequence of the 2004 drought and locust plague in some areas of Niger, farmers practicing FMNR in the village of Dan Saga did not need food assistance. They "were able to meet their own needs through selling firewood and non timber forest products"(11). It is a common experience for Nigerian FMNR practitioners to become much more resilient to adverse events such as droughts and insect attacks, as they now have additional resources to draw from in times of need. By default, FMNR is a powerful tool for climate change adaptation.

2.7.3 ADOPTION AND RAPID SPREAD OF FMNR, ETHIOPIA

In 2004, World Vision Australia and World Vision Ethiopia initiated a forestry-based carbon sequestration project as a potential means to stimulate community development, while engaging in environmental restoration efforts. An innovative partnership with the World Bank, the Humbo community–based Natural Regeneration Project, involving the regeneration of 2728 ha of degraded native forests, brought social, economic, and ecological benefits to the participating communities. Within just two years of operation, communities were collecting wild fruits, firewood, and fodder. At the same time, it was reported that wildlife were returning to the regenerated area, and erosion and flooding had been reduced due to the FMNR efforts. In addition, the communities are now receiving payments for the sale of carbon credits through the World Bank.

The factors contributing to the Humbo success story include securing legally binding forest user rights from the government; organization of community members into cooperatives; delineation of forest areas to the cooperatives; provision of training on forest management; FMNR; book-keeping and leadership; and community ownership of, and responsibility for, implementing the management plan and upholding agreed by-laws. In particular, securing user rights gave community members the confidence that they would benefit from their investment of time and effort in restoring the forest.

Following the success of Humbo, workshops were held during 2009 and 2010 in Mekele, the capital of the Tigray Region in northern Ethiopia, and experience-sharing visits to Niger and Humbo were organized. Within eight months of the second workshop, 273 people, including departmental heads, development agents, sector specialists, administrators, NGO staff, and model farmers, had been taken on exposure and training visits within Tigray. Some 20,000 ha of land and 10 ha-FMNR model sites in each of 34 subdistricts were set aside for research and demonstration.

---

[7]Millions fed. Proven successes in agricultural development. Chapter 7, Re-greening the Sahel: Farmer-led innovation in Burkina Faso and Niger, Chris Reij, Gray Tappan, and Melinda Smale. http://www.ifpri.org/publication/millions-fed.

FIGURE 2.7.8 AND FIGURE 2.7.9

Above: Rapid forest restoration through a community-managed FMNR program in Humbo, Ethiopia.

The Tigray regional government is the first government in the world to institutionalize FMNR by including it in the normal annual planning and implementation cycle of the Department of Agriculture and charging the Tigray Agricultural Research Institute with the task of researching and promoting best-practice FMNR for Tigray.

The federal government of Ethiopia is also building on the success and impact of FMNR, as demonstrated in Humbo. In April 2011, the government committed to reforest 15 million ha of degraded land as part of its Climate Resilient Green Economy Strategy and national target to become a carbon-neutral economy by 2025.

Project outcomes at Humbo on the FMNR movement across Ethiopia and internationally have been enormous. A constant stream of individuals and groups have either heard of or visited Humbo independently; many of these people then apply what they learned on their own land. In addition to articles and press releases published by the World Bank and World Vision, the Humbo project has received widespread coverage in the media and environmental journals (12).

FIGURE 2.7.10 AND FIGURE 2.7.11

Above: Following an FMNR conference in Ethiopia in 2010, the regional government of Tigray facilitated intensive workshops, field training, and experience-sharing visits and set aside over 20,000 ha for experimentation and demonstration of FMNR in communities.

2.7.4 FMNR IN GHANA: FROM DESPAIR TO "LIFE AND JOY"

In 2008, communities in Talensi, in northern Ghana, felt like hopeless victims of climate change (13). They stated that in recent years, their crops increasingly suffered from regular droughts, insect pests, floods, and damaging windstorms. There was a shortage of firewood, wildlife had long left the area, water sources were dry for much of the year, people and animals alike were regularly hungry, and poverty was increasing. Indiscriminate clearing of the forest and annual burning of bushland contributed to the high rate of deforestation and environmental destruction in the district.

FIGURE 2.7.12 AND FIGURE 2.7.13

Above: Just two years into the Talensi project, nine communities have protected 203 ha of land from bush fires and begun practicing FMNR. In that short time, fodder and firewood have become available, wild fruits and wildlife have returned, and people feel confident that they are not hopeless victims of climate change and desertification, but rather that they can do something to reverse the environmental degradation surrounding them and create a better future for themselves and their children. FMNR has now spread to 2000–3000 ha of farmland.

An FMNR project was implemented in Talensi in 2009. Within two years, everybody, from local people to government leaders, was amazed that their previously burned out and barren landscape was now coming back to life with a healthy forest of 1–3-m-tall trees without planting a single tree. The communities are extremely thankful for all the benefits they are already experiencing because of FMNR, including wild fruits being abundantly available, firewood close at hand, fodder for livestock, and the return of some wildlife. This is an example of a community that had directly

Yameriga, site 1

Wakii

FIGURE 2.7.14 AND FIGURE 2.7.15

Forests and Water Officer, Aboubacca Cidibe."Thousands of projects have come through here but this FMNR, there is no comparison, if we are the judges. We have nothing but our environment. Since we started working with FMNR we have already started seeing the benefits that we have not seen with any other project. The type of benefits we see pushes me sometimes to leave my home and just walk through my field to appreciate the trees and environment. When things get to where they need to be, we will see more yields and the path will be clear." Female lead farmer, Thiapy, Senegal.

contributed to forest destruction and subsequently felt helpless to stop it. Following the introduction of FMNR, the community became forest protectors and rehabilitators.

After studying who held power in the community, the project targeted leaders and authorities—chiefs, traditional land custodians, and local government authorities—to ensure that they understood and were convinced of the value of FMNR. The leaders in turn convinced the communities to set aside pilot sites to test FMNR on and to participate in the activities. World Vision provided training and regular follow-up. By mid-2012, the total forest area under FMNR was 203 ha, with up to an additional 2000–3000 ha of FMNR on farmland. Because of the enormous impact that FMNR is having on the participating communities and its acceptance and endorsement by traditional and contemporary authorities, World Vision Australia approved a proposal for the extension of the current project and an expansion into three new districts of northern Ghana.

2.7.5 FMNR IN SENEGAL: APPRECIATING THE ENVIRONMENT

World Vision Senegal has implemented two FMNR projects that sought to promote restorative environmental practices in the regions of Kaffrine and Diourbel. These projects have changed peoples' attitudes and practices. Previously, "people did not care about the environment and if they saw someone cutting a tree they would not stop them as it was not their problem. Now, anybody seeing trees being cut will respond and try to stop it."[8] Within just five years, trees have been regenerated on 62,000 ha of previously cleared farmland.

The success of FMNR in this project and its acceptance by authorities, communities, and NGOs has led World Vision Australia to approve funding for extension of the existing scheme. Therefore, the new plan will reach 100% of the World Vision program area in the Kaffrine region, expand FMNR into the regions of Tamba, Kolda, and Fatick, and engage with national ministries and networks to institutionalize FMNR and other sustainable land use practices. Additionally, 100,000 ha of farmland are expected to be revegetated within the next three years and changes in national forestry policy and practice are expected to affect millions of hectares of land. In that the Senegalese government is one of the chief proponents of the Great Green Wall of the Sahara and Sahel Initiative,[9] its adoption of FMNR as the primary means of forest restoration is expected to have a spillover effect on all the Sahelian Great Green Wall signatory countries.

In a significant development, participants of the Second African Drylands Week, convened by the Africa Union in Ndjamena in Chad, August 25–29, 2014, made the following declaration: "(We) RECOMMEND AND PROPOSE that the drylands development community, through the African Union, and all collaborating and supporting organizations, commit seriously to achieving the goal of enabling EVERY farm family and EVERY village across the drylands of Africa to be practicing FMNR and Assisted Natural Regeneration by the year 2025."[10]

2.7.6 BENEFITS OF FMNR

While the promotion of FMNR has positive impacts on sectors and cross-cutting themes as diverse as climate change, poverty, income generation, biodiversity, gender, nutrition, and migration, its impact on food production alone gives sufficient justification for wide-scale promotion and adoption. Today,

---

[8] Forests and Water Officer, Aboubacca Cidibe, Kaffrine, Senegal.

[9] http://www.thegef.org/gef/great-green-wall.

[10] The full communiqué can be found at http://rea.au.int/en/content/second-africa-dry-land-week-n%E2%80%99djamena -chad.

millions of hectares of land around the world lie idle or are not performing optimally due to degradation; as a result, there are people who are going hungry unnecessarily.

In Niger, farmers have reported that land that had become too degraded to successfully grow crops has been restored through FMNR. In some cases, such as in the Tahoua region, through soil and water conservation measures such as half-moons, planting pits, and FMNR, bare, lateritic soils that had become too hardened to cultivate have been restored to productivity. In some years, the presence of trees has resulted in the survival of interplanted crops while crops in the open have desiccated. Preliminary research results by the Senegal Agricultural Research Institute indicate that within two years of FMNR implementation, crop yields increased from an average of 296 kg/ha to 767 kg/ha (Figure 2.7.16).

FIGURE 2.7.16

Millet yield, with and without FMNR (Rendement = Yield of Millet). Above left: Preliminary findings of the Senegal Agricultural Research Institute show that practicing FMNR on farmland can result in a doubling of crop yields. Right: This photo was taken during a dry spell in the rainy season and shows superior growth and survival rates of crop plants growing close to trees compared to those in the open.

2.7.7 PRECONDITIONS FOR THE SCALE-UP OF FMNR

Experience gained from the four FMNR movements described previously highlights the preconditions for scale-up. Observations of the original movement in Niger show that factors inherent to FMNR, such as its low cost and simplicity, enabled the rapid spread of the method. The movements that followed,

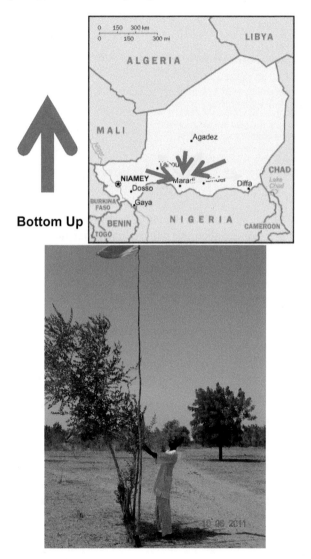

FIGURE 2.7.17

Above left: FMNR in Niger was a largely bottom-up process, with FMNR spreading directly from farmer to farmer. Above right: Enthusiastic farmer in Senegal by his own initiative, spread the word on FMNR by erecting a flagpole beside a market road and a sign post. Women freely shared information on FMNR among themselves.

which were initiated by World Vision projects in its program areas, demonstrated that project staff and communities could draw on a range of methods appropriate to the context to best foster the subsequent widespread adoption of FMNR.

The inherent contributing factors to the rapid spread of FMNR in Niger include:

- The simplicity, visibility, and intuitive nature of FMNR, which enabled peer-to-peer informal promotion so that farmers learned from and imitated other farmers. The spread of FMNR in Niger from the SIM project based in Maradi was largely by word of mouth from farmer to farmer. SIM sent "FMNR champions," namely, SIM staff and farmer practitioners, to various parts of Niger. In addition, NGOs, faith-based groups, Peace Corps volunteers, farmers' groups, and government foresters were welcomed to come and learn about FMNR in Maradi. Yet, the very rapid and large-scale adoption of FMNR appears to be largely a result of farmer-to-farmer promotion. Many early adopters would have simply observed FMNR when they passed through farmlands and, due to the intuitiveness of the method, began practicing it upon returning home without any additional external support.

- A Nigerian farmer once told the principal author, "If you as a foreigner convince me to try a new innovation and I implement it, but it doesn't work, I will suffer, but you can fly home with no consequences. Even if your Nigerian project staff shares this information with me, I will be skeptical because that person is paid to give those instructions, while it is me who bears the risk and the consequences. However, if another farmer tells me to try something, I know it is genuine because their livelihood also depends on the efficacy of what they are promoting." Farmers are much more likely to be convinced about the value of a new practice by their peers than by external agents.

- Building on established relationships in the community, assistance with food for work programs during a food crisis, and encouragement of pioneers and "early adopters" resulted in a critical mass of farmers adopting FMNR. The 1984 famine and the subsequent SIM-managed food for work program followed on from over 50 years of SIM presence and relationship building in Niger. Normally, in this very conservative society, individuals are averse to being singled out for being or acting differently to long-standing, accepted social norms. The food for work program enabled a critical mass of people to experience FMNR firsthand, in a social environment where it would not be criticized.

- A response to a crisis—Farmers had their backs to the wall.

If traditional farming practices were meeting the basic needs of the community, it is unlikely that FMNR would have spread so rapidly. The fact was that farmers were scared. All they had learned from their forebears about farming was failing them. In average years, farmers could not grow enough food for their families; and in exceptional years, people suffered greatly and even starved. The need to adapt to environmental, climatic, and demographic changes pushed people to move beyond norms of behavior and adopt new practices, which clearly contributed to resilience, survival, and even prosperity.

It should be noted that the adoption of FMNR in some nearby regions from 1984 onward seems to have been spontaneous and, as far as the authors can tell, not related to events in the Maradi region. For example, in the state of Zinder, some 1 million ha of parklands, dominated by the native tree species, *Faidherbia albida*, have emerged in recent decades.[11] Whereas *Faidherbia* is a component species in the FMNR of Maradi, other species more suitable for fuel wood and traditional construction, such as *Piliostigma reticulatum* and *Guiera senegalensis*, are dominant. In the village of Dan Saga, Aguie

---

[11]Reij, C. 2013, Senior Fellow, World Resources Institute, Washington DC, personal communication with Tony Rinaudo.

Department, farmers returning late from Nigeria, after the rainy season had already started, did not have time to conduct traditional slashing of trees regrowing from tree stumps. They noticed that while neighbors' crops suffered from severe windstorms in the early 1980s, their crops were afforded protection from the "bushes" growing in their fields; and from this observation, the practice of FMNR was adopted widely in the village. By 2007, 170 villages were involved and 53 village committees had been established, each encompassing three or four villages. Some 130,000 ha were being managed under FMNR, and fields that were practically treeless in 1984 were covered with 103 to 122 trees per hectare (16). However, even here, there may have been some contact with SIM Maradi, through the food for work program conducted in 1984 (17).

World Vision Australia began promoting FMNR through grant projects in West and East Africa in 2005. Since that time, the number of projects and the associated locations has grown exponentially, with some $13.5 million invested throughout 11 countries today.

Major lessons learned from the introduction of preconditions in the project context for widespread uptake of FMNR include:

- Committed and skilled national staff members drive the project—Where there has been no history of FMNR projects within a World Vision National office, FMNR has usually been introduced by support office staff through seminars and workshops. Once funding is procured and projects have started, annual follow-ups, monitoring, and encouragement visits are made by support office staff. An FMNR champion is selected from within the target country, and the best progress has been observed where senior management have supported and encouraged project implementation. The bulk of the extension work is done by farmer volunteers who first demonstrate their commitment to FMNR on their own farms before engaging others to practice it. This approach of using peer-to-peer extension can be highly effective. Trust and close working relations between WV National staff and communities are essential for initial acceptance and uptake of FMNR. Increasingly, WV National Offices are learning from and being inspired by each other's adoption of FMNR.
- Overcome farmers' perceived risks.
- Fear of reduced crop yields as a result of competition and shading from trees.
- Education, such as site visits and demonstrations, as well as incentives early on in the project, such as those offered in the food for work program in Niger, enables demonstration of the benefits of FMNR, such as increased crop yields. In turn, such positive demonstrations entice the local community to try the FMNR approach.
- Fear of losing management rights over trees on their land; regenerated trees being stolen or burned.
- In some cases, tree ownership and management rights are not clear at the outset. A precondition for the uptake of FMNR is negotiation and clarification to recognize these rights. Also, in some cases, the Forestry Department may inadvertently discourage FMNR uptake by the enforcement of regulations requiring permits for harvesting and transporting wood. In these cases, diplomatic advocacy for reforms on the part of the Forestry Department is necessary. Often it is initially simpler to work with district-level forest officers who tend to be more flexible in their approach. Communities can be supported to form binding agreements with the government Forestry Department, giving them ownership (or at the very least, user rights) of the trees and a greater measure of protection from theft. Developing such agreements requires understanding of the relevant laws and cultural standards and advocating for enabling laws and, where necessary, new cultural norms for reforestation to occur. While this process may take some years, it is a crucial step for effective FMNR uptake.

- The possibility of an initial drop over two to three years in available fuel wood as tree regrowth is first protected.
- Communities should be forewarned of the possibility of reduced fuel wood availability in the early FMNR implementation stages and contingency plans drawn up. Fuel-efficient stoves can assist in reducing household demand for firewood.
- Farmers inform farmers—Sending farmers (both men and women) to visit farmers where FMNR has already been introduced is an important step to convince early adopters of the method. Experienced local FMNR champions can also travel to new communities to demonstrate the method and share information.
- Recruit women in the process, as well as men—Women should be included in training as FMNR champions from the outset. Women are often highly motivated and are good communicators who are able to spread the news rapidly through their well-established networks. Engaging female farmers enables men and women in communities to work together to achieve the same goals. Equally, men need to be brought on the journey. There are instances of men (who are usually the decision makers on land issues) preventing their wives from practicing FMNR.
- Apply multiple approaches for communicating behavioral change—Behavioral change is a gradual process requiring many inputs that helps FMNR to become a movement, by becoming a whole-of-community topic of interest, not just the domain of farming household heads. Examples of various approaches include:
  - Dynamic full-time staff in each location spend most of their time engaged in coaching and encouraging community partners.
  - Nominating village FMNR champions are given in-depth training and assisted with transport and may, depending on the context, be given a monthly stipend for achieving work targets. Village champions then frequently visit farmers, providing encouragement, updates on information, and advice.
  - Providing workshops and conducting exchange visits.
  - Creating an enabling legal and cultural environment. An enabling legal environment gives farmers the security of knowing that if they invest their energy in FMNR, they will benefit. This provides a very great incentive to not only practice FMNR in the first place, but to protect their trees from threats such as theft, fire, and animals. An enabling legal environment simply means that farmers have (and know that they have) the legal rights of ownership—or at the very least, user rights—to the trees they manage. To bring such laws into effect at a national level may meet with resistance and inertia and take years to achieve. In the interim, it may be prudent to create memorandums of understanding with sympathetic local officials. An enabling cultural environment gives farmers the security to know that they will not be ostracized or ridiculed for practicing FMNR. This is an environment in which FMNR becomes normal behavior and to practice it does not raise eyebrows. In fact, in Niger, once FMNR was well on the way to becoming standard practice, the tables turned and those who did not practice it were in danger of being ostracized.
  - Changing the perceived role of government forestry service staff. Too often, forestry staff are trained primarily as forest protectors and enforcers of the law and they may not readily see the value of supporting farmers to adopt FMNR. Greater emphasis on extension, outreach, and enabling of the adoption of FMNR is needed in forestry departments.

- Including FMNR and environmental education at school so children can help on their farms and discuss ideas with their parents.
- Conducting awareness-raising interviews on radio programs.
- Engaging local leaders in discussions and negotiations, involving faith leaders in the promotion of environmental protection and involving women as well as men.
- Creating promotional caravans that entertain and inform.

2.7.8 FROM THE GRASSROOTS TO A GLOBAL MOVEMENT

There is a danger of approaching the subject of FMNR scale-up from a perspective in which it is only necessary to understand the drivers and preconditions of adoption of FMNR in rural communities, and that donors, NGOs, policy makers, and UN agencies have fully understood, embraced, and promoted FMNR from the outset. In fact, the spread of FMNR has largely occurred under the radar of national and international organizations, and even as more evidence came to light, acceptance of FMNR into mainstream thinking has occurred at a glacial pace at best.

At some levels, FMNR could be perceived as an affront to more conventional, complex, and expensive approaches to combating deforestation and desertification. A common perspective is that such complex, long-standing, and extensive problems surely require complex, expensive, long-term solutions. The Niger FMNR regreening phenomena begs to differ.

The bottom line is that presumably illiterate, poor, risk-averse farmers with everything to lose if they implement a technology that does not work are adopting FMNR in droves. The authors believe that FMNR is being freely adopted by tens of thousands of farmers in dozens of countries throughout Africa and Asia because it is a low-cost, rapid, flexible, and accessible tool that delivers multiple benefits. FMNR enables farmers to respond quickly to their ever-changing economic, environmental, and social reality. They adapt this flexible tool, happily sacrificing optimum output for the much more desirable outcome of yield and income stability.

Resource-poor, risk-averse farmers have to survive and want to thrive in a highly risky social-environmental-economic reality. Failure can literally mean disaster, or even death, so they opt for the stability of yield/income over time rather than maximum yield in some years. Dr. Richard Stirzaker of the Commonwealth Scientific and Industrial Research Organization (CSIRO) wrote the following:

> I have followed the development of Farmer Managed Natural Regeneration (FMNR) since its very beginning in Niger during the early 1980s. Thirty years later, independent scientists have hailed FMNR as contributing to the greatest positive transformation of the Sahel. I agree.

FMNR is a counter-intuitive idea. Traditional agroforestry has always tried to specify the ultimate tree-crop combination and arrangement that maximises complementarily. FMNR is based on a naturally regenerating suite of tree species, each growing where they are because they have demonstrated an ability to best exploit a specific niche and overcome prevailing constraints. The farmer then thins and selects from this "template" that nature has produced. Farmers derive their livelihoods from cropping around the trees, cutting browse for animals, producing construction poles and firewood. The contribution each of these options make towards food security depends on current trees density, rainfall, availability of labour, and the prevailing prices for the different products, providing food and income stability in a very variable environment.

I do not think that any research programme, no matter how well funded, would have come up with the idea, because it expertly combines the subtleties of location specific tree selection with farmer specific opportunities and constraints. (18)

In many ways, FMNR can be considered a "no regrets" technology. During 30 years of FMNR promotion, the social, economic, and environmental benefits of FMNR have become apparent while very few negative impacts have come to light, despite in-depth evaluations in several West African nations and Ethiopia. It has been said that if you have nothing to lose and everything to gain, then by all means, go for it. FMNR is a technology that we can and should confidently promote in every opportunity open to us. In the 1980s, FMNR was promoted at the village level, and it took root and spread. In the early 2000s, organizations such as World Vision and the Africa Regreening Initiative promoted FMNR at district and national levels, and from that initiative, FMNR has taken root and is spreading. This stepwise progression showed that even with an incremental increase in effort, an exponential increase in adoption could be achieved.

In April 2012, this realization led World Vision Australia, in partnership with the World Agroforestry Center and World Vision East Africa, to host an international conference in Nairobi called *Beating Famine*, in order to analyze and plan how we could improve food security through the use of FMNR and Evergreen Agriculture (19). The conference was attended by more than 200 participants, including world leaders in sustainable agriculture, five East African ministers of agriculture and the environment, ambassadors and other government representatives from Africa, Europe, and Australia, and leaders from nongovernment and international organizations such as the Food and Agriculture Organization (FAO) of the United Nations.

Two major outcomes of the conference were:

- The desire to build a global network of key stakeholders
- Country, regional, and global level plans for scale-up

The conference acted as a catalyst for regular media coverage in some of the world's leading outlets and a noticeable increase in momentum for an FMNR global movement. Media has included features in distinguished outlets such as *Der Spiegel* (a German weekly paper), the Australian Broadcasting Corporation's *Lateline* (a TV program), ZDF TV News (a news broadcast in Germany), *The Guardian* (a newspaper in the UK), and *PBS NewsHour* (a news program on US public television). This heightened awareness of FMNR has created an opportunity for its exponential spread worldwide to assist with food security, increase rural incomes, develop social capital and capacities, and build environmental resilience against climate change.

As a result, the tide is turning, and FMNR is now supported and actively promoted by an increasing number of international organizations. Since evidence of the significant impacts and spread of FMNR in Niger have come to light, largely through the combined efforts of World Vision, The World Agroforesty Centre (ICRAF), the African Regreening Initiative, and the World Resources Institute, other significant organizations are taking up the cause. Increasingly, particularly since the Beating Famine Conference in 2012, ICRAF has championed FMNR and actively sought partnerships with World Vision and other NGOs, lobbied governments and major donors, pressed for favorable policy changes at the governmental and intergovernmental levels, and contributed to building an evidence base for FMNR. The World Resources Institute is actively promoting FMNR and documenting its uptake and impact. The United States Agency for International Development (USAID) has listed FMNR as one of the best-practice interventions of choice for natural resource management (NRM) (20) and resilience projects. The World Bank and others, in the quest for more "climate-smart agriculture," have also recognized the important role of agroforestry and the success of FMNR and EverGreen Agriculture in Niger (21).

In 2010, FMNR won the Interaction[12] Best Practice and Innovation Initiative award in recognition of its high technical standards and effectiveness in addressing the food security and livelihood needs of small producers in the areas of NRM and agroforestry. In 2011, FMNR won the World Vision International Global Resilience Award for the most innovative initiative in the area of resilient development practice and natural environment and climate issues. In 2012, FMNR won the Arbor Day Award for Education Innovation (22). In addition, *The Guardian* listed FMNR third in its list of the top innovations transforming Africa (23). In 2013, World Vision Australia was awarded equal second place for the United Nations Convention to Combat Desertification (UNCCD) Land for Life Awards for its work in promoting FMNR.

The "Beating Famine" conference and subsequent global recognition and support of FMNR set the scene for a new approach to spread FMNR—namely, the engagement of all stakeholders simultaneously. If FMNR could spread in Niger as a largely bottom-up movement across 5 million ha in just 20 years, what would be possible if all stakeholders—government policy makers and extension services, civil society, NGOs, faith-based organizations, UN agencies, donors, communities, and individuals—worked together toward achieving the same goals? Technically, there is no reason why numerous countries could not achieve regreening rates of 5 million ha in 5 years simultaneously.

The rapid uptake of FMNR cannot fully be explained by its simplicity, low cost, or quick rewards. Individuals and communities that had lost hope and felt like helpless victims of poverty and climate change are being empowered to address their situation. FMNR practitioners realize that they can do something very tangible and within their means to address these serious issues.

Normally, farming is considered drudgery. People do not typically go to their farms "just to appreciate the trees and environment." The deep sense of pride and satisfaction that FMNR has triggered in Senegalese lead farmer (see **FMNR in Senegal – appreciating the environment**) is being replicated wherever FMNR has been introduced.

---

[12]InterAction is a US umbrella organization for NGO.

FMNR does not create dependency; it is a bottom-up approach that puts individuals and communities firmly in the decision-making seat. In a very real sense, FMNR is giving people back their dignity and sense of belonging. Participants in the Humbo Community Managed Natural Regeneration project were asked how they felt about the changes they were experiencing. Their responses capture their sense of restored pride, joy, and dignity that has been gained through using FMNR to restore their mountain:

> "Very, very happy. When we went to Hobicha market [in the past] when we felt tired there was no shade, but now we go through the shade."
>
> "Very good. Previously this place was destroyed by the people, only rock [was left]. There was a big problem with soil erosion. The trees protect the hillside from erosion, even the fallen trees. People were using it previously for charcoal production. I feel happiness when I look at the hill now."
>
> "When it was rock, we did not have rain. I feel very proud. I feel proud not only for me but for the village, the *woreda*. This project is known in the whole world."
>
> "I am so happy when I feel the fresh air and moisture from the hillside."

In turn, the change in people's view of themselves and their place in the world has generated a genuine and very effective enthusiasm to share the good news of FMNR with others.

In response to the shifting scale of the FMNR movement, World Vision has developed something called the "FMNR Hub"[13] to move forward from the existing piecemeal approach to FMNR scale-up to a globally coordinated one that does the following:

- Capitalizes on the current momentum and opportunities for global scale-up
- Supports our current projects (which will inevitably increase)

The work of the FMNR Hub involves:

- Building an FMNR global network
- Creating awareness through conferences, workshops, exchange visits, and accessible information on the Web and in written and video form
- Advocating for favorable NRM policies and market development
- Training of FMNR trainers and champions
- Maintaining the FMNR Hub website (http://www.fmnrhub.com.au)

---

[13]The FMNR Hub is an FMNR center of excellence, managed by a small team in World Vision Australia. It leads and fosters the development of FMNR globally through coordination, collaboration, communication, building evidence and science, and project fund-raising. In particular, it initiates projects in new regions, acts as a communications hub and knowledge bank, coordinates the network of partners and collaborators for a global FMNR movement, provides technical support to projects, and advocates and raises funds for FMNR projects.

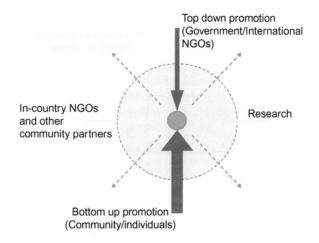

FIGURE 2.7.18

Above: Diagrammatic representation of engagement with all stakeholders via the FMNR Hub for the rapid promotion and uptake of FMNR.

Table 2.7.1 FMNR Benefits and Impacts

Economic	Social	Environmental
Increases income through improved crop yields, sale of tree products, and improved livestock production Reduces expenditures and increases consumables, such as food, fodder, firewood, poles, edible fruits and leaves, etc. Increases household assets Offers new income opportunities via carbon credit revenues Reduces risk from flooding and drought Economic flow-on effects such as employment and greater purchasing capacity Development of business models to increase income	Fosters realization, acceptance, and the resolve to change Creates an enabling environment, fosters and reinforces a positive outlook, enables a resolve to further improve one's well-being, and shifts from passive acceptance to mobilization for change Builds collaboration, networks, and partnerships Fosters tree ownership and land tenure security for farmers Increases education and training Increases empowerment for women Creates community advocates Increases food security, health, and resilience Improves the environmental comfort of rural communities Gives rise to hope and optimism, which improves adaptive capacity Community capacity building to deal with local, regional, and national governments and regulators	Widespread adoption of FMNR: Restores tree cover Increases biodiversity Reduces erosion Enriches soils Increases water availability (through reduced soil-moisture evaporation and increased water infiltration and groundwater recharge) Reduces wind and temperatures Increases climate change adaptation and mitigation through carbon sequestration by regenerated trees

Continued

Table 2.7.1 FMNR Benefits and Impacts—cont'd		
Economic	**Social**	**Environmental**
	Improved environmental governance through clarification of land and resource rights, and adoption of locally enforced rules governing the use of trees on farms, movement of livestock, control of bush fires, and other measures to benefit the community and prompt additional investments in FMNR Community development (including reduced migration of young people and men to cities) Better opportunities for medical treatment, children's education, nutrition and clothing, etc.	

FMNR fits the classic sustainable development paradigm affecting economic, social, and environmental aspects of people's lives.
Francis, R., Weston, P., and Birch, J., 2015, The Social, environmental and economic benefits of FMNR, World Vision Australia,
Melbourne, available at http://fmnrhub.com.au/fmnr-study/.

After one year of operation, the FMNR Hub has achieved the following:

- Winning the UNCCD 2013 Land for Life Award in Windhoek, Namibia, in September 2013. The prize of $30,000 will be used to support the development of FMNR in southern Africa.
- Establishing a website at www.fmnrhub.com.au to share news, project information, reports, and research globally.
- Conducting scoping trips and multiregional workshops in southern Africa, India, and Haiti.
- Becoming a founding member of the Evergreen Agriculture Network, hosted by the World Agroforestry Center in Nairobi. This led to partnering in various sustainable agricultural projects in East Africa.
- Producing an FMNR literature review, which identifies 24 key social, environmental, and economic impacts and will influence ongoing research into FMNR.
- Producing a fully referenced Wikipedia entry, available at: http://en.wikipedia.org/wiki/Farmer -managed_natural_regeneration.
- Presenting at the Caux Dialogue on Land and Security, which explores the human connections between poverty, conflict, and environmental degradation.[14]
- Making a presentation to the Education Concerns for Hunger Organization (ECHO) Conference in the Dominican Republic, and raising awareness of FMNR with major organizations in Latin America for the first time in 2013.
- Coordinating a media visit to Timor-Leste for a feature article published in *The Melbourne Age and Sydney Morning Herald Good Weekend Magazine* April 26th, 2014.

---

[14]See more on this event at http://caux.iofc.org/en/Dialogue2014#sthash.X1TtdByM.dpuf.

2.7.9 CONCLUSIONS

FMNR continues to be, principally, a movement of simple farming practices to regenerate trees that is largely being spread by farmers themselves. Recognition and support for this movement has recently taken hold within international organizations, which is making a significant contribution to global widespread uptake of the method.

Understanding the preconditions for the rapid uptake of FMNR among all stakeholders is not merely an academic exercise. The authors believe that continued support for the FMNR movement can protect vulnerable people from famine and improve their livelihoods.

The recent food crisis in East Africa declared in mid-2011 provides a stark reminder that the need for solutions addressing root causes of hunger is greater today than ever before. In 2010, Roland Bunch, an international agricultural consultant, did a study of six African nations (Zambia, Malawi, Kenya, Uganda, Niger, and Mali). He wrote about his findings in a chapter of the 2011 State of the World Report entitled "The Coming Famine" (24). He concluded that four major factors are on a collision course in a sort of "perfect storm" that will almost surely result in an African famine of unprecedented proportions, probably within the next 4 to 5 years. It will most heavily affect the lowland semiarid to sub-humid areas of Africa (including the Sahel, parts of eastern Africa, and a band from Malawi across to Angola and Namibia), and he calculated that unless the world does something dramatic, 10 to 30 million people could die of famine between 2015 and 2020.

Victor Hugo once wrote, "Nothing is as powerful as an idea whose time has come." For the sake of millions of small farmers daily living precariously close to disaster, let's hope for and work toward making now the time for FMNR.

SOIL, WATER, AND ENERGY—THE RELATIONSHIP TO LAND RESTORATION

COMPUTATIONAL POLICY SUPPORT SYSTEMS FOR UNDERSTANDING LAND DEGRADATION EFFECTS ON WATER AND FOOD SECURITY FOR AND FROM AFRICA

3.1

Mark Mulligan

King's College London, Strand, London, UK

3.1.1 LAND DEGRADATION POLICY SUPPORT

3.1.1.1 WHAT IS POLICY SUPPORT?

Policy support systems (PSSs) are an extension of the ubiquitous and highly variable decision support systems (DSSs), which can range from simple flowcharts to sophisticated geographical information system (GIS)–based simulation tools. They combine geospatial data and models of process in order to examine baseline conditions or the impacts of scenarios or policy interventions. DSSs are usually intended to assist decision making around a specific issue, such as whether to implement a specific land management intervention (where to add check dams, where to permit irrigation, where to build terraces, etc.). PSSs, on the other hand, assist decision making around the design of much broader policies such as adaptation of agriculture to climate change, land use planning, and land use incentive schemes, all of which might include a range of individual management actions, addressing complex issues and with impacts on many processes. Designing and understanding the impacts of proposed policies for restoring degraded land constitute a typical PSS use case.

3.1.1.2 WHICH POLICY MAKERS ARE INVOLVED?

DSS and PSS are usually targeted at technical assistants to policy makers and form only part of the information input to the policy-making process. They are not designed to advise on which policy to adopt, but rather to act as a digital testing place to better understand the likely implications of adopting various policies. They thus add to the weight of evidence in favor of—or against—a particular policy. The availability of such a testing ground is particularly important where the policy involves a variety of landscapes, ecosystems, socioeconomic activities, and stakeholders such that the usual expert evaluations and conceptual scenario analysis may not identify all positive and negative outcomes of a particular policy. The policy may thus yield unintended and unhelpful surprises upon implementation.

Where landscapes are spatially heterogeneous, temporally variable, or both, as is very much the case for Africa, spatially explicit, data-based simulation tools can help handle and communicate the outcomes of the resulting complexity over particular administrative or biophysical regions. Such tools combine relatively generic rules for the operation of biophysical and socioeconomic processes with highly specific, spatially explicit data on biophysical and socioeconomic properties. PSS thus enables individual learning and colearning of stakeholder groups and can also provide project-specific advice for the implementation of more robust and better-tested policy.

3.1.1.3 WHICH POLICIES SHOULD BE USED?

Careful land use and land management policy are fundamental to sustainable agriculture and development in marginal environments. To be of any real benefit, land use strategies and incentives must be sustainable and profitable in the long term, and land management options must also be effective at avoiding land degradation for the long term. The impacts of both land use and land management strategies must also be understood for the range of socioeconomic and biophysical conditions that exist in the region of influence of the proposed policy since the response will not be uniform across variable landscapes and those applying policy should not assume that it will be. Policies relevant to the restoration of degraded lands that need to be tested in this way include:

- Investment or incentivization of particular crop choices or land management techniques, including irrigation, terracing, contour ploughing, ploughing, and slope reforming
- Infrastructural investments such as dams, water transfers, and desalinization facilities
- Conservation and protection schemes such as designation of protected areas, deforestation, buffer strips, check dams, vegetation, and soil restoration

All of these items involve complex interactions with biophysical processes (and thus stores and fluxes of water, sediment, soil, and plant productivity) and human actions such as crop choice and land management. Over spatially and temporally explicit landscapes, the only effective way to manage the knowledge base around such complexities is through the application of data and process-based, spatially explicit modeling (Mulligan, 2009).

3.1.2 INFORMATION NEEDS FOR LAND RESTORATION

3.1.2.1 STATIC DESERTIFICATION ASSESSMENTS VERSUS DYNAMIC PSS

Understanding the geography of both global and African desertification to date has focused on the application of GIS- and remote sensing–based assessments of desertification risk or of observed land degradation. Examples of the former include Sommer et al. (2011) and Spinoni et al. (2014) and of the latter include Bai et al. (2008) and Helldén and Tottrup (2008). While these assessments are useful as geographical prioritization exercises, they usually indicate where there is risk from desertification or where land degradation is occurring, but say little about *why* it is occurring and thus are of limited use in prevention, adaptation, mitigation, or land restoration. To facilitate land restoration, we need to understand:

- Where and why degradation has occurred
- How to reduce the pressures that led to desertification
- Understanding where, what, and how to restore
- Understanding how much to restore
- Identifying the priority regions that will have the greatest positive environmental and societal impact through restoration

We will tackle these questions in turn in this discussion, starting with where and why degradation has occurred. We use the WaterWorld PSS throughout the analysis, which is a fully distributed, process-based hydrological model that utilizes remotely sensed and globally available data sets for application to supporting hydrological analysis and decision making, and is especially useful in ungauged and data-poor environments, since it can run entirely from globally available data sets. If users have local data, they can be uploaded and used in WaterWorld analysis, but even if they do not, simulations are still possible with global data sets delivered with the model.[1] WaterWorld simulates a hydrological baseline as a mean during the period 1950–2000, and it can be used to calculate hydrological scenarios of climate change, land use change, land management options, impacts of extractives (oil and gas and mining), and impacts of changes in population and demography, as well as a combination of these. It is fully documented online at www.policysupport.org/waterworld and described in Mulligan (2013a).

In addition to water quantity, its seasonality, and soil erosion and deposition, WaterWorld uses an index of the extent to which water is affected by human activities: the so-called human footprint (HF) on the water quality index (Mulligan, 2009b). The HF essentially assumes that if rain falls on land that has low human impact, then it will generate clean runoff, whereas if it falls on human land uses that may form point (mines, oil wells, roads, urban) or nonpoint (unprotected cropland and pasture) contaminant sources, then the runoff generated is not clean (i.e., has a human footprint). The HF index (expressed as a percentage) at a point is thus the proportion of water at that point that fell as rain on upstream human-impacted areas and is an indicator of the potential level of contamination of water. All land covers are assumed to have the same unit impact, though users can change this if there is reason to believe, for example, that particular areas of cropland have a higher or lower pollutant release than others. The impacts on water quality are thus the magnitude and distributions of human land uses upstream in relation to where the rain falls.

3.1.2.2 AFRICA SUBJECT TO RECENT LAND DEGRADATION

In Figure 3.1.1, we use a number of the global data sets available in and used by WaterWorld to map (i) observed degradation in Africa according to the remote sensing analysis of Bai et al. (2008) and a series of potential drivers for land degradation through the distribution of (ii) croplands (Ramankutty et al., 2008), pastures (Ramankutty et al., 2008), and deforestation (Hansen et al., 2013). Much of the observed degradation seems to be centered on the Congo Basin, which has relatively little current pressure from cropland and pasture, though it does have significant deforestation. Other areas include eastern South Africa, which is experiencing significant cropping.

---

[1]The policy support system is accessible through a Web browser (and runs on a server, requiring no local installation). It is available at www.policysupport.org/waterworld. The model version used here (v2.x) runs simulations on entire continents, 10-degree-square tiles, countries, or major basins at 1-km^2 resolution or on 1-degree-square tiles at 1-ha resolution.

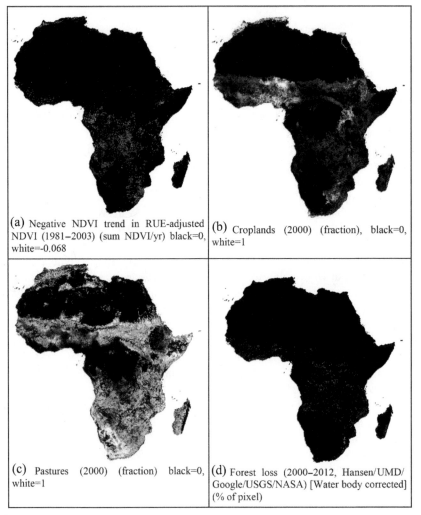

FIGURE 3.1.1

Potential indicators and drivers of current land degradation in Africa.

The analysis of Bai et al. (2008) shows a negative trend in rainfall use efficiency (RUE)–adjusted normalized difference vegetation index (NDVI) between 1981 and 2003 using time series Global Inventory Modeling and Mapping Studies (GIMMS) NDVI data from a range of satellites at 8-km resolution. A negative trend in NDVI may indicate land degradation, but it is not necessarily land degradation. The RUE-adjusted values remove the impact of temporal variation in rainfall, but negative trends may also indicate land use change over the period. Land use changes generating negative trends in NDVI tend to be associated with agriculturalization. The anthropogenic maintenance of lower NDVI is likely to lead to reductions in soil organic matter, and thus land degradation in the long term.

Figure 3.1.1(a) shows the distribution of degradation as indicated by the data set of Bai et al. (2008) alongside the distribution of (b) croplands, (c) pastures, and (d) deforestation. The croplands and pastures maps of Ramankutty et al. (2008) were used. Ramankutty's data combines agricultural inventory data and satellite-derived land cover data to map land use to form the only comprehensive characterization of the distribution of global agriculture in the year 2000. In order to better capture pressures resulting from the cattle density, for some of the statistics, we use livestock densities as well as pasture fractions. Livestock densities are calculated from Robinson et al. (2013). Wildland grazers include cattle, buffalo, goats, and sheep as head-counts per square km. For deforestation, we use the Hansen et al. (2013) data set, which used Landsat imagery from 2000–2012 to characterize annual deforestation. On the basis of these 12 years of 30-m data, we correct for the presence of water bodies and calculate an annual average deforestation rate at 1-km resolution.

Degradation appears highest across parts of central Africa, whereas existing croplands are most extensive in West and East Africa and pastures are most extensive throughout sub-Saharan Africa (outside the Congo Basin). Deforestation is highly localized and most common throughout central Africa in some of the same areas where degradation is observed. Table 3.1.1 indicates that the countries with the greatest intensity of degradation (i.e., mean degradation per unit area) are Gabon, Angola, and the Democratic Republic of Congo (DRC), but that the countries with the greatest total degradation (i.e., the sum of degradation over the national territory) are DRC, Angola, and South Africa. Cropland fraction is not highest in the countries that show the greatest degradation, though one of these, Angola, does appear in the top 10 countries for fractional cover of pastures. Similarly, livestock headcounts are also highest per unit area in countries that are not among those in the top 10 for mean degradation (with the exception of South Africa). However, a number of countries in the top 10 for mean degradation are

Table 3.1.1 Top 10 African Countries for Degradation and Intensity of Associated Land Use Drivers

Rank	Mean Degradation (highest)	Total Degradation (highest)	Cropland Fraction (highest)	Pasture Fraction (highest)	Livestock Headcount (highest)	Deforestation Rate (2000–2012) (highest)
1	Gabon	DRC	Nigeria	South Africa	Tunisia	Côte d'Ivoire
2	Angola	Angola	Uganda	Somalia	Nigeria	Liberia
3	DRC	South Africa	Ghana	Eritrea	Morocco	Benin
4	Congo	Zambia	Benin	Mozambique	Burkina Faso	Sierra Leone
5	Zambia	Tanzania	Morocco	South Sudan	Ethiopia	DRC
6	South Africa	Gabon	Côte d'Ivoire	Namibia	Senegal	Mozambique
7	Sierra Leone	Namibia	Burkina Faso	Botswana	South Sudan	Ghana
8	Tanzania	Congo	Cameroon	Tanzania	Sudan	Tanzania
9	Zimbabwe	Ethiopia	Tunisia	Zimbabwe	Kenya	Zambia
10	Namibia	Zimbabwe	South Africa	Angola	South Africa	Guinea

also in the top 10 for mean deforestation, such as Sierra Leone, DRC, United Republic of Tanzania, and Zambia. This suggests that deforestation is a major driver of degradation, as measured by Bai et al. (2008). This may reflect the influence of land use change on the metric in this study. Though the land use change observed here (2000–2012) is largely outside the period measured by Bai et al. (2008) (1983–2003), there is likely to be some correlation between historic and current deforestation geographies.

While these agroecological factors are potential drivers of current and recent land degradation, examining the parts of Africa at greatest risk of future land degradation requires looking at broader agroclimatic factors, as well as broader human pressures and dependencies on the environment than those associated with agriculture alone.

3.1.2.3 AFRICA AT RISK OF FUTURE DEGRADATION

Figure 3.1.2 shows a series of agroclimatic risk factors mapped at 1-km resolution for Africa. The aridity index[2] (a) is clearly most arid (closer to 0) in Saharan Africa, western South Africa, and the Horn of Africa. Aridity is an important driver for land degradation since it is a primary determinant of the productivity (and thus the carrying capacity) of the land (Dregne, 1985; Geist and Lambin, 2004). Rainfall seasonality is often also considered critical as a risk factor since strong rainfall seasonality can lead to seasonal shortages of water for agricultural and domestic use and significant seasonality in productivity. The Walsh and Lawler (1981) seasonality index is calculated by WaterWorld using the monthly rainfall data from Hijmans et al. (2005) at 1-km resolution and representative of the period 1950–2000. The seasonality index (Figure 3.1.2b) shows a strongly seasonal regime throughout much of Africa, especially south of the Sahara and throughout southern Africa. Dry matter productivity (DMP) calculated from the SPOT-VGT product[3] (Figure 3.1.2c) shows very high values in central Africa, where high rainfall supports lush tropical forests, but particularly low values in Saharan Africa, western South Africa, and the Horn of Africa, in line with the distribution of aridity. Mean crop suitability[4] under rainfed conditions and low inputs (FAO and IIASA, 2012; Mulligan, 2013b) (Figure 3.1.2d) shows similar patterns of productivity. Low suitabilities can be considered a risk factor for land degradation because of the potential for overcropping, while high suitabilities can be considered a driver because of the likelihood of intensive agriculturalization.

Table 3.1.2 shows that when these metrics are averaged by country, the countries with the highest agroclimatic risk factors include Egypt, Western Sahara, Mali, Tunisia, Zambia, and Sudan. Egypt is most arid and most seasonal, with the lowest productivity and lowest crop suitability of all African countries on average, nationally. This reflects the extreme concentration of productivity in Egypt along the Nile and the significant reliance on irrigation for much of the country's remaining landmass under hyperarid conditions.

---

[2]This represents P/PET, where P = precipitation and PET = annual total potential evapotranspiration. P is derived from Hijmans et al. (2005) and PET is calculated by WaterWorld (Mulligan, 2013a)

[3]Mulligan (2009a) developed a global data set for mean DMP. This is calculated from analyses of DMP data made available by VITO (http://www.geoland2.eu/) every 10 days over the 10-year period (1998–2008) from 1-km resolution SPOT-VGT NDVI data. Dry matter productivity represents the daily growth of standing biomass.

[4]This is calculated as the mean crop suitability for all 49 crops modeled in the IIASA GAEZ exercise under rainfed, low-input conditions.

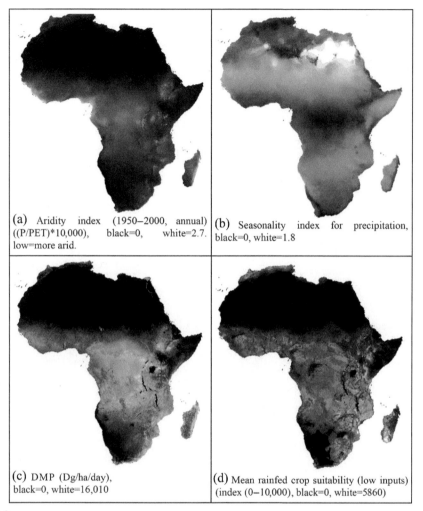

FIGURE 3.1.2

Agroclimatic risk factors for future land degradation in Africa.

As is the case globally, population[5] (Figure 3.1.3a) in Africa is highly clustered around urbanization. We show only population densities of fewer than 1000 persons/km^2 to better highlight rural populations, and these are clearly most extensive in West Africa, the Great Lakes region, and the Nile basin. The distribution of population considered poor [from Elvidge et al. (2009) (Figure 3.1.3b)] is broadly in line with the distribution of population itself. The previously discussed agroclimatic risk factors have to be examined

[5]Population data are taken from LandScanTM Global Population Database, 2007. For Landscan, sub-national level census counts for periods around 2007 were distributed within a 1-km resolution grid based on likelihood coefficients generated from proximity to roads, slope, land cover, nighttime lights, and other data sets. It is the most spatially detailed global population data set available.

Table 3.1.2 Top 10 African Countries for Agroclimatic Desertification Risk Factors

Rank	Aridity Index (most arid)	Rainfall Seasonality (most seasonal)	DMP (lowest)	Crop Suitability (lowest)
1	Egypt	Egypt	Western Sahara	Egypt
2	Zimbabwe	Sudan	Libya	Western Sahara
3	Zambia	Niger	Mauritania	Mauritania
4	Western Sahara	Chad	Niger	Libya
5	Tanzania	Senegal	Algeria	Algeria
6	Uganda	Mali	Egypt	Niger
7	Tunisia	Mauritania	Eritrea	Somalia
8	South Africa	Libya	Sudan	Eritrea
9	Somalia	Zambia	Tunisia	Mali
10	Sierra Leone	Burkina Faso	Mali	Sudan

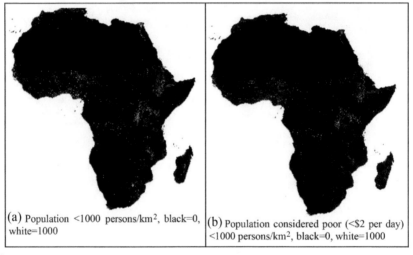

(a) Population <1000 persons/km^2, black=0, white=1000

(b) Population considered poor (<\$2 per day) <1000 persons/km^2, black=0, white=1000

FIGURE 3.1.3

Human risk factors for future land degradation in Africa.

within the context of these population pressures since they determine both pressure on the environment and the local and downstream population demand for an agriculturally productive environment.

3.1.2.4 GLOBAL CHANGE RISK FACTORS

Having examined current land, agroclimatic, and human risk factors for land degradation, we now turn our attention to the distribution of global change risk factors in Africa. We examine both land use change (deforestation) and climate change.

3.1.2.5 DEFORESTATION SCENARIOS

While data sets and tools for understanding recent global deforestation have improved significantly in recent decades (e.g., Hansen et al., 2013), there are few high-resolution, globally applicable deforestation models that can be used to produce future projections. The models that do exist on a global scale are usually of relatively coarse resolution and require complex application within the context of broader global change models, such as Dynamic integrated climate-economy (DICE) (Nordhaus, 1992), Policy analysis of the greenhouse effect (PAGE) (Hope et al., 1993), and Integrated Model to Assess the Greenhouse Effect (IMAGE) (Alcamo et al., 1998). Our deforestation scenarios are produced with the QUICKLUC (v2.0) deforestation model, which forms part of WaterWorld, and its companion tool, Co$tingNature, which is described next.

QUICKLUC (v2.0) projects deforestation forward on the basis of recently measured rates that can be assessed from one of three data sets: global forest change (GFC) (Hansen et al., 2013), Terra-i (Reymondin et al., 2012) and Forest Monitoring for Action (FORMA) alerts (Hammer et al., 2014), or all three combined (see Figure 3.1.4). It is an equilibrium model that projects measured rates forward and then allocates the deforested pixels on the basis of distance-based rules. Recent rates of deforestation are assessed per regional administrative areas according to FAO (2014). The resulting regionally averaged rates can be multiplied by the user to increase or decrease rates across the study area to reflect global economic, population, and market conditions. Rates can be added to the measured base rate in order to "seed" deforestation in regions where it recently has not occurred. If there is reason to believe that the measured deforestation data set is an overestimate, then fractional forest cover losses below a given number can be ignored to remove noise (partial pixel loss). The resulting regionally averaged rates are used to project future deforestation for the number of years specified. For each region, the number of cells to be deforested over the period is calculated from the rate multiplied by the number of years. The allocation of these pixels is calculated on the basis of Euclidean proximity to existing deforestation fronts (according to the chosen historic data set) and accessibility to population centers (according to Uchida and Nelson, 2009).

If the user opts to allocate by agricultural suitability, then allocation is also controlled by normalized mean agricultural suitability (rainfed, low inputs) for 49 crops based on the International Institute for

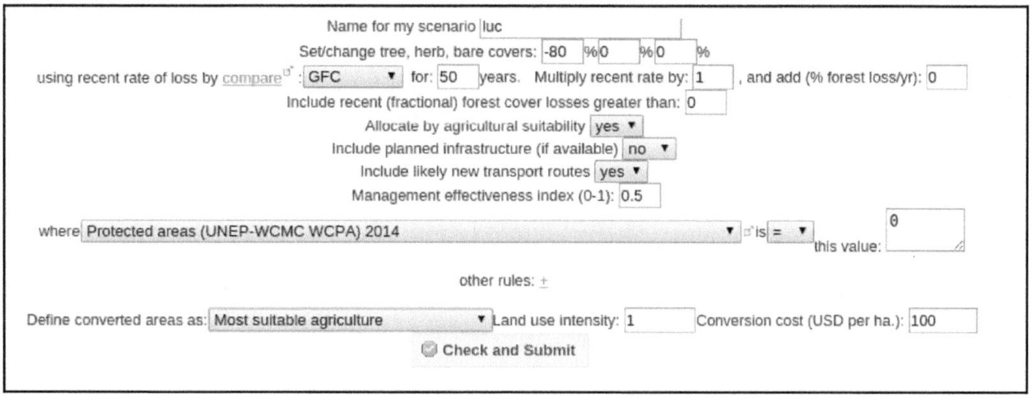

FIGURE 3.1.4

Setup for typical QUICKLUC deforestation scenario.

Applied Systems Analysis (IIASA) Global Agroecological Zones (GAEZ) analysis (Food and Agriculture Organization (FAO), 2012). In this case, the allocation surface combines proximity to recent deforestation, accessibility to population centers, and crop suitability, all normalized within the study area. If the user chooses to also "include planned (transport) infrastructures," which are not yet built and thus are not represented in the accessibility field (and if a data set for their location is available), these roads are added to the deforestation fronts map. The same is true for "likely new transport routes," which, if selected, use a map that connects each urban center from the Schneider et al. (2009) map with linear transport connections and adds these to the deforestation fronts map before proximity is calculated.

The user may also opt to exclude deforestation occurring in certain areas (for example, in protected areas) and set an effectiveness by which this occurs. An effectiveness of 1 excludes all allocated deforestation, while an effectiveness of 0 excludes none of it. Deforestation is further restricted from taking place in areas with tree cover of less than 10% or where recent deforestation has already taken place according to the data set chosen. Despite these constraints, which affect only the allocation within the region but not the rate of change within the region, deforestation rates calculated for each region are maintained. Tree cover in the deforested pixels is reduced according to the user input and replaced by herb and bare cover according to their original ratios in the given pixel. Land use type is changed according to the user specification: to a single use, to the most suitable use, or to the most common use.

The scenario applied here (Figure 3.1.4) removes 80% of tree cover (leaving isolated trees) at rates according to Hansen et al. (2013) for 50 years into the future. Allocation of land is done as a function of agricultural suitability and includes likely new transport routes. Deforestation is excluded from protected areas (with a 50% efficiency), and deforested land is converted to the most suitable of cropland or pasture. The resulting scenario for Africa (Figure 3.1.5a) shows the continuation of deforestation throughout West Africa, Central Africa, and East Africa in particular. Different rules will, of course, produce quite different scenarios. As is the case for greenhouse gas (GHG) emission scenarios, the final products are projections, not predictions.

3.1.2.6 CLIMATE CHANGE SCENARIOS

We use WaterWorld to compare baseline (WorldClim 1950–2000; see Hijmans et al., 2005) climate conditions with a special report on emissions scenarios (SRES) A2a climate scenario downscaled to 1-km resolution using the delta method relative to the WorldClim climate baseline (see Mulligan et al., 2011a). The Special Report on Emissions Scenarios (SRES) (Nakicenovic et al., 2000) A2a scenario represents high growth and a global 3.5 °C warming relative to 1990 by 2100. Data from this scenario were used as monthly, downscaled GCM output (Ramirez and Jarvis, 2010) for temperature and precipitation.

We use an ensemble mean of the 17 available GCMs (see Mulligan et al., 2011a) as a means of reducing the potential bias associated with using a single GCM (given significant uncertainty between GCMs). The resulting scenario shows decreases in rainfall (Figure 3.1.5b) in coastal areas of North Africa, southeast Africa, and southwest Africa, with increases elsewhere, especially in the Sahara. These combine with changes in temperature (Figure 3.1.5c) of warming throughout Africa, but with most warming in the western part of Saharan Africa and throughout southern Africa.

When these metrics are averaged by country, we see different countries in the top 10 for each metric (Table 3.1.3), though Malawi, Uganda, Nigeria, and Sierra Leone head the list for mean population and mean poverty, while the West African countries and Mozambique are highest for deforestation.

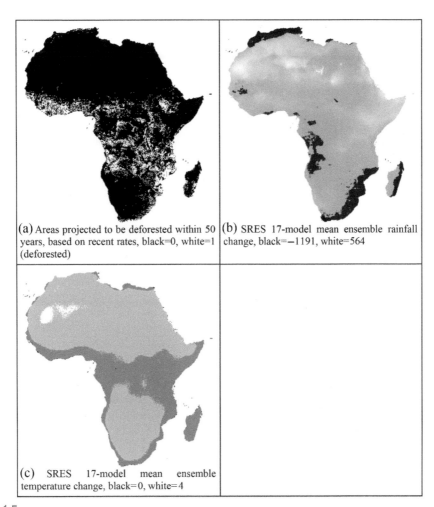

Environmental change risk factors future land degradation in Africa.

Rank	Rural Population <1000 (highest)	Rural Population Considered Poor (highest)	Deforestation Projected (highest)	SRES Precipitation Change (greatest drying)	SRES Temperature Change (highest)
	Table 3.1.3 Top 10 African Countries for Human and Environmental Change Risk Factors				
1	Malawi	Malawi	Côte d'Ivoire	Morocco	Mauritania
2	Uganda	Uganda	Mozambique	Mozambique	Mali
3	Nigeria	Nigeria	Liberia	South Africa	Algeria
4	Sierra Leone	Sierra Leone	Ghana	Madagascar	Niger
5	Ghana	Ethiopia	Zambia	Tunisia	Botswana
6	Ethiopia	Ghana	Guinea	Angola	Burkina Faso
7	Benin	Burkina Faso	Sierra Leone	Congo	Zimbabwe
8	Burkina Faso	Benin	Benin	Senegal	Chad
9	Senegal	United Republic of Tanzania	Angola	Zimbabwe	Sudan
10	Kenya	Senegal	Uganda	Malawi	Morocco

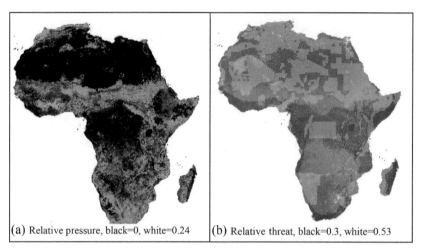

(a) Relative pressure, black=0, white=0.24   (b) Relative threat, black=0.3, white=0.53

FIGURE 3.1.6

Integrated pressure and threat factors.

The greatest drying is projected for Morocco and Mozambique, while the greatest warming is projected for Mauritania and Mali. It is important to recognize that different emission scenarios produce geographically different projections, as do different combinations of GCMs, so while these projections are useful in the context of examining potential risk, they are not predictions of the climatic future of these countries.

3.1.2.7 INTEGRATING PRESSURES AND THREATS

Finally we integrate a range of these risks into two single indices: current pressure and future threat, as calculated by WaterWorld's companion tool, Co$ting Nature (www.policysupport.org/costingnature). Pressure is an index of a range of current human pressures on the land and water and combines normalized mapped values for population, wildfire frequency, grazing intensity, agricultural intensity, dam density, and infrastructure (dams, mines, oil and gas, urban) density, all equally weighted. Future threat is an index of the potential for pressure to increase in the future and combines normalized values for accessibility, proximity to recent deforestation Moderate Resolution Imaging Spectroradiometer (MODIS), projected change in population and gross domestic product (GDP), projected climate change, current distribution of nighttime lights, mining, and oil and gas concessions. The resulting maps indicate the highest pressure (Figure 3.1.6a) is throughout sub-Saharan Africa and East and West Africa. The highest threat tends to be outside the areas of highest pressure and into the currently less affected Sahara and Congo areas.

3.1.2.8 HUMAN DEPENDENCY

In addition to understanding where agroclimatic risks and human pressures on the environment are greatest, we must also understand where human dependency on the environment is greatest to understand the impacts of potential land degradation. While local dependency is at least partly captured by the population and poor population metrics described previously, there are also significant

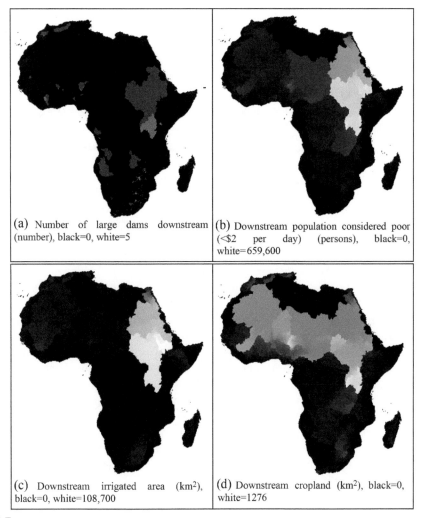

(a) Number of large dams downstream (number), black=0, white=5

(b) Downstream population considered poor (<$2 per day) (persons), black=0, white=659,600

(c) Downstream irrigated area (km²), black=0, white=108,700

(d) Downstream cropland (km²), black=0, white=1276

FIGURE 3.1.7

Downstream (hydrologic) dependencies.

"downstream" hydrological and supply chain dependencies that create teleconnections[6] between environments and populations in different geographical locations. Figure 3.1.7 describes some of these dependencies. For instance, Figure 3.1.7(a) describes, for each pixel, the number of large dams downstream, based on Mulligan et al. (2011b) and indicates areas where enhanced soil erosion and transportation will have impacts on reservoirs or hydroelectric-power (HEP) turbines downstream.

---

[6]A teleconnection is a relationship between a variable or variables over large distances (typically thousands of kilometers).

Table 3.1.4 Top 10 African Countries for Aggregate Pressure, Threat, and Downstream Dependence

Rank	Current Pressure	Future Threat	Dams Downstream	Poor Population Downstream	Irrigated Area Downstream	Cropland Downstream
1	South Africa	Mali	Uganda	Uganda	Uganda	Uganda
2	Ghana	Namibia	South Sudan	South Sudan	South Sudan	South Sudan
3	South Sudan	Tunisia	Sudan	Sudan	Sudan	Niger
4	Nigeria	Nigeria	Tanzania	Egypt	Egypt	Sudan
5	Mozambique	Sudan	Ethiopia	Ethiopia	Ethiopia	Mali
6	Somalia	Ethiopia	Kenya	Tanzania	Tanzania	Chad
7	Morocco	Eritrea	Côte d'Ivoire	Chad	Eritrea	Nigeria
8	Eritrea	South Sudan	Zambia	Eritrea	Kenya	Algeria
9	Côte d'Ivoire	Mauritania	Angola	Niger	Guinea	Mauritania
10	Zimbabwe	Zambia	Eritrea	Kenya	Mali	Burkina Faso

Figure 3.1.7(b) shows, for each pixel, the downstream poor population and indicates the number of people that may thus benefit from ecosystem services provided by that pixel (or be affected by its desertification). Figure 3.1.7c describes downstream irrigated areas, and like (a) and (b), the upper Nile shows up very strongly as one of only a few areas in Africa that supplies significant downstream irrigated areas. Finally Figure 3.1.7(d) shows the areas of downstream cropland and highlights the dependence of western and eastern African croplands on water from the Nile, Niger, and Chad basins (given their size and downstream agriculture).

Table 3.1.4 shows the mean value for these indices by country and indicates area-average current pressure to be greatest in South Africa and future threat greatest in Mali. Uganda, South Sudan, and Sudan top the list for all the downstream metrics (given their location at the head of the Nile). Avoiding land degradation and restoring desertified lands in these countries are important for the countries themselves, but also for all the populations, agricultures, and dams downstream of them.

Downstream dependency, however, is not just hydrological. Since all countries in Africa trade agricultural commodities with other countries in Africa and throughout the world, there are also supply-chain teleconnections between African countries and those elsewhere that are downstream in the supply chain.

3.1.2.9 SUPPLY CHAIN TELECONNECTIONS (ALL COMMODITIES)

The WaterWorld supply chain module was used to analyze countries benefiting from agricultural production in Africa.[7] The WaterWorld supply chain module uses the UN Comtrade database (United Nations Commodity Trade Statistics Database, 2013) of export values (in US dollars) for the years

---

[7]COMTRADE data were not available for the following countries: Angola, Benin, Botswana, Cameroon, Chad, Comoros, Democratic Republic of the Congo, Equatorial Guinea, Eritrea, Gabon, Guinea-Bissau, Guinea, Lesotho, Liberia, Libya, Mali, Morocco, Mozambique, Republic of Congo, Sao Tome and Principe, Seychelles, Sierra Leone, Somalia, Swaziland, and Western Sahara.

2007–2011 to map the key flows of agricultural products from the study area as a means of highlighting international dependencies on agriculture in these areas. COMTRADE uses the Global Trade Analysis Project (GTAP) commodity classes.[8]

The greatest African agricultural exports for the period 2007–2011 were for sugarcane and beet ($14 billion), bovine cattle, sheep and goats, horses ($7.3 billion), Food products n.e.c.[9] $500 million), wool, silkworm cocoons ($120 million), vegetables, fruit, nuts ($80 million), leather products ($64 million), oilseeds ($53 million), bovine meat products ($46 million), sugar ($26 million), dairy products ($22 million), processed rice ($22 million), cereal grains n.e.c. ($20 million), animal products n.e.c. ($18 million), meat products n.e.c. ($18 million), crops n.e.c. ($16 million), plant-based fibers ($12 million), vegetable oils and fats ($10 million), wheat ($3.7 million), and paddy rice ($3.5 million).

Figure 3.1.8(a) shows the export volumes and destinations of food products n.e.c for Africa. Some of the trade is within Africa, with destinations for African food products to most African countries, but there are also significant markets in Europe, the United States, and South Asia. A similar pattern emerges for crops n.e.c (Figure 3.1.8b). Desertification in Africa thus has considerable potential to affect food markets within other countries in Africa and around the world.

Having exemplified the role of geospatial policy support tools in identifying risks, dependencies, and potential impacts of land degradation at the continental scale in Africa, we now move to using the same tools to better understand the likely impact of restoration scenarios at the national scale.

3.1.3 RESTORING AFRICA

So, how might we restore the parts of Africa already degraded and prevent further degradation in those areas identified as facing agroclimatic and socioeconomic risk? The analysis given up to now in this section indicates that two key components of restoration could involve afforestation/reforestation and (sustainable) agricultural intensification, since both deforestation and intensive agriculture have an influence on the potential for, and impact of, land degradation. Afforestation may protect against soil erosion and degradation, though at some evaporative loss cost compared with grazing land and cropland, and also at some opportunity cost to the avoided agriculture on reforested lands. Sustainable agricultural intensification through increased ecoefficiency (reducing inputs while maintaining productivity) may reduce the water quality impacts of agriculture while sustaining production, at minimum cost in fertilizer and associated energy. Here, we use Gabon as a case study for analyses of such scenarios at the national scale since Gabon has the highest national average rate of land degradation in recent decades, according to Bai et al. (2008).

3.1.3.1 THE DESERTIFICATION BASELINE FOR GABON

Figure 3.1.9 shows the Co$ting Nature pressure and threat maps for Gabon, normalized at the national scale. They indicate a country under relatively little human pressure over much of its territory, but concentrated pressures in the southeast, south, and parts of the northwest. Future threat is more evenly

---

[8]https://www.gtap.agecon.purdue.edu/databases/contribute/detailedsector.asp.
[9]Not elsewhere classified—i.e., not incorporated into other commodity classes.

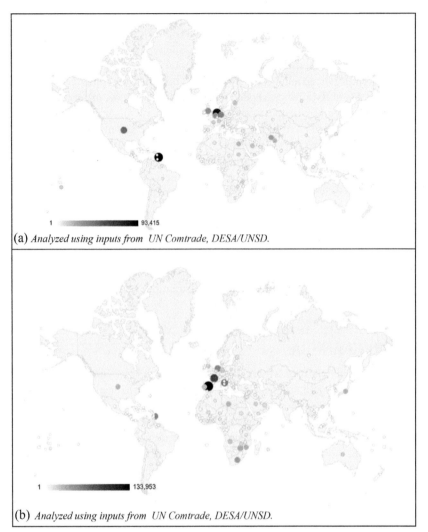

FIGURE 3.1.8

Export volumes from Africa for (a) food products n.e.c. and (b) crops n.e.c., by destination country (in thousands of US dollars per year).

distributed across the country, with particularly high values toward the west. The lack of pressure is surprising given the apparent level of land degradation, according to Bai et al. (2008).

In this context, we now run a baseline analysis with WaterWorld. The baseline represents the mean hydrological situation from 1950–2000. Figure 3.1.10(a) and (b) show water balance by pixel and region, respectively. They indicate a generally positive water balance, with significant spatial variability

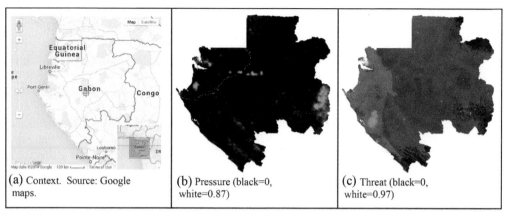

(a) Context. Source: Google maps.

(b) Pressure (black=0, white=0.87)

(c) Threat (black=0, white=0.97)

FIGURE 3.1.9

Pressure and threat for Gabon, locally indexed.

at the pixel scale and between regions, with the driest areas on the southern coast and farthest east. Seasonality, according to the Walsh and Lawler (1981) index, is also highest at the coast (Figure 3.1.10c,d). The highest DMP (Figure 3.1.10e,f) does not coincide with the highest water balance but occurs farther inland where water balance is lower but seasonality of water balance is also less. A closer look at the Bai et al. (2008) index for Gabon (Figure 3.1.10g) indicates potential degradation throughout the country (30% of the territory), except the coastal zone. Since the (Hansen et al., 2013) data indicate that most of this area is under forest rather than agriculture, the Bai et al. (2008) figure seems an overestimate.

In order to examine the potential of reforestation for restoring degraded lands and recovering ecosystem services for local and downstream beneficiaries, we implement an afforestation scenario in WaterWorld. For this, we use the WaterWorld land use change scenario tools. Our scenario (Figure 3.1.11) adds tree cover to all pixels where current tree cover is less than 60% and local slope is greater than 3 degrees. The scenario is thus targeted on steeper, deforested, or sparsely vegetated lands. Forested areas are converted to natural cover, and thus any previous agricultural land use is replaced. As a result of this intervention, tree cover in Gabon increases by 2% and 1,595,600 ha are converted, but croplands decrease by around 30,000 ha and pastures decrease by around 60,000 ha. Wildland grazer headcount falls by 4000. If active reforestation were costed at, say, $100 per hectare, this intervention would cost $159.56 million.

Afforestation leads to increases in evapotranspiration throughout the afforested area, but also leads to increased fog interception (*sensu* Bruijnzeel et al., 2011) in the more mountainous areas. As a result, afforested areas show decreases in water balance in the lowlands and increases in the uplands (Figure 3.1.12b). Since we replace agricultural land with natural cover, the human footprint on water quality (a measure of the potential for agroindustrial contamination of water) decreases throughout the afforested areas and their downstream rivers (Figure 3.1.12c). Finally, the increased forest cover also decreases soil erosion in all regions where significant afforestation occurs (Figure 3.1.12d). However, for water balance, erosion, and water quality, this intervention leads to no change over 90%–95% of the national territory since the intervention does not cover and is not upstream of those areas.

In our second scenario, we implement an ecoefficient agricultural intervention (Figure 3.1.13) in which we convert a randomized clustered 50% of agricultural land on slopes greater than 2 degrees to have a reduction of input of 50%, changing the "land use intensity" from 1.0 to 0.5. While this has an associated technological cost, it also reduces expensive agricultural inputs, so we estimate the net cost as $50/ha.

This scenario only affects the human footprint (water quality) index and changes land management over some 364,800 ha (which would cost some $18.2 million @ $50/ha) (Figure 3.1.14). Water quality improves as a result, but by a very small value (−0.0.047) at the national scale and is not affected at all

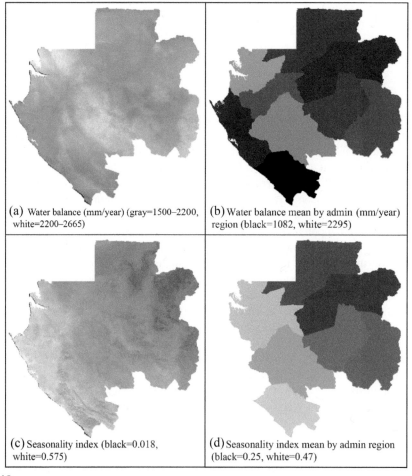

(a) Water balance (mm/year) (gray=1500–2200, white=2200–2665)

(b) Water balance mean by admin (mm/year) region (black=1082, white=2295)

(c) Seasonality index (black=0.018, white=0.575)

(d) Seasonality index mean by admin region (black=0.25, white=0.47)

FIGURE 3.1.10

The baseline land degradation context for Gabon.

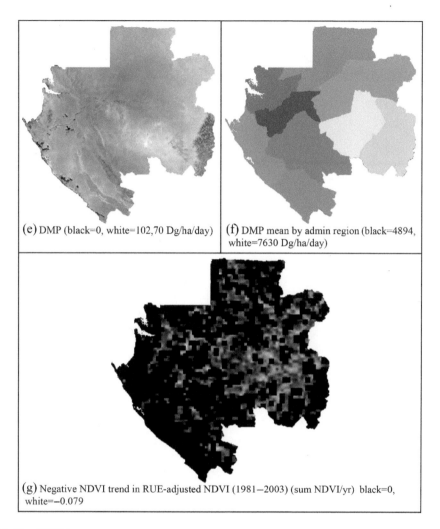

(e) DMP (black=0, white=102,70 Dg/ha/day)

(f) DMP mean by admin region (black=4894, white=7630 Dg/ha/day)

(g) Negative NDVI trend in RUE-adjusted NDVI (1981−2003) (sum NDVI/yr) black=0, white=−0.079

FIGURE 3.1.10, CONT'D

Name for my scenario aff

Set/change tree, herb, bare covers: +100 %0 %0 % for approx: 100   per-cent ▼ of land, cluster, scale: 0.3

where Slope gradient (degrees)   ▼ σ is >= ▼   this value: 3

other rules: -

... and ▼ where Cover of tree-covered ground (MODIS 2010)   ▼ σ is < ▼   this value: 60

Try it. ᵘ

Define converted areas as: All Natural   ▼ Land use intensity: 1   Conversion cost (USD per ha.): 100

◎ Check and Submit

FIGURE 3.1.11

An afforestation scenario.

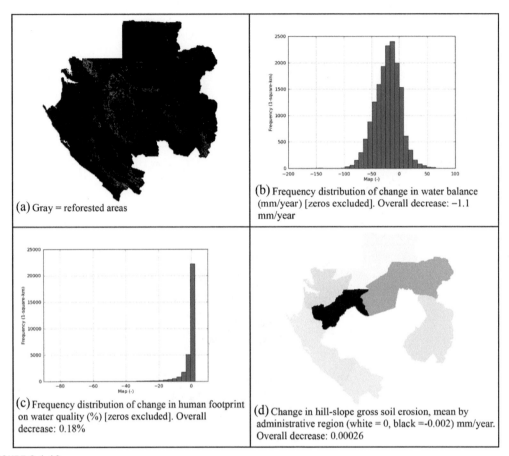

(a) Gray = reforested areas

(b) Frequency distribution of change in water balance (mm/year) [zeros excluded]. Overall decrease: −1.1 mm/year

(c) Frequency distribution of change in human footprint on water quality (%) [zeros excluded]. Overall decrease: 0.18%

(d) Change in hill-slope gross soil erosion, mean by administrative region (white = 0, black =-0.002) mm/year. Overall decrease: 0.00026

FIGURE 3.1.12

Impacts of an afforestation scenario.

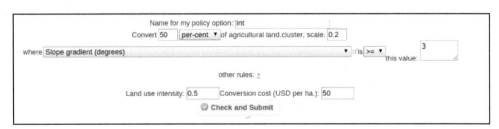

FIGURE 3.1.13

A sustainable agricultural intensification scenario.

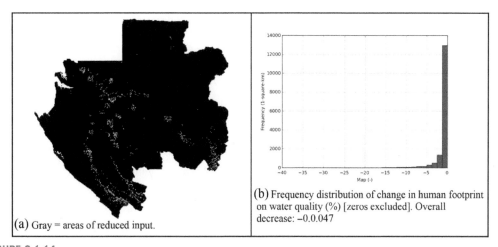

(a) Gray = areas of reduced input.

(b) Frequency distribution of change in human footprint on water quality (%) [zeros excluded]. Overall decrease: −0.0.047

FIGURE 3.1.14

Outcomes of the ecoefficient agriculture scenario.

over some 95% of the national territory, decreasing pollution only over the remaining 5%. This modest success would come at a considerable economic cost, as well as some opportunity cost.

3.1.4 CONCLUSIONS

There is no silver bullet for mapping and monitoring desertification at regional to global scales. A number of useful remote sensing analyses exist, but within these, it is difficult to distinguish degradation from land use and climate variability. We show some evidence at the continental scale for association between deforestation and remotely sensed land degradation metrics. However, it seems more likely that the metrics are affected directly by deforestation rather than by actual degradation subsequent to deforestation. Combining remotely sensed land degradation assessment, spatial mapping of land use, as well as mapping of climatic and human risk factors alongside human dependencies on local and upstream water and land, can help to understand key risks.

However, managing those risks and understanding the kind of restoration activities that will have positive impacts required much more than satellite monitoring and static spatial risk analysis. Dynamic policy support systems allow the assessment of baselines for risk, local, downstream, and supply-chain dependency at scales from the national to the local. Understanding the environmental baseline helps one to understand the likely risk areas and factors for land degradation, but also to design policies that may restore landscape function and ecosystem services.

Using some example interventions, we showed in this section that even large interventions have little impact at the national scale, and recovering even 5% of landscapes in a typical country is likely to cost hundreds of millions of dollars. Thus, interventions have to be very cheap in terms of implementation costs and opportunity costs in order to make a worthwhile difference at the very large scale that land degradation is occurring. Every hectare of land farmed provides an economic and livelihood return to farmers and their families. It is, therefore, not surprising that farming shapes entire landscapes.

If restoration activities are to compete for landscapes with farming, the economic and livelihood returns to restoration have to be clear, significant, and accessible to farmers. Benefit sharing mechanisms and markets for ecosystem services that promote restoration have their work cut out for them to compete with the relentless conversion of land to agriculture and its progressive degradation.

REFERENCES

Alcamo, J., Leemans, R., Kreileman, E., 1998. Global change scenarios of the 21st century. Results from the IMAGE 2.1 Model, Elsevier, Amsterdam. Available at http://www.fao.org/nr/gaez/en/.

Bai, Z.G., Dent, D.L., Olsson, L., Schaepman, M.E., 2008. Global assessment of land degradation and improvement 1. identification by remote sensing. GLADA report 5. Version 5.

Bruijnzeel, L.A., Mulligan, M., Scatena, F.N., 2011. Hydrometeorology of tropical montane cloud forests: Emerging patterns. Hydro. Proc. 25 (3), 465–498.

Dregne, H.E., 1985. Guarded optimism for the future of arid lands: Aridity and land degradation. Environ.: Sci. Policy Sustain Devel 27 (8), 16–33.

Elvidge, C.D., Sutton, P.C., Ghosh, T., Tuttle, B.T., Baugh, K.E., Bhaduri, B., Bright, E., 2009. A global poverty map derived from satellite data. Comp. Geosci. 35, 1652–1660.

Food and Agriculture Organization (FAO), 2014. Global Administrative Unit Layers (GAUL). Digital dataset, available online at http://www.fao.org/geonetwork/srv/en/metadata.show?id=12691.

Food and Agriculture Organization (FAO) and International Institute for Applied Systems Analysis (IIASA), 2012. Agro-ecological suitability and productivity. Crop suitability index. Digital dataset, available online at http://www.fao.org/nr/gaez/about-data-portal/agricultural-suitability-and-potential-yields/en/.

Geist, H.J., Lambin, E.F., 2004. Dynamic causal patterns of desertification. Biosci. 54 (9), 817–829.

Hammer, D., Kraft, R., Wheeler, D., 2014. Alerts of forest disturbance from MODIS imagery. Intl. J. Appl. Earth Obs. Geoinform. 33, 1–9.

Hansen, M.C., et al., 2013. High-resolution global maps of 21st-century forest cover change. Science 342 (November 15), 850–853.

Helldén, U., Tottrup, C., 2008. Regional desertification: A global synthesis. Global Planet. Change 64 (3), 169–176.

Hijmans, R.J., Cameron, S.E., Parra, J.L., Jones, P.G., Jarvis, A., 2005. Very-high-resolution interpolated climate surfaces for global land areas. Intl. J. Clim. 25, 1965–1978.

Hope, C., Anderson, J., Wenman, P., 1993. Policy analysis of the greenhouse effect. Energy Pol. 23, 327–338.

LandScan™ Global Population Database, 2007. Oak Ridge, TN: Oak Ridge National Laboratory. Available at http://www.ornl.gov/landscan/.

Mulligan, M., 2009. Integrated environmental modelling to characterise processes of land degradation and desertification for policy support. In: Hill, J., Roeder, A. (Eds.), Remote Sensing and Geoinformation Processing in the Assessment and Monitoring of Land Degradation and Desertification. Taylor and Francis.

Mulligan, M., 2009a. Global mean dry matter productivity based on SPOT-VGT (1998–2008). http://www.ambiotek.com/dmp.

Mulligan, M., 2009b. The human water quality footprint: Agricultural, industrial, and urban impacts on the quality of available water globally and in the Andean region. In: Proc. Intl. Conf. Integ. Water Res. Manage. Climate Change, Cali, Colombia. Available online here http://www.ambiotek.com/publications/CINARA_Industry_and_mining.pdf.

Mulligan, M., 2013a. WaterWorld: A self-parameterising, physically-based model for application in data-poor but problem-rich environments globally. Hydrol. Res. http://dx.doi.org/10.2166/nh.2012.217.

Mulligan, M., 2013a. SimTerra: A consistent global gridded database of environmental properties for spatial modelling. http://www.policysupport.org/simterra [based on FAO (2012) Agro-ecological suitability and productivity. Crop suitability index (class).].

Mulligan, M., et al., 2011a. The nature and impact of climate change in the challenge program on water and food (CPWF) basins. Water Intl. 36 (1), 96–124. http://dx.doi.org/10.1080/02508060.2011.543408.

Mulligan, M., Saenz, L. van Soesbergen, A., 2011b. Development and validation of a georeferenced global database of dams. Digital dataset available online at http://www.ambiotek.com/dams.

Nakicenovic, N., et al., (Eds.), 2000. Special Report on Emissions Scenarios. Cambridge University Press, Cambridge, UK.

Nordhaus, W., 1992. An optimal transition path for controlling greenhouse gases. Science 258, 1315–1319.

Ramankutty, et al., 2008. Farming the planet: 1. Geographic distribution of global agricultural lands in the year 2000. Global Biogeochem. Cycles 22. http://dx.doi.org/10.1029/2007GB002952. GB1003.

Ramirez, J., Jarvis, A., 2010. Disaggregation of Global Circulation Model outputs. Decision and policy analysis working paper no. 2. Available at: http://climate.org/media/ccafs_climate/docs/Disaggregation-WP-02.pdf.

Reymondin, L., et al., 2012. Terra-i: A methodology for near real-time monitoring of habitat change at continental scales using MODIS-NDVI and TRMM. CIAT-Terra-i. http://terra-i.org/dms/docs/reports/Terra-i-Method/Terra-i%20Method.pdf.

Robinson, T.P., et al., 2013. Mapping the global distribution of livestock. PLoS One. 9 (5). http://dx.doi.org/10.1371/journal.pone.0096084. e96084.

Schneider, A., Friedl, M.A., Potere, D., 2009. A new map of global urban extent from MODIS data. Environ. Res. Lett. 4. article 044003.

Sommer, S., et al., 2011. Application of indicator systems for monitoring and assessment of desertification from national to global scales. Land Deg. Develop. 22 (2), 184–197.

Spinoni, J., Vogt, J., Naumann, G., Carrao, H., Barbosa, P., 2014. Towards identifying areas at climatological risk of desertification using the Köppen–Geiger classification and FAO aridity index. Intl. J. Clim. 35 (9), 2210–2222.

Uchida, H., Nelson, A., 2009. Agglomeration index: Towards a new measure of urban concentration. Background paper for the World Bank's World Development Report. Available online at http://siteresources.worldbank.org/INTWDR2009/Resources/4231006-1204741572978/Hiro1.pdf.

United Nations Commodity Trade Statistics Database, 2013. Department of Economic and Social Affairs/Statistics Division, 2013. COMTRADE database. http://comtrade.un.org/db/.

Walsh, P.D., Lawler, D.M., 1981. Rainfall seasonality: Description, spatial patterns, and change through time. Weather 36, 201–208.

THE VALUE OF LAND RESTORATION AS A RESPONSE TO CLIMATE CHANGE

3.2

Guy Lomax

The Nature Conservancy, Oxford, UK

3.2.1 ECOSYSTEMS AND CLIMATE CHANGE

The biosphere plays a fundamental role in Earth's carbon cycle. Earth's terrestrial ecosystems and soils together contain between 2 and 3 trillion tons of organic carbon (2000–3000 GtC; IPCC, 2013). This is 2.5–3.5 times the amount of carbon present as CO_2 in the atmosphere and substantially more than that present in known fossil fuel reserves (IPCC, 2013). More than 100 billion tons of atmospheric carbon is captured by photosynthesis and cycled through these ecosystems every year, and they currently remove and permanently sequester more than a quarter of the 10 billion tons of excess CO_2 entering the atmosphere each year as a result of human activity (IPCC, 2013; Le Quéré et al., 2009).

Human activities have historically released hundreds of billions of tons of this carbon into the atmosphere, but net emissions from deforestation and other land use changes have fallen in recent years to about 900 million tons per year in 2011, only 9% of all humanity's CO_2 emissions (IPCC, 2013). This apparently small contribution may give the impression that land has only marginal importance for responding to the threat of climate change relative to cutting emissions from fossil fuels.

However, this review of current research reveals that land, ecosystems, and agriculture are central to many different facets of humanity's response to climate change that go far beyond its contribution to current emissions, and that land, therefore, merits a much more central place in global efforts to address climate change than it has at present. Furthermore, the climate case for land restoration is intimately linked to wider land goals such as food security, economic and development opportunities, sustainability of water resources, and preserving biodiversity. The emerging climate case is, therefore, also an opportunity to drive forward action on land restoration for environmentally and socially sustainable development.

While land use change is currently a net source of CO_2, land differs from fossil fuels in that some of the carbon released historically can be restored through known management practices, and such restoration often brings economic and environmental benefits (Victoria et al., 2012; Lal, 2004; Smith et al., 2008). When this restoration potential is also considered, land becomes an opportunity to tackle rising CO_2 levels that is both globally significant and economically attractive. Sections 2–4 explore this opportunity, as well as the wider benefits of restoring carbon to soils, in more detail.

Land Restoration. http://dx.doi.org/10.1016/B978-0-12-801231-4.00019-7

And land is equally fundamental to efforts to adapt to climate change, which is the subject of section 5. Land directly mediates many of the anticipated impacts of climate change on human welfare, and maintaining resilient agricultural systems and natural ecosystems is central to avoiding the most severe consequences of the climate change that is already inevitable (IPCC, 2014a; Colls et al., 2009).

Finally, land also provides humankind with more than 98% of its food supply (FAO, 2013) and must also meet rising demands for clean energy resources through the development of bioenergy over the century. Such demands are likely to place increasing pressures on existing agricultural systems and natural ecosystems and risk driving further degradation and land conversion (Smith et al., 2013). Section 6 emphasizes the importance of restoring degraded land to productivity, sustainable intensification, and multifunctional agriculture to meet this demand without further clearance of natural systems (Smith et al., 2013; FAO, 2009).

Restorative management of land and agriculture has a uniquely important place in protecting humanity and the natural world from the impacts of climate change. Reconciling this need with the ever-increasing demands that we are placing on production from land will certainly not be easy, but emerging paradigms for land management offer many opportunities to do so effectively. Section 7 concludes by briefly reviewing these needs and what this new focus on land as a climate response might entail for wider land restoration efforts.

3.2.2 RESTORING TERRESTRIAL CARBON STOCKS

Over the history of human civilization, however, deforestation, agricultural practices, and land degradation have driven the release of a substantial fraction of the carbon once in this system now into the atmosphere. Since 1750, an estimated 180 billion tons of stored carbon have been released through deforestation and soil carbon losses, compared to 375 billion tons from fossil fuel burning and other industry (IPCC, 2013). Looking further back, another 50–300 billion tons may have been released through land clearance since the dawn of agriculture around 8000 years ago (Ruddiman, 2013). Of this, on the order of 50-100 billion tons have been released through degradation of the soils themselves (Lal, 2004), with the rest mostly from deforestation.

This loss of carbon from ecosystems and soils has been driven by a diverse set of practices, including land clearance, burning, overgrazing of pastures and grasslands, crop cultivation practices encouraging carbon loss, and drainage of wetlands for agriculture (IPCC, 2013; Lal, 2004; Smith et al., 2007, 2008; Parish et al., 2008; FAO, 2000). Fortunately, there are also a wide range of known practices and land management approaches that can restore a large fraction of this carbon lost from soils and vegetation and preserve stocks that are still present (Smith et al., 2007).

In general terms, increased carbon sequestration will result from those practices that stimulate establishment and growth of perennial plants (including trees), increase overall productivity, raise organic matter inputs to the soil, stimulate a healthy soil ecosystem, and reduce losses of carbon from soils and biomass (Victoria et al., 2012; Smith et al., 2008; Eagle et al., 2012). Some changes in management bring added benefits to greenhouse gas (GHG) mitigation by reducing the use of fuels or fertilizers that entail emissions of CO_2 and the powerful GHG nitrous oxide (Lal, 2004; Smith et al., 2008; Eagle et al., 2012). Examples of specific approaches to restore carbon to degraded lands include the following (Victoria et al., 2012; Lal, 2004; Smith et al., 2008; Eagle et al., 2012; Branca et al., 2013; Conant, 2010; Schoeneberger et al., 2012; Teague et al., 2011):

- Adding manures, composts, and other organic mulches to the land surface
- Increasing year-round plant cover on cropland with cover crops
- Diversifying crops or rotations with nitrogen-fixing species, improving fertility
- Improved water management and erosion control
- Changing livestock management to eliminate overgrazing, stimulate grass growth, and restore ecosystem function
- Integrating livestock into cropping land or rotating cropland with perennial pasture
- Planting perennial crops, grasses, or trees instead of or in addition to annual crops
- Revegetation, adding organic material, and facilitating natural regeneration of extremely degraded soils
- Restoring wetlands that have been drained or burned for agriculture

The most appropriate mix of approaches for a particular context in the real world depends on geographical and environmental factors, the agricultural context, and the needs of the land user. The amount of carbon that a given practice might restore similarly varies widely with the soil type, the climate, the current state of the land and soil, the history of land use, and other local considerations, as well as the details of how a practice is implemented (Lal, 2004; Branca et al., 2013). Our understanding of the ecological processes governing such variations is still limited and is an active area of soil science research (Eagle et al., 2012; Stockmann et al., 2013), but we can draw a few general conclusions about what these methods can achieve.

Many changes in practices (e.g., introducing cover crops or livestock management) can restore soil carbon without entailing any major changes to the agricultural system or loss of production. In some cases, they may even reduce the labor requirements and costs of managing a particular piece of land (Victoria et al., 2012). These moderate changes in practices tend broadly to stimulate between 0.1 and 1 ton of net carbon drawdown from the atmosphere per hectare of land per year (0.1–1.0 tC/ha/year) (Smith et al., 2008; Eagle et al., 2012; Branca et al., 2013; Conant et al., 2001; Lal et al., 2004). Greater changes in land use, such as conversion of annual cropland to perennial crops, pasture, or forestry, can yield higher rates of sequestration of 1–5 tC/ha/year and can more rapidly restore carbon and soil health to land that has been degraded by intensive practices like arable agriculture (Smith et al., 2008; Eagle et al., 2012; Nabuurs et al., 2007). Where these land use changes lead to loss of production, however, these may reduce a farmer's income if compensation is not available (Eagle et al., 2012).

Where land degradation is most severe (to the point where fertility and productivity for land users is greatly reduced), this is invariably associated with great losses of soil carbon of around 30–50 tons per hectare (Lal, 2004; FAO, 2000). Restoring such degraded lands is, therefore, a crucial opportunity, allowing sequestration rates of several tons of carbon per hectare per year in the course of increasing ecosystem services and productivity (Smith et al., 2008; Lal, 2009). For context, the annual emissions of an average European Union (EU) citizen were 2.2 tons of carbon per year (tC/year) in 2009, including the footprint of imported products (Eurostat, 2013).

While our understanding of the relationships between land management and carbon is far from complete, it is clear that there are many possible approaches to resequester a significant amount of carbon into soils and ecosystems.

3.2.3 THE RESTORATION OPPORTUNITY IN CONTEXT

How might this local sequestration potential translate into a global response to rising CO_2 emissions? While the achievable scale is an extremely difficult question to answer quantitatively, there have been many efforts to place order-of-magnitude bounds on the global biophysical potential to sequester carbon through restoring lands and soils. The most comprehensive estimates have appeared in the work of the Intergovernmental Panel on Climate Change (IPCC), which has attempted to scale up average per-hectare figures from a wide review of different practices across available land areas, breaking down potentials by land type and broad climate zone (Smith et al., 2007; 2008; 2014).

Excluding forest planting and management, they identified a total global potential of around 1.5 billion tons of carbon per year (GtC/year) abatement across the earth's 1.5 billion ha of cropland, 3.5 billion ha of grazing land, hundreds of millions of hectares of severely degraded and desertified land and tens of millions of hectares of cultivated wetlands, the majority of which represented sequestration.

Of this, about 1 GtC/year was deemed "economic potential" by 2030, costing less than $100/tCO_2 abated to implement, and about one-quarter of this was possible through restoring the most degraded lands. Around the same level of mitigation may be viable through reforestation, forest management, and reducing deforestation (Nabuurs et al., 2007).

Let us put this in a global context. While the particular figures are uncertain, these estimates show that between them, better land and forest management have the potential to mitigate anthropogenic emissions on the order of 2 GtC/year—one-fifth of all humanity's emissions and comparable to emissions from electricity generation worldwide (IPCC, 2014).

It is even possible that this is an underestimate. The nonforestry figures, for example, are based on scaling up the average sequestration rates observed across all studies the authors reviewed, typically for single practices (Smith et al., 2007). It is likely that applying suites of several beneficial practices on the same land could bring greater benefits than apparent from such single-practice studies (Eagle et al., 2012; Branca et al., 2013). Additionally, these average figures include those results where practices were least effective for a given context and led to little or no increased sequestration. An improved understanding of the ecological and soil processes acting on soil organic matter may allow us to better tailor such approaches to maximize sequestration (Stockmann et al., 2013; Gougoulias et al., 2014; Teague et al., 2013).

The economic costs also make land carbon sequestration attractive in the context of global efforts to tackle climate change. The costs stated are more than $100/tCO_2 sequestered or avoided by 2030, with at least one-third of the total scale available at below $20/tCO_2. This compares favorably to the IPCC's estimated net costs per ton of CO_2 avoided for several electricity sector mitigation technologies in 2030: $0–50/tCO_2 for wind energy, $20–100/tCO_2 for carbon capture and storage technology, and more than $100/tCO_2 for solar generation (IPCC, 2007). As land management is dominated by sequestration of CO_2, it is notable that these costs are also well below many estimates for emerging technologies aiming to capture and sequester CO_2 from the atmosphere itself, such as direct air capture, enhanced mineral weathering, bioenergy with carbon capture and storage (Bio-CCS), or ocean liming (McLaren, 2012). And where some such proposals have raised fears of unintended consequences (e.g., the effects of mineral weathering on water chemistry), land restoration is associated with proven cobenefits.

The economic costs also hide a key fact that is clear from many other contributions to this volume: many practices that sequester carbon within existing agricultural systems, especially those that improve degraded land, can bring benefits well beyond CO_2 abatement. As we will see in section 4, soil carbon is itself central to healthy soil function, and in the right circumstances, such restoration can pay for itself through improved productivity, reduced costs of inputs, and protection of farmers' livelihoods from risk (Lal, 2004; Branca et al., 2013). If these benefits can be captured and internalized, a substantial part of this global potential may be economically attractive for land users, even without a carbon price.

Restoring degraded ecosystems and soils is, therefore, a global opportunity for carbon sequestration that is of comparable importance to mitigating climate change as major branches of clean energy. Once the potential economic benefits of carbon sequestering practices in degraded soils are considered, the opportunity is even more attractive.

3.2.4 THE IMPORTANCE OF SOIL CARBON

Sequestering carbon in land is not solely motivated by climate change, but it is an integral part of improving the health and function of degraded soils. In many cases, therefore, managing land for climate benefits is closely aligned with managing for wider goals, such as sustainable food production, providing ecosystem services, and resilience to weather events.

Unlike carbon stored in a geological reservoir, of course, carbon is not stored as an inert substance. One major pool of aboveground biomass carbon represents the growth of trees, shrubs, and other plant life, and the soil carbon pool represents plant root development, soil life such as fungi and bacteria, and soil organic matter (SOM). *SOM* is a general term for the complex and diverse products of the decomposition of biological tissue in soils, as well as a large group of organic compounds secreted by plant roots and other soil life (Victoria et al., 2012; Stockmann et al., 2013). SOM is composed of about 50% soil organic carbon (SOC).

This organic material and the soil ecosystem it supports are both fundamental to healthy soil function and are the basis of a productive aboveground ecosystem. The functions of a healthy SOM pool in the soil include the following (Victoria et al., 2012; Lal, 2011, 2014; Liu et al., 2006):

- A natural store of key plant nutrients
- Improved retention of nutrient additions
- Improved nutrient availability to plants through increased ion exchange capacity
- Improved water infiltration and retention, increasing water availability in dry periods and reducing flooding, erosion, and nutrient leaching during heavy rainfall
- An energy source for beneficial soil biota and the diverse services they provide
- A buffer against soil pH changes
- Improved physical structure of soil and resistance to erosion
- A substrate for microbial breakdown of contaminants, agricultural chemicals, and potential GHGs, such as methane

The macroscopic result of these soil-level processes is that soils with high levels of SOC tend to have higher fertility, more consistent soil moisture, more resilience to weather events and other disturbances, lower leaching of nutrients and erosion of soils, and lower requirements for agricultural inputs than soils with low SOC. Current research points to the existence of a threshold of 1–2% SOC, or 2–4% SOM (depending on the soil and climate), below which soil fertility and productivity rapidly collapse (Stockmann et al., 2013). The most severely depleted agricultural soils, however, can have

SOC levels as low as 0.1% SOC (Lal, 2011). In addition to the global climate benefit, restoring carbon to such soils is, therefore, a key component in improving their local fertility, resilience, and ecological function.

3.2.5 LAND AND CLIMATE CHANGE ADAPTATION

These central functions of SOM in forming healthy, resilient soils are one component of the role of land in the second pillar of climate change response: adaptation.

Among the most severe risks of climate change to human welfare are its potential to exacerbate issues of water availability and quality, food security, and degradation of both vital terrestrial and coastal ecosystems. These impacts account for four of the eight key risks of climate change identified by the IPCC (2014), and all are directly mediated by land management. In low-latitude and dryland agriculture, especially, climate change poses substantial risks to food and water security through heat-related reductions in yield, crop damage from extreme weather events, erratic rainfall and increased drought risk, and loss of species and natural ecosystems that provide food and water cycling services (IPCC, 2014a).

Ecosystems can also indirectly mitigate the impacts of the remaining four key risks (coastal storm surges and flooding, inland flooding, breakdown of key infrastructure through extreme weather, and heat-related morbidity and mortality) through acting as protective buffers, improving water management, and affecting local microclimate (IPCC, 2014a; Colls et al., 2009).

Restoration of degraded soils and improving resilience of agricultural systems at all levels will determine how well food systems will be able to adapt to this threat. Maintaining healthy levels of SOM in soils is a central element of resilience to water stress, flooding, and erosion risk (Victoria et al., 2012), as discussed in section 4, and so actions that encourage drawdown of carbon into soils often also have benefits for adaptation.

Yet this is only one component of land-based adaptation, and there are many other important practices that can improve the ability of land users to respond to climate change. Climate risks can also be controlled through targeted agricultural systems for rainfall harvesting and watershed management to reduce flooding, drought, and erosion risks and buffers and structures to protect crops and soils from damage and erosion (Colls et al., 2009; Howden et al., 2007; Anwar et al., 2012).

A second element in land-based adaptation is diversification of species, agricultural systems, and income sources. Increasing species or genetic diversity among crops or livestock and using species best adapted to the local conditions improve resistance to disease, disturbance, and climatic variability (Colls et al., 2009; Howden et al., 2007). Integrating different farming systems and income sources, such as crops and livestock, similarly increases the resilience of land users' livelihoods.

Finally, systems such as agroforestry can harness particular ecological functions and relationships to mitigate the impacts of climate on productivity. For example, the use of trees and shrubs may give shade to crops and livestock, stabilize eroding soils, improve water infiltration, and support beneficial organisms (Schoeneberger et al., 2012; Branca et al., 2011; Scherr et al., 2012). Similarly, in natural, nonagricultural ecosystems, fire management, restoration of drained wetlands, creation of natural ecosystem buffers, conserving biodiversity, and providing migration routes for species can all help to preserve vital ecosystem services in the face of climate change (Colls et al., 2009; Parish et al., 2008; Schoeneberger et al., 2012).

Land is arguably even more central to climate change adaptation than it is to climate change mitigation. Fortunately, those practices that restore carbon to soils are often the same practices that improve the resilience of agricultural and natural systems. However, focusing on soil carbon will not necessarily be enough. Successful adaptation can include many other practices, such as diversification of farmers' livelihoods and landscape-level management of water. And those regions that are most vulnerable to climate change stresses, such as drylands, may not coincide with the regions with the most potential to sequester carbon (IPCC, 2014a; Branca et al., 2013). While mitigation and adaptation have historically tended to be separate activities, we will need to begin integrating both these elements into landscape-level plans and policies to achieve climate goals (Harvey et al., 2014).

3.2.6 MEETING THE RISING DEMANDS ON LAND

Land provides us with almost all human food, along with materials and energy. Demand for all of these is projected to increase in the next decades (FAO, 2009), placing higher stresses on cultivated and natural ecosystems even in the absence of climate change. The challenge from a climate perspective will be to meet and manage these demands without accelerating degradation of current agricultural land or driving clearance or overexploitation of carbon-dense natural systems.

In 2050, the world population is expected to have risen to 9–10 billion (Smith et al., 2013), and feeding this growing population has been a concern for many years. The Food and Agriculture Organization (FAO) estimate that under current trends of population growth and rising incomes, food production will need to be 70% higher in 2050 than today, which is expected to result from a projected 75% increase in meat consumption as populations become wealthier (FAO, 2009).

On top of this rising demand for food and feed, there are also ongoing efforts to encourage large-scale expansion of bioenergy to displace fossil fuels in key sectors of the energy system (IPCC, 2014b; IEA, 2011, 2012). Even with efforts to phase out the use of food crops for biofuels (IEA, 2011), this will still intensify the demands that humankind is making of Earth's terrestrial production. International Energy Agency (IEA) projections anticipate more than a threefold increase in bioenergy use by 2050, yielding 1 billion tons of further net carbon emissions abatement per year and involving 8–11 billion tons of biomass harvested annually for bioenergy ends (IEA, 2011, 2012). For context, all biomass harvested globally through crops, grazing, and forestry in the year 2000 amounted to 12 billion tons (Krausmann et al., 2008). And the latest economic models for climate change mitigation increasingly rely on even larger harvests by the end of the century to fuel a pairing of Bio-CCS technologies for the purpose of reducing the amount of CO_2 in the atmosphere in absolute terms (Fuss et al., 2014).

Can such demands be met? Over the last 60-70 years, dramatically increased crop yields and higher system efficiency have allowed food supply to support a tripling of population with very little increase in the global land area under cultivation (Smith et al., 2013), yet these rates of increase have been falling more recently[10], and such efforts to intensify have also led to land degradation and loss of soil carbon (Lal, 2011, 2014). Expansion of agriculture has also historically been the dominant driver of deforestation and loss of grassland soil carbon (FAO, 2000). The challenge facing the food system is now to reconcile the three imperatives of meeting rising food and energy demand, sustainable land management, and preservation of vital ecosystems.

This will require efforts across the whole agricultural and food systems to contain rising demand, sustainably increase productivity of existing agricultural land, and bring currently degraded land back into production. This must incorporate three approaches:

- *Land-sparing practices*, or sustainable intensification practices, which improve the production of food and other goods from existing agricultural land without degrading it, thereby meeting rising demand without expansion of agricultural area into new ecosystems. At the most basic level, this could mean improved fertility and yields, such as through better nutrient management, crop varieties, and SOC levels (Smith et al., 2013; Branca et al., 2013). Other options include increasing the number of crops in rotations, integrating different crops or products into a single piece of land (e.g., through agroforestry), and converting wastes or residues into energy products where appropriate (Smith et al., 2013; Eagle et al., 2012).
- *Sustainable agricultural expansion*, which allows new or degraded land to contribute to production while maintaining carbon stocks and ecological function. Land restoration is the most important element, bringing marginal land or previously degraded soils back into productive use and enabling expansion of the productive area without encroachment into natural ecosystems (Smith et al., 2013; Lal, 2009, 2011). Also in this category are land-sharing agricultural systems that enable significant production of food, fiber, and energy from carbon-rich natural ecosystems without driving carbon loss, such as paludiculture in peatlands, sustainable grazing systems on grasslands, or aquaculture (Victoria et al., 2012; FAO, 2009; Parish et al., 2008; Teague et al., 2013), as well as landscape-level planning to maintain ecological function in agricultural systems (Scherr and McNeely, 2008).
- *Reducing demand for food*, through reducing waste and encouraging dietary change, easing pressures on land use and potentially freeing current agricultural areas for land-demanding practices of reforestation, wetland regeneration, or simply set-asides that sequester high levels of carbon, or for production of bioenergy (Smith et al., 2013; Eagle et al., 2012; Powell and Lenton, 2012). While these can be difficult to bring about in practice, in theory, these demand-side changes could unlock far greater potential for global carbon sequestration than is possible through a focus only on the land management (Smith et al., 2013).

Historic increases in yields and intensification have allowed us to start to reduce emissions from land conversion in the last decades, even amid rapidly rising food demands. However, humanity will need to find new ways of sustainably expanding supply, reducing waste, and managing demand to prevent fresh acceleration of these losses in the coming decades. Restoring unproductive degraded land will form an essential part of reconciling these competing needs. Ultimately, ensuring that our existing land will continue to meet human needs over this century is as important for the climate as it is for food security.

3.2.7 CONCLUSION

Historic degradation of soils and clearance of forests have led to the release of hundreds of billions of tons of carbon into the atmosphere, part of which can be restored over the coming decades. Techniques for resequestering carbon in ecosystems are available and economically attractive relative to many clean energy technologies, and the global potential to do so is great. Yet equally important is adaptation for managing land to safeguard our food supply and ecosystem services in a future threatened by climate risks. Encouraging carbon sequestration, facilitating adaptation, and increasing sustainable productivity are equally critical and mutually supportive elements of land's role in mitigating the effects of anthropogenic climate change. Climate-oriented policies must begin to integrate all of these elements in global and local-scale planning if they are to be effective (Scherr et al., 2012; Harvey et al., 2014; Scherr and McNeely, 2008). Efforts to restore degraded land worldwide are a particularly outstanding opportunity for the climate, bringing substantial benefits through all three of these elements, as well as restoring economic value and ecosystem services.

However, the complexity of the biophysical and human dimensions of land issues makes them particularly difficult to make progress on, and efforts to encourage sustainable land management and improved agricultural methods over the last decades have demonstrated the scale of the challenges. These challenges will not be resolved straightforwardly simply through adding the climate change lens or implementing a carbon price for biological sequestration. However, the growing realization of the value of ecosystem-based responses to climate change demonstrates that they deserve a more central place in climate change mitigation and adaptation discourse than they have today and represents a huge opportunity to refocus international attention and investment toward driving forward sustainable management and land restoration efforts.

REFERENCES

Anwar, M.R., et al., 2012. Adapting agriculture to climate change: A review. Theor. Appl. Climatol. 113 1–2, 225–245. http://dx.doi.org/10.1007/s00704-012-0780-1.

Branca, G., et al., 2011. Climate-Smart Agriculture: Mitigation of Climate Change in Agriculture Series 3. Food and Agriculture Organization, Rome, Italy.

Branca, G., et al., 2013. Food security, climate change, and sustainable land management. A review, Agron. Sustain. Dev. 33 (4), 635–650. http://dx.doi.org/10.1007/s13593-013-0133-1.

Colls, A., Ash, N.N., Ikkala, N., 2009. Ecosystem-Based Adaptation: A Natural Response to Climate Change. International Union for the Conservation of Nature (IUCN), Gland, Switzerland.

Conant, R.T., 2010. Challenges and opportunities for carbon sequestration in grassland systems. Food and Agriculture Organization, Rome, Italy.

Conant, R.T., Paustian, K.E.T., Elliott, E.T., 2001. Grassland management and conversion into grassland: Effects on soil carbon. Ecol. Appl. 11 (2), 343–355. http://dx.doi.org/10.1890/1051-0761(2001)011[0343:GMACIG]2.0.CO;2.

Eagle, A.J., et al., 2012. Greenhouse Gas Mitigation Potential of Agricultural Land Management in the United States: A Synthesis of the Literature. Nicholas Institute, Duke University, Durham, NC.

Eurostat, 2013. Carbon dioxide emissions from final use of products. Available at: http://ec.europa.eu/eurostat/statistics-explained/index.php/Carbon_dioxide_emissions_from_final_use_of_products. [Accessed 16 November 14].

Food and Agriculture Organization (FAO), 2000. World Soil Resources Report: Land Resource Potential and Constraints at Regional and Country Levels. Food and Agriculture Organization, Rome, Italy.

Food and Agriculture Organization (FAO), 2009. How to Feed the World in 2050. Food and Agriculture Organization, Rome.

Food and Agriculture Organization (FAO), 2013. FAO Statistical Yearbook 2013. Food and Agriculture Organization, Rome.

Fuss, S., et al., 2014. Betting on negative emissions. Nat. Clim. Chang. 4 (10), 850–853. http://dx.doi.org/10.1038/nclimate2392.

Gougoulias, C., Clark, J.M., Shaw, L.J., 2014. The role of soil microbes in the global carbon cycle: Tracking the below-ground microbial processing of plant-derived carbon for manipulating carbon dynamics in agricultural systems. J. Sci., Food Agric 94 (12), 2362–71. http://dx.doi.org/10.1002/jsfa.6577.

Harvey, C.A., et al., 2014. Climate-smart landscapes: opportunities and challenges for integrating adaptation and mitigation in tropical agriculture. Conserv. Lett. 7 (2), 77–90. http://dx.doi.org/10.1111/conl.12066.

Howden, S.M., et al., 2007. Adapting agriculture to climate change. Proc. Natl. Acad. Sci. U S A. 104 (50), 19691–19696. http://dx.doi.org/10.1073/pnas.0701890104.

Intergovernmental Panel on Climate Change (IPCC), 2007. Climate Change 2007: Mitigation. Contribution of Working Group III to the Fourth Assessment Report of the Intergovernmental Panel on Climate Change. Cambridge University Press, Cambridge, UK, and New York.

Intergovernmental Panel on Climate Change (IPCC), 2013. Climate Change 2013: The Physical Science Basis. Contribution of Working Group I to the Fifth Assessment Report of the Intergovernmental Panel on Climate Change. Cambridge University Press, Cambridge, UK, and New York.

Intergovernmental Panel on Climate Change (IPCC), 2014a. Climate Change 2014: Impacts, Adaptation, and Vulnerability. Part A: Global and Sectoral Aspects. Contribution of Working Group II to the Fifth Assessment Report of the Intergovernmental Panel on Climate Change. Cambridge University Press, Cambridge, UK, and New York.

Intergovernmental Panel on Climate Change (IPCC), 2014b. Climate Change 2014: Mitigation of Climate Change. Contribution of Working Group III to the Fifth Assessment Report of the Intergovernmental Panel on Climate Change. Cambridge University Press, Cambridge, UK, and New York.

International Energy Agency (IEA), 2011. Technology Roadmap: Biofuels for Transport. OECD Publishing, Paris. http://dx.doi.org/10.1787/9789264118461-en.

International Energy Agency (IEA), 2012. Technology Roadmap: Bioenergy for Heat and Power. OECD Publishing, Paris.

Krausmann, F., et al., 2008. Global patterns of socioeconomic biomass flows in the year 2000: A comprehensive assessment of supply, consumption and constraints. Ecol. Econ. 65 3, 471–487. http://dx.doi.org/10.1016/j.ecolecon.2007.07.012.

Lal, R., 2004. Soil carbon sequestration to mitigate climate change. Geoderma. 123 (1–2), 1–22. http://dx.doi.org/10.1016/j.geoderma.2004.01.032.

Lal, R., 2009. Carbon Sequestration in Saline Soils. J. Soil Salin. Water Qual. 1 (1–2), 30–40.

Lal, R., 2011. Sequestering carbon in soils of agro-ecosystems. Food Policy 36, S33–S39. http://dx.doi.org/10.1016/j.foodpol.2010.12.001.

Lal, R., et al., 2004. Managing soil carbon. Science 304, 393.

Le Quéré, C., et al., 2009. Trends in the sources and sinks of carbon dioxide. Nat. Geosci. 2 (12), 831–836. http://dx.doi.org/10.1038/ngeo689.

Liu, X., et al., 2006. Effects of agricultural management on soil organic matter and carbon transformation—A review. Plant Soil. 52 (12), 531–543.

McLaren, D., 2012. A comparative global assessment of potential negative emissions technologies. Process Saf. Environ. Prot. 90 (6), 489–500. http://dx.doi.org/10.1016/j.psep.2012.10.005.

Nabuurs, G.J., et al., 2007. Forestry. In: Metz, B. et al. (Ed.), Climate Change 2007 Mitigation. Contribution of Working Group III to the Fourth Assessment Report of the Intergovernmental Panel on Climate Change. Cambridge University Press, Cambridge, UK, and New York.

Parish, F.A., et al., 2008. Assessment on Peatlands, Biodiversity and Climate Change: Main Report, Global Environment Center, Kuala Lumpur & Wetlands International, Wageningen.

Powell, T.W.R., Lenton, T.M., 2012. Future carbon dioxide removal via biomass energy constrained by agricultural efficiency and dietary trends. Energy Environ. Sci. 5 (8), 8116. http://dx.doi.org/10.1039/c2ee21592f.

Ruddiman, W.F., 2013. The anthropocene. Ann. Rev. Earth Planet. Sci. 41 (1), 45–68. http://dx.doi.org/10.1146/annurev-earth-050212-123944.

Scherr, S.J., McNeely, J.A., 2008. Biodiversity conservation and agricultural sustainability: Towards a new paradigm of "ecoagriculture" landscapes. Philos. Trans. R. Soc. Lond. B Biol. Sci. 363 (1491), 477–494. http://dx.doi.org/10.1098/rstb.2007.2165.

Scherr, S.J., Shames, S., Friedman, R., 2012. From climate-smart agriculture to climate-smart landscapes. Agric. Food Secur. http://dx.doi.org/10.1186/2048-7010-1-12.

Schoeneberger, M., et al., 2012. Branching out: Agroforestry as a climate change mitigation and adaptation tool for agriculture. J. Soil Water Conserv. 67 (5), 128A–136A. http://dx.doi.org/10.2489/jswc.67.5.128A.

Smith, P., et al., 2007. Agriculture. In: Metz, B., Davidson, O.R., Bosch, P.R., Dave, R., Meyer, L.A. (Eds.), Clim. Chang. 2007 Mitigation. Contrib. Work. Gr. III to Fourth Assess. Rep. Intergov. Panel Clim. Chang. Cambridge University Press, Cambridge, MA.

Smith, P., et al., 2008. Greenhouse gas mitigation in agriculture. Philos. Trans. R. Soc. Lond. B Biol. Sci. 363 (1492), 789–813. http://dx.doi.org/10.1098/rstb.2007.2184.

Smith, P., et al., 2013. How much land-based greenhouse gas mitigation can be achieved without compromising food security and environmental goals? Glob. Chang. Biol. 19 (8), 2285–2302. http://dx.doi.org/10.1111/gcb.12160.

Smith, P., et al., 2014. Agriculture, forestry and other land use (AFOLU). In: Climate Change 2014: Mitigation of Climate Change. Contribution of Working Group III to the Fifth Assessment Report of the Intergovernmental Panel on Climate Change. Cambridge University Press, Cambridge, UK, and New York.

Stockmann, U., et al., 2013. The knowns, known unknowns, and unknowns of sequestration of soil organic carbon. Agric. Ecosyst. Environ. 164, 80–99. http://dx.doi.org/10.1016/j.agee.2012.10.001.

Teague, W.R., et al., 2011. Grazing management impacts on vegetation, soil biota and soil chemical, physical and hydrological properties in tall grass prairie. Agric. Ecosyst. Environ. 141 (3–4), 310–322. http://dx.doi.org/10.1016/j.agee.2011.03.009.

Teague, W.R., et al., 2013. Multi-paddock grazing on rangelands: Why the perceptual dichotomy between research results and rancher experience? J. Environ. Manage. 128, 699–717. http://dx.doi.org/10.1016/j.jenvman.2013.05.064.

Victoria, R., et al., 2012. The Benefits of Soil Carbon: Managing Soils for Multiple Economic, Societal, and Environmental Benefits, in: UNEP Year B. United Nations Environment Programme, Nairobi.

ECONOMICS, POLICY, AND GOVERNANCE OF LAND RESTORATION

THE IMPORTANCE OF LAND RESTORATION FOR ACHIEVING A LAND DEGRADATION–NEUTRAL WORLD

4.1

Luca Montanarella

European Commission, DG Joint Research Centre (JRC), Ispra, Italy

4.1.1 INTRODUCTION

Anthropogenic and environmental pressures on the earth's ecosystems have led to substantial land degradation in all parts of the world throughout the centuries. The scale of this degradation and the complex socioecological drivers behind it have only been acknowledged in recent years.

The 1977 United Nations Conference on Desertification (UNCOD) originally focused on land degradation that mainly occurred in the Sahel region, due to extensive drought phenomena (UNCOD, 1977; Verstraete et al., 2009). Unfortunately, the UNCOD did not highlight that the natural resources of most other biomes were also degraded due to various socioecological drivers, including overexploitation. The establishment of the United Nations Convention to Combat Desertification (UNCCD) in 1994 was a huge step forward toward globally acknowledging the severity of the impact of land degradation on ecosystem health and human livelihoods. Yet, the convention was only targeted at drylands, overlooking the fact that ecosystem degradation is a global challenge and not limited only to certain biomes. Drylands are still the main target of the UNCCD, which demonstrates the limited focus of the convention on arid, semiarid, and dry subhumid areas.

In recent years, the global attention to land degradation has broadened as updated definitions of this term indicate. For example, the Global Environment Facility (GEF) defines *land degradation* as "any form of deterioration of the natural potential of land that affects ecosystem integrity either in terms of reducing its sustainable ecological productivity or in terms of its native biological richness and maintenance of resilience" (GEF, 2013a). Recalling that we are in the Anthropocene era (Crutzen and Stoermer, 2000), which is marked by the increasing ability of human beings to shape our planet, land degradation has been recognized as a human-driven process affecting terrestrial ecosystems under all climatic conditions and limiting the extent of ecosystem services. In this sense, land degradation should be defined as follows: "Any process that limits the ability of a landscape to deliver ecosystem services."

Nevertheless, the formal definition of land degradation, as adopted by all parties to the UNCCD, reads as follows: "a reduction or loss, in arid, semiarid, and dry subhumid areas, of the biological or

Land Restoration. http://dx.doi.org/10.1016/B978-0-12-801231-4.00020-3

economic productivity and complexity of rainfed cropland, irrigated cropland, or range, pasture, forest, and woodlands resulting from land uses or from a process or combination of processes, including processes arising from human activities and habitation patterns, such as soil erosion caused by wind and/or water; a deterioration of the physical, chemical, and biological or economic properties of soil; and a long-term loss of natural vegetation" (UNCCD, 1994).

We will base the remaining discussion on this definition because it has been adopted by nearly all countries; the UNCCD is the most ratified Multilateral Environmental Agreement (MEA), with 195 parties. Sticking to the agreed, legally binding definition allows the discussion to make progress in a measurable and objective way while avoiding the lengthy debates on the definition of land degradation and desertification. Though these debates are scientifically pertinent, they have prevented a more effective implementation of the convention on the ground. Following from the aforementioned definition of land degradation, we can define *land restoration* as follows: "an increase, in arid, semiarid, and dry subhumid areas, of the biological or economic productivity and complexity of rainfed cropland, irrigated cropland, or range, pasture, forest, and woodlands resulting from sustainable land management practices, such as a reduction of soil erosion caused by wind and/or water; a restoration of the physical, chemical, and biological or economic properties of soil; and a long-term increase of natural vegetation."

It is important to note that, of the three criteria adopted for the definition of land degradation, two are strictly related to soil properties, namely: (i) soil erosion and (ii) soil quality, which is defined as a combination of physical, chemical, and biological or economic properties of soil. Interestingly, these two criteria are also the two soil-related indicators of the Organisation for Economic Co-operation and Development (OECD) for regular reporting on the agricultural environment (OECD, 2013). Yet, as shown in Figure 4.1.1, current data on soil erosion by water reported by the OECD indicate large eroded areas in countries that the UNCCD does not recognize as affected by land degradation. This demonstrates the limited focus of the convention on arid, semiarid, and dry subhumid areas, while the call for a land degradation neutral world, endorsed at the 2012 Rio+20 Conference, addressed a global issue extending far beyond the UNCCD.

In this discussion, we will therefore consider land restoration on a global scale, without a specific focus on arid, semiarid, and dry subhumid areas.

4.1.2 DEFINITION AND ACCOUNTING OF LAND DEGRADATION NEUTRALITY

The definition of land degradation neutrality needs to be consistent with the adopted definitions of legally binding agreements and conventions. Therefore, we need to take into account the definition provided by the UNCCD and the related definition of land restoration, as previously discussed. In an ideal world, as described in the outcome document of the Rio+20 Conference, "The Future We Want" (UN, 2012), we should stop the increasing trend of land degradation by balancing ongoing degradation processes with equivalent restoration activities. The result should be a stable status quo of land degradation on a global scale (i.e., a land degradation–neutral world). We need to define a baseline that is equivalent to the status of global land degradation in 2012 (when the Rio+20 conference occurred) and implement an accounting system to measure any increase in degradation against this baseline. Positive increments of land degradation must then be neutralized by equivalent efforts at land restoration (Chasek et al., 2015).

In practical terms, this means assessing the area of land affected by (i) soil erosion; (ii) deterioration of the physical, chemical, and biological or economic properties of soil; and (iii) long-term loss of natural vegetation, and implementing an effective monitoring system to detect any changes in degradation or restoration to these areas.

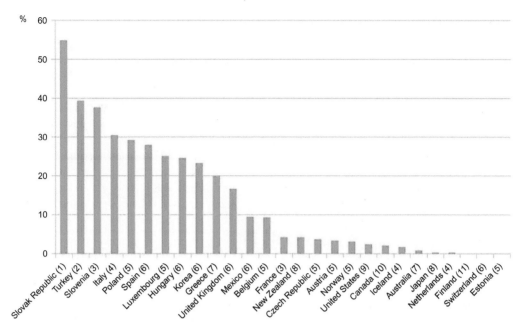

Countries are ranked in terms of highest to lowest % share of agricultural land at risk to water erosion.
1. Data for Slovak Republic refer to 2003-04.
2. Data for Turkey refer to 1990-94.
3. Data for France and Slovenia refer to 2006-07.
4. Data for Iceland, Italy, and Netherlands refer to 1995-99.
5. Data for Austria, Belgium, Czech Republic, Estonia, Luxembourg, Norway, and Poland refer to 2009-10.
6. Data for Hungary, Mexico, Spain, Switzerland, and United Kingdom refer to 2000-02 and Korea refer to 2002.
   Soil erosion data for Spain includes agriculture and forestry land. For Switzerland the total agricultural area includes summer pastures (alpine pastures). For Mexico, the area of risk is the sum of moderate + severe + extreme erosion categories.
7. Data for Australia and Greece drawn from OECD (2008). Data for Greece covers all land, Including agricultural land.
8. Data for Japan and New Zwaland refer to 1985-89.
9. Data for United States refer to 2007-08.
10. Data for Canada refer to 2005-06, values for cultivated cropland.
11. Data for Finland refer to 2001.

Sources: OECD (2008), *Environmental Performance of Agriculture in OECD Countries Since 1990*, http://www.oecd.org/agriculture/; Joint Research Centre, European Union; Unpublished Estimates Pan-European RUSLE Model JRC, 2011; and national sources.

FIGURE 4.1.1

Agricultural land areas classified as having moderate to severe water erosion risk, OECD countries, 1990–2010. Risk of water erosion greater than 11 tons/ha/year of soil loss, as a percentage share of total agricultural land area (OECD, 2013).

Land degradation neutrality should be, therefore, defined as follows: *a state whereby the amount of degraded land remains stable within specific temporal and spatial scales.*

Since it is unrealistic to expect land degradation processes to halt completely, given the predicted increase in resource consumption in the near future, we must focus on effective land restoration strategies in order to neutralize these trends. The following sections address land restoration as a key element in achieving land degradation neutrality.

4.1.2.1 LAND DEGRADATION

It is estimated that 25% of the world's land area is either highly degraded or undergoing high rates of degradation. Land degradation is especially severe in developing countries. It is estimated that two-thirds of African land is already degraded to some degree and that land degradation affects at least 485 million people (65% of the entire African population). By the 2050s, 50% of agricultural land in Latin America will be subject to desertification (UNCCD, 2014). Moreover, in the context of developing countries, the issue of gender cannot be ignored, as (usually) poor women are most vulnerable to land degradation: "It is the world's poorest inhabitants that are most affected by negative alteration to the natural environment and climate change, and of the world's poor some 60 percent are female. … Because of their dependency on natural resources, rural women in developing countries are the group that is being hit the hardest by the effects of environmental degradation and depletion of natural resources." (UNU-LRT, 2013, p. 5). Also, women often have less control over and less decision-making power regarding land. A land degradation neutral world needs to be people centered because poor and marginalized populations are usually the most affected:

- *Population growth*: There is an absolute increase in the quantity of resources consumed required to sustain and guarantee the basic needs of a growing global population. This exerts pressure on land and other natural resources, which renders them less available and less productive. Thus, arable land rapidly degrades with implications for food security. Moreover, the high population growth in developing countries, particularly within groups whose livelihoods depend on the exploitation of ecosystem services, concentrates environmental stress and degradation in regions that are most reliant on those services (UN, 2012).
- *Poverty*: Low income and education levels are frequently associated with overexploitation of resources that leads to environmental degradation. Moreover, natural shocks (droughts and floods, among others) further affect fragile ecosystems and destroy the vegetation and soil, reducing the ability of the land to support lives and provide ecosystem services (UN, 2012).
- *Unsustainable human activities*: Poor irrigation schemes, excess use of mineral fertilizers, and overgrazing lead to salinization of land, soil acidification, and destruction of vegetation, respectively. These consequences ultimately diminish the ability of land to recover its environmental services naturally.
- *Land tenure systems*: Unequal, gender-biased, and insecure land tenure prevents people from making necessary investments and decisions to protect land from becoming degraded. As a result, poor and vulnerable people, particularly women, frequently find themselves trapped in a cycle of poverty within which they are unable to take ownership of the land and its productivity (FAO, 2012).
- *Poor governance*: The top-down approach that is common in the developing world has led to environmental degradation, while inclusive governance would lead to better results.

In an interconnected world, influence and causality regarding land degradation occur between countries and regions. Indirect Land Use Changes (ILUCs) have been extensively documented:

- European countries are largely dependent on food, fiber, and biomass produced on soils in other parts of the world. Such dependence has been also described as "virtual soil imports" (SERI, 2011). Therefore, policy decisions in Europe may trigger substantial land use changes in other parts of the world. Important EU policies, like the Common Agricultural Policy (CAP) and the Energy Policy, have documented impacts in developing countries (Kretschmer et al., 2013).
- Degradation impacts in the developing world spill over to the developed world, as the increasing migratory pressure on developed countries is documented every day (UNCCD, 2013).

4.1.3 LAND RESTORATION

Extensive literature and documentation are available regarding ecological restoration (see, e.g., SER, 2004), which has become a point of international debate—especially within the framework of the Convention for Biological Diversity (CBD) since the adoption of the Aichi targets in 2010. Particularly relevant is target 15: "By 2020, ecosystem resilience and the contribution of biodiversity to carbon stocks have been enhanced, through conservation and restoration, including restoration of at least 15 percent of degraded ecosystems, thereby contributing to climate change mitigation and adaptation and to combating desertification" (CBD, 2011). According to the CBD (2010), by some estimates, two-thirds of the planet's ecosystems are degraded. The global potential for forest landscape restoration alone is estimated to be on the order of 1 billion ha, or about 25% of the current global forest area.[1] As we have defined land degradation, restoration activities to limit soil erosion, improve soil quality, and increase natural vegetation should be considered options to achieve land degradation neutrality.

Restoration activities are ongoing in many parts of the world, with extensive restoration projects being implemented in both developed and developing countries. Large projects are, however, not necessarily more successful then smaller community-based activities. The local socioeconomic and historical context needs to be taken into account when developing restoration plans. For example, highly developed industrialized countries have been focusing most land restoration resources on recovering sites degraded by industrial activities and urban and infrastructure development. Europe has a heritage of abandoned mining and industrial sites (approximately 3 million potentially contaminated sites) that require huge investments for full restoration of original soil functions (Van Liedekerke et al., 2014).

In other parts of the world, the dominant degradation processes may be different, which would require tailored approaches to the local conditions. The recent assessment by Bao et al. (2014) reports that land degradation hot spots cover about 29% of global land area and occur in all agroecologies (agricultural production system ecologies) and land cover types. Land degradation is especially important in grasslands, which host approximately 3.2 billion people. A particularly relevant area to the UNCCD is the Sahel region, historically considered the area most affected by desertification from as early as the 1970s. A relatively recent report (Reij and Smaling, 2008) shows an increasing improvement of the vegetation cover in this area, which demonstrates the effectiveness of restoration strategies being developed and implemented on local scales. It is clearly demonstrated that local communities can effectively restore degraded areas by implementing relatively simple and effective management practices (see, e.g., Web Alliance for Re-Greening Africa, 2014). On the other hand, large investments in the same region aimed toward regreening the Sahel still have to prove their effectiveness. An important example is the recently established initiative to build a 7000-km-long, 15-km-wide "Great Green Wall" (GGW) from Dakar to Djibouti to act as a green barrier of protection against desert advancement. The GEF has granted US 100.8 million to the GGW-participating countries to expand sustainable land and water management (SLWM) and adoption of these practices in targeted landscapes and in climate-vulnerable areas in west African and Sahelian countries (GEF, 2013b).

Several positive examples of successful land restoration activities exist:

- Sandwatch education program in the Cook Islands involved schools in managing beach environments. By integrating activities into schools' curricula, education programs do not require

---

[1]http://www.cbd.int/doc/strategic-plan/targets/T15-quick-guide-en.pdf.

the extra commitment and time of dedicated educators. This is replicable in other countries and can be very effective in reducing beach degradation.

- The "Adopting the Living Well in Balance and Harmony with Mother Earth" principle in Bolivia has seen enhanced resource management. This principle, if based on national policies, recognizes and respects natural rights.
- Antidesertification initiatives supported by governments and the private sector have restored 450,000 ha of land through agroforestry in China.

It will be interesting to assess the restoration effects that these large-scale initiatives will deliver over the next years.

4.1.3.1 RESTORING ERODED SOIL

Soil erosion is a widespread degradation process that naturally occurs on sloping land, in the case of water erosion, or in windy areas with scarce vegetation. It is not inherently a degradation process; it only becomes "degradation" if a measurable effect of human-induced erosion can be detected. Since naturally occurring erosion rates are variable according to local conditions, it is not possible to define a universally valid value for the definition of soil erosion processes for land degradation assessments. However, in OECD countries, a threshold value of 11 tons/ha/year of soil loss has been proposed as a possible value above which soil erosion should be considered as a human-induced degradation process (Figure 4.1.1).

Restoration of areas affected by soil erosion has been extensively documented in many parts of the world, starting with the famous "dust bowl" in the midwestern United States, which triggered the adoption of the US Soil Conservation Act in 1935. This act has driven extensive restoration programs in the United States with a remarkable result: a consistently decreasing area of eroded soils (USDA, 2013).

Other very well-documented examples of effective restoration practices on land affected by erosion exist in many parts of the world: in Iceland, the world's oldest soil conservation service has restored soil on eroded land for more than 100 years (Aradottir et al., 2013); in Australia, the extensive application of the Landcare program has effectively reversed soil erosion trends in many areas (Commonwealth Intergovernmental Working Group for the UNCCD, 2002); and in many other parts of the world, effective soil erosion mitigation measures have been documented (WOCAT, 2007).

Impressive restoration programs on areas affected by wind erosion have been implemented in China (Hu et al., 2009) with substantial effects in limiting the impact of wind erosion on the human population. The Three-North Shelterbelt Program is the largest afforestation program in the world. It aims to establish 35 million ha of shelterbelt forests between 1978 and 2050. By then the forest cover in the Three-North Region of China will increase from 5 to 14%. So far, over 13 million ha of plantations have been established; 4.33 million ha of mountainous and highly degradable land were closed and protected from grazing and fire, and 0.4 million ha of forests were established by aerial seeding. As the Three-North Shelterbelt Program progresses, only the land located in more arid and infertile sites remains available for afforestation and rehabilitation. Among the established plantations, 60% are owned by individuals, 30% by the state, and 10% by collectives (FAO, 2002).

4.1.3.2 RESTORING SOIL QUALITY

The second main process identified by the UNCCD as a driver of land degradation is the "deterioration of the physical, chemical, and biological or economic properties of soil," which can be summarized as loss of soil quality. Soil quality has been extensively discussed both within the scientific community and among policy makers (Tóth and Németh, 2011) and a large amount of research exists to demonstrate the importance of soil quality for sustainable development. Soil quality is usually defined in relation to specific functions: good soil can perform certain functions, like producing biomass (agriculture, forestry, grasslands, etc.), filtering water, or storing biodiversity and carbon. The multifunctionality of soils is also well understood in the policy-making framework and has thus been fully incorporated into the EU Soil Thematic Strategy and related legislative proposals (Toth et al., 2007).

Restoring soil quality is essentially equivalent to restoring soils' full multifunctionality. Usually, restoration efforts limit themselves to restoring only some soil functions, depending on future land use plans. For example, restoration in contaminated sites is not usually aimed at full restoration of soil capacity to deliver good quality food, but is limited to restoring its hydrological functions to prevent groundwater contamination, as well as occasionally restoring additional functions, such as providing areas for construction of housing and infrastructure or natural areas for biodiversity and carbon storage. Restoration of contaminated sites remains a very costly activity and is progressing only in limited areas. Regular reporting in Europe about the restoration of contaminated sites is compiled from data provided by national authorities to the European Commission (Panagos et al., 2013). The results of the latest report (Van Liedekerke et al., 2014), however, indicate that restoration is progressing very slowly in Europe.

4.1.3.3 RESTORING NATURAL VEGETATION

As defined previously, the UNCCD's third factor of land degradation is the long-term loss of natural vegetation. Thus, restoration activities should aim at restoring natural vegetation in degraded areas. It seems obvious that these restoration activities will have to exclude all areas dedicated to biomass production, like agricultural areas (e.g., croplands, pastures, etc.) and managed forest areas, as natural vegetation restoration implies full ecosystem recovery in its natural condition. Typically, those areas dedicated to natural vegetation restoration are (or will be) nature protection areas. These areas are particularly relevant for achieving the Aichi Target 15 of restoration of at least 15% of degraded ecosystems by 2020. In the EU, these areas may well coincide with already identified nature protection areas within the NATURA 2000 network.

4.1.3.4 BUILDING MOMENTUM IN LAND RESTORATION

The developed world can help solve these issues by providing financial backing for land restoration projects. It can also spearhead legal reforms in the developing world and advocate for peace, as well as being mediators and advisors in areas of political instability, which may otherwise cause irrevocable harm to the natural environment, as has been seen in some war-torn countries. The developed world can also share some of its successful strategies and help the developing nations implement these strategies.

Several opportunities are available for land restoration in the framework of South-South cooperation:

- Enhancing and strengthening local/traditional-level resource management mechanisms.
- Implementing strategies to eliminate poverty and give people alternative livelihoods to reduce their dependence on natural resources, which would enable the land to sustainably support them without degrading.
- Educating and creating awareness campaigns regarding the causes and effects of land degradation. This can also include training resource managers who can then pass on knowledge at the grassroots level. Education should also start at the elementary school level.
- Encouraging collective action by those affected by land degradation. This can enable people to feel control over their resources and for their voices and concerns to be taken seriously. Through collective action, restoration can be achieved and degradation reduced.
- Enabling an extensive interdisciplinary approach to land restoration. To restore land to its initial state is too large a task to be carried out by a single sector, or even by a single nation. Thus, the developing world should collaborate, create connections among resource sectors, and seek cross-border support from other countries.
- Devolving resources to the local level will also help reduce population movements to urban centers and thus reduce land degradation in these areas; urbanization is among the major drivers of environmental deterioration.
- Controlling population growth will help reduce degradation and achieve restoration.
- Enhancing equity to land resources so that all residents feel important and act as custodians of land resources.
- Implementing legal reforms to remove restrictions on women's rights to land ownership, including titling; which refers to reforming family laws to allow women to inherit land.
- Mainstreaming gender into land resource management policies will also help restore land in the developing world, where most societies are patriarchal.
- Compelling industries involved in degradation activities (e.g., mining and quarrying) to ensure the land is restored. Encouraging payment for environmental services can also do this.
- Documenting and sharing land restoration success stories.
- Including sustainable land management into national decision making and policies.

4.1.4 CONCLUSIONS

Realistically, achieving a land degradation–neutral world is only possible by implementing an extensive soil restoration program that focuses on areas affected by soil erosion, loss of soil quality, and loss of natural vegetation. Achieving political commitment and global consensus on such a target requires adherence to agreed definitions. Even so, the original definition of land degradation, as it is in the UNCCD text, may be questionable from a scientific point of view; however, it remains a major achievement that nearly all countries of the world have accepted this definition and have ratified the convention. To build on such a wide consensus, a solid political basis must be formed to make substantial progress in combatting desertification by restoring degraded lands, as defined in the convention text. Setting clear, measurable targets is possible, as demonstrated by the regular reporting on soil erosion and soil quality by the OECD. Both these indicators already cover most of the substantial elements of land degradation, as defined by UNCCD. Concerning the third element, the long-term loss of natural vegetation, a wide range of remote sensing tools exist for assessing the extent of the degradation process and for monitoring restoration progress in degraded areas. The planned World Atlas of Desertification will provide the necessary baseline for such future monitoring systems (Zucca et al., 2014).

REFERENCES

Aradóttir, Á.L., Petursdottir, T., Halldorsson, G., Svavarsdottir, K., Arnalds, O., 2013. Drivers of ecological restoration: lessons from a century of restoration in Iceland. Ecol. and Society 18 (4), 33.

Bao Le, Q., Nkonya, E., Mirzabaev, A., 2014. Biomass productivity-based mapping of global land degradation hotspots. ZEF Discussion Papers on Devel, Policy 193, 57.

Chasek, P., Safriel, U., Shikongo, S., Futran Fuhrman, V., 2015. Operationalizing zero net land degradation: The next stage in international efforts to combat desertification? J. of Arid Env. 112 (A, January), 5–13.

Commonwealth Intergovernmental Working Group for the UNCCD, 2002. National report by Australia on measures taken to support implementation of the United Nations Convention to Combat Desertification. Government of Australia, Department of the Environment, Canberra, Australia.

Convention on Biological Diversity (CBD), 2011. Strategic plan for biodiversity 2011–2020: Further information related to the technical rationale for the Aichi biodiversity targets, including potential indicators and milestones. UNEP, Nagoya, Japan.

Convention on Biological Diversity (CBD), 2010. COP 10 Decision X/2. strategic plan for biodiversity 2011–2020.

Crutzen, P.J., Stoermer, E.F., 2000. The "Anthropocene." Global Change Newsletter 41, 17–18.

FAO, 2012. Voluntary Guidelines on the Responsible Governance of Tenure of Land, Fisheries and Forests in the Context of National Food Security.

FAO, 2002. Project "Afforestation, Forestry Research, Planning and Development in the Three North Region of China"—GCP/CPR/009/BEL.

Global Environmental Facility (GEF), 2013a. Land Degradation. Available from: http://www.thegef.org/gef/land_degradation (accessed 24.11.14).

Global Environmental Facility (GEF), 2013b. The Great Green Wall Initiative. Available from: http://www.thegef.org/gef/great-green-wall (accessed 01.10.14).

Hu, Y., Zhao, B., Ma, B.L., 2009. New concept and control measures for wind erosion in agricultural land: A case study in northern China. In: Fernandez-Bernal, A., de la Rosa, M.A. (Eds.), Arid Environments and Wind Erosion. Nova Science Publishers, Hauppauge, NY.

Kretschmer, B., Allen, B., Kieve, D., Smith, C., 2013. The sustainability of advanced biofuels in the EU: Assessing the sustainability of wastes, residues and other feedstocks set out in the European Commission's proposal on Indirect Land Use Change (ILUC). Biofuel ExChange briefing No 3. In: Institute for European Environmental Policy (IEEP), London, UK.

Organisation for Economic Co-operation and Development (OECD), 2013. OECD Compendium of Agri-environmental Indicators. OECD Publishing, Paris.

Panagos, P., Van Liedekerke, M., Yigini, Y., Montanarella, L., 2013. Contaminated sites in Europe: Review of the current situation based on data collected through a European network. J. Env. Pub. Health. 2013.

Reij, C.P., Smaling, E.M.A., 2008. Analyzing successes in agriculture and land management in sub-Saharan Africa: Is macro-level gloom obscuring positive micro-level change? Land Use Policy 25, 410–420.

Sustainable Europe Research Institute (SERI), 2011. Europe's global land demand: A study on the actual land embodied in European imports and exports of agricultural and forestry products. Available from http://seri.at/.

Tóth, G., Németh, T., 2011. Land Quality and Land Use Information - in the European Union – 399 pp. – EUR 24590 EN– Scientific and Technical Research series – ISSN 1831–9424, ISBN 978-92-79-17601-2.

Tóth, G., Stolbovoy, V., Montanarella, L., 2007. Soil quality and sustainability evaluation—An integrated approach to support soil-related policies of the European Union. JRC Position Paper EUR 22721.

United Nations (UN), 2012. The Future We Want. Available from: http://www.uncsd2012.org/thefuturewewant .html (accessed 01.01.15).

UN Convention to Combat Desertification (UNCCD), 1994. Elaboration of an international convention to combat desertification in countries experiencing serious drought and/or desertification, particularly in Africa. UNCCD, New York.

UN Convention to Combat Desertification (UNCCD), 2013. White paper 1: Economic and Social Impacts of Desertification, Land Degradation and Drought. In: Second UNCCD Scientific Conference. April 9–12.

UN Convention to Combat Desertification (UNCCD), 2014. The Land in Numbers. Secretariat of the UNCCD, Bonn, Germany.

UN Conference on Desertification (UNCOD), 1977. Desertification: Its Causes and Consequences. Pergamon Press, Oxford, UK.

United Nations University, Land Restoration Training Programme (UNU-LRT), 2013. Gender Equality Policy. Agricultural University of Iceland, Reykjavík, Iceland.

US Department of Agriculture, 2013. Summary Report: 2010 National Resources Inventory. Natural Resources Conservation Service, Washington, DC, and Center for Survey Statistics and Methodology, Iowa State University, Ames, Iowa.

Van Liedekerke, M., Prokop, G., Rabl-Berger, S., Kibblewhite, M., Louwagie, G., 2014. Progress in the Management of Contaminated Sites in Europe. JRC Ref. Rep. EUR. 26376.

Verstraete, M.M., Scholes, R.J., Stafford Smith, M., 2009. Climate and desertification: looking at an old problem through new lenses. Frontiers in Ecol. and the Env. 7, 421–428.

Web Alliance for Regreening in Africa, 2014. Available from: http://w4ra.org/ (accessed 01.10.14).

World Overview of Conservation Approaches and Technologies (WOCAT), 2007. Where the Land Is Greener: Case studies and analysis of soil and water conservation initiatives worldwide. In: Liniger, H., Critchley, W. (Eds.), CTA, the Netherlands; FAO, Rome; UNEP, Nairobi; and CDE, Bern.

Zucca, C., et al., 2014. The role of soil information in land degradation and desertification mapping: A review. In: Kapur, S., Erşahin, S. (Eds.), Soil Security for Ecosystem Management: Mediterranean Soil Ecosystems 1. Springer International Publishers, Cham, Switzerland, pp. 31–59.

TRANSFORMING LAND CONFLICTS INTO SUSTAINABLE DEVELOPMENT: THE CASE OF THE TAITA TAVETA OF KENYA

4.2

Ednah Kang'ee

Armed Conflict and Peace Studies Program, Department of History and Archaeology, University of Nairobi, Kenya

4.2.1 INTRODUCTION

Land conflicts are indeed a widespread phenomenon, which can transpire at any time and in any place. Both the rise in population and the pressure on the environment caused by human activities have aggravated the perception of land as a dwindling resource, cementing the connection between land and violent conflict. Land ownership and land use rights in Africa often result in land disputes that cause negative effects on the certainty of land markets, tenure and food security, economic production, and reduction of poverty. Overdependence on a single or very small number of resources often triggers land conflicts (Collier and Hoeffler, 2000).

Sustainable development, which is development that meets the needs of the present without compromising the ability of future generations to meet their own needs, composes two key concepts: that of needs, in particular the essential needs of the world's poor, to which overriding priority should be given; and the idea of limitations imposed by the state of technology and social organization on the environment's ability to meet present and future needs (WCED, 1987).

In Kenya, land ownership, access, and use have always been entwined. One of the most critical challenges Kenya faces today is the management of land and natural resources. The 1954 Swynnerton Plan accorded secure individual land titles to African farmers. The plan was further fortified by the 1959 Native Lands Registration Ordinance, which was replaced after independence by the 1963 Registered Land Act and the 1968 Land Adjudication Act. Though the registration process appeared to increase tenure security for many land owners, it also created new forms of disputes, such as provocations over registered land and conflicts over land sales (Shipton, 1988). Furthermore, the high cost of registration hindered the streamlining of registrations after land transactions, such as inheritance and sales.

This text investigates the nature and causes of land conflicts in Kenya using Taita Taveta as a case study. It also explores options that can be pursued to transform these conflicts into sustainable development.

Land Restoration. http://dx.doi.org/10.1016/B978-0-12-801231-4.00021-5

4.2.1.1 TAITA TAVETA COUNTY

The county of Taita Taveta borders Makueni, Kitui, and Tana River to the north, Kajiado to the north-west, Kilifi and Kwale to the east, and Tanzania to the southwest (GOK, 2005). It has a population of 284,657, as cited in the Republic of Kenya Ministry of Planning and National Development's 2009 population and housing census results. It covers an area of 16,975 km^2. The land is classified as arid and semiarid (hereafter ASAL). The bulk, made up of 11,100 km^2, is within Tsavo east and Tsavo west national parks. There are also water bodies, such as Lakes Chala and Jipe, and the hilltop forests. The people of Taita Taveta, the Wadawida and Watuweta, occupy the remaining 5,876 km^2. They speak *kidawida* and *kituweta*, which are Bantu languages. Bantus first moved to the area of Taita Taveta in approximately 1000–1300 A.D. (Vogt and Wiesenhütter, 2000). They migrated to Kenya through Tanzania in five groups, each group settling in different parts of the present Taita Taveta (Bravman, 1987).

Land conflicts in Taita Taveta have been consistent since Kenya gained independence from Britain in 1963. These conflicts have imposed a heavy toll in terms of destruction of social infrastructure and loss of property and human lives. The economic, cultural, and socioeconomic change in Taita Taveta chiefly depends on land. The population continues to have difficulty conforming to the modern agrarian economy and in coping with land degradation, low agricultural output, and exacerbated conflicts over access to and control of land. Intermittent conflicts over land have caused major social and economic instability. For instance, on October 13, 2010, retired teachers in the area demonstrated in the streets of Voi town against the Kenya National Union of Teachers over an alleged deal to sell 100 acres of prime land, which belonged to the teachers' investment company and had been acquired in 1980 (Nation Correspondent, 2010).

4.2.1.2 THE SOCIAL AND POLITICAL ORGANIZATION OF THE WADAWIDA AND WATUWETA

The Wadawida and Watuweta were organized in patrilineal clans, which were referred to as *vichuku*. Lineage was referred to as *kivalo*, or *kichuku*, and generally had a span of four generations. The nuclear unit in the series of kinship group was the patrilocal extended family known as *kinyumba*. The *kinyumba*'s main functions were to maintain the close relationships of members and give them a sense of community. Furthermore, it was the principal land and cattle owning unit of which the head of the extended family was the legal representative (Mkangi, 1983).

Precolonial Taita Taveta territory was divided into administrative *malolo* (singular: *ilolo*), which translates as "districts." The boundaries were strips of no-man's lands, which were made up of high, bare, uninhabitable plateau, too steep slopes, or the like. In more densely populated parts, one could find small rivers without natural vegetation in their beds. Prinns (1952) claims that religious ties existed among the people and were of great importance. Kinship ties also formed an integral part of relationships, but neither were they peculiar of this unit. The main principles of this kinship were common territory and political identity, rights over land, and conformity of culture; a vague expression shared by the Wadawida and Watuweta as a basis for unity. However, the district was the smallest unit in which being culturally distinct was both felt and expressed (Prinns, 1952).

Rights over land during precolonial Taita Taveta were held by other units and by kin groups in general and specifically by the *kinyumba* whose leader held the land rights and owned gardens, on its behalf, within the *ilolo* and could acquire any of the available cultivatable parts within the

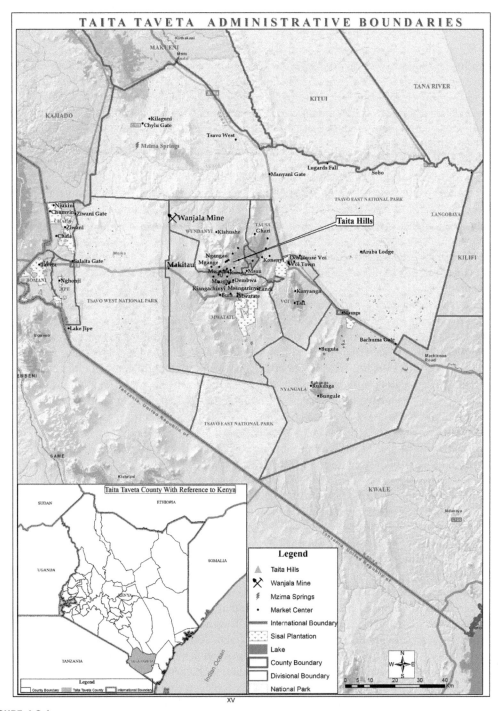

FIGURE 4.2.1

Taita Taveta Administrative Boundaries

boundaries. According to Paola Fleuret, structural aspects of the social organization at the community level influenced access to the management of land, water, livestock, and labor. At the household level, gender-based division of responsibilities and expectations among men, women, and children influenced the use of individual and household resources (Fleuret and Fleuret, 1991). Resource management strategies were thus embodied by variation in household structure, by the interests of men and women, which were not necessarily congruent or interdependent, and by suprahousehold relations. Married women obtained rights to use land through their husbands, but husbands could allocate land they owned to others, for example, their brothers and cousins and other male kin members. Thus, decisions about the division of plots were largely made by husbands/men. Unmarried women had lesser and insecure access to land through their fathers or brothers and generally had to request renewal of land use rights on an annual or seasonal basis from their brothers or fathers.

During the precolonial period, the geographical location of the people of Taita Taveta determined the type of economic activity. The people were distributed in their five administrative divisions. Those located in the highlands were mainly involved in commercial and subsistence farming, with the main cash crops being bananas, beans, cotton, sugarcane, maize, sisal, coffee, cassava, sweet potatoes, exotic and tropical fruits (especially mangoes and avocados), and many other horticultural products. In addition, beekeeping and bee honey production was an important economic activity in areas like Voi and Wundanyi.

This was different during the postcolonial period. The upper zones of the Wundanyi division suffered acute food deficits mainly because of high levels of soil erosion. Farmers were more inclined to grow high value horticultural crops, most of which were sold in neighboring towns, such as Mombasa. However, most farmers sold their produce to middlemen instead of selling directly at Kongowea, a wholesale market in Mombasa, because the middlemen from neighboring communities had adequate vehicles to handle tough terrains, as well as storage facilities to preserve the agricultural goods. As a result, the Wadawida and Watuweta became vulnerable to manipulation by middlemen when prices were set far below prevailing market prices in Mombasa. People could not manage the type of irrigation infrastructure required to propagate land in such areas. Though mining activities yielded high returns, they also came with high risks, such as huge open casts left behind, which posed a risk to the residents, especially children. Effluent draining into the ground and contaminated water sources, as well as flying stone pieces, were other risks involved. This was not received well at all by the locals.

Voi town, which was also a home to the Maungu Buguta Settlement Scheme, was strategically located along the junction of the railway lines from Nairobi to Mombasa and Taveta, which presented infinite opportunities for potential entrepreneurs and a big market for agricultural produce. The road network in Voi also links the town to other major towns like Mombasa and Nairobi. Public bus and van services are available day and night. However, the people of Taita Taveta often engaged in conflict over the settlement scheme due to the fact that they were squatters on their own land. Squatting in the area could be traced from 1892, when the Catholic Holy Ghost Fathers Missionaries arrived in the area and established the Bura Mission on 1,000 acres. The Catholics believed in both religious enlightenment and vocational training, such as material well-being. They appropriated land for building their centers and for other uses, particularly agriculture (Gichia Njogu, 2004). The inhabitants resented the loss of this land and, as a result, conflict emerged between them and the missionaries. Other lands acquired by missionaries included Wusi mission and Mbale mission. In many cases, the land acquired by the missionaries was more fertile than other land nearby, and people were forced to live as squatters on the low land. Some people also lived as squatters at sisal estates. However, they encountered forceful evictions by sisal estate owners, who even used firearms and destroyed the property of the squatters. Others took refuge near the park, as they claimed that the land allocated to the park was more fertile. This was very dangerous and resulted in the deaths of both humans and animals, as they competed for space.

Tausa and Mwambirwa were the least populated divisions because they are located in the lowlands with inadequate rainfall and were only suitable for range activities. A dominant problem in the two divisions was the poor road network, which led not only to overpriced food and nonfood items, but also to a general lack of investments due to transportation dilemmas. There was also low farm production due to lack of arable land and adequate food reserves.

4.2.1.3 LAND CONFLICTS OVER THE ESTABLISHMENT OF TSAVO NATIONAL PARK

Tsavo National Park (20,720 km^2) is one of the oldest and largest parks in Kenya. It was named after the Tsavo River, which flows from west to east through the national park. It is located partly in Taita Taveta and occupies an area of about 11,747 km^2. It was opened in April 1948 and is divided into east and west sections on the A109 road by the Mombasa-Nairobi railway. It borders the Chyulu Hills National Park and the Mkomazi game reserve in Tanzania. The park is very popular with tourists due to the diversity of wildlife including the famous "big five" (lion, black rhino, Cape buffalo, elephant, and leopard). The park is also home to a great variety of bird life, such as the black kite, crowned crane, lovebird, and the African sacred ibis. Tsavo east is the larger of the two. It is generally flat, with dry plains across which the Galana River flows. Tsavo west is more mountainous and wet, with swamps, Lake Jipe, and the Mzima Springs. It is known for its bird life, large mammals, and its black rhino sanctuary (World Database on Protected Areas, 2000).

The indigenous people became squatters on their land, as the park occupies at least 65% of Taita Taveta. This caused land conflicts due to the scramble for land between the locals and the government ministry in charge of tourism, forests, and wildlife in 1962. The conflicts arose over park boundaries, human destruction of wildlife habitat, destruction of crops and farms by wildlife, and the loss of flora and fauna biodiversity (Cobb, 1976).

The existence of Tsavo National Park between Taita and Taveta districts restricted the free movement of people and goods and hampered development of the districts. Before its establishment, the area was a concurrence zone where trade took place between the Wadawida and Watuweta and their neighbors. Inhabitants cultivated the land for subsistence farming. However, they were rendered landless and left to be squatters on their own land after the government allocated the land to Tsavo National Park. This was met by resistance from the Wadawida and Watuweta who felt cut off from accessing their once common ground for trade (Wright, 2005). The earliest records of human-elephant conflict in Tsavo are from 1916, when the district commissioner of Voi asked permission from the government administration to allow the local people to kill elephants that were damaging crops. One local man recalled killing several elephants to protect his maize and other crops and to sell the ivory in the early 1940s (Kasiki and Smith, 2000).

There was a motion in support of Taita Taveta district during a parliamentary debate in 1981, which suggested that state- and foreign-owned land should be given to the squatters (Kenya National Assembly Official Records, 1981). Members of Parliament argued that in normal circumstances animals lived in the bushes while humans lived on fertile land for agriculture and for their dwellings. However, in the case of Taita Taveta, animals had been given the fertile land and human beings were barely surviving on the arid land. One of the components of the greed and grievances theory entails that faster economic growth reduces risk, likely because it raises the opportunity cost of joining a rebellion. The government's vision to allocate a large part of land to Tsavo National Park can be viewed as the potential for faster economic growth in terms of the park's appeal for tourists. This therefore raises the opportunity costs of the people of Taita Taveta joining a rebellion to fight for their land, largely occupied by the park, while they are squatters. According to Collier and Hoeffler (2000), the higher the dependence on primary commodities, such as natural resources, the greater the risk of conflict. This is due to the resources being a main financial component of rebel groups and a weak governmental structure. For example, overdependence on the Tsavo National Park will lead to lower-than-average wages, which will contribute to conflicts, as workers will be dissatisfied. There is therefore a need to diversify the economy.

However, transformation of this conflict can be achieved if communities become involved in decision-making processes that affect the coexistence between them and the park and if they are given the opportunity to benefit from the park (e.g., through job creation).

The proposed Kenya Natural Resources Benefit sharing bill of 2014 could aid in the transformation of conflict over the Tsavo National Park. Part four of the bill specifies that 20% of the revenue collected from natural resources should be paid into a sovereign wealth fund established by the national government. It also stipulates that 60% be shared between national and county government in the ratio of 60% to the national government and 40% to the county government.

4.2.1.4 LAND CONFLICTS BETWEEN WORKERS IN THE MINES AND OWNERS OF THE MINES

Mining activities have very high yields, despite the very high risks to the environment and life. In Taita Taveta, mining has culminated in a struggle, which has led to disputes, displacements, and killings. For example in January 2009, in Kamtonga Village, Mwatate, which is a mineral-rich district, a family was evicted from their home by more than 400 villagers who raided the piece of land to excavate gemstones. This was after word circulated that the home was sitting on deposits. The attackers were workers from nearby large-scale mining companies (Anonymous Respondent, 2009a-e). Later in September 2009, a Scottish gemstone miner, Campbell Bridges, was killed over mining tussles in the Kabanga, Mwatate district. The assailants were armed with arrows, spears, and machetes. His killing was due to an apparent dispute over mining rights (Odula, 2009).

Most of the large-scale mining companies in Taita Taveta hail from outside the area in Nairobi. Examples include companies like the Sanghani/Wanjala Mines and Devki Steel Mills, which have invested billions of shillings. Devki Steel Mills, which had acquired a 300-acre plot at Mbulia Group Ranch in Tausa Division in 2010, affirmed that the construction mining industry would be completed in three phases and would have the capacity to process 250,000 tons of iron ore annually when fully operational. The firm intended to spend 16 billion shillings in the first and second phases of the project and 20 billion shillings in the third stage (Mnyamwezi, 2010). However, the Wadawida and Watuweta did not acknowledge the news of the investment by Devki Steel Mills. They asserted that the Wundanyi MP, Thomas Mwadeghu, had colluded with some Taita Taveta county council officers to issue a mining license to an investor who was extracting iron ore from the land without consulting its residents.

Unlike Collier and Hoeffler, Keen stresses that two forces interact so that greed generates grievances and rebellion, which in turn legitimizes further greed (Keen, 2000). This explains the family eviction in Kamtonga village and killing of the geologist-Campbell Bridges by the greedy, grieving, rebellious groups. Further, we see how economic opportunities and political competition combine to shape the likelihood, character, and duration of conflicts.

Land conflicts over mining rights can be transformed into sustainable development through the sharing of natural resources with the communities from which they are extracted. Kenya's proposed Natural Resources Benefit sharing bill of 2014 suggests that mining and oil exploration companies are expected to sign new benefit agreements with counties that will be statutory and deposited with the Senate if a new wealth sharing bill is adopted into law. For example, this would be done by contributing profits to the community and the environment and fostering community and stakeholder participation early in the process (Zani, 2014). The mining projects must meet the needs of all its stakeholders including shareholders and other financial stakeholders; employees; the community; the environment; and local, state, and federal governments. The firms will have a year to sign the benefit-sharing agreements with respective county governments that will also include nonmonetary benefits after the bill becomes law.

The bill further suggests the formation of a County Benefit Sharing Committee to negotiate with an affected organization on behalf of the county. The committee shall be composed of the county executive committee member responsible for finance, the chairperson of the committee of the respective county assembly responsible for matters relating to natural resources, and five persons elected by the local community where the resource is located.

In addition to the bill, the enactment of policies to safeguard the interests of the workers in the mines provides for a criterion on mining investment and regulation of mining activities. The enforcement of royalties to the communities where minerals are exploited will boost their economic development. The bill further advocates for mining workers' compensation schemes to be revised to avoid exploitation.

4.2.1.5 LAND CONFLICTS BETWEEN WORKERS AND LARGE-SCALE FARM OWNERS

Sisal farming is one of the main economic activities. It was started by colonial settler, Ewart Grogan, who in the early 1920s began to convert what was then seen as wasteland into a productive agricultural region. However, this method of land use conflicted with the local people's subsistence farming and herding of cattle and depleted the grass they had used for grazing. Sisal farming blossomed largely because of the high demand for the "white gold," as the processed sisal was known during World War I and II. A major source of conflict was the diversion of water pipes to sisal estates instead of the area settled by the people. The youth and men often engaged in physical fighting with the sisal estate owners. They would attempt to destroy the water pipes supplying water to the estate. The formation of sisal estates in Taita Taveta often led to the displacement of people. The establishment of the Voi Sisal Estate, for example, also led to the displacement of local cultivators. The local cultivators curbed eviction and fought to secure their livelihoods. The Kenya National Assembly official records show that, on March 12, 1983, squatter houses and property were burnt down by personnel from the Taveta sisal estates. This included six homesteads, comprised of 22 huts situated in Kimala and Jipe (Kenya National Assembly Official Record, 1983).

The conflict here can also be explained by rebel aspirations to sources of funding for rebellion that would provide the would-be insurgents effortless access to natural resources. The youth and workers in the sisal estates represented the rebels.

This conflict can be transformed through the incorporation of development ethics to determine the rights and responsibilities of various stakeholders, the empowerment of the displaced to avoid victimization, and allowing stakeholders to share equitably in benefits of the land resource. This will be achievable if the government enforces more interventions to secure legal guarantees.

4.2.1.6 LAND CONFLICTS OVER INFRASTRUCTURE

Wadawida and Watuweta have recorded poor infrastructure as a cause of conflict in the area. The civil society and professionals from the area have petitioned the government to address pertinent issues affecting the local community's infrastructure. Both economic infrastructures, such as public utilities, and physical infrastructure, such as transport networks in the area, are poor. As a result, in November 2010 Voi residents took to the streets and barricaded the Mombasa highway to protest against the poor state of roads in Taita Taveta. Protesters crippled transport between Mombasa and Nairobi for the better part of the day as travelers were left stranded (Manyindo et al., 2010). There are grievances among the people of Taita Taveta when it comes to economic development of the area. The people feel marginalized, which is why they resort to sporadic demonstrations on the streets, which further paralyzes economic activities in the area and in neighboring areas, resulting in heavy losses. Grievances take a heavy toll in this case, which results in conflict.

This conflict can be transformed if both local and national governments budget to improve infrastructure such as road transport systems and community buildings such as city markets, social halls, street lighting, and schools, among others. These would contribute to sustainable development. For example, improvement of roads will enhance trade activities as farmers will be able to transport their produce easily and will be able access markets even outside the counties.

4.2.1.7 LAND CONFLICTS BETWEEN WIDOWS AND THEIR IN-LAWS

In Taita Taveta, often due to the scarcity and competition for land, widows have been dispossessed of their deceased husband's land. They have, therefore, engaged in conflict with brothers of the departed husbands regarding rights of ownership and inheritance. The view concerning widows' deprivation in such instances is that they can always get married to other men who will give them property, such as land. Thus, brothers of the deceased take advantage and grab the land, depriving the widow of her rights.

In the Bura area in 2009, for example, three widows were forcefully evicted from their matrimonial homes by brothers of their deceased husbands. One widow in protest colluded with her brothers, along with a group of about eight young men from her neighboring village, to fight her two brothers-in-law and force them out of her home. The fight ended with one of the in-laws being hospitalized at the Voi district hospital in critical condition. However, her brothers-in-law had succeeded in forcefully evicting her. The matter had been reported to the chief and is waiting to be resolved (Anonymous Respondent, 2009a-e). As stipulated by Collier, combatants' desire of the deceased brothers land for their self-enrichment emerges as another cause of the conflict, in accordance with the greed versus grievance theory.

This type of conflict can be transformed into sustainable development through increasing awareness of the Law of Succession and Inheritance Act of 1981. Such awareness could ensure that acts of injustices on widows be avoided and allow them access to land. This will enable them to contribute to the well-being of their families by engaging in farming activities to sustain their livelihoods, both for subsistence and commercial use.

4.2.1.8 LAND CONFLICTS BETWEEN TAITA TAVETA AND THEIR NEIGHBORS OVER AREAS OF JURISDICTION

During a leaders' meeting at Wundanyi in 2009, there was no agreement on boundaries between Taita Taveta and the neighboring districts. The boundaries were unclear to the residents, the county council of Taita Taveta, and the district and provincial administrations. Elders from the area have complained that their neighbors had encroached upon their districts each time there was a conflict. Various meetings attempting to determine the district boundaries were held between elders, councilors, members of Parliament, district and provincial commissioners from Taita Taveta district and coast province, and their counterparts from Kibwezi district and eastern province. For example, the town of Mtito Andei was claimed by Taita Taveta's municipal council who had collected revenue from it since independence. As a result, there were demonstrations by the Kamba and Taita ethnic groups in September 2009 following a dispute at the highway township of Mtito Andei. Taita Taveta county council had put the border at the Mtito Andei River, which meant that the Mtito Andei town was under the jurisdiction of Taita Taveta. The root cause of the dispute between Makueni and Taita Taveta was the revenue collection from the busy highway businesses at Mtito Andei (Correspondent, 2009).

There have been other unresolved issues of the boundaries between Taita Taveta and Kwale district. Since independence, the Wadawida and Watuweta have had boundary disputes with the Mijikenda of neighboring Kwale. According to the Wadawida and Watuweta, the boundary stretches up to Taru Hill, while the Mijikenda in Kwale claimed that the border was the Miasenyi trading centre in Voi District. The leaders also accused the government of failing to address the boundary disputes between Taita Taveta and its neighbors in the Kibwezi and Kajiado districts. However, leaders from both Kwale and Taita Taveta have used the boundary disputes to their advantage in order to win more votes to their sides (Anonymous respondent, 2009a-e). The Wadawida and Watuweta have voted for their leaders with the hope of resolving the boundary problems as they had been promised, but to no avail.

In 2003, President Mwai Kibaki established Taveta district, carved out from what used to be Taita Taveta district. According to Mwandawiro Mghanga, the problem was that the decision to create the new district was made without consulting the people of the districts and without determining the boundaries between Taveta and the new Taita district. This added yet another source of conflict. Furthermore, the residents of the two districts constantly called for the removal of the park from the districts. An administration chief affirmed that each time he patrolled the area he perceived to be his jurisdiction, he would meet with his fellow colleagues doing the same (Anonymous Respondent, 2009a-e). He added that every fortnight there was a conflict between him and his colleagues and the people they represented, for there were no clear boundaries. His predecessor had engaged in a fight after a confrontation with a colleague over their areas of jurisdiction and the people they represented, which is how he acquired his position. This further shows that even those in the local administration were not sure of their own areas of jurisdiction, thereby paving the way for the clans to easily engage in land conflict.

Incentives of conflicts initiated by greed in this case are manifested in multiple ways, including economic gain through control of goods and resources, such as land or by increased political power.

There is a need to shift communities toward peaceful coexistence with their neighbors as opposed to scrambling for land resources and fertile land. The loss incurred during conflict between communities is massive when compared with the gains made when communities adjust to peaceful coexistence. Communities can engage in transformation of arid and semiarid land to arable land through irrigation systems, which will help address rural poverty. Facilitating farmers in these areas with agricultural skills of horticultural produce will enable them to produce enough to feed themselves and generate income.

Peaceful coexistence is a prerequisite for sustainable development. We have seen how land conflicts impact negatively on a community in conflict. However, when parties in conflict choose to peacefully coexist, social cohesion is attainable and people are able to focus on promoting economic development and eradicating poverty. This will place countries on the path to achieving sustainable growth and development.

4.2.2 CONCLUSION

While the theoretical approach of greed versus grievances may not offer an understanding of land conflicts in Taita Taveta, it does shed some light on some of these conflicts, and therefore the model cannot totally be disregarded. This is because many factors come into play with conflict, which cannot be ascribed only to greed versus grievance.

Land conflicts in Taita Taveta can be traced back to the history of settlement in the region. There were a number of factors that contributed to the conflict. For instance, political factors, such as the entrenchment of the colonial administration in Kenya, which led directly to inequality in land ownership and use, landlessness, squatting, land degradation, and the resultant poverty and Africans' resentment of the white settlers. The struggle for survival involved fighting over scarce resources (Callinicos, 2010). It is evident from the conflict that the community was not involved in the decision-making process in the area, therefore causing grievances and an environment for conflict. The people have blamed their leadership for the land problems they have encountered.

Land conflicts have led to loss of biodiversity and other forms of land degradation. This has resulted in decreasing yield growth, hence increased food insecurity at the global level. However, food security can be achievable in both developed and developing countries by dealing with constraints that limit higher yields in such regions. Such actions include increased investment in agricultural research and addressing market conditions and rural services, reducing postharvest losses by investment in processing and storage in developing countries, and increasing public awareness in developed countries to change food consumption habits that lead to food losses. Improved water productivity is also vital to increase yield in the areas where water productivity is low. Bioenergy studies have proved that the usage of second generation feedstock provides some potential for liquid bioenergy that does not compromise food security and biodiversity (Nkhonya, et al., 2012). Community involvement in the decision-making process over natural resources reduces land conflicts because the local people become more accountable and engaged in the management of the natural resource for the common benefit of the community. As a result, benefits obtained from the natural resource contribute to the wages of community workers sustaining livelihoods and increasing food security.

Similarly, respect for women's land rights allows for the women to gain access to land use and ownership. This paves the way for women accessing credit from formal sources. It also increases women's interests in land conservation. The money obtained can be used to purchase farm equipment and enable the women to venture into commercial and subsistence farming, which will go a long way to educate their children and sustain their families while bringing about development in the society.

REFERENCES

Anonymous Respondent, 2009a. Oral Interview. December 2, Bura, Kenya.

Anonymous Respondent, 2009b. Oral Interview. December 3, Voi, Kenya.

Anonymous Respondent, 2009c. Oral Interview. December 3, Wundanyi, Kenya.

Anonymous Respondent, 2009d. Oral Interview. December 4, Mwatate, Kenya.

Anonymous Respondent, 2009e. Oral Interview. December 4, Jipe, Kenya.

Bravman, B., 1987. Making Ethnic Ways: Communities and Their Transformations in Taita, Kenya, 1800 to 1950. James Currey, Oxford, UK..

Callinicos, A., 2010. The Revolutionary Ideas of Karl Marx. Bloomsbury, London.

Cobb, S., 1976. The Distribution and Abundance of the Large Herbivores Communities of Tsavo National Park (East), Kenya. Ph.D Thesis, Oxford University, Oxford, UK.

Collier, P., Hoeffler, A., 2000. Greed and Grievance in Civil War. The World Bank Policy Research Working Paper 2355, Washington DC.

Daily Nation Correspondent, 2010. Retired Teachers and Knut Officials wrangle over Land. Daily Nation. Nairobi, Kenya.

Fleuret, P., Fleuret, A., 1991. Social organization, resource management, and child nutrition in the Taita Hills, Kenya. American Anthropologist, New Series 93 (1), 91–114.

Gichia Njogu, J., 2004. Community-Based Conservation in an Entitlement Perspective; Wildlife and Forest Bio-diversity Conservation in Taita, Kenya. Afr. Studies Center, Leiden, Kenya.

GOK, Ministry of Planning and National Development, 2005. Taita Taveta District Strategic Plan 2005-2010 for the Implementation of the National Population Policy for Sustainable Development. National Coordination Agency for Population and Development.

Kasiki, S.M., Smith, R.J., 2000. A Spatial Analysis of Human-Elephant Conflict in the Tsavo Ecosystem Kenya. Gland, Switzerland.

Keen, D., 2000. Incentives and Disincentives for Violence. In: Berdal, M., Malone, D. (Eds.), Greed and Grievance: Economic Agendas in Civil Wars. Lynne Rienner Publishers, Boulder, CO, pp. 19–43.

Kenya National Assembly Official Record (Hansard), 1983. Burning of Squatters Houses and Property in Taveta. March 15–July 12.

Kenya National Assembly Official records (Hansard), 1981. Taita/Taveta District State and Foreign Owned Land to be Given to the Landless. Nairobi, Kenya.

Manyindo, J., Kitimo, A., Bocha, G., 2010. Kenya: Residents Block Busy Highway to Protest at Poor Road Network. Daily Nation. November 9.

Mkangi, G.C., 1983. The Social Cost of Small Families and Land Reform. A Case Study of the Waitaita of Kenya. Pergamon Press Ltd, Oxford, UK.

Mnyamwezi, R., 2010. Taita to Host New Steel Plant. The Standard. November 19.

Nairobi Chronicle Correspondent, 2009. Conflict Between Kambas and Taitas Brewing. The Nairobi Chronicle On-line J. September 13.

Nkhonya, E., et al., 2012. Sustainable Land Use for 21st Century. United Nations.

Odula, T., 2009. Campbell Bridges Dead: Mob Kills Famous Geologist In Kenya: Police Report. Huffington Post. http://www.huffingtonpost.com/2009/08/13/campbell-bridges-dead-mob_n_258346.html.

Prinns, A.H.J., 1952. The Coastal Tribes of the Northeastern Bantu Pokomo. Nyika, Teita.

Shipton, P., 1988. The Kenya Land Tenure Reform: Misunderstandings in the Public Creation of Private Property. In: Downs, R.E., Reyna, S.P. (Eds.), *Land and Society in Contemporary Africa*, Hanover: University Press of New Hampshire, London: University Press of New England.

Vogt, N., Wiesenhütter, J., 2000. Land Use and Socio-Economic Structure of Taita-Taveta District (S-Kenya) – Potentials and Constraints. October.

World Commission on Environment and Development (WCED), 1987. Our Common Future. Oxford University Press, Oxford, UK. p. 43.

World Database on Protected Areas, 1948. Tsavo East National Park. Available from: http://protectedplanet.net/sites/752. Accessed on October 2, 2010.

Wright, D.K., 2005. New Perspective of Early Regional Interaction Networks of East African Trade: A View from Tsavo National Park Kenya. Afr. Archaeological Rev. 22(3).

Zani, A., 2014. The Natural Resources Benefit Sharing Bill. Nairobi, Kenya.

CASE STUDY: TARANAKI FARM REGENERATIVE AGRICULTURE. PATHWAYS TO INTEGRATED ECOLOGICAL FARMING

4.3

Tom Duncan

Home Ecology & Brooklet Farm Education, Brooklet, Australia

4.3.1 CASE STUDY: INTRODUCTION

Ben Falloon (Figure 4.3.1) is a fourth-generation farmer on a 400-acre property in Woodend, Central Victoria, Australia. He is pioneering a polyculture planned grazing system within a holistic management framework. The polyculture system includes cattle, chickens, sheep, pigs, and ducks while vegetables and fruits are cultivated in zones that benefit from passive irrigation, and natural fertilizer is provided by on-farm animals, composted manure, and biomass residues. The relationship between the cattle and chickens is dynamic, cattle graze cells intensively for 1–5 days, and chickens then graze that cell in a mobile chicken barn called an *eggmobile*, feeding off insect larvae around cattle manure. This provides a protein-rich diet for egg production while also fertilizing the pasture with chicken manure.

Irrigation tanks called *K-line irrigation pods* are laid out on pasture land in a snaking pattern within the chicken grazing cell to maximize the surface area coverage of drip irrigation pipe. Irrigation is scheduled for nighttime to reduce evaporation and conserve precious water resources. Mimicking nighttime rainfall boosts grass pasture production significantly as a result of less evaporation and higher volume of water available to grass roots. Enhancing crop water use efficiency is just one aspect of boosting pasture production and reducing waste. The beneficial effect of irrigating pasture covered by chicken manure is rapid regrowth of pasture due to soluble nitrogen being fertigated into the grass pasture, providing soluble nitrogen to the grass ecosystem of roots, fungal, and bacterial rhizosphere associates. Rapid uptake of nitrogen by the diverse grass pasture not only provides rapid regrowth of grass, it may also reduce the amount of nitrogen gas released into the atmosphere from chicken manure. The process of irrigating at night immediately after chickens lay manure on the pasture liquefies some components of the manure and drip-feeds the nutrients to the grass roots, where nitrogen is rapidly taken up by the grass ecosystem. Chicken manure left to dry on pasture without irrigation speeds up the process of ammonia becoming volatile and releasing nitrogen gas to the atmosphere, where it acts as a climate change gas (albeit not a significant source). The climate benefits if rapidly irrigating pastured chickens requires further study to quantify climate change benefits from reduced nitrogen

Land Restoration. http://dx.doi.org/10.1016/B978-0-12-801231-4.00022-7

FIGURE 4.3.1

Ben Falloon and his daughter, Maya, in the "Eggmobile" attached to a tractor. The eggmobile houses chickens and provides egg-laying roosts and egg collection areas. The eggmobile is a movable barn for chickens, following the food source provided by protein-rich insects that colonize cells recently grazed by cattle.

Credit: Copyright © 2016 Home Ecology and Brooklet Farm Education.

release to atmosphere. Deeper root growth in the grassland which then senesces leaves significant stores of carbon in the soil, further adding to the climate benefit of the integrated ecological agriculture approach, as compared to conventional grassland management where cattle and chickens are not grazed holistically or in cells. This is suggested as a research topic for field researchers and experimental agriculture scientists to further quantify the benefit of nighttime irrigation of chicken cells and deeper root growth from extra soluble nutrients entering the grassroot zone.

The nighttime K-line irrigation management practice makes the pasture suitable for grazing again in a far shorter time than conventional monocultural grazing systems. This results in a higher rate of production than most other farms in the district. Central Victoria farming has been marginal for since the 1980's due to its changing rainfall pattern resulting in less rain (Figure 4.3.2), while the relatively small land holdings and the demise of sheep farming on small to medium-sized farms has meant many farms are no longer productive land systems. Rainfall is approximately 600–800 mm annually, and summer comes late compared to surrounding areas due to the higher altitude of 500 m of elevation in the Macedon Ranges.

4.3.2 DECLINE OF FAMILY FARMS

The traditional mixed family farm business has been declining in Australia since the 1980s due to issues including deregulation of the dairy, wheat, and wool boards, relatively high bank interest rates compared to other industrialized nations, and various free market initiatives that reduce tariffs on imported

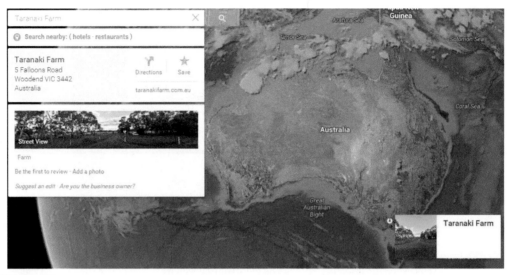

FIGURE 4.3.2

The location of Taranaki Farm in Central Victoria, adjacent to large land areas to the north that have suffered desertification. Central Victoria is at risk of desertification as well, but farms like Taranaki Farm are leading examples of best-practice agriculture management against desertification.

Credit: Copyright © 2016 Home Ecology and Brooklet Farm Education.

produce. This process of decline enabled the consolidation of small cattle, sheep, and wheat farm holdings into large corporate-owned farms and also contributed to the fragmentation of farms into low-productivity properties on smaller plot sizes. For example, the author's third-generation organic permaculture farm was originally part of Taranaki Farm's land, then owned by Jack Falloon, and was sold to the author's grandfather when economic conditions were unfavorable for chicken farming. This fragmentation into a smaller farm made it even more difficult to earn a living because competitiveness has been closely linked to economies of scale and size of land holdings.

Australia's climate is also extremely variable, with long periods of drought, sometimes lasting 12 years or more. This is often followed by several years of flooding; in the last decade, this pattern has occurred over more than 67% of the continent. These extreme climate events have led to a decline of family farms (Figure 4.3.3) because many farms were not resilient in the face of changing climate and shifting market dynamics.

Against this general trend of family farm decline, new ways of managing the soil, water, animals, pasture, and crops have enabled some farmers to go beyond these challenges and thrive. This case study focuses on the use of a combination of holistic management, permaculture, and Keyline design at Taranaki Farm (Figure 4.3.2.1). These management practices, principles, methodologies, and tools ensure water supply year-round for livestock, irrigation, and soil development, bringing about a significant increase in pasture productivity. The practices were developed with input from such specialists as Darren J. Doherty of Regrarians Ltd., a non-for-profit company based in Victoria, Australia, dedicated to the 'regenerative enhancement of the biosphere's ecosystem processes' providing training courses globally.

FIGURE 4.3.2.1

Satellite image of Taranaki Farm illustrating the coupling of road layout with Keyline drains and dam delivering water to pasture and animals. Approximately nine Eggmobiles can be seen centre-left following cattle cell movements.

Credit: Copyright © 2016 Home Ecology and Brooklet Farm Education.

FIGURE 4.3.3

Multigeneration family farm with Maya Falloon and a sheepdog. Eggmobiles can be seen in the distance with pastured chickens.

Credit: Copyright © 2016 Home Ecology and Brooklet Farm Education.

4.3.3 THE RISE OF RESILIENT FARMS—KEYLINE DESIGN

The need to design farms to be resilient to droughts, fires, and floods has produced two Australian methods of holistic design and management, which are now at the basis of sustainable land management and agriculture: Keyline and permaculture. Keyline design was popularized in P. A. Yeomans's book *Water for Every Farm* (1965). Yeomans pioneered the use of larger farm dams for irrigation in Australia, as well as designing, manufacturing, and supplying chisel plows and subsoil aerating rippers. He developed the Keyline comprehensive system (Figure 4.3.4), consisting of amplified contour ripping to control rainfall runoff and enable controlled flood irrigation of undulating land without the need for terracing. In many regions of the world, undulating land is terraced to control water and nutrients in cropping and animals systems; however, the process of making terraces involves significant labor or machinery, as well as above-ground earthworks, which can result in significant loss of soil, nutrients, soil flora, and fauna, contributing to waterway and marine environment pollution. The Keyline system's use of contour ripping enhances water infiltration into pasture and crop land; the subsequent buildup of soil structure—in combination with holistic management and planned grazing—prevents erosion or nutrient and soil particulate export from the property. Additionally, Keylines aerate the soil with Keyline deep rippers, which enhance water-holding capacity and movement of humus-building microorganisms and soil-building fauna.

The core of Yeomans's Keyline concept is to improve soil quality quickly, through targeted intervention via strategic soil and water management. Figure 4.3.5 demonstrates Keyline design developed by Darren Doherty (Doherty and Jeeves, 2015) of Regrarians, being combined with both the holistic management decision-making toolkit for planned grazing, developed by Allan Savory (1999), as well

FIGURE 4.3.4

Keyline ripper lines following 'Keyline Pattern Cultivation' guidelines can be seen in these photos. Keyline water conservation channels with overflows are not enough for the modern farm; instead, K-line irrigation pods combined with holistic management inspired the movement of cattle and chickens rapidly from cell to cell, boosting productivity significantly with low water use and no pollution.

Credit: Copyright © 2016 Home Ecology and Brooklet Farm Education.

FIGURE 4.3.5

Elements of design include Keyline design, holistic management, and planned grazing across the landscape with cattle and chicken eggmobiles. The chicken eggmobiles move across the landscape, and their previous positions are clearly illustrated by the new green grass growth in patches, where the K-line irrigation pods have provided nighttime irrigation on chicken-manured pasture.

Credit: Copyright © 2016 Home Ecology and Brooklet Farm Education.

as permaculture animal polycultures, a design developed by Bill Mollison (Mollison, 1988) and David Holmgren (2002).

The result of this combination is a holistic and regenerative agricultural system that includes the following:

- The acceleration of soil-building activity
- An increase in nutrients
- An increase in soil water-holding capacity through an increase in carbon matter
- A larger total biomass output than simple linear farming systems
- No water pollution
- Multiple income streams from a variety of livestock animals

The use of Keylines also often incorporates dams, which are linked by Keyline swales (Figure 4.3.6). These water conservation and distribution channels often involve a 1:400 fall (i.e., every 400 m, the Keyline channel descends 1 m in the landscape), which dramatically slows the flow of rainfall over the landscape. During large rainfall events, Keyline channels deliver the excess water to the dams, which may fill up and overflow or backfill the drains and overflow over designated gates and banks into pastures or crop areas according to farming or rotation-cropping activities. Thus, the Keyline channels guide excess water to prevent damage to surrounding land and to ensure optimal water use.

This concept of water storage through Keyline dam use was outlined by Yeomans in 1954. His objective was to provide water for every farm and to pioneer resilient and regenerative farm

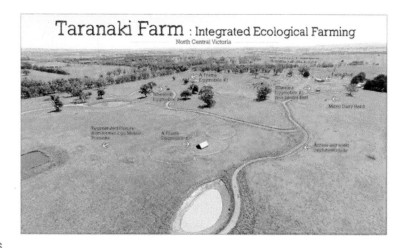

FIGURE 4.3.6

Dams provide water for the K-line irrigation pods, and in large rainfall events, water spills onto pasture, passively irrigating the grass. K-line irrigation shows grass regrowth from former eggmobile locations. The overall pattern follows the Keyline design. Moving from pattern to details, polycultures of mixed pastures, cattle, chickens, and K-line irrigation are incorporated to boost productivity and reduce waste.

Credit: Copyright © 2016 Home Ecology and Brooklet Farm Education.

infrastructure. Creating a climate-resilient water infrastructure is not new—as evidenced by the use of terrace paddies in Asia and chinampas in the Americas over the last two millennia—but Keyline's framework and design methodology is unique in that it is paired with specific deep rippers and subsoil aerators that use the soil as the primary store of water, by increasing soil carbon from senescing grass roots. This pairing of technology and management techniques is often the domain of large corporate conglomerates due to the required intellectual and capital inputs; it is rare to find both a design and management system and high-tech equipment and tooling to implement the systems in small-, medium-, and large-scale sustainable agriculture methodologies.

Darren Doherty of Regrarians whom designed the Taranaki Farm Keyline system has recently improved upone the Keyline SuperPlow that Ben Falloon originally developed in 2010. The new Keyline SuperPlow developed by Darren not only rips the soil deeply and sows diverse grass species, it allows deeper grass root growth in one pass instead of over several years by injecting biochar, bonechar and biofertilisers into the soil, creating opportunities for long organic carbon chains to form, all of which dramatically increase soil based water storage and rapid grass root growth downwards into the aerated soil soils that have been ripped by the plow. Doherty states "the Keyline SuperPlow prototype now deep cultivates, injects & sprays compost tea / biofertilisers and sows cocktail / shotgun biological subsoilers and pastoral species all in a single pass." Best known for applying permaculture at the broadacre scale after Bill Mollison requested Doherty take on a leadership role, he has created a new broadacre farm management method called Paddock2Paddock - whereby animal bones are turned into bonechar and animal viscera is turned into biofertiliser for integration back into the soil. Combining bonechar with it's rich calcium phosphate and calcium carbonate content, with the nitrogen rich biofertiliser and injecting this mix into the soil via a Keyline SuperPlow on contour into the landscape. The rapid regenerative management method increases soil carbon and

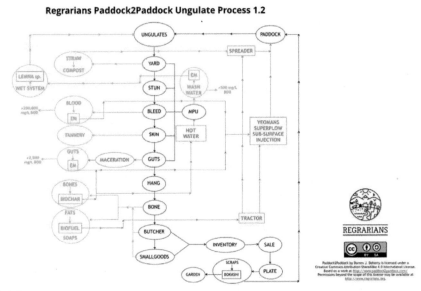

FIGURE 4.3.6.1

The Paddock2Paddock process closes the nutrient gap that Paddock2Plate and also conventional agriculture production and consumption models have, as a result of most bones and nutrients in the viscera going offsite.

Credit: Copyright © 2016 Home Ecology and Brooklet Farm Education.

nutrients dramatically improving pasture growth and animal productivity. Home Ecology and Brooklet Farm Education (Duncan, 2015) have partnered with Doherty at Regrarians to provide the conversion technologies to small and medium scale regenerative farms, that enable farms to turn animal bones into bonechar suitable for integration into pasture sub-soil, and also biofertiliser production utilizing Black Soldier Fly larvae colonies to convert viscera into hygienic nutrients for combining with bonechar and sub-soil injecting via the Keyline SuperPlow. The new Paddock2Paddock methodology as seen in the above Figure 4.3.6.1 will provide a new wave of productivity on small to medium scale farms that slaughter their own animals and sell directly, such as Taranaki Farm and others around the world. It may also provide one of the most cost effective methods to close the nutrient loop in farming, whilst increasing on farm productivity and becoming cost competitive against large abattoirs and meat retailers.

4.3.4 PERMACULTURE—A DESIGN SCIENCE

Permaculture was developed by ecologist and former professor Bill Mollison at the University of Tasmania and his mentee/collaborator David Holmgren who was completing a degree in Environmental Design at the College of Advanced Education in Hobart. Mollison taught structural ecology and fundamentals of systems analysis. He encouraged his students to develop agricultural solutions with a holistic view to generating maximum positive impact on human civilization. He conceived his theory of permaculture when surveying Tasmania's vast old-growth forests and witnessing the complex relationships between plant species and the various trophic levels and layers from soil-based roots and tubers to grasses, shrubs, understory trees, canopy trees, and vines that stretched from bottom to top. Mollison thought the "web of life" that sustains complex peak forest ecosystems could be replicated on farms: diverse cultivated ecologies could mimic nature's use of polycultures to support the needs of the web's various parts for inputs, such as sunlight, water, nutrients, minerals, and fauna.

On field projects as a consulting ecologist and permaculture designer in Africa and the Asia-Pacific, Mollison observed farmers running cattle with birds such as chickens and fowl, a learned tradition stemming from existing symbiotic relationships between birds and cattle as evident in nature. Allan Savory, who coined the terms *holistic management* and *planned grazing*, also observed this predator-prey movement with attendant birds and other animals grazing on herbivore-related insects and parasites. Birds not only eat parasitic ticks that otherwise could paralyze and infect cattle with human diseases such as rickettsia and Lyme disease, but they also feed on fly and other insect larvae, which colonize livestock manure. Thus, the cultures that Mollison observed integrated this knowledge into livestock management, benefiting from healthier livestock and healthier birds, which in turn resulted in better-quality meat, better-quality bird eggs and meat, and less disease by eliminating such factors as ticks and fleas.

This symbiosis supports and reinforces beneficial relationships with net beneficial outcomes to all and at little cost and no ecological harm. Thus, improved agricultural designs can better tackle the problems facing farms with climate extremes, cost constraints, and a lack of scalable technology for sustainable agriculture, as discussed in *Permaculture: A Designers' Manual* (Mollison, 1988).

Holmgren (2002) further developed the concept of permaculture in *Permaculture: Principles and Pathways Beyond Sustainability*, which develops an incisive and comprehensive structural analysis consisting of 12 core permaculture principles of design. Holmgren has implemented his teachings and principles at his home, Melliodora Farm, in central Victoria. Melliodora Farm is a 1-ha farmlet and one of the longest continually running permaculture-designed systems in the world, having been established in 1985. It is comprehensively designed to provide food, fuel, and fiber.

Small family farms can ill afford great capital or equipment expenditures. Thus, Ben Falloon uses a combination of permaculture, Keylines, and holistic management planned grazing, after the example of Joel Salatin of Polyface Farm in Swoope, Virginia, in the United States, which pioneered a combination of these systems into a viable and productive family farm. Falloon initiated his Keyline water design under the guidance of Darren J. Doherty, an approved Keyline designer. In conjunction with Salatin's approach, holistic management and planned grazing was implemented in such a way that cattle are run on the 400-acre farm following the Keyline pattern; chicken flocks are then run in eggmobiles that move around after the cattle, eating the insect larvae that colonize the manure the animals leave behind. K-line irrigation pods are integrated into the eggmobile system for nighttime irrigation, which is water efficient and enables rapid pasture growth and fertilization. Mimicking nighttime rainfall has corresponding water efficiency benefits because lower nighttime temperatures result in less evaporation, and therefore more infiltration and plant uptake versus daytime irrigation. The addition of K-line irrigation pods that can deliver thousands of liters of water to a chicken-grazing area during nighttime is a significant innovation that pushes productivity beyond most farm systems.

Figure 4.3.7 illustrates the Keyline water conservation channels connecting to dams across the landscape. Falloon's eggmobile is shown in the center, with a clearly delineated circle, which is the fence line within which chickens live and graze for approximately 1 week (depending on the season) before moving to the next patch along the Keyline pattern. K-line irrigation occurs via the eggmobile and mobile water carts, which enhance pasture growth; when combined with chicken manure nutrients,

FIGURE 4.3.7

Keyline conveyance swales linking to dams, which provide water for the K-line irrigation pods integrated with the chicken eggmobiles.

Credit: Copyright © 2016 Home Ecology and Brooklet Farm Education.

soil structure and pasture regrowth are accelerated for the planned grazing herd. The K-line irrigation system innovates beyond Keyline because it is very targeted to chicken-mobile manure sites, dramatically increasing pasture growth and productivity.

Following the Keyline pattern has a purpose. In large rainfall events, dams and water conservation and distribution channels overflow at certain points; as the water flows over and through the layers of chicken manure immediately down-grade of the swale, it carries the nutrients in the manure to the rest of the pasture. This promotes passive pasture fertigation and inhibits the intensive spread of manure, which may be washed into local waterways during large rainfall events (which itself is a significant contributor to algal blooms and fish deaths in receiving waterways, lakes, and marine environments).

Falloon's regenerative agriculture innovations have been improved in several ways, including by following cow herds with mobile milking equipment (Figure 4.3.8). This eliminates the waste and pollution that other dairy farms face in concentrating milking cows in a confined space twice daily: the management of nutrients and waste on conventional dairy farms often requires the production of unhealthy wastewater lagoons that spill over into waterways during large rainfall events or, in some cases, require expensive biodigesters (though these do generate power and heat from biogas). It should be noted that low-tech family biodigesters can be built in the developing world for under US$50 with design examples provided by the United Nations Food and Agriculture Organization (FAO). Farmers face cost difficulties in industrialized nations with biodigesters because of extensive environmental impact assessments and engineering for design and construction is required, generally pushing up costs

FIGURE 4.3.8

Mobile dairy cow milking unit that is attached to a tractor and wheeled through grasslands.

Credit: Copyright © 2016 Home Ecology and Brooklet Farm Education.

for biodigesters. Pasteurization and homogenization are consequences of the industrial dairy farm model that packs as many cows into a milking barn as possible, with close proximity and accumulation of feces and disease-carrying bacteria. Conversely, the accumulation of waste and disease and the corresponding waterway pollution rarely happen in nature because the predator-prey relationship keeps herds moving. Savory noted this fact and embedded it within the planned grazing methods.

Falloon's solution enhances grasslands carbon, which in turn increases nutrient- and water-holding capacity, which in turn boosts cattle and chicken production. This method further prevents disease from spreading through the herd and to humans, as well as preventing animals from overgrazing large sections and compacting the soil extensively, which causes soil pugging and anaerobic conditions where disease or harmful soil bacteria can flourish, including foot and mouth disease.

4.3.5 COMPLEXITY AND CHAOS INTO ORDER, FROM PATTERNS TO DETAILS

Permaculture proposes that farm designs follow a scientific approach that understands and observes holism; it should move from the macro to the micro view in a cumulative developmental and nonlinear process to allow the periods and cycles of the seasons to be included. Permaculture should be designed from the pattern down to details, with patterns including seasons; climate; animal migration and breeding cycles; plant growth and seeding; rainfall and catchments; human society patterns of food and nutrition requirements; local community demands for local, fresh, sustainably, and ethically produced food; and local market access and logistics. These patterns set the stage for a self-organizing farm,

FIGURE 4.3.9

Eggmobile following Keyline swales and dams.

Credit: Copyright © 2016 Home Ecology and Brooklet Farm Education.

whereby the design parameters are encapsulated within a set of variables that seek to eliminate pollution, waste, poisons, antibiotics, inefficiencies, animal cruelty, and marketing falsehoods.

Were factory farms to apply holistic management, permaculture, and Keyline design practices (Figure 4.3.9), the use of regenerative and restorative agriculture would increase dramatically, which paves the way toward sustainable food production on a grand scale. The nature of complex cultivation systems is that total biomass output can be greater than simple linear systems that characterize industrial factory farms with their associated wasteful byproducts. The yield gap has narrowed for many organic agroecologically farmed crops, compared to industrial or genetically modified organism (GMO) transgenic types of farming systems as published by proceedings of the Royal Society of London. Having equal yield, no poisons, and most fertilizer and inputs generated on site or locally is delivering better food to consumers at a lesser cost of production while improving the land and water catchment.

The complex model developed for Taranaki Farm exemplifies the ways in which a holistic management approach may deliver a self-organizing system that emits no pollution, waste, poisons, or antibiotics and that entails no inefficiencies or unethical animal management. Some of the system's parameters are embedded within the financial framework of having a viable family farm and strongly integrated with a community-driven demand for local, fresh, sustainable, and ethically produced food. Communities must be educated on the use and benefits of regenerative farming, especially when compared with conventional or industrial farming.

4.3.6 TARANAKI FARM—LOCAL MARKETS FOCUS WITH FINANCIALLY SUSTAINABLE COMPLEX SYSTEMS

Falloon did not start out to develop a self-organizing system that combines holistic management, Keyline design, and permaculture design science. Instead, he simply asked the following question: what is the most regenerative, sustainable, productive, and profitable way to farm and raise animals?

The market power of large megafarms and corporate supermarket chains has made it necessary for family farms to develop their own set of complex supply chains, connected through social media and

FIGURE 4.3.10

Buying clubs enable Falloon to invest in infrastructure and acquire inputs to underwrite his farming and animal-raising activities.

Credit: Copyright © 2016 Home Ecology and Brooklet Farm Education.

including multiple and seasonal product types. Community support agriculture (CSA) is a leading emergent model that provides food to consumers within a 100- or 200-mile radius. This arrangement enables local farms to invest in seasonal activities, such as sowing seeds, harvesting hay, installing irrigation, and ordering packing machinery and boxes, with the confidence that its products will be sold. While this model has not been fully developed at Taranaki Farm, Falloon can invest in future farm activities based on strong community support from the Taranaki Farm Buying Club (Figure 4.3.10).

Ultimately, it is Falloon's buying club and the retailers that sell his products that underwrite his regenerative agriculture enterprise and the financial well-being of his family. An on-farm facility has been built to enable the community to visit the farm and pick up goods. This will significantly improve the economic sustainability of the farm and further build the Taranaki regenerative agriculture brand.

Regenerative farming systems generally need to establish a direct relationship with the consumer in order to attract the best possible price for the products being sold. Competing against conventional produce with large retail conglomerates or food chains is part of the problem, in that it can force farmers to try and cut corners with regard to sustainable and ethical soil, water, and animal management practices. The economics of conventional, industrial, or GMO farming are generally pushing profit margins so low that farmers may resort to unsustainable practices. Falloon overcame these limitations by developing a model where he can send his livestock to the local abattoir and sell the packaged meat in his on-farm store. The resulting increase in overall livestock income is approximately 200%–500% above previous income when sold at the wholesale market (Ben Falloon, personal communication, December 2014).

Significant administrative hurdles to local government issuing a permit for the on-farm store delayed the building of the store (and hence the overall farming enterprise) by several years. This bureaucracy inherently reduces the likelihood of other farmers contemplating the bold move of selling direct from their farm. Local, state, and federal government policy makers need to support on-farm stores and abattoirs if they want to see regenerative land management be taken up by mainstream farmers and boost the rural economy. Cutting red tape and allowing farmers to sell direct to the public and butcher their own animals is a logical step to unlocking the true value of the agriculture sector and rural economy. The slow pace of reform in local, state, and federal policy-making circles for on-farm stores and abattoirs is preventing market competition from small to medium-size farms against large retailers. Government agencies such as Primesafe that regulate small farm abattoirs in the state of Victoria have been described by farms near Taranki Farm as draconian and anti small farm abattoirs. Primesafe are themselves creating a market distortion with their heavy handed-ness which is inherently anti-competitive, to the benefit of large abattoirs and supermarket chains. It is logical to argue that this type of legal and planning bureaucracy, which prevents small to medium-sized farms opening an on-farm store or abattoir, is an anticompetitive legal constraint that is holding back economic growth in rural areas, by preventing local food production and consumption, and therefore the laws and policies need to be vastly simplified and empower farmers and consumers with market choice. To consider otherwise is to support the current anticompetitive duopolies operating the retail market segment in Australia.

Increased income with direct sales to local consumers makes farming attractive to the younger generation and heralds the renaissance of small- to medium-scale sustainable farming. If farms are permitted by local governments and health departments to have an on-farm abattoir or utilize local mobile abattoirs, the financial returns are further increased to a level that excites many young farmers contemplating a return to the land. Removing the go-betweens from the financial equation, such as the local or regional abattoir and the retail outlet, results in farmers and graziers receiving almost all of the profit and value chain from the animals. This is a good outcome for farmers and, importantly, good for the regeneration of the land, soil, and water that the management system and animals thrive within.

Social media platforms also enable transactions to be made without the overhead associated with multiple customers; in some cases, farms can serve up to 400 different homes and retail and wholesale customers by organizing through social media outlets. Complex financial and marketing systems that support complex farms are codependent in developing a restorative and regenerative agricultural sector and business model. The Open Food Foundation (OFF) is an example of an open-source online food hub that enables farmers and consumers to develop their own trading relationship. While this is not a new concept, it is on a larger scale than previously experienced. Falloon may be able to leverage such a platform to enable online purchase aggregation (Taranaki Farm's Facebook page has over 25,000 followers and growing).

4.3.7 ETHICS AND RESTORATIVE AGRICULTURAL ECONOMY

The question of how to avoid pollution, animal cruelty, and overuse of antibiotics within the context of permaculture, Keyline, and holistic management farming has been answered on Taranaki Farm through smart biomimicry design and its use of mobile milking and eggmobile units (Figure 4.3.11). The former

FIGURE 4.3.11

Mobile cow milking unit.

follows the herd around the farm in its migration across the cells in the Keyline design, with the latter following. Such mobile technologies are desirable for many reasons, one of which is the adaptive and nonlinearity of its application and the achievement of multiple holistic management filters. And with mobile phone technologies allowing farmers to connect to customers, we can increasingly say that mobility is a principle of regenerative agriculture. Biodiversity benefits on regenerative farms are increased through the retention of native vegetation, native habitat corridors as part of a bioregional approach, and encouragement of native grasslands, which foster a healthy population of native animals. The corresponding production benefits of having local wildlife include better pollination of crops and higher amounts of beneficial pest management species that control invasive or parasitic species. The result of increased on-farm native biodiversity is enhanced crop and animal production rates and protection of unique biodiversity.

Designing new mobile technologies such as eggmobiles, mobile dairy milking units, and movable solar-electrified fences can improve farm productivity and overall output. The seasonal and yearly observation of mobile cow and chicken systems can be measured in terms of positive impacts on the farm, including production rates, soil carbon increases, and nutrient-holding capacity. Water-holding capacity is also a key metric that can be observed by visual inspection of grasslands and pastures, according to how long it stays green after the last rainfall. This can be accomplished through visual inspection and increasingly augmented through the use of mobile, airborne drones that take seasonal photos. These data can supply information as to the effectiveness of the management system, relating to holistic management's planned grazing, and support permaculture's principle to observe and interact at a systems level. The movement of animal herds, nutrients, and water can be tracked and productivity metrics accumulated to feed back into harvest, sowing, milking, and selling schedules, which are connected to online customer acquisition platforms. All of this supports the emergence of complex, holistically managed agricultural systems that are part of a restorative and regenerative agricultural enterprise and economy.

Regenerative agriculture offers a rich and rewarding avenue for family farms and community co-operative farms to transition to a more sustainable future underwritten by the community. It also offers a method by which the desertification of the world's grasslands may be reversed, leading once more to productive marginal lands. Currently, the regenerative farming branding is an emergent factor in producer branding, but increasingly, as customers learn more about how their food is grown, regenerative agriculture farms will benefit from better brand recognition and customers voting with their dollars.

A key factor in the growth of the regenerative agriculture sector is the development of a comprehensive design manual to guide farmers toward a truly regenerative management system that is holistic and based on proven and effective solutions. Darren Doherty who co-designed and developed some aspects of Taranaki Farm's Keyline system is active in the production of the Polyface Farm documentary with his family and, with Andrew Jeeves the 'Regrarians Handbook' that will provide a very significant integration of permaculture design, Keyline, and holistic management (Doherty and Jeeves, 2015). The intent of the documentary and manual is to shift animal management and family farming toward sustainability, holism, and high productivity. The manual will be a succinct and sequential outline of over 300 integrated methodologies and techniques that have been proven over years of application to work toward regenerating production landscapes. It will offer a rich supply of regenerative agriculture knowledge and practices that are acknowledged by land managers around the world as the key factors to success in their regenerative agriculture enterprises. Taranaki Farm and Regrarian designed farms demonstrate the sustainable and safe way for the world to feed itself by achieving environmental, social and economic objectives in an ethical manner.

Family farm gathering. Credit: Copyright © 2016 Home Ecology and Brooklet Farm Education.

REFERENCES

Doherty, D.J., Jeeves, A., 2015. Regrarians Handbook. Regrarians Ltd., Bendigo, Australia. Accessed February 2015 from http://www.regrarians.org/regrarian-handbook/.

Duncan, T., 2015. Regrarians #Paddock2Paddock Design with Home Ecology and Brooklet Farm Education. Accessed August 2015 from http://www.homeecology.net/regrarians-paddock2paddock-design-with-home-ecology-and-brooklet-farm-education/.

Holmgren, D., 2002. Permaculture: Principles and Pathways Beyond Sustainability. Holmgren Design Services, Hepburn, Victoria.

Mollison, B., 1988. Permaculture: A Designers' Manual. Tagari Publications. Hobart, Tasmania.

Savory, A., 1999. Holistic Management: A New Framework for Decision Making. Island Press, Washington, DC.

Yeomans, P.A., 1965. Water for Every Farm: A Practical Irrigation Plan for Every Australian Property. K.G. Murray Publishing Company Pty Ltd, Sydney, Australia.

REGENERATING AGRICULTURE TO SUSTAIN CIVILIZATION

4.4

Allan Savory[1], Tom Duncan[2]

Savory Institute, Boulder, Colorado[1] Home Ecology & Brooklet Farm Education, Brooklet, Australia[2]

4.4.1 INTRODUCTION

Agriculture, the production of food and fiber, is truly the foundation of civilization: it has enabled humanity to move beyond the constant quest for food and develop such areas of social life as education, writing, scientific research, specialization of skills, business, government, nation building, warfare, and urban life. However, agriculture has at times lead to the downfall or complete collapse of civilizations over long periods of time, as it has rarely been practiced in a sustainable manner and has functioned more similarly to the mining industry (along a linear pathway as opposed to circular); in that it extracts nutrients from the ground but fails to replenish them. Sustaining civilizations has been a constant problem for thousands of years. Throughout history, many civilizations have failed in all regions of the world. Where historically armies have changed civilizations, those civilizations have continued, albeit in changed form, while those that failed because of environmental degradation due to their agriculture practices ceased to exist (Costa, 2012; Lowdermilk, 1942). Today, with more than half the world's population living in cities (UN report), this historical regional pattern cannot be continued. To sustain civilization as urban land overtakes the rural land available for farming, agriculture needs to become more efficient and, therefore, must be based on scientific teachings—mainly those of the biological sciences.

The threat posed by accelerated climate change stems not only from fossil fuels, but also from current agricultural practices and land uses, as well as global desertification. Realizing that climate change is likely to continue even in a post–fossil fuel world (Hare, et al., 2011), Savory began (in the 1990s) to write, lecture, and produce videos stating the problem was not sustainable agriculture, but rather how to sustain civilization. So long as people and media focused on the need for sustainable agriculture, most of the population in cities would think of climate change as an agricultural problem and not one of great concern to cities and nonfarmers. He believed the focus needed to shift toward sustaining cities through a new environmentally regenerative form of agriculture, replacing the current agricultural practices that endanger civilization.

While agriculture is destroying the biodiversity and proper functioning of rivers, lakes, and oceans (this includes the economic functions as well as the environmental functions) as a result of excess nutrients, algal blooms, chemicals, and sometimes inappropriate livestock access to fragile aquatic ecosystems, this section confines itself to an analysis of the state of the land (Duncan, 2009) where agricultural production is more intensely dependent on soil life and quality than on any other factor. Both

Land Restoration. http://dx.doi.org/10.1016/B978-0-12-801231-4.00023-9

the living soil and the plant and animal life needed to sustain civilizations require water, nutrients, and carbon. It is in the soil that complex life ensures the balanced, healthy cycling of these three essentials.

Allan Savory, a teacher of agricultural policy development, often poses this question: "If we were limited to only one measurement every year to evaluate the stability and health of any piece of land—farm, ranch, forest, or national park, for instance—what would that be?" The answer that eventually emerges is that it would need to measure, at the most basic level, the quality of the water running off a piece of land. This would inform as to the amount of water lost from runoff and what that water carried (i.e., soil particles, chemical, other pollutants, etc.). As Savory humorously concludes, if civilization is truly to be sustained, farmers (or perhaps consumers or policy makers) should be asked to drink the water running off farming land from which they consume the produce.

With both droughts and floods increasing worldwide (Cai et al., 2014; Savory, 1999), landscapes' water retention is increasingly the focus of both current forms of agriculture (organic and chemical/industrial). Thousands of years ago, the Nabataeans (CE 37 to about 100), whose cities spread (according to some historians, including Gibson, 2004) from Yemen to Damascus, and from western Iraq into today's Sinai desert, and the Charcoans (CE 828–1126), who lived on land that is today the United States, attempted to sustain their cities by retaining water on their landscapes through sophisticated water runoff harvesting methods. They succeeded for a time, but were eventually overrun by the desertification of the areas surrounding their water harvesting and irrigation projects.

The water, nutrients, and carbon upon which modern agriculture depends are influenced by land managers, including individual farmers and communities over entire watershed areas, river catchments, even vast regions spanning nations, as well as by the agricultural policies of governments and international agencies.

This subsection argues that a new agricultural philosophy is needed if modern civilization, with its urban focus, is to survive the threats of an increasing population, accelerating climate change, and far-reaching desertification. The following subsection expands on the need for such a new philosophy, paying close attention to the different challenges facing different types of environments, specifically the challenge posed by desertification and the need to use rainfall effectively to prevent or reverse this phenomenon. The third subsection outlines various new agricultural methods that have been developed to use water more effectively, including Keyline design, Permaculture, and land reclamation. The fourth proposes Holistic management as the best method available to serve the needs of modern agriculture. The fifth expands on how and why Holistic management must be employed worldwide. The final subsection concludes.

4.4.2 THE NEED FOR A NEW AGRICULTURAL PHILOSOPHY

Civilization is, by definition, urban. However, it survives only if there is sufficient agricultural production. Where some civilizations have failed altogether, environmental destruction brought about by agriculture is often to blame; for example the Mayan and the early civilizations of Mesopotamia (Costa, 2012; Diamond, 2005; Lowdermilk, 1942). Perhaps the longest-known civilization is that at the mouth of the Nile, where civilization has persisted over some 10,000 years, although changing with military conquests, while, along the Nile, civilizations with similar agricultural methods have failed. These varying success levels are associated with the silt that erodes from soil in the upper reaches of the river: the annual silt load of the Nile flows past cities, along the river, depositing on farms at the mouth, enriching the soil every year, which enables continuous production of food. Historically, other long-lasting civilizations have been sited on seashores or large navigable rivers, surviving not off their own agricultural production, but off food brought in from vast and distant lands. This unsustainable use of land is compounded by the use of fossil fuels to transport and store food, as well as waste disposal that pollutes land and oceans.

New thinking is now more essential than ever.

4.4.2.1 BROAD ENVIRONMENTAL DIFFERENCES

It has long been recognized that arid and semiarid environments are most vulnerable to degradation (Báez et al., 2013; Cherwin and Knapp, 2012), culminating in the formation of man-made deserts, which lead to failed civilizations (Lowdermilk, 1932; Savory, 1999). In studying land degradation, Savory & Bingham (1988) concluded that all environments are fragile and affected by our actions and that the belief that only the more arid environments are subject to desertification is an oversimplification. Savory notes that the fate of water on the land differs not with the rainfall level, but rather with the annual distribution of humidity (Ravi et al., 2004; Savory, 1999). Where the distribution of humidity is more even, desertification does not occur, even with relatively low rainfall, regardless of the agricultural practices; however, in regions with high rainfall, but dry seasons, land degradation occurs rapidly, again regardless of the agricultural practices. Further, Savory (1988) notes that abandoned civilizations in perennially humid regions are found under fully recovered soils and vegetation, while those in seasonally dry regions are found in deserts, still expanding thousands of years later.

Believing that, in all environments and over millennia, soil life, plants and animals evolved together; not one before another (Axelrod, 1985; Frank and Groffman, 1998; Mack and Thompson, 1982; Milchunas et al., 1989). Savory (1988) noted that in more humid environments, woody vegetation dominates, most herbivores are insects, and large predators hunt their limited larger prey singly; meanwhile, in seasonally dry environments, perennial grass plants play a more significant role and high numbers of large herbivores prevailed, accompanied by pack-hunting predators, which influenced herbivore behavior and movement. The soils and soil life appear to be influenced by the plant and animal populations on the land. Thus, the great grain-growing regions of the world are former grasslands that supported large herbivore populations, where deep carbon-holding soils developed, as opposed to the humid more forested regions (Aguilar et al., 1988; Savory, 1999; Wright and Wimberly, 2013).

Where mainstream science sees arid and semiarid environments as being prone to desertification and terms them "fragile," Savory (1988), recognizing that all environments are fragile and that desertification occurs in high rainfall areas, coined the term "brittle" for seasonally humid or dry environments (in the more humid environments, dead plant parts softly crumple without snapping, while, in the more seasonal environments, dead twigs, grass leaves, and stems snap when crushed by hand). He developed a simple one to 10 brittleness scale, with perennially very humid environments at the lower end and seasonally humid environments with many prolonged dry periods at the higher end, regardless of the amount of rainfall.

4.4.2.2 DESERTIFICATION

Desertification, the change of productive land into desert, is the final stage of land degradation brought about by agriculture. This process begins with available rainfall. Whereas it is often thought that agriculture is dependent on total rainfall (Turner, 2004; Wallace, 2000), Savory (1988) recognizes that agriculture is reliant on *effective* rainfall, a term coined by him in the 1960s.

Desertification and soil erosion has long been regarded as a problem, but it is, in fact, a symptom of ineffective rainfall, resulting from a loss of biodiversity and soil structure (Chapin III, et al., 2000; Rasmussen et al., 2001; Savory, 1999). Thus, the problem that agriculture needs to address is not desertification, but rather those practices and policies that result in the available rainfall becoming less effective, as life in and on the soil decreases.

Effective rainfall is rain that falls and infiltrates the soil, only leaving the soil through plant transpiration or subsurface flow to rivers, wetlands, and aquifers. Ineffective rainfall is rain that either flows off the soil surface or that infiltrates the soil, but subsequently leaves by evaporation from bare soil surfaces. In arid climates when the rate of precipitation exceeds the rate of water infiltration the resultant runoff to waterways can contain significant volumes of nutrients and soil particles (Duncan, 2009). This further degrades soil biodiversity, grasslands, and crop production. Simultaneously, rainfall runoff can degrade instream aquatic ecosystems and the economic values inherent in clean water used for drinking water and irrigation, in addition to contributing to damage or destruction of fish spawning habitat, reefs and marine waters (Duncan, 2009). Excess nutrients, herbicides and pesticides can severely limit or destroy marine water's ability to support complex life. Duncan with Melbourne Water and EPA Victoria demonstrated that pollutants generated on farm land can contain significant nutrients, herbicides and pesticides. In the PortsE2 and Filter model for Port Philip and Westernport Bay it was demonstrated that approximately 4,000 – 10,000* tonnes of nitrogen (*depending on wet or dry year) can run-off a 1.3 million hectares of catchment where approximately 50% is farm land in a temperate climate with approximately 800mm of rain annually (EPA Victoria & Melbourne Water, 2009).

While a number of factors govern the effectiveness of available rainfall, the greatest is the nature of the soil surface: whether it is bare and exposed between plants, how closely spaced the plants holding dead plant material or litter in place are, and the nature of the soil materials in regards to particle types (e.g., sand, clay, loam, rock, etc.). The rate of soil infiltration depends upon the rate of application and the porosity of the soil's top millimeters. If the exposed soil has its structure sealed or capped with remnant plant matter that cakes and seals the soil, the rate of infiltration decreases (USDA NRCS South Dakota, 2013). If the soil between plants is covered with dead plant material or litter but does not seal or cap the soil, which can often make it difficult to actually see where the soil surface begins, infiltration tends to be rapid (Savory, 1999). If heavy rainfall (high rate of application) is hampered by close plant spacing holding dead litter in place as it flows across the surface, the actual rate of application at the surface is decreased, which allows greater infiltration in larger rainfall events. If, following precipitation, the soil surface is bare, exposed, or sealed with a thin skin of primitive life forms and without plant root systems penetrating the earth; then much of that rainfall evaporates out of the top of the soil in the following days.

Desertification does not occur in regions ranking low on the brittleness scale (good distribution of humidity throughout the year), because the soil is not exposed for long periods of time over billions of hectares, even if farmed with poor agricultural practices. For example, as Savory observed, pastures in England with relatively low rainfall that have been overgrazed heavily for centuries, show little bare soil and do not desertify, despite such bad management. In such environments, any large area of exposed soil is rapidly recovered by nature, as any gardener would know (Doneen and MacGillivray, 1943; Evans and Etherington, 1990; Ravi et al., 2004). However, in regions high on the brittleness scale—even with high rainfall—agricultural practices have led, for centuries, to a high percentage of exposed soil over prolonged time and, subsequently, to less effective rainfall. This, in turn, leads to desertification and its many symptoms: increased frequency and severity of man-made droughts and floods, poverty, social breakdown, violence, mass urban emigration, and climate change. Even natural droughts and floods become more severe where rainfall is largely ineffective; such as early civilizations of Mesopotamia (Diamond et al., 1999). While Savory was a government research officer in the 1960s, he first noted a large area in Africa where Zimbabwe, Botswana, and South Africa meet. The governments of the three countries proclaimed a serious drought at the same time as the International Red Cross was collecting money on nearby city streets for the flood victims in Mozambique on the same river system.

The importance of effective rainfall becomes apparent by looking at a recent demonstration of rainfall infiltration on YouTube (USDA NRCS South Dakota, 2013). In this demonstration, 25 mm of rainfall was applied to three adjacent similar soil sites in a seasonally humid/dry environment: site A was healthy grassland with high plant diversity and soil cover between plants, site B was land with less plant diversity and soil cover between plants, and site C was cropland recently planted in the soil of site B. On site A, the water fully infiltrated the soil in 10 seconds; on site B, it infiltrated in 7 min; while site C infiltration took over 30 min. Infiltration was 40 times faster on healthier rangeland, and 180 times faster than on unhealthy cropland. Therein lies the story of most droughts and floods afflicting the severely desertifying the western United States.

In the most problematic and violent regions, from North Africa to India and up into China, about 95% of the land used for agriculture is rangeland, high on the brittleness scale, not cropland (World Bank, 2014). As previously discussed, if only one measure was taken every year to assess the land's health, it would be the quality of the water flowing off the land. If this water carries soil sediment, it may indicate unhealthy soil structure that lacks carbon matter, fungi, bacteria, and soil fauna; insufficient plant coverage or biodiversity, with the possible exceptions of true deserts and the cutting of the outer banks of a meandering river. Globally, agriculture today results in more than 75 billion tons of dead eroding soil annually (Pimentel et al., 1995). Such figures are derived mainly from the world's croplands, often ignoring the greater erosion from the remaining approximately 80% of land, and even low figures amount to more than 10 tons annually per capita, worldwide (Savory, 1999).

4.4.3 WATER MANAGEMENT: AGRICULTURAL PRACTICES AND POLICIES

In recent years, crop farmers engaged in chemical/industrial agriculture have begun using no-till methods by direct seeding without plowing and using herbicides to control weeds. While this represents a move in the desired direction, with regard to improving the effectiveness of rainfall and increasing carbon matter in soils, this type of farming is still monoculture chemical agriculture. However, farmers who keep soil covered, do not turn the soil over, and who seed their main crop with cover crops demonstrate high yields. Recent studies have demonstrated that organic agroecological systems can yield the same productivity ($\pm5\%$) as monoculture industrial farming practices or genetically modified organism (GMO) crop systems (Ponisio et al., 2014). The studies demonstrate that multicropping and crop rotations substantially reduce the yield gap (to $9\pm4\%$ and $8\pm5\%$, respectively) when the methods were applied in only organic systems (Ponisio et al., 2014). Some farmers' successes, like those of Joel Salatin of Polyface Farm in the Shenandoah Valley, Virginia, and Gabe Brown of Brown's Ranch near Bismarck, South Dakota, are gaining increasing recognition and the farmers have also incorporated livestock to ensure the biological breakdown of crop residues, while keeping the soil covered.

This section examines other effective practices for managing water.

4.4.3.1 KEYLINE—WATER FOR EVERY FARM

Water management methods aimed at increasing the water-holding capacity of soils include deep ripping Yeomans plows and Keylines, as developed by P. A. Yeomans in *Water for Every Farm,* (Yeomans, 1973). Keylines are dug slightly off contour as conveyance swales (temporary water-holding and or redirecting earthworks) that collect water and convey harvested water to dams, which hold it for irrigation, whilst slowing the water moving through the landscape and allowing for maximum infiltration into soil and groundwater. Keyline conveyance swales may also have overflow zones that allow controlled flood irrigation onto pastures and crops in large rainfall events. Some Keyline conveyance swales also have infiltration capacity, depending on the amount of clay and sand in the swale soil strata. They are often placed at key points in the landscape, where convex landforms meet concave ones, minimizing the earthworks and soil disturbance, while also allowing for easier conveyance of water back onto the land via off-contour swales and irrigation by gravity to the land below. In principle, controlling water flow on contour holds water on the land longer, allowing greater infiltration, which is the Permaculture swale methodology. The decision to utilize Keyline of Permaculture swales will depend on the desired outcome and also factor in soil types, climate, rainfall, crops, animals, and other factors.

Keyline ripping, using deep-ripping Yeomans ploughs, aerate compacted soils, allowing better water infiltration, and builds up soil carbon in rip lines from plant colonization in what was previously compacted or barren soil. These rip lines also allow for navigation by earthworms, ants, and beetles into parts of the soil substrata that previously were inaccessible or too compacted. The deep ripping can be used in place of swales and dams with regard to pasture production for livestock. However, if cropping is required, the swales and dams provide a more reliable source of water for crop irrigation, which can lead to vastly improved crop productivity. In developing nations where there is insufficient machinery for swale and dam construction; but there is enough machinery for ripping or chiselling, Keyline ripping may significantly enhance the water-holding capacity of cropping and pasture lands.

Taranaki Farm has utilized Keyline ripping as seen in Figure 4.4.1, with ripped contours showing higher levels of pasture growth in thin strips where the Keyline ripper has provided aeration, space for roots to grow deeper, and the provision of increased water holding capacity. The figure provides an example of a farm that has embedded Holistic Management, Keyline, and Permaculture into the whole farm planning, design, and management, and then gone beyond these frameworks to a highly evolved biomimicry and integrated ecological agriculture approach. Benefits of using the management frameworks, methodologies, design principles, and technologies in combination, and cumulatively from the outset, are that the farm has little to no pollution, no antibiotics, no biocides, and that the total productive output is equal to conventional monocropping in some products. Pasture productivity is far above a conventional monoculture pasture grazing operation due to the integration of chicken eggmobiles and rapid chicken manuring of pasture, mobile Keyline irrigation pods that mimic nighttime irrigation onto chicken-manured zones, and rapid movement of cattle from cell to cell on a daily basis to stimulate pasture growth according to the Holistic Management and Planned Grazing framework.

The Keyline scale of permanence used when considering the overall whole farm planning design exercise, as developed by Yeomans, has the following components:

- Climate
- Land shape
- Water supply

FIGURE 4.4.1

Keyline ripping on contour seen from above, integrating K-line irrigation pods with nighttime irrigation at chicken eggmobiles. Holistic Management and Planned Grazing across the landscape in a polyculture context.

Credit: Copyright © 2016 Home Ecology and Brooklet Farm Education.

- Farm roads
- Trees
- Permanent buildings
- Subdivision fences
- Soils

More recently, in 2013, Darren J Doherty of Regrarians (who was trained by Yeomans) proposed an update of these components into what is now been developed in the *Regrarians Handbook* and platform (Doherty & Jeeves, 2015):

- Climate
- Geography
- Water
- Access
- Forestry
- Buildings
- Fences
- Soils
- Economy
- Energy

The *Regrarians Handbook* brings together a significant body of work, modernizing Keyline and Permaculture design with a long list of improvements utilizing the latest scientific action research and field-tested innovations from farms around the world (Doherty, 2015).

Doherty makes an excellent point - that soil health is a foundational element of ecological farming – yet in regards to livestock and grazing, farms are exporting minerals at an unsustainable rate. Animal bones contain significant amounts of calcium and phosphate, the fate of which is usually exported off farm. The mineral deficit cumulatively impacts soil and plant health and ultimately can impact human health if eating minerally depleted food. Additionally, the impacts of mining for phosphates and calcium products has an environmental and social impact in the process of extraction and distribution, with phosphates also being indicated as aggravating algal blooms from farm run-off. Doherty has developed a unique approach called Paddock2Paddock which complements the now popular Paddock2Table concept – whereby animals are processed on farm, animal bones are charred in a biochar system, whilst animal viscera is processed into biofertiliser. The combining of nitrogen ricj biofertiliser with bonechar which is 33% calcium phosphate and 35% calcium phosphate, provides an ideal regular soil fertility program, all of which can be incorporated back into the soil with Doherty's Keyline SuperPlow. The concept in Keyline with deep ripping enhanced water infiltration and soil based carbon through the extra growth of grass roots and their senescing. Paddock2Paddock promises to close the nutrient and mineral loop in farming, whereby nutrients and minerals do not leave the farm and are incorporated on-site and locally. Duncan and Doherty have developed a comprehensive suite of technologies to close the nutrient and mineral loop, developing biochar and biofertiliser production systems that are modular, dock-able and easy to use by farmers whilst processing animals on-site (Duncan, Home Ecology and Brooklet Farm Education, 2015).

4.4.3.2 PERMACULTURE AND SWALES

Permaculture is a design science developed by Bill Mollison and David Holmgren (1978) and later Mollison and Reny Slay (1988); it offers a set of principles, strategies, and methods to conduct farm planning and is often focused on self-sufficiency and sustainable human settlements, while also emphasizing the sustainable bioeconomy at local and regional scales. Design methods emphasize the importance of observation and interaction with systems over a cyclical period to understand the periodicity of the area in question including the land, soil, water, animal, plant growth, biodiversity, and catchment conditions that yield insights into productivity opportunities, suitability of various species for incorporation or removal from the overall farm plan, and, importantly, the identification and management of risk.

Permaculture as practiced and taught by Duncan (Figure 4.4.2) incorporates cultural, environmental and economic considerations in its implementation as a result of the complexity and differences that different geographies, cultures, and climates experience. Permaculture provides a decision support system and many hands-on tools, including composting methods, nutrient management, settlement design, water harvesting, and infiltration design with swales and dams, polycultural kitchen gardens for food security, integration of animal systems into tree cropping and perennial systems, and methods for enterprise and community to collaborate around sustainable food, fiber, and fuel production. Farmers who have completed Permaculture design certificates with Duncan, develop new knowledge, skills and personal empowerment with practical tools to make their farming operations sustainable,more self-sufficient and economic. Permaculture design, landscaping and farm management enterprises are many and varied with Permaculture experts offering a variety of services and products including soil fertility management, compost, biodiversity restoration, waterway repair services through to intensive vegetable and fruit production and animal management.

FIGURE 4.4.2

A Permaculture kitchen garden in Hebei, China.

Credit: Copyright © 2016 Home Ecology and Brooklet Farm Education.

In the context of drylands and lands suffering from desertification, communities experiencing food insecurity have succeeded in developing small plots of land, integrating Permaculture swales on contour to help water infiltrate the soil and convey water to dams by backfilling once swales are filled during rainfall events. The Permaculture principle of "catch and store energy" (Holmgren, 2002) is characterized by Permaculture swales and Keyline drains and dams. Storage of water for dry periods is often an essential step to take for arid and desert communities to enable water supply to kitchen gardens and self-sufficiency cropping in the dry seasons. Passive irrigation doesn't require pumps or power, therefore it is considered an appropriate technology and design approach that can be implemented by hand using simple tools such as hoes and spades. Permaculture swales are on contour, whereas Keyline drains operate most commonly on a 1:1000 or 1:500 ratio, meaning that the infiltration and conveyance drains only drop 1 m over 1000 m, slowing the passage of water across the landscape dramatically and increasing water holding capacity of the landscape, which is an important aspect for arid regions with sandy soils and very little carbon matter in soils. Keyline drains and Permaculture swales may also have overflow gates that allow water to slowly and passively flood or inundate designated areas in rainfall events, often in desirable places, such as cropping areas or zones slated for grazing.

Attempts to integrated Permaculture swales in highly arid environments have demonstrated the effectiveness of the technique at the small scale for food security, kitchen gardens, market gardens, and food forests with complex ecologies and mixed species of food, fiber, medicine, fuel, and habitat production. The cyclical nature of Permaculture design, productivity observations, and farm management can be labor intensive and is suited to farmers with the ability to input their own labor or who can access tools or machinery.

4.4.3.3 LARGE-SCALE LAND RECLAMATION

Perhaps the best known large-scale land reclamation involving the redirection of runoff rainfall is the Chinese success widely publicized by John Liu, director of the Environmental Education Media Project, in his film *China's Sorrow—Earth's Hope* (Liu, 1995). Faced with severe silting of China's Yellow River, the Loess Plateau Watershed Rehabilitation Project embarked on a restoration project in northern China, with funding provided by the World Bank and the International Development Association (IDA). This project demonstrated significant increased agricultural production and improved people's lives over approximately 3.5 million ha, using large machinery and labor to harness water runoff and contour the landscape, while planting trees and crops and removing livestock where appropriate. It is both recognized notable because of the scale of this Chinese effort. We do not know today if the Nabataeas efforts to sustain their cities by harnessing runoff and planting trees and crops collectively reached a similar scale. This excellent example in China is being widely touted as *the* solution, but China's desertification continues to expand at alarming rate and Beijing is sometimes inundated with up to quarter of a million tons of sand dumped on it.

Coauthor Tom Duncan worked in a China-Australia five year partnership between AusAID and China's department of water resources in the Water and Agriculture Management in Hebei project (Figure 4.4.3), which also reached a similar scale. This innovative and holistic project assisted 30 million farmers in Hebei province, from arid inner Mongolian plateau down to the regions of central China plains, that are experiencing desertification and complete exhaustion of shallow water tables with waterways running dry from overextraction. Hebei also encircles Beijing and Tianjin, both of which are

FIGURE 4.4.3

Water harvesting and distribution in Hebei, China.

megacities that have experienced severe water shortages and degraded water quality. Strategic objectives included ensuring safe water supply to the megacities of China by working with farming communities to sustainably irrigate their farms without exhausting the water table or extracting from environmental flows in waterways.

The project developed integrated farming and water management systems in cooperation with the Department of Water Resources, including the updating of geographic information systems (GIS) and aerial information systems to reflect hydrogeological data that could quantify the amount of water recharge in the catchments and a sustainable water harvest for the province of Hebei that allowed farmers and urban water users a fair share, without compromising the aquatic ecological health of waterways, lakes, and groundwater dependent ecosystems. The department also carried out quantification of environmental flows to Beijing and Tianjin water supply reservoirs in cooperation with Australian experts (Figure 4.4.4). Capacity building at the ministerial and advisor levels through to the agricultural extension and training officers meant that a holistic project design could achieve a holistic outcome.

Duncan developed the information and training materials for capacity building at a ministry and departmental level, and simultaneously fed back information from farmer-level surveys and Women's Development Association surveys to reflect the feedback in water and agriculture policy development at the departmental level. This was a novel process for the central planning approach, to instead have feedback from the ground up to the department level, with facilitation of farmer and policy maker combined roundtables to rapidly accelerate the level of change in sustainable water and agriculture management in Hebei province. Facilitation of simultaneous development of village master plans

FIGURE 4.4.4

Community water tank stores harvested water and schedule irrigation to orchards, crops, and animal water troughs. The village chief identifies the next steps to a World Bank consultant assisting in the water planning process.

Credit: Copyright © 2016 Home Ecology and Brooklet Farm Education.

FIGURE 4.4.5

Village meeting to plan soil types, water, earthworks, irrigation, cropping, trees, and water moisture monitoring probes.

developed by farmers themselves (Figure 4.4.5) for the purpose of sustainable water and agriculture management objectives delivered concise and accurate plans that were then implemented and fed back into policy target documents describing the hydrogeological water balance of Hebei province and the corresponding benefits to water dependent ecosystems and urban water users in Beijing, Tianjin, and Hebei's capital city, Shijiazhuang.

Project innovations were spread across the Hebei province region of 18.7 million ha and included large-scale community water harvesting on terraces and swales, integrated ecological agriculture methods such as livestock cell grazing, irrigation scheduling, composting integration, polyculture tree crops with advanced grafting techniques, integrated pest management, stakeholder surveying, education programs, village agriproduction and water master planning, GIS, and hydrogeological modeling. Water harvesting and irrigation scheduling techniques were developed in consultation with all communities, which resulted in each village forming a community water user association to manage water resources effectively, finance the installation of pumps, and pay for operations and maintenance. Solar pumps and conventional pumps provided header tanks with water and were installed collectively instead of individual shallow water wells that had long since run dry from overextraction—and the community-based header tanks had enough water to boost production by significant rates resulting in income generation above and beyond expectation, along with the attendant land restoration benefits that accrue from more sustainable practices.

Duncan's work emphasizes the community grassroots upward method of project development, whereby the farming communities, including Women's Development Association representation,

FIGURE 4.4.6

Women's Development Association village survey in Hebei, China; part of the participatory village master-planning process.

Credit: Copyright © 2016 Home Ecology and Brooklet Farm Education.

identify their needs in terms of crop, animal, water, and soil management (Figure 4.4.6), and are then provided training and materials resources to achieve the goals. This resulted in the production of Duncan's lunar calendar, which rural farmers are familiar with (lunar calendars are the dominant type of calendar used by farmers), that incorporated the innovation of integrated farming and water management systems information on a monthly and weekly basis in the agricultural lunar calendar (Figure 4.4.7) using visual aids (as most farmers were functionally illiterate). Cultural context is an important aspect to defining project success benchmarks and gaining a cultural license to operate a project within communities. Conversely, agricultural extension officers who were traditionally trained in central universities would supply farmers with pages of written notes; however due to functional illiteracy many farmers would use the paper to cook an evening meal on their home-cooking fire. Furthermore, conventional calendars were also used for the home fires, because they did not observe the traditional Chinese culture of the lunar calendar and intricacies including the Chinese year revolving around the new moon cycle (often in February or March). The new moon cycle that occurs in transition from winter to spring indicates the correct time of year for sowing or harvesting crops according to observations from agronomists and hydrologists over a period of several thousand years of Chinese civilization.

The lunar calendar could also be folded up and put in a farmer's pocket so he or she could take it into the fields to carry out more complex tasks, such as monitoring water moisture levels in preexisting gneiss rock strata tree crop terraces or graft walnut trees at the suitable time of year on the terraces. Precise photo instructions accompanied the lunar calendar for tree grafting, water monitoring with gyprock probes, compost integration into soils for optimum carbon and nutrient rates in soils which also increased water holding capacity in the arid, degraded soils. Many of the terrace lands were classified as wastelands with regard to soil quality and land use capacity. Subsequently, after the project was implemented, the agricultural output and sustainability of water harvesting from the compost

FIGURE 4.4.7

The participatory village master-planning process incorporated training material development. A lunar calendar that integrated farming and water systems information was identified as the most useful training tool.

Credit: Copyright © 2016 Home Ecology and Brooklet Farm Education.

integration, water harvesting, and community irrigation systems exceeded the existing categorization system of agricultural production categories and delivered a result that can provide guidance to other large-scale projects (Figure 4.4.8).

4.4.4 HOLISTIC MANAGEMENT

Holistic Management (Savory, 1999) is not simply a different technique, but involves a major paradigm shift developed by Savory to address the complexity inherent in management. It is applicable to any management situation, although this section examines it in the context of agriculture. Every farm is unique: it has its own management, clients, suppliers, land, and local economy. Each of these components is constantly changing and, thus, they are unique every year. As a result of this uniqueness and of the social, environmental, and economic complexities faced by each farm, there can be no "best management practice" applicable to every farm. The social, environmental, and economic outcomes or results are, however, the direct results of management actions.

Holistic Management recognizes that management at any level simply cannot avoid social, cultural, political, environmental, and economic complexities. Therefore, the context for any objective needs to be understood holistically. Generic or universal management can be described as "reductionist": it reduces the context for objectives to need, desire, profit, or address a single specific problem. The contexts for our actions are too simplistic, without care for the complexities involved, which results in unplanned and

FIGURE 4.4.8

Large-scale water harvesting in an arid inner Mongolia plateau, Hebei, China.

Credit: Copyright © 2016 Home Ecology and Brooklet Farm Education.

unintended consequences (whether they be immediate or eventual). For example, a crop of corn grown by applying recommended fertilizers, insecticides, and herbicides will sell profitably; however, unintended consequences will likely follow: a drinking well may be polluted, people may react to chemicals used, and soil life and the soil's ability to cycle nutrients and hold water and carbon may decrease. Even such seemingly harmless practices as Keyline plowing may lead to unintended consequences due to a lack of concern for the context of the action: perhaps the ranch was overcapitalized (e.g., too much money was tied up in land, buildings, or machinery relative to income from low cattle numbers) and the action exacerbated this by investing yet more money on part of the land; additionally, perhaps the runoff water was not genuine excess surface flow from effective rainfall, but was a symptom of ineffective rainfall.

Any action—no matter how good it may appear or how much research and past results support it—still needs to be situated within a local holistic context. Management actions should be considered first and foremost as opposed to technological innovations that are capital intensive.

A holistic context is defined by people determining how they want their lives to be, tying that desired life to the preservation of the healthy state of land (environment) far into the future in such condition that future generations can live similar lives. This holistic context then provides the context within which all major management decisions and actions must be taken, ensuring that they socially, environmentally, and economically support the short- and long-term desires for that land.

Throughout history, agriculture has led to the failure and abandonment of many cities in all regions of the world. In *The Watchman's Rattle: Thinking Our Way Out of Extinction,* Rebecca Costa (2012) concludes that those cities did not fail because of their agriculture alone, but because their societies could not address the complexity of rising population and deteriorating environment.

Prior to Napoleon's time, it was assumed that major blunders experienced in political, social, and environmental realms were the result of amateurism—people could buy or inherit their positions as heads of companies, armies, etc. However, the philosopher John Ralston Saul discovered that major blunders increased as organizations began to be led by professionally trained experts. He suggests that this is due to experts' training in narrow silos of knowledge: "The reality is that the division of knowledge into feudal fiefdoms of expertise has made general understanding and coordinated action not simply impossible but despised and distrusted" (Saul, 1992. p. 8). He continues, "In many ways the differences between various languages today are less profound than the differences between the professional dialects within each language. Any reasonably diligent person can learn one or two extra tongues. But the dialect of the accountant, doctor, political scientist, economist, literary historian or bureaucrat is available only to those who become one" (p. 112).

Whereas Costa sees the solution to addressing complexity as using advancing technology to speed the frontal part of the human brain that deals with complexity, Saul sees the need to move beyond silo-style education and organizations. However, there is a shorter, simpler way to address complexity in a practical manner through Holistic Management.

4.4.4.1 SYSTEMIC PROBLEMS

Systemic problems are caused by one common source, as opposed to many special or individual sources. In 1987, Savory attended a conference in Sweden addressed by Gro Harlem Brundtland, formerly prime minister of Norway, special envoy to the UN, and vice chair of the international group (Carlsson, et al., 1987). Brundtland outlined many problems facing the international community, each costing millions or billions to fix, but all worsening. She appealed to scientists to find connections in the hope that this could lead to greater successes. What Savory concluded was that, rather than connections, the problems stemmed from one systemic common cause. That common cause was the genetically embedded behavior that humans use when making conscious decisions: underlying even the most complicated decisions, we develop objectives and then use some tool to achieve them, based on many factors.

However, there are two simple flaws in this method that affect both complexity and agriculture. First, man-made desertification, which was initiated thousands of years ago, was misunderstood: livestock was blamed and there were no tools to prevent or reverse this phenomenon. Secondly, an adequate context was missing for the decisions to be made holistically and thus address the inescapable complexity.

4.4.4.2 TOOLS TO ADDRESS AGRICULTURE, DESERTIFICATION, AND CLIMATE CHANGE

It has long been thought that humans have far more tools at their disposal to manage the environment than they really do. In fact, humanity only currently has three tools with which to manage the environment at large; these are technology (such as machinery, chemicals, etc.), fire, and the concept of resting the land to allow biodiversity to restore itself. We do use small living organisms, such as in making cheese and wine, or use genetic engineering to create new organisms, but not to manage the environment at large.

None of these three tools can sustain biodiversity in most of the vast seasonal rainfall regions of the world, which include mainly grasslands, savannahs, and former grasslands that are now desert. To maintain the annual biological cycle—birth, growth, death, and decay—in environments such as these, where grass plants provide the main stability and soil cover, the presence of large numbers of herbivores is required. Without these herbivores and their symbiotic partners, gut microorganisms and pack-hunting predators, decay is impaired. Grasslands tend to shift to environments dominated by woody plants in higher rainfall areas and to lands with smaller woody plants and crusted soil cover in lower rainfall areas. Traditionally, fire was used to burn dying grass material to try and sustain healthy grasslands and their soils. Recognizing that rested rangelands shifted from grasslands to woody vegetation areas or desertified areas, range scientists in the 1950s recommended large machines, such as the Dixon Imprinter, to greatly disturb rangelands by crushing down dead plant material and breaking sealed soil surfaces, roles formerly performed by vast populations of animals (Imprinting Foundation, 2014; Dixon and Carr, 1999).

If agriculture is to sustain megacities, available rainfall must once more be made effective and, thus, reverse human-made desertification. Worldwide, roughly one-third of the land where humidity is near perennial, desertification does not occur; thus, simply resting the land restores biodiversity and full ecosystem function. However, many lands across the world experience seasonal humidity and dry periods; here resting the land either totally or partially as happens when there are too few large animals with changed behaviour on the land (Savory, 1999) disrupts soil and plant life leading to a loss of biodiversity. In fact, in seasonal humidity grasslands, only properly managed livestock can today reverse desertification over billions of hectares while also feeding people.

4.4.4.3 PROPERLY MANAGED LIVESTOCK

In the 1960s, it became clear that livestock that run free on the land rather than being grain-fed in feedlots could contribute to reversing desertification. Thousands of years of grazing animals in herds as pastoralists have done and many are doing today claim that mob grazing caused large and expanding human-made deserts in seasonal (and especially low-rainfall) regions. A century of modern rangeland science, fencing, and many rotational and other grazing systems accelerated desertification, as first observed in Africa and confirmed in the United States (Savory, 1999; Sinclair, Fyxell, 1985).

Andre Voisin (1959), a French pasture scientist, proposed that the disappointing results of rotational grazing in Europe's more evenly humid environment were due to a misunderstanding of overgrazing. For centuries, overgrazing has been thought to be the result of too many animals grazing. Voisin suggested that it instead results from plants being exposed to grazing for too many days and/or reexposed after insufficient time for root recovery, regardless of animal numbers. He recognized that some planning was needed to address an innate complexity, replacing rotational and other grazing systems.

However, Voisin's simple "rational grazing" planning process, which uses a calendar, notebook, and land plans, is inadequate to deal with the greater complexity in the world's more seasonal environments and especially where rainfall is less. This includes most of the grasslands of the Americas, Australia, and Africa, and right through to China. The greater complexity in such vast regions includes more erratic rainfall, prolonged dry years, more diversity of wildlife, cultures, plants, soils, topography, and more, as well as conflicting land uses. In the 1960s, Savory solved this problem by researching all professions to see who had ever dealt with anything even nearly as complicated. He found what he sought in hundreds of years of accumulated military experience in Europe. Simply taking their planning processes designed for immediate battlefield situations and doing the planning on a chart on which four dimensions and many considerations could be easily plotted worked immediately. Over the next few years he and farmers in five countries that worked with him were achieving noteworthy success in reversing desertification using livestock as the tool. However, they then began experiencing erratic results, including complete failures. Analyzing the failures, Savory concluded none were due to the grazing planning process, but all were his fault because he had failed to take into account all of the social, cultural, and economic complexity that farmers faced daily. This eventually led to discovering the systemic problem common to all management.

4.4.4.4 PLANNED GRAZING: HOLISTIC MANAGEMENT AND HOLISTIC PLANNED GRAZING

Any management action will always involve not only nature's complexity, but also social, cultural, and economic complexities—which is the reason that the word *holistic* came into what was originally simply a grazing planning process and results became consistently achievable.

Any farmer would determine if livestock should be run at all when initial decisions are made in a holistic context. For example, if farming in say Brazil's tropical humid forests, any farmer managing holistically would always realize that grazing cattle could never be justified socially, environmentally, or economically. Once it is determined that livestock are necessary in the relevant holistic context, then the holistic Planned Grazing process should be used. While most newly promoted rotational grazing systems like mob grazing are a welcome step forward because they recognize the need for greatly increased animal impact periodically applied in grasslands and do generally produce better results in the more reliably humid environments, they fail to address the complexity and needs in the seasonally humid/dry environments where desertification is occurring.

Through the Savory Institute, a global network of affiliated locally led and managed learning/training hubs is expanding. Through these hubs, farmers, ranchers, pastoralists, universities, environmental organizations, and government agencies are beginning to collaborate, managing holistically, and providing professionally trained advisors, as well as gathering data concerning results.

4.4.5 POLICY AND DEVELOPMENT PROJECTS

As agriculture is a big driver of climate change and as climate change will continue even in a post–fossil fuel world unless a new regenerative agriculture system is developed, it is important to understand that the use of Holistic Management at the level of the farm, ranch, or forest alone will not save modern cities. Change must be made at the policy level in all countries and at the development project level in developing countries. These two levels are similar in that both policies and development projects only arise to address specific problems and universally today that problem provides the context for objectives and the actions that are taken.

To succeed, the objectives of policy and development projects need to meet the following three criteria:

- They need public support.
- They need to address the underlying cause of the problem, not merely the symptoms.
- They need to have a context greater than the problem to avoid unintended consequences.

Today, it would be difficult indeed to find any development project or agricultural policy that has full public support, does not deal only with symptoms, and does not lead to unintended consequences. Although no government has yet done so at a higher level, several government and international agencies, including the US Department of Agriculture, the World Bank, United States Agency for International Development (USAID), and US universities, have engaged Savory in workshops, training staff to analyze existing policies and devise fresh policies within the holistic framework. In all, more than 2000 officials and faculty members have undergone such training using their own policies as examples. All policies used were found to have a high likelihood of failure, and most would lead to unintended damaging consequences, because the context for policy objectives was too narrow for the complexity of the problem. By using a holistic context, a workable policy could be developed. Such training in the use of the holistic framework has also been successful with smaller groups of officials in India and Lesotho, as well as with a large nongovernmental organization (NGO) in Kenya with similar results.

Duncan specializes in developing large scale project plans and has experience in delivering plans within Australia, China, Malaysia, and Laos with some significant results that demonstrate that the Holistic Management approach is a driver for sustainable change that yields multiple benefits across government, farming, environment, economy, and societal sectors.

4.4.6 IMPROVING MANAGEMENT

Progress has, unfortunately, been institutionally stalled by 50 years of rejection of the concept of using livestock as a tool to reverse desertification (Briske et al., 2008). This is somewhat understandable, as livestock has been deemed the cause of desertification for centuries. It is counterintuitive and paradigm changing to accept that only properly managed livestock can reverse this problem.

While it is possible to gather data and evidence of the results of Holistic Management, it is difficult to subject any decision-making process or the Planned Grazing process to experimental scientific protocols; as Holistic Management often involves changing multiple variables at the same time. However, performance data, despite resulting in ranchers winning many good stewardship awards and a constant repetition of success when managing holistically, has at times been rejected as anecdotal by experimental range scientists. This could be overcome if a large enough data set was gathered that was able to take into account all the variables involved in Holistic Management. The inability of some academic authorities to understand that no management process using all the available science of the day can be subjected to experimental protocols is perhaps best described by Thomas Kuhn in his 1962 book, *The Structure of Scientific Revolutions* and is a paradigm problem—not any lack of intelligence or wrong motive. Perhaps it helps to use the analogy of World War II. In that mighty and complicated conflict the allied civilian leaders used a constant decision making and planning process, using all available science—and they won. No one would call that anecdotal or believe it could be or needed to be subjected to experimental protocol in order to either accept or to understand the science. That is almost exactly what Holistic Management and its holistic Planned Grazing involves—people using a holistic context to aid constant decision making and planning using all available science.

Despite institutional stalling caused by the irrelevant studies which purport to prove that rotational grazing systems do not support any of the claims made for Holistic Management and holistic Planned Grazing thousands of people, farmers from organic to mainstream, ranchers, scientists, and now some organizations are beginning to practice holistic management. The practice has spread to millions of hectares on six continents. Increasingly, institutional scientists from universities, large environmental NGOs, and government agencies are becoming involved in gathering social, economic, and environmental performance data based on results, which can only help accelerate the shift in public perception about livestock and hopefully facilitate institutional acceptance of the holistic framework in policy and project development worldwide.

REFERENCES

Aguilar, R., Kelly, E.F., Heil, R.D., 1988. Effects of cultivation on soils in northern Great Plains rangeland. Soil Sci. Soc. Amer. J. 52 (4), 1081–1085.

Axelrod, D.I., 1985. Rise of the grassland biome, central North America. Bot. Rev. 51 (2), 163–201.

Báez, S., Collins, S.L., Pockman, W.T., Johnson, J.E., Small, E.E., 2013. Effects of experimental rainfall manipulations on Chihuahuan Desert grassland and shrubland plant communities. Oecologia 172 (4), 1117–1127.

Briske, D., et al., 2008. Benefits of rotational grazing on rangelands: an evaluation of the experimental evidence. Range. Ecol. Manage. 61, 3–17.

Cai, W., et al., 2014. Increasing frequency of extreme El Niño events due to greenhouse warming. Nature Clim. Change 4 (2), 111–116.

Carlsson, I., Brundtland, G.H., Ramphal, S.S., 1987. Towards Sustainable Development: Fourteen Case-Studies Prepared by African and Asian Journalists for the Nordic Conference on Environment and Development at Saltsjöbaden, Stockholm, 8–10 May 1987: with Overview Chapters. Panos Institute, London, ©1987.

Chapin III, F.S., et al., 2000. Consequences of changing biodiversity. Nature 405 (6783), 234–242.

Cherwin, K., Knapp, A., 2012. Unexpected patterns of sensitivity to drought in three semi-arid grasslands. Oecologia 169 (3), 845–852.

Costa, R., 2012. The Watchman's Rattle: Thinking Our Way out of Extinction. Vanguard Press, Jackson, TN.

Diamond, J., 2005. Collapse: How Societies Choose to Fail or Succeed. Penguin Press.

Dixon, R.M., Carr, A.B., 1999. Land Imprinting for Restoring Vegetation in the Desert Southwest. The Future of Arid Grasslands: Identifying Issues, Seeking Solutions. The Imprinting Foundation Tucson AZ, 325.

Doherty, D.J., Jeeves, A., 2015. Regrarians Handbook. Regrarians, Australia. Accessed on February XX, from, http://www.regrarians.org/regrarian-handbook/.

Doneen, L.D., MacGillivray, J.H., 1943. Germination (emergence) of vegetable seed as affected by different soil moisture conditions. Plant Physio. 18 (3), 524.

Duncan, T., 2009. Floating Biofilter Conference Paper. International Water Association and Stormwater Industry Association Victoria, Melbourne, online http://www.stormwater.asn.au/component/content/category/75-conference-papers.

Duncan, T., Home Ecology., Brooklet Farm Education., 2015, Regrariand Paddock2Paddock Design With Home Ecology And Brooklet Farm Education, Accessed on August 18, from http://www.homeecology.net/regrarians-paddock2paddock-design-with-home-ecology-and-brooklet-farm-education/.

Evans, C.E., Etherington, J.R., 1990. The effect of soil water potential on seed germination of some British plants. New Phytol. 115 (3), 539–548.

Frank, D.A., Groffman, P.M., 1998. Ungulate vs. landscape control of soil C and N processes in grasslands of Yellowstone National Park. Ecology 79 (7), 2229–2241.

Gibson, D., 2004. The Nabataeans: Builders of Petra. Xlibris Corporation, Bloomington, Indiana.

Hare, W.L., Cramer, W., Schaeffer, M., Battaglini, A., Jaeger, C.C., 2011. Climate hotspots: key vulnerable regions, climate change and limits to warming. Regional Envir. Change 11 (1), 1–13.

Holmgren, D., 2002. Permaculture: Principles and Pathways Beyond Sustainability. Holmgren Design Services, Victoria, Australia.

Imprinting Foundation, 2014. The Imprinting Foundation: New Hope for Damaged Lands, Tucson, Arizona. Accessed September 25, 2014, from, http://imprinting.org/.

Kuhn, T.S., 1962. The Structure of Scientific Revolutions (1st ed.). University of Chicago Press. LCCN 62019621.

Leblanc, M.J., et al., 2008. Land clearance and hydrological change in the Sahel: SW Niger. Glob. Planet. Change 61 (3), 135–150.

Liu, J., 1995. China's Sorrow—Earth's Hope. Environmental Education Media Project in China (EEMPC) onlinehttp://eempc.org/, © 2015.

Lowdermilk, W.C., 1942. Conquest of the Land Through Seven Thousand Years. U. S. Department of Agriculture, Soil Conservation Service, Washington.

Mack, R.N., Thompson, J.N., 1982. Evolution in steppe with few large, hooved mammals. Amer. Nat. 757–773.

Milchunas, D.G., Lauenroth, W.K., Chapman, P.L., Kazempour, M.K., 1989. Effects of grazing, topography, and precipitation on the structure of a semiarid grassland. Veget. 80 (1), 11–23.

Mollison, B., Holmgren, D., 1978. Permaculture One: a perennial agricultural system for human settlements. Transworld Publishers, Melbourne, A Corgi Book.

Mollison, B., Slay, R.M., 1988. Permaculture: A Designers' Manual. Tagari Publications, Tasmania, Australia.

Pimentel, D., et al., 1995. Environmental and economic costs of soil erosion and conservation benefits. Science-AAAS-Weekly Paper Ed. 267 (5201), 1117–1122.

Ponisio, L., et al., 2014. Diversification practices reduce organic to conventional yield gap. Proc. Royal Soc. London. doi:10.1098/rspb.2014.1396.

Rasmussen, K., Fog, B., Madsen, J.E., 2001. Desertification in reverse? Observations from northern Burkina Faso. Global Env. Change 11 (4), 271–282.

Ravi, S., D'Odorico, P., Over, T.M., Zobeck, T.M., 2004. On the effect of air humidity on soil susceptibility to wind erosion: the case of air-dry soils. Geophys. Res. Lett. 31(9).

Reij, C., Tappan, G., Belemvire, A., 2005. Changing land management practices and vegetation on the Central Plateau of Burkina Faso (1968–2002). J. Arid Environ. 63 (3), 642–659.

Saul, J.R., 1992. Voltaire's Bastards: The Dictatorship of Reason in the West. Simon and Schuster, New York.

Savory, A., 1999. Holistic Management: A New Framework for Decision Making. Island Press, Washington, DC.

Savory, A., Bingham, S., 1988. Holistic Resource Management, 1718. Island Press, Covelo, CA.

Sinclair, A.R.E., Fryxell, J.M., 1985. The Sahel of Africa: ecology of a disaster. Canad. J. Zool. 63 (5), 987–994.

Kuhn, Thomas S., 1962. The Structure of Scientific Revolutions. Harper, Collins, Chicago.

Turner, N.C., 2004. Agronomic options for improving rainfall-use efficiency of crops in dryland farming systems. J. Exper. Bot. 55 (407), 2413–2425.

United Nation, 2014. World's Population Increasingly Urban with More than Half Living in Urban Areas. Available from: http://www.un.org/en/development/desa/news/population/world-urbanization-prospects-2014.html. Accessed on July 10, 2014.

USDA NRCS South Dakota, 2013. Grassland Soil Health: Infiltration. Available from: http://youtu.be/IqB4z7lGzsg. Accessed April 3 2015.

Victoria, E.P.A., 2009. Better Bays and Waterways. Water Quality Improvement Plan, Melbourne, Australia.

Voisin, A., 1959. Grass Productivity. Philosophical Library Inc., New York.

Wallace, J.S., 2000. Increasing Agricultural Water use Efficiency to Meet Future Food Production. Agri. Ecosys. Env. 82 (1), 105–119.

World Bank, 2014. Permanent Cropland (% of land area). Accessed on September 20, 2014, from, http://data.worldbank.org/indicator/AG.LND.CROP.ZS.

Wright, C.K., Wimberly, M.C., 2013. Recent land use change in the Western Corn Belt threatens grasslands and wetlands. Proc. Natl. Acad. Sci. 110 (10), 4134–4139.

Yeomans, P.A., 1973. Water for Every Farm: A Practical Irrigation Plan for Every Australian Property. K.G. Murray Publishing Company Pty Ltd, Sydney, Australia.

LAND DEGRADATION: AN ECONOMIC PERSPECTIVE

4.5

Hannes Etter

Economics of Land Degradation Initiative, Bonn, Germany

Land is the foundation of the production of food and other valuable goods. All humans depend on the availability of fertile soils and related benefits that land provides, and as such, it is one of the most precious resources on our planet. Despite its critical role in human sustenance and survival, land is increasingly being degraded and ultimately lost due to overuse and unsustainable land management practices such as fertilization or intensive agriculture (ELD Initiative, 2013b). Land degradation threatens global food security, particularly when coupled with climate change effects. These include increasing risks of famine and drought and the consequent reduction of available fertile land, which is paired with an already constrained supply of food and other land-related resources (Smith et al., 2007) and further aggravated by increasing food demands from population growth. It has been noted that "serious doubts have been raised on the capacity of land to meet the demands of a human population rapidly increasing to 9 billion" (Godfray et al., 2010). As a result, smallholder farmers, who represent the majority of the global poor, are directly endangered by the loss and degradation of land (Barbier and Hochard, 2014).

In addition to the negative impact on food security and local livelihoods from reduced land productivity, it is also a serious threat to economic development on local, national, and global scales: 24% of usable land on Earth is already degraded, a loss estimated around US$40 billion per annum (ELD Initiative, 2013b). In Niger, for example, about 8% of gross domestic product (GDP) is being lost due to land degradation, which causes nutrient depletion and salinization, and negatively affects one of the most important sectors for national food security: the production of staple crops (Nkonya et al., 2011). Elsewhere, the agricultural sector of the United States loses $44 billion annually, amounting to $247/ha of cropland or pasture (Eswaran et al., 2001). These economic losses are mirrored around the world in all places where land degradation is occurring.

Protecting this natural resource in a manner that allows for continuous future use and stability requires sustainable land management (SLM; ELD Initiative, 2013b; Smith et al., 2007; Godfray et al., 2010; Barbier and Hochard, 2014; Nkonya et al., 2011)[1]. Approaches to SLM are manifold and cover a wide range of technological or chemical measures, but also include land tenure reform and changes in resource governance (Eswaran et al., 2001). Global initiatives such as World Overview of Conservation

---

[1] SLM is defined as "the adoption of land use systems that enhance the ecological support functions land management: of land with appropriate management practices, and thus enable land users to derive economic and social benefits from the land while maintaining those of future generations. This is usually done by integrating socio-economic principles with environmental concerns so as to: maintain or enhance production, reduce the level of production risk, protect the natural resource potential, prevent soil and water degradation, be economically viable, and be socially acceptable."

Land Restoration. http://dx.doi.org/10.1016/B978-0-12-801231-4.00024-0

Approaches and Technologies (WOCAT; Liniger and Schwilch, 2002), World Resources Institute (WRI; https://www.wocat.net/), Partner Initiative for Sustainable Land Management (PISLM; http://www .sids2014.org/index.php?page=view&type=1006&nr=2389&menu=1507), and others (World Bank, 2006) demonstrate the availability of a large base of SLM knowledge. However, this knowledge is not always applied, so the adoption of SLM practices remains insufficient to reverse or even halt global land degradation. This is due to a variety of obstacles, including lack of financial resources to switch to SLM and other economic, technical, legal, political, social, cultural, and environmental barriers (ELD Initiative, 2013b). However, from an economic point of view, there is a strong need for adopting SLM, as well as increasing evidence that it is actually more affordable over the long term: as highlighted by the ELD Initiative (2013b), SLM approaches often come at a lower cost than that of degradation effects on productivity and other land-based services. Later, they yield further benefits to society: in a recent ELD study in Botswana by Favretto et al. (2014), communal land management was found to be more sustainable compared to private cattle ranches, because it provided higher values of ecosystem services.

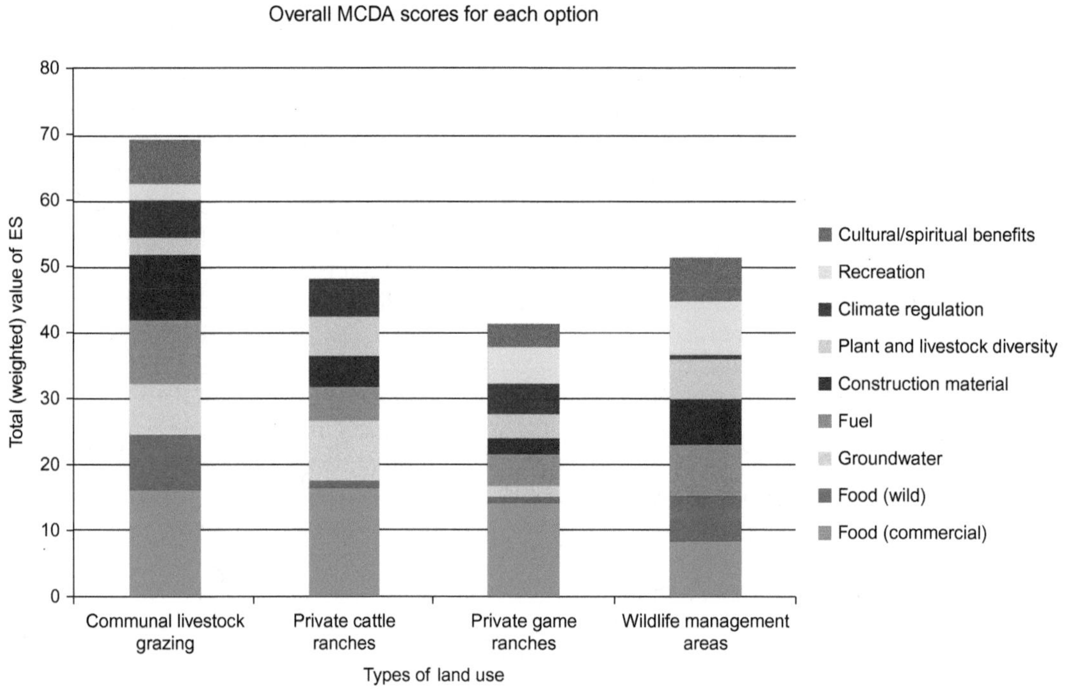

FIGURE 4.5.1

Multi-criteria decision analysis (MCDA) scores for different land use types in Botswana (Favretto et al., 2014)

In the light of uncertain future developments and trends (e.g., climate change and natural disasters), land resource management should be adapted toward resilience in ecosystems, productivity, and livelihoods. Concepts of ecosystem-based adaptation (Uy and Shaw, 2012) or ecosystem-based disaster risk reduction (Renaud et al., 2013) have proven useful to frame such transitions and should be included in the consideration and adoption of SLM. Figure 4.5.2 depicts possible land management pathways,

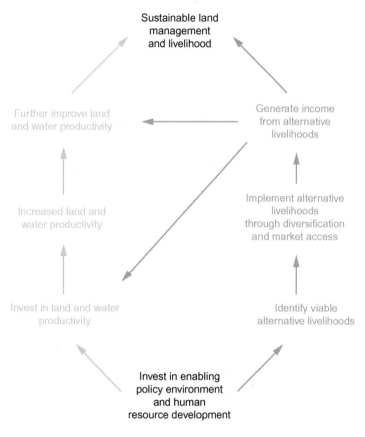

FIGURE 4.5.2

Pathways to sustainable land management (ELD, 2013b)

underlining much-needed choices that must be based on sound information and facts while being comprehensible for decision makers and stakeholders (ELD Initiative, 2013b).

4.5.1 THE ECONOMICS OF LAND DEGRADATION INITIATIVE

Against this background, and in light of the exacerbating effects of continued land degradation, the need for clear metrics as a foundation to select potential and accessible options was identified, with the goal of facilitating an SLM-enabling environment. The Economics of Land Degradation (ELD) Initiative was thus founded as "an initiative for a global study on the economic benefits of land and land based ecosystems" (http://www.eld-initiative.org/index.php?id=23). It focuses on the value and benefits of SLM while providing a scalable approach for analyses of the economics of land degradation based on the cost and benefits of land and land-based ecosystems. A primary goal of the ELD Initiative is to integrate the economics of land degradation in policy strategies and decision making through increased political and public awareness.

Within this framework, three stakeholder groups were identified as strongly interconnected and influential in regards to the practice and implementation of SLM. They represent the core ELD target groups:

- Political decision makers
- The private sector
- The scientific community

Recognizing the different needs of these stakeholders at various governance, spatial, and temporal scales is crucial for applicable, relevant results. Drivers of land degradation and barriers to the adoption of SLM differ at the regional, national, local, and household levels. For example, increasing changes in land use to agriculture as a result of the increasing human population is a global driver of degradation, whereas conflict can create pockets of land degradation (e.g., the abundance of land mines in Cambodia in the 1990s that rendered much land unsafe for any use, including food production). As these levels of land management are interconnected and depend on processes within the panarchy of different interacting resource governance scales (http://www.sustainablescale.org/ConceptualFramework/UnderstandingScale/MeasuringScale/Panarchy.aspx), it is important to assess the impacts of land degradation and the benefits that holistic and long-term land management provides. Recognizing the challenges of a multiscale and multistakeholder approach, the ELD Initiative adapted a six-step approach originally developed by Noel and Soussan (2010). Each step can be disaggregated to meet the specific demands of a particular setting. They are as follows (ELD Initiative, 2013b):

- **Inception**: Identification of the scope, location, spatial scale, and strategic focus of the study, based on stakeholder consultation and the preparation of background materials on the socioeconomic and environmental context of the assessment.
- **Geographical characteristics**: Assessment of the quantity, spatial distribution, and ecological characteristics of land cover types, categorized into agroecological zones and analyzed through the use of a geographical information system (GIS).
- **Types of ecosystem services**: Analysis of ecosystems services stocks and flows for each land cover category, based on the ecosystem service framework.
- **Role of ecosystem services and economic valuation**: The role of the assessed ecosystems services in the livelihoods of communities living in each land cover area and also the role of overall economic development in the study zone. This implies estimating the total economic value of these services to estimate the benefits of action or the cost of inaction.
- **Patterns and pressures**: Identification of land degradation patterns, drivers, and pressures on the sustainable management of land resources, including their spatial distribution and the assessment of the factors causing the degradation. This is to inform the development of scenarios for cost-benefit analysis. The following substeps can be taken to choose the appropriate valuation method under available data, resources, local capacity, and specific objective to be achieved: (i) deciding the type of environmental problem to be analyzed; (ii) reviewing which valuation method is appropriate for that problem and the type of environmental value to be captured (use value or total economic value); (iii) considering what information is required for the identified environmental problem and chosen valuation method; and (iv) assessing what information is readily available, how long it would take to access it, and at what monetary cost.

- **Cost-benefit analysis and decision making**: The assessment of sustainable land management options that have the potential to reduce or remove degradation pressures, including the analysis of their economic viability and the identification of the locations for which they are suitable.

The ELD Initiative identified an additional step as an addendum to the methodology, which is to take action with the most economically feasible choice, ensuring that it is actually implementable. This also likely requires the dissolution of the aforementioned barriers and obstacles to SLM practices (ELD Initiative, 2013b). To further that aim, a recent publication of the initiative, *Economics of Land Degradation Initiative: Practitioners Guide* (ELD Initiative, 2014a), provides a user-friendly, step-by-step guide that facilitates uptake of SLM practices by practitioners in the field, based on sound cost-benefit analyses and robust, custom assessments. As practitioner involvement is essential for the implementation of SLM, this guide "provides practitioners and decision-makers with the skills necessary to make an economic case for preventing or reversing land degradation and to adopt more sustainable land management options" (ELD Initiative, 2014a).

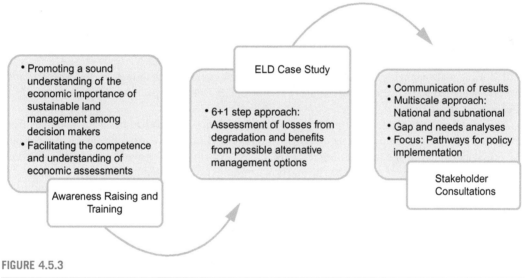

FIGURE 4.5.3

Dissemination of economic valuation of SLM (Hannes Etter)

4.5.2 FROM SCIENTIFIC KNOWLEDGE TO ACTION: IMPLEMENTATION OF ECONOMIC VALUATION

In order to ensure the application and implementation of results from these assessments on a broader scale, as well as facilitate their upscaling to larger levels, the ELD Initiative promotes an iterative, multiscale dialogue between relevant parties involved in land management. Management of land can sufficiently influence the practice of resource use, including the three noted target groups (scientific community, private sector, and policy and decision makers). Figure 4.5.3 provides a schematic overview of this process, organized along three components: (i) an inaugural training workshop, (ii) the ELD assessment, and (iii) subsequent stakeholder consultations. The training of high-level decision makers (e.g., on the ministerial political level or among members of a company's supervisory board) provides a sound understanding of economic assessments, while also raising awareness of the importance of sound resource management. This also sets the frame and focus for a subsequent ELD assessment, ensuring results that ultimately meet the demands and needs of decision makers and their constituents. Results from economic assessments of the benefits from action then directly feed into stakeholder consultations at different administrative scales. These consultations are currently being used to communicate the potential of SLM, as well as identify feasible pathways and tools for the implementation of desired management options. For instance, successful consultations by the ELD Working Group on *Options and Pathways to Action* in Kenya concluded with clear recommendations to enable the implementation of SLM on a national and subnational level (ELD Initiative, 2014b). Potential tools facilitate the financing of SLM, such as payments for ecosystem services, taxes, subsidies, and the establishment and use of new markets such as carbon offsets (ELD Initiative, 2013b) will be adapted to context-specific needs. While pathways to action are mostly informed by the scientific community and steered by political actors, they also often have a strong impact on and are influenced by the private sector, connecting all these stakeholder realms (ELD Initiative, 2013b). As such, and also in light of their influence on the use of natural resources, it is prudent to include the private sector as decision makers and relevant stakeholders when performing cost-benefit analyses of SLM.

It is clear that managing land more sustainably, which includes rehabilitation and restoration of formerly degraded soils, does not only relate to ecological or social perspectives, but also matters from an economic standpoint.

An increasing body of literature provides a strong foundation to mobilize financial resources from governments as well as the private sector (ELD Initiative, 2013a). However, different challenges persist. Full understanding of ecosystem services and the benefits that they provide is still difficult to calculate, scaling management approaches up and down remains arduous, and multistakeholder processes can incite dormant conflicts and power inequalities. The application of scientifically robust and practical SLM requires a holistic conceptual foundation to select among the proposed management solutions. The landscape approach, defined as a way "to conceptualize and implement integrated multiple objective projects" (Sayer et al., 2013), is often proposed to "enhance the sustainable management and rehabilitation of production capacities" (Hett et al., 2013). In this context, an economic assessment of benefits of a particular action can not only provide a better understanding of the status quo and the importance of the present value of land as a resource base, but also contribute to a sound understanding of the long-term benefits of SLM. This includes rehabilitation and restoration activities and ultimately helps to unify stakeholders and support human safety and security and ecological sustainability and conservation efforts in a rapidly changing and challenging world.

REFERENCES

Barbier, E.B., Hochard, J.P., 2014. Land Degradation, Less Favored Lands and the Rural Poor: A Spatial and Economic Analysis. A Report for the Economics of and Degradation Initiative. Department of Economics and Finance. University of Wyoming, Laramie, WY. http://www.eld-initiative.org/fileadmin/pdf/ELD __Assessment.pdf.

ELD Initiative, 2013a. Opportunity Lost: Mitigating Risk and Making the Most of Your Land Assets. An Assessment of the Exposure of Business to Land Degradation Risk and the Opportunities Inherent in Sustainable Land Management. Available from, www.eld-initiative.org (accessed 08.08.15).

ELD Initiative, 2013b. The Rewards of Investing in Sustainable Land Management. Interim Report for the Economics of Land Degradation Initiative: A Global Strategy for Sustainable Land Management. Available from, www.eld-initiative.org (accessed 08.08.15).

ELD Initiative, 2014a. Principles of Economic Valuation for Sustainable Land Management Based on the Massive Open Online Course, the Economics of Land Degradation. Practitioner's Guide. Available from, www.eld-initiative.org (accessed 08.08.15).

ELD Initiative, 2014b. Support Towards the Economics of Land Degradation Initiative. Report on the ELD Kenya Stakeholders Consultations April 2014. Available from: http://www.eld-initiative.org/fileadmin/pdf/Reports_WG_Stacey/1_ELD_Kenya_Report.pdf (accessed 08.08.15).

Eswaran, H., Lal, R., Reich, P.F., 2001. Land Degradation: An Overview. In: Bridges, E.M. et al., (Ed.), Responses to Land Degradation. Proc. 2nd. International Conference on Land Degradation and Desertification, Khon Kaen. Thailand. Oxford Press, New Delhi, India.

Favretto, N., et al., 2014. Assessing the Socio-Economic and Environmental Dimensions of Land Degradation: A Case Study of Botswana's Kalahari. Report for the Economics of Land Degradation Initiative. Leeds, UK. Available from, http://www.see.leeds.ac.uk/research/sri/eld/ (accessed 08.08.15).

Godfray, H.C.J., et al., 2010. Food security: the challenge of feeding 9 billion people. Science 327 (5967), 812–818.

Hett, C., et al., 2013. Landscape Mosaics Maps as a Basis for Spatial Assessment and Negotiation of Ecosystem Services and their Trade-Offs at the Meso-Scale: Examples from Laos, Madagascar and China. Available from, http://www.cde.unibe.ch/v1/CDE/pdf/HETT.pdf (accessed 08.08.15).

Liniger, H., Schwilch, G. Enhanced decision-making based on local knowledge. The WOCAT method of sustainable soil and water management. Mount. Res. Develop. 22(1), 14–18.

Nkonya, E., Gerber, N., von Braun, J., De Pinto, A., 2011. Economics of Degradation. The Costs of Action Versus Inaction. Available from, http://www.ifpri.org/sites/default/files/publications/ib68.pdf (accessed 08.08.15).

Noel, S., Soussan, J., 2010. Economics of Land Degradation: Supporting Evidence-Base Decision Making. Methodology for Assessing the Costs of Degradation and Benefits of Sustainable Management. Paper commissioned by the Global Mechanism of the UNCCD to the Stockholm Environment Institute (SEI), Global Mechanism, Bonn Germany.

Renaud, F.G., Sudmeier-Rieux, K., Estrella, M., 2013. Why Do Ecosystems Matter in Disaster Risk Reduction? In: Renaud, F.G., Sudmeier-Rieux, K., Estrella, M. (Eds.), The Role of Ecosystems in Disaster Risk Reduction. UNU-Press, Tokyo, Japan, pp. 3–26.

Sayer, J., et al., 2013. Ten principles for a landscape approach to reconciling agriculture, conservation, and other competing land uses. Proc. Natl. Acad. Sci. U. S. A. 110 (21), 8349–8356.

Smith, P., et al., 2007. Agriculture. In: Metz, B. et al., (Ed.), Cambridge University Press, Cambridge, UK, and New York.

Uy, N., Shaw, R., 2012. Overview of Ecosystem-Based Adaptation. In: Uy & Shaw, Ecosystem-Based Adaptation. pp. 3–19.

World Bank, 2006. Sustainable Land Management Sourcebook. Available from, https://innovationpolicyplatform.org/sites/default/files/rdf_imported_documents/eBook.pdf. Emerald Group Publishing Limited, Bradford, UK.

FOUR RETURNS, THREE ZONES, 20 YEARS: A SYSTEMIC APPROACH TO SCALE UP LANDSCAPE RESTORATION BY BUSINESSES AND INVESTORS TO CREATE A RESTORATION INDUSTRY

4.6

Willem H. Ferwerda

Rotterdam School of Management, Erasmus University, Rotterdam, the Netherlands; IUCN Commission on Ecosystem Management, Gland, Switzerland; Commonland Foundation, Amsterdam, the Netherlands

4.6.1 INTRODUCTION

"We have a choice to make during our brief visit to this beautiful blue and green living planet: to hurt it or to help it."

Ray Anderson, founder and chairman of Interface Inc. (May 2009)

According to scientists at the Stockholm Resilience Centre, humanity is rapidly approaching the nine boundaries of the productive ecological capacities of the planet (Rockström et al., 2009).[1] Recently, an international team of scientists concluded that four of these nine planetary boundaries (climate change, loss of biosphere integrity, land-system change, and altered biogeochemical cycles—i.e., phosphorus and nitrogen) have been crossed as a result of human activity (Steffen et al., 2015). In fact, society is said to have entered a new geological period—called the *Anthropocene*[2]—in which human activity,

---

[1]The planetary boundaries framework defines a safe operating space for humanity based on the intrinsic biophysical processes that regulate the stability of the Earth system, which include the following: 1. Stratospheric ozone depletion; 2. Loss of biosphere integrity (biodiversity loss and extinctions); 3. Chemical pollution and the release of novel entities; 4. Climate Change; 5. Ocean acidification; 6. Freshwater consumption and the global hydrological cycle; 7. Land system change; 8. Nitrogen and phosphorus flows to the biosphere and oceans; and 9. Atmospheric aerosol loading.

[2]This term was first proposed by the Nobel Prize–winning atmospheric chemist Paul Crutzen and Eugene F. Stoermer (Crutzen and Stoermer, 2000).

Land Restoration. http://dx.doi.org/10.1016/B978-0-12-801231-4.00025-2

particularly traditional business practices, significantly affects and influences the Earth's natural events and ecosystems (Steffen et al., 2007).

Environmental degradation is a threat to human well-being, the global economy, international trade, and society as a whole.[3] Annually, an estimated US$4.3–20.2 trillion is lost due to the conversion of natural ecosystems to other uses (Constanza et al., 2014).[4] The World Resources Institute (WRI), the University of Maryland, and the International Union for Conservation of Nature (IUCN) have calculated that 2 billion ha (equivalent to the areas of China and the United States combined) are degraded or destroyed (WRI, 2014). Thus, the restoration of natural capital forms an essential part of the outcomes of 2012's Rio+20 UN conference (UN, 2012) and of the indicators of UNEP's Greening Economy Initiative (UNEP, 2011).

The Stockholm Resilience Centre's nine planetary boundaries provide a physical and biological basis for understanding the interconnection of the world's different global environmental threats. The relationship between the environmental ceiling of each planetary boundary and the 11 dimensions of human well-being, identified through government priorities for Rio+20, is shown in Figure 4.6.1.

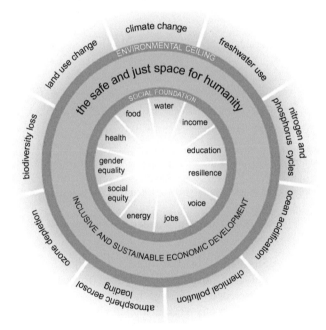

FIGURE 4.6.1

A safe and just space for humanity is based on the social foundation of humanity, consisting of the 11 dimensions of human well-being and keeping well within the environmental ceiling, formed by each of nine planetary boundaries (after Kate Raworth).

---

[3]See, e.g., Convention on Biological Diversity (2011), Sukhdev et al. (2014), Millennium Ecosystem Assessment (2005), Johansson et al. (2012), IUCN (2014) IUCN and CEM (2014), and McKinsey & Company (2011).
[4]In comparison, the 2013 gross world product (GWP) is estimated at US$74.31 trillion (Index Mundi, 2014).

Given this interconnectedness among global environmental issues, no single issue can be resolved without understanding its interaction with the others. In a restoration model, the rationale for people's actions should be connected to what is scientifically and technically possible, while also focusing on what is urgently needed from an ecological point of view. While human activities have caused many of today's problems, we can reverse this damage. This section discusses potential future actions toward a systemic solution.

4.6.1.1 RETHINKING OUR RELATIONSHIP WITH NATURE

Solving the ecological crisis requires more than technical innovation; it also requires the integration of the knowledge and experiences of different stakeholder groups who have established a clear vision of a sustainable future, wherein economic activity operates within the functional boundaries and capabilities of the planet. Barriers and institutional silos must be broken down, and human society must accept its interdependence and interconnectedness with ecosystems and their functions. In economic terms, this requires a shift of both theory and practice from a linear understanding of finance and business activity to a cyclical one (incorporating feedback loops), which is based on a solid understanding of how natural and anthropogenic systems work. It is imperative that we restore the mutually beneficial relationship between human society and nature, while moving toward a circular economy—that is, one that is not dependent on infinite growth. This type of economy provides a coherent framework for systems level redesign and, as such, offers an opportunity to harness innovation and creativity to enable a positive, restorative economy (Ellen MacArthur Foundation, 2014). The Triple Bottom Line (otherwise known as "Triple P," for Planet, People, and Profit) philosophy[5] encourages the idea that reducing one's impact is sufficient, which has led to important steps in sustainability. However, the rate of biodiversity loss and the conversion of natural ecosystems, which together constitute major degradation, continue at an increased rate. The conclusion, therefore, must be that the Triple Bottom Line model does not go far enough to preserve and restore ecosystems. We need to develop this idea into one of restoration and protection of "planetary resilience" and understand that "ecosystems are economics" (Figure 4.6.2).

This suggests that companies should work on not only lowering their unsustainable impacts but also scaling up their positive impacts through ecological restoration in partnership with key stakeholders.

4.6.1.2 OBSTACLES LIMITING BUSINESS INVOLVEMENT IN RESTORATION

To scale up one's positive impacts, scientists, practitioners, and local communities must work together to understand the complexity of ecological and socioeconomic drivers. Obstacles between stakeholder groups must be removed to allow essential knowledge sharing and collaboration. Designing and implementing effective, efficient, and engaging restoration projects and programs will enable businesses and investors to scale up ecosystem restoration efforts in a cost-effective manner. Figure 4.6.3 depicts the most important reasons for businesses and investors not to become active in restoration activities.

---

[5]With sustainable development having been articulated by the World Commission on Environment and Development (Brundtland Commission, 1987), the "triple bottom line" was first discussed by Freer Spreckly (1981) and then further developed by John Elkington (1997).

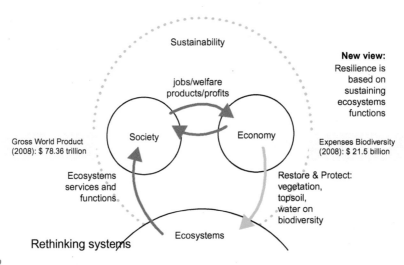

FIGURE 4.6.2

A more complete view of sustainability. The blue arrows are the basis of our unsustainable economic system; we need to organize the green arrow with all stakeholders, including business, to sustain ecosystem functions.

Obstacles to involve business in ecosystem restoration

Opportunities to address it

	Obstacles	Opportunities
1	Lack of long term thinking Time frame needs to expand from 2-3 years to 20+ years	Focus on family companies and pension funds (patient capital, long term)
2	Economic value of ecosystems is poorly understood and externalities not accounted for	Show that it works in existing scalable projects
3	Local communities continue pattern of behaviors, which are unknowingly detrimental	Focus on Income, Jobs and Education
4	Solutions are often presented overly complex while simple proven tools and techniques exist	Use a Language that everyone understands…and film it!
5	Silo thinking approach Many stakeholders working in isolation projects; well intentioned but not additive.	Holistic (Systemic) thinking approach We need people with an internal approach of connectivity or purpose.

SOURCE: Willem Ferwerda, Nature Resilience, ecological restoration by partners in business for next generations. RSM, IUCN CSV (2012), McKinsey team analysis (2012), Common land Foundation (2014)

FIGURE 4.6.3

Overview of the five most important obstacles and opportunities to involve business and investors in ecosystem restoration.

BOX 4.6.1 THE LOESS PLATEAU WATERSHED REHABILITATION PROJECT (CHINA)

The Loess Plateau Project started in 1994 with US$252 million provided by the World Bank's International Development Association and the Chinese government. With a total budget of approximately US$491 million and a project area covering **1,56** million ha (or **15,6000** km^2, approximately half the size of Belgium). The following outcomes provide insight into what can be achieved through cooperation and long-term focus:

- Sediment flow into the Yellow River decreased by more than 53 million tons and continues on this downward trend.
- A network of small dams store water for use by towns and farmers when rainfall is low, increasing the vegetation cover which reduces the risk of flooding.
- Replanting and grazing bans increased the perennial vegetation cover from 17% to 34%.
- Local food supply increased.
- More than 2.5 million people in four of China's poorest provinces—Shanxi, Shaanxi, Gansu, and the Inner Mongolia Autonomous Region—were lifted out of poverty, with the poverty rate reducing from 59% to 27%.
- Farmers' incomes increased from around US$70 per year, per capita, to around US$200.
- Substantial downstream benefits from the reduced sedimentation and global improvements through carbon sequestration.

The project's principles have been adapted and replicated widely throughout China by the Green for Grain programme. It is estimated that as many as 20 million people in China have benefited from such an approach. The continuing aim is to restore the entire Loess plateau, which is approximately the size of France.

In order to achieve significant results in scaling up ecosystem restoration programs, the following lessons from the Loess Plateau must be taken into account:

- The scientific definition of "restoration and rehabilitation" needs to be clear and concise before the start of the project.
- Intergenerational sustainable profit models with a long-term time frame (20-40 years) is a key success factor.
- Incorporating local farmers into the rehabilitation of the land helps gain favour for the project and increases the project's longevity by making them future custodians. This helps generating income for the farmers.
- The use of a common language that all stakeholders understood, simplified understandings of complicated restoration concepts and enabled farmers and experts to work together.
- A wide range of technical, hydrological, ecological, and agricultural tools was made accessible through government support.

Stakeholders include governments, businesses, farmers, landowners, and local communities, all needing to cooperate to understand long-term intergenerational projects instead of focusing on short-term activities that achieve few sustainable impacts (Van Andel and Aronson, 2012). The Chinese Loess Plateau Watershed Rehabilitation Project is an existing silo-breaking project and a practical example of a massive landscape-rehabilitation project (Box 4.6.1, Figure 4.6.4).

FIGURE 4.6.4

Rehabilitated hills of the Loess Plateau at Gao Xing Zhuang village, China, where, before 1995, the landscape was heavily eroded. Photo credit: Kosima Weber Liu, EEMP.

4.6.1.3 TARGETS

Accelerating restoration means developing a common target that enables all stakeholders to reach out from their respective silos (be they academic, business, personal, etc.) and work together. The Bonn Challenge (GPFLR, 2013a) reflects the best agreement so far to address a restoration agenda, with its core commitment being to restore 150 million ha of lost forests and degraded lands worldwide by 2020. However, both Action 2020, launched in 2013 by the World Business Council on Sustainable Development, and the Global Alliance on Climate Smart Agriculture, launched by John Kerry, United States Secretary of State and Sharon Dijkstra, minister of the Netherlands, at the Climate Summit on September 23, 2014, in New York, address ecosystem restoration as well.

Although there are enormous opportunities for increasing food production, biodiversity, water security, and the accumulation of biomass in the topsoil by recovering lost functionality in production landscapes, not one global initiative or consortium has succeeded in involving the business sector in large-scale restoration of degraded lands and biodiversity. This is particularly serious given the urgent need to scale up and accelerate ecosystem restoration, as it is connected to alleviating poverty in many developing countries (UNEP, 2010).

4.6.1.4 SYSTEMIC MODEL

Scaling up ecological restoration with business partners is not about controlling complexity, but rather about distributing the weight of complexity among partners. It should be practical (e.g., by using common practices) and replicable (e.g., by using a common language and working in partnerships), while actively searching for economical viable restoration projects based on a combination of sustainable business cases (such as forestry, agriculture, tourism, etc.). Ecological restoration is both a philanthropic endeavor and a core economic issue. It requires establishing new partnerships between companies and local organizations to work together on long-term projects that give priority to tailor-made business cases. Over the last 10 years, there has been an increasingly larger group of participants from the business, finance, scientific, and civil society spheres coming together to pave the way for designing and implementing practical solutions. The next generation of business leaders should be educated in business schools on ways to integrate and emphasize the importance of "natural capital." Working with nature will not only strengthen the long-term technical and strategic functioning of businesses but also boost the morale and passion of its employees. One in four corporate titans worldwide view biodiversity loss as a threat to their business growth (TEEB – The Economics of Ecosystems and Biodiversity for Business – Executive Summary 2010a). The challenge is to provide companies and investors with a model that brings different interests together based on a sustainable business model and a systemic approach (Box 4.6.2).

4.6.2 ECOSYSTEM RESTORATION: THE ECONOMY RELIES ON ECOLOGY

The four types of ecosystem services first recognized by the Millennium Ecosystem Assessment (2005) are established among scientists and policy makers:

BOX 4.6.2 A SYSTEMIC APPROACH FOR ECOSYSTEM RESTORATION BRINGS ALL INTERESTS TOGETHER

A systemic approach is characterized by the following elements:

- Addressing the interests of companies and investors, along with research institutions, business schools, civil-society organizations, local governments, and farmers
- Selecting system thinkers—that is, attract people in the business community, NGOs, and scientists who are committed to a new way of achieving socioeconomic and ecological sustainability based on systems thinking[6]
- Using all available technology—that is, making use of a tailor-made innovative toolbox (including practical and financial tools) for ecosystem restoration and sustainable agriculture. Practical examples such as innovative devices for tree planting without irrigation, permaculture, removal of invasive species, as well as innovations within the financial sector, such as green bonds and habitat banking.
- Educating future business leaders by creating a direct relationship between business schools and restoration projects to influence the new generation of business leaders, which ensures that economic and business activities protect and restore ecosystems through business models that are built to make restoration a viable investment proposition
- Complementing this approach with the existing efforts of farmers, landowners, governments, NGOs, and scientists

[6]Systems thinking is understanding how factors/inputs influence one another within a whole. For example, in nature, ecosystems are composed of various elements (air, water, movement, plants, and animals) that depend on one another to survive or perish. Likewise, organizations consist of people, structures, and processes that work together to succeed. Systems thinking is defined as problem solving by viewing problems as parts of an overall system. The antithesis to systems thinking is reacting to specific parts, outcomes, or events, which may contribute to the development of unintended consequences (Senge, 1990).

- **Provisioning services**: Food (including seafood and game), crops, wild foods, and spices; water; pharmaceuticals, biochemicals, and industrial products; and energy (hydropower and biomass fuels)
- **Regulating services**: Carbon sequestration and climate regulation, waste decomposition and detoxification, purification of water and air, crop pollination, and pest and disease control
- **Supporting services**: Nutrient dispersal and cycling, seed dispersal, primary production, and infrastructure and housing
- **Cultural services**: Cultural, intellectual, and spiritual inspiration; recreational experiences (including ecotourism); and scientific discovery

To provide these services, which form the basis of wealth creation, ecosystems must be functional. Given this premise (that ecosystem restoration is beneficial to the economy), degrading and destroying a primary asset to make money in the short term is counterproductive. Logically, one should save or conserve such an important asset's value (at least) and seek to improve or enhance its condition, subsequent worth, and continued productivity (at best). For example, any factory owner would avoid sacrificing his or her production equipment for the sake of a single product's development. However, this is precisely how the world economy manages ecosystems and landscapes: natural capital stocks are sacrificed for the sake of what they enable us to produce. It is a short-term solution to the problem of income generation, in essence. Nkonya et al. (2011) describe the stakes: "Awareness of environmental risks has moved to the forefront of global consciousness during the past 25 years. However, this awareness has not translated into comprehensive action to address the problem of land degradation, which poses a serious threat to long-term food security. This inaction is primarily the result of limited knowledge of the costs related to land degradation and of insufficient institutional support" (Nkonya et al., 2011).

Among most scientists, and increasingly among members of the business community, it is widely accepted that healthy ecosystems form the basis of a sound and sustainable economy. The Economics of

Ecosystems and Biodiversity (TEEB) provides important insights into the relationship between ecosystem degradation and its costs to the global economy and society.[7] These are huge: TEEB estimates the economic loss due to degradation to be US$21–72 trillion per year, and Costanza et al. (2014) estimate that US$4.3–20.2 trillion has been lost due to the conversion of ecosystems (for plantations, agricultural land, mining, oil exploration, urban areas and infrastructure) annually. By comparison, in 2012, the gross world product (GWP) was approximately US$84.97 trillion. TEEB concludes that the restoration and conservation of ecosystems is no longer an issue to be tackled solely by nongovernmental organizations (NGOs) and other charitable organizations or by donor-funded development projects such as public-private partnerships (P3s).

TEEB intends to produce a global study on the economic benefits of land-based ecosystems by highlighting the value of sustainable land management and providing a global approach for the economic analysis of land degradation. It aims to integrate these economic analyses into policy strategies and decision making by increasing the political and public awareness of the costs and benefits of land-based ecosystems. Together with TEEB, a genuinely holistic view of the issues at stake could be established. TEEB for business identifies seven steps[8] that allow companies to better account for the value of natural capital, which should lead to involvement in ecosystem restoration. However, this must be financially attractive to ensure that businesses act on this information.

4.6.3 RESTORING ECOSYSTEM FUNCTIONS IS RESTORING OUR ECONOMY

"The nation that destroys its soil destroys itself."

Franklin Delano Roosevelt (February 26, 1937)

Several companies increasingly are realizing that investing in environmental sustainability is highly profitable in the medium and long terms due to lower costs and higher revenues. Many have pursued actions on the basis of corporate social responsibility (CSR) and environmental impact-reduction strategies.

4.6.3.1 BEYOND IMPACT REDUCTION

Good initiatives abound in direct investment projects and through the introduction of environmentally (and socially) friendly production processes, including those that embrace new certification schemes and introduce participatory processes with all relevant stakeholders. Such certification initiatives based on supply chains include the Forest Stewardship Council, Marine Stewardship Council, Sustainable Trade Initiative, the soy and palm oil roundtables, Utz Certified, Rainforest Alliance, and International Organization for

---

[7]TEEB (2010b), The Economics of Ecosystems and Biodiversity: Mainstreaming the Economics of Nature: A Synthesis of the Approach, Conclusions and Recommendations of TEEB.

[8]These seven steps are (TEEB, 2010a):

- Identify the impacts and dependencies of your business on biodiversity and ecosystem services (BES).
- Assess the business risks and opportunities associated with these impacts and dependencies.
- Develop BES information systems, set SMART (specific, measurable, achievable, relevant, and time-bound) targets, measure and value performance, and report your results.
- Take action to avoid, minimize, and mitigate BES risks, including in-kind compensation (offsets) where appropriate.
- Grasp emerging BES business opportunities, such as cost efficiencies, new products, and new markets.
- Integrate business strategy and actions on BES with wider CSR initiatives.
- Engage with business peers and stakeholders in government, NGOs, and civil society to improve BES guidance and policy.

BOX 4.6.3 GLOSSARY ON ECOSYSTEM RESTORATION AND REHABILITATION

The variety of the definitions that are key to this topic is the result of decades of scientific debate. It is clear that humanity cannot go back to the natural state of ecosystems that existed prior to its interference. As time wears on, more people than ever will exist, living off the land, which means there will be more pressure on the world's resources. In order to successfully scale up restoration efforts with business and investors, a universal understanding of the key terms is vital.

Standardization (ISO) 26000, almost all of which are related to commodities for the international market (coffee, cocoa, soy, palm oil, and timber) where the consumer pays a premium for responsibly managed supply chains. While reducing environmental impacts is the focus, it is insufficient on its own to properly protect ecosystems: new business models have to move beyond certification and environmental impact assessments (EIAs). Ecosystem restoration is an important approach to reversing the depletion of natural capital; unfortunately, it has been underutilized and underfunded. There are many reasons for companies to act, however, including the ethical, financial, and competitiveness reasons (Nidulomu et al., 2009).

A challenge to restoration stems from the existence of different scientific views and perspectives on landscape and ecosystem restoration. These dominate the scientific debate and often slow action on the ground. This so-called novel ecosystem concept is constantly in debate (Murcia et al., 2014), particularly regarding key definitions (SER, 2004; FAO, 2005a; Lamb and Gilmour, 2003); see Box 4.6.3.

4.6.4 RESTORATION AND REHABILITATION

Ecological restoration is defined as "[t]he process of assisting the recovery of an ecosystem that has been degraded, damaged, or destroyed" (SER, 2004). It is often used broadly to mean returning a site or system to "pre-disturbance conditions" and implies connecting an ecosystem, as it occurred and developed in the past, to its future potential to evolve and adapt. The notion of "historical continuity" is relevant and useful because it gives people living in degraded ecosystems a better understanding of the ecosystems services in the past.

Ecological rehabilitation seeks to reestablish the productivity of a natural ecosystem, as well as some, but not necessarily all, of the plant and animal species thought to originally[9] inhabit that site, though "pre-disturbance" conditions are not always possible to achieve. In time, the protective function of that ecosystem and many of its ecological services may be reestablished. Emphasis is generally put on restoring ecosystem processes and functions to increase the flow of services and benefits to people (FAO, 2005b; Clewell and Aronson, 2013).

4.6.4.1 DIFFERENT LANDSCAPES, DIFFERENT APPROACHES

"By locating forests, pastures and water bodies within the larger ecological, social and economic context, landscapes open up to meet a range of objectives that could address demands of preservation, conservation and exploitation."

Jagdeesh Rao, CEO Foundation for Ecological Security, India (Accessed on 24 August, quote on website www.commonland.com)

---

[9]For ecological or economic reasons, the new habitat might also include species not originally present at the site.

Landscapes are often defined as sets of overlapping ecological, social, and economic networks within a specific area, making them ideal units for planning and decision making, as they allow for the integration of various sector plans and programs within a spatial context. Each landscape calls for its own kind of restoration. The resilience of ecosystem functions is the starting point in restoring ecosystem functions, which will increase landscape biomass, biodiversity, and organic matter. It will further increase ecosystem services, such as pollination, retention of water, soil fertility, and human well-being. Landscapes are created where an increase in biodiversity and vegetation cover contribute to newly developed agricultural lands. Within those mosaic landscapes, ecological, sustainable agricultural, and economic zones coexist, as they sustain the natural resilience of the ecosystem.

Although many use "ecosystems" and "landscapes" interchangeably, the most complete definition of "landscape restoration" is provided by the GPFLR (2013b): "[T]urn barren or degraded areas of land into healthy, fertile, working landscapes where local communities, ecosystems and other stakeholders can co-habit, sustainably." On a national level, political and practical steps are taken are being taken in Rwanda, El Salvador (Figure 4.6.5), and Ethiopia to restore large areas of degraded ecosystems. These countries' leaders increasingly understand that restoring ecosystem functions equally restores the economy.

4.6.4.2 CREATING A RESTORATION INDUSTRY

Although some individual companies may contribute to carbon compensation schemes or support individual restoration projects, a wider global initiative (consortium, mechanism) to engage business is urgently needed, because the business community has many of the essential capabilities required for success, including a hands-on approach, the ability to mobilize local communities, and the resources to finance projects.[10]

FIGURE 4.6.5

Degradation in Bajo Lempo, El Salvador. The government of El Salvador is working on a national restoration program (Programa Nacional de Restauración de Ecosistemas y Paisajes, PREP) to transform the agro-sector and restore ecosystems that are critical for a sustainable, secure, and productive landscape over 50% of the country (1 million ha). Photo credit: author Willem Ferwerda.

---

[10]The need for business involvement is underlined by the 2011 call to action at the World Conference on Ecological Restoration by the Society of Ecological Restoration, and it was reinforced in 2013 at the 5th World Conference on Ecological Restoration. The State of the World's Land and Water Resources for Food and Agriculture (FAO, 2011).

Estimates of costs and benefits of restoration projects in different biomes

	Biome/Ecosystem	Typical cost of restoration (high scenario)	Estimated annual benefits from restoration (avg. scenario)	Net present value of benefits over 40 years	Internal rate of return	Benefit/cost ratio
		U$$/ha	U$$/ha	U$$/ha	%	Ratio
1	Coral reefs	542,500	129,200	1,166,000	7%	2.8
2	Coastal	232,700	73,900	935,400	11%	4.4
3	Mangroves	2,880	4,290	86,900	40%	26.4
4	Inland wetlands	33,000	14,200	171,300	12%	5.4
5	Lakes/rivers	4,000	3,800	69,700	27%	15.5
6	Tropical forests	3,450	7,000	148,700	50%	37.3
7	Other forests	2,390	1,620	26,300	20%	10.3
8	Woodland/shrubland	990	1,571	32,180	42%	28.4
9	Grasslands	260	1,010	22,600	79%	75.1

FIGURE 4.6.6

Estimated returns from ecological restoration (UNEP, 2010, after The Economy of Ecosystems and Biodiversity, TEEB, 2009).

Several studies have found a significant profit motive to funding restoration activities. For example, an IUCN analysis (IUCN, n.d.) shows that, once restored, 150 million ha would pump more than US$80 billion into national and global economies and close the climate change "emissions gap" by 11%–17%. Based on TEEB data (2009) (Figure 4.6.6), a mean investment per hectare of US$2390 is needed; to restore 200 million ha, US$478 billion, or about EUR450 billion, is needed over 20 years. That's approximately US$24 billion or EUR 22.5 billion a year. De Groot et al. (2013) find that, even in a worst-case scenario (i.e., discount rate of 8%, 100% of the maximum cost, and a restoration benefit of 30% of the total economic value), investing in restoration still breaks even or provides a financial profit (in total economic value) in six ecosystem types.

4.6.5 A TOOLBOX OF PROMISING SOLUTIONS

Experiences from many ongoing or completed projects have already created a toolbox to tackle future projects. These tools include several promising technical solutions, as well as social- and stakeholder-management tools that are important for successful restorations.

4.6.5.1 SOCIAL TOOLS

While economic drivers are key to success, they are not the only drivers. Social participatory skills are also an important prerequisite to achieve success in restoration, as is the social structure among stakeholders. Many lessons can be learned from development organizations, conservationists, and farmers. Living Lands, a nonprofit organization for conserving and restoring living landscapes, integrates the

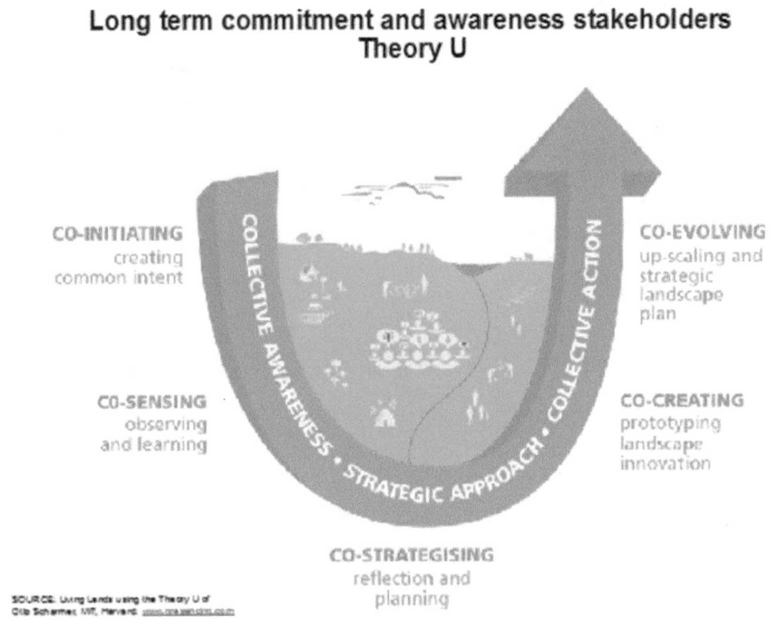

FIGURE 4.6.7

Bottom-based stakeholder engagement process and Theory University, which was developed by Otto Scharmer (MIT and Harvard). It is now used successfully by Living Lands in South Africa to give land users and landowners a deeper understanding of their interaction with the land and develop its potential. Figure design: Living Lands.

"U" methodology (Scharmer and Kaufer, 2013; Senge et al., 2004), transdisciplinary research, and eco-system approach. The approach provides opportunities for all stakeholders to identify and create viable community-based responses (Figure 4.6.7). "Theory U" proposes that the quality of the results that we create in any kind of social stakeholders system is a function of the quality of awareness, attention, or consciousness from which the participants in the system operate.

4.6.5.2 TECHNICAL TOOLS

Several examples have shown that it is possible to regreen eroded areas (Figure 4.6.8). Once stakeholders are able to combine greening with successful economic activities, a business model will emerge. Low-tech solutions can create biomass in dry, degraded lands through permaculture techniques, water storage systems,[11] and Biochar,[12] a soil enhancer that can hold carbon, boost food security, and increase soil biodiversity. Many more such low-tech innovative ideas exist and are in the process of being scientifically tested and analyzed with promising results. They are often the result of citizens' initiatives, entrepreneurial inspiration, or individual or collective responsibility.

---

[11]Land Life Company (2014) and Groasis (2014) specialize in developing tree-planting devices without irrigation.
[12]International Biochar Initiative (2014).

Many proven technologies exist

Waterboxx – Life Land Box

Permaculture

Waterworks

Holistic Livestock Management

Dune stabilization

Reclamation of mining sites

SOURCE: Nature Resilience, ecological restoration by partners in business for next generations. RSM, IUCN CEM (2012), Deere WUR, GPFLR (FAO, WRI, IUCN, WUR), Wetlands International, Savory Institute, RWS AG.

FIGURE 4.6.8

Many proven technologies exist.

4.6.6 BUSINESS SCHOOLS: PREPARING MANAGERS FOR A RESTORATION INDUSTRY

"One of the biggest things that needs to change is the educational system. Universities are still teaching a system to students that destroys the biosphere."

Ray Anderson, CEO Interface In: Bloomberg Magazine Online Extra: "Stop Destroying The Biosphere" July 18, 2004 http://www.bloomberg.com/bw/stories/2004-07-18/online-extra-stop-destroying-the-biosphere

Demand is increasing for producers and consumers to understand and reduce their ecological footprints, including the natural resource footprint, throughout the entire value chain. TEEB studies with business cases on ecosystem restoration and biodiversity conservation could fill this niche in the agenda and curricula of business schools. Additionally, organizations like the World Business Council for Sustainable Development (WBCSD), the World Resources Institute, and the IUCN have developed a Corporate Ecosystem Services Review (Hanson et al., 2008) and a Guide to Corporate Ecosystem Valuation (Hanson et al., 2012). These tools assist businesses in evaluating their impacts and dependency on ecosystems and in determining the risks and opportunities of their current operations. Despite a wide array of methods and frameworks, none of the present valuation tools, which should be simple in practice, are easily applicable by the business community; often, they are complex, presented in a manner and using language that are not immediately relevant to decision making in the private sector.

Some valuation frameworks help businesses understand and identify the material (tangible) risks and benefits of ecosystem services. However, while the Corporate Ecosystem Valuation is an important step forward, it does not provide sufficient incentives for companies to restore natural capital and, therewith, agricultural systems that, in the long run, rely on ecosystem functionality. A global standard for the assessment and valuation of landscapes is needed for the private sector to incorporate restoration activities into decision-making frameworks.

A disconnect exists between business schools' curricula and the growing recognition among government bodies and within academic circles of the importance of healthy ecosystems for the survival of this planet. Case studies that show how businesses implement and finance ecosystem restoration projects are excellent examples and invaluable resources. The onus is on groups like the WBCSD and business biodiversity networks, such as Leaders for Nature and UN Global Compact, to find new ways of disseminating the science and research that they have developed to the business communities.

4.6.7 CLOSING THE GAP BETWEEN BUSINESS AND ECOSYSTEM RESTORATION

"The Corporation 2020 is the firm of the future. It produces positive benefits for society as a whole, rather than just its shareholders. It encourages positive social interactions among workers, management, customers, neighbours, and other stakeholders. It is a responsible steward of natural resources. It invests in the productivity of its workers through training and education. It strives to produce a surplus of all types of capital, including financial, natural, and human capital. [It] can be best characterised with four terms—goal alignment, community, institute, and 'capital factory.'"

Corporation 2020 (n.d.)

Businesses are now actively seeking ways in which they can make positive contributions regarding environmental degradation. There is evidence that they are deepening their understanding and awareness of ecosystem impacts and dependencies such as the 2010 agreement biodiversity between the Dutch Confederation of Industries and conservation organizations as part of the Leaders for Nature network (Wensing, 2012). Many mechanisms are now in place for the private sector to contribute to ecosystem restoration, while numerous successful projects present the opportunity to scale up efforts. In addition, a large body of knowledge on how to achieve ecosystem restoration has been accumulated. Despite the possibility of gaps in our knowledge, implementation of large-scale restoration is the obvious next step, and further learning must come from doing.

4.6.7.1 THE ROLE OF BUSINESS

Not only has the value of businesses' contributions to ecosystem restoration been established, but the various ways in which businesses can contribute have also been broken down and defined. The United Nations Convention to Combat Desertification (UNCCD) has outlined a role for the private sector that includes the following actions (adapted from UNCCD, 2012):

- Engage in investments that increase efficient land use and the resilience of related ecosystems functions and services and reduce or mitigate risks.

- Invest in research and development on sustainable land-use management.
- Establish and implement public-private partnerships that ensure social inclusiveness.
- Support the development of information-sharing mechanisms, especially at the local level, with a focus on sustainable land-use management and related goods and services.
- Within a CSR framework, engage in reporting, at the national and international levels, on actions taken toward halting degradation and on best practices, lessons learned, and management models that are in use and suitable for attaining such targets.

4.6.7.2 REMOVING OBSTACLES THAT PREVENT PRODUCTIVE PARTNERSHIPS

Although motivation and awareness within the private sector are increasing, and despite the steady influx of new project initiatives to regreen the planet and restore natural capital, the net-positive-impact action on ecosystems remains scarce. This lack of engagement is largely due to the significant barriers that exist between businesses and those organizations and communities involved in ecological initiatives, including local communities, NGOs, farmers, ecologists, economists, and policy makers. These barriers range from a lack of networking across groups to differences in the use of language and a lack of trust. At the same time, new alliances must be forged based on common understandings of what must and can be done. In other words, private sector involvement depends on the following:

- Intersectoral and interinstitutional collaboration, necessitating the breakdown of institutional silos
- A simple global standard ecosystem-service valuation tool that is backed by science
- International, widely accepted guidelines, tools, and technologies for ecosystem restoration, including a means of reintroducing sustainable agricultural practices
- A wiki database of ecosystem and landscape restoration projects, which can provide models for replication and scaling-up implementation
- A smart and simple broker mechanism, which is regionally replicable and endorsed by leaders in the field, to engage companies in major restoration projects
- A commitment by all participating companies, scientists, governments, NGOs, and local communities to a long-term approach and perspective on this undertaking

4.6.7.3 THE NEED FOR A TRUSTWORTHY DEAL MAKER

An interinstitutional framework or mechanism must be established to build the necessary trust and connections between the business community and stakeholders (civil society organizations, governments, and educational institutions, among others). This must be based on ecological science so these barriers can be broken down and productive collaboration on major projects realized.

The partnerships created as a result of this framework would hold immense promise for ecosystem restoration; the costs and benefits would be distributed proportionately (i.e., equitably and justly) and take into account long-term goals. Serious attention would need to be given to TEEB recommendations.

Companies could expect to benefit from these partnerships in a number of ways, such as developing new tools and insights into sustainable decision making, experiencing working with different sectors, developing new networks, and developing a positive brand and reputation.[13]

---

[13]See the IUCN's Business Engagement Strategy (2012) and Operational Guidelines for Private Sector Engagement (2009), for example.

4.6.7.4 AVOIDING LAND GRABS AND GREEN WASHING

It is not always problematic for wealthy companies to invest in ecosystem restoration to create new agricultural lands in developing countries for commercial use. However, occasionally local people are forced from the land (i.e., land grabs) or less food is grown. Recent data indicate that, since 2001, at least 80 million ha of land deals have been land grabs (Oxfam, 2012). Ecosystem restoration partnerships should cooperate with local organizations and use accepted international restoration guidelines to avoid these practices. Additionally, green washing, which is used to promote the perception that a company's aims and policies are environmentally friendly, may be used to manipulate popular opinion to support otherwise questionable aims. Working in long-term partnerships with businesses will curtail these unethical activities.

4.6.8 CREATING ECOSYSTEM RESTORATION PARTNERSHIPS

The idea that maximization of return on investment (ROI) per hectare in a relatively short period (over a few decades) leads to almost worldwide degradation, loss of biodiversity and topsoil, water scarcity, and loss of security is termed the *degradation industry*. Unfortunately, this concept is not widely accepted. While using natural resources with abandon created prosperity for many years, it is now a severe threat to the well-being of humans, the global economy, trade, and society. By creating partnerships with businesses, knowledge and complementary expertise can more easily be shared (by breaking down knowledge silos), allowing businesses to become leaders based on a long-term purpose. The ambition, as formulated in this section, calls for an international mechanism to ensure the productive involvement of the private sector in ecological-restoration efforts.

4.6.8.1 ECOSYSTEM RESTORATION PARTNERSHIPS

By design, partnerships to tackle ecosystem restoration should be flexible enough to be adapted to different circumstances around the world, scaled up to meet the demands of the largest initiatives, or both. In the approach and operations, these partnerships should draw on best-practice examples from other mechanisms that have proven successful at creating interinstitutional collaboration and contribution, many developed by the private sector. Such an approach will lead to the creation of ecosystem restoration partnerships (ERPs). ERPs should be neutral and independent agencies that bring together existing networks of businesses and business schools, scientific institutions, governments, and local development partners. They should be empowered, endorsed, and financed by committed private-sector institutions.

4.6.8.2 CRITICAL SUCCESS FACTORS

Ecosystem restoration projects should be funded through a variety of finance and incentive mechanisms, including social investment funds that may pay off dividends in the form of quantities of carbon sequestered, groundwater recharged, or increases in agricultural production. The critical success factors of ERPs are the following:

- *Focus*: Restoring hectares of degraded landscape and seascape based on ecosystem science.
- *Endorsement* from ecosystem scientific institutions and civil-society organizations.

- *Connecting companies and implementing partners* through business networks and business schools and implementing partners (NGOs, landowners, farmers).
- *Business cases* should be made based on successful examples from agriculture, carbon sequestration, and biofuel production and be ethical and responsible.
- *Long-term commitment*: One generation (20 years, which could be divided into four 5-year periods).
- *Results*: A wide range of result data should be gathered, including the number of consortium projects, the number of business schools involved, the number of investors, the amount of ecosystem services that are restored, and the increase of local agricultural production and income.
- *Organization*: A smart, small-broker mechanism should be developed that is supported by partners.
- *Income models* should be developed that elucidate the return potential of projects.
- *Replicable*: The chosen model should be regionally replicable.
- *Ambitions* in line with the Bonn Challenge, Action 2020, and Vision 2050 of the WBCSD and international UN agreements.
- *Investors* should be sought among family companies and social entrepreneurs who understand the importance of a long-term vision and investment (on the order of 20 years) and whose companies will play a part in this vision; multilateral and bilateral institutional investors should also be invited to join, including the World Bank, Global Environment Facility, regional development banks, and development financial institutions.
- *Local stakeholder networks and field activities*: ERPs should operate within learning networks, through which all stakeholders (including farmer associations, local entrepreneurs, and civil-society organizations) can connect; examples include such networks as those created by the GPFLR, Landscapes for People, Food, and Nature Initiative, and the Alliance for Climate-Smart Agriculture, which use a "blended approach" (live meetings, social media, and web-based learning support tools), which may be complemented with vocational training centers.
- *Science*, as developed and endorsed by recognized international institutions and bodies, should form the basis for projects.
- *Governments* should play a role by confirming their commitment to land restoration.[14]
- *Knowledge transfers, communication and learning* through practical examples, case studies, and news will be vital to future projects' success.

An overview of the stakeholders and the role of a restoration business developer/deal maker is shown in Figure 4.6.9.

---

[14]For example, in 2011, Rwanda announced the Forest Landscape Restoration Initiative to reverse the degradation of soil, water, land, and forest resources by 2035 and to use ecosystem restoration to create jobs. Additionally, in 2014, Ethiopia pledged to restore 15 million ha (1/6 of the country's total area) of degraded and deforested land by 2025 (see Minnick et al., 2014). Latin America has also been active in this area, with new legislation in Mexico, Argentina, and Colombia that aims to promote sustainable development, reduce climate change, and alleviate poverty.

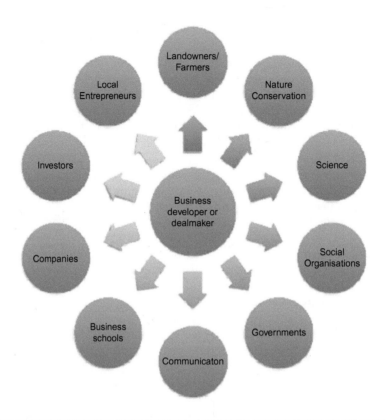

FIGURE 4.6.9

A restoration business developer acts as a deal maker to put together the best available combination of stakeholders for implementing large-scale restoration projects based on a business proposition. An additional advantage is that business schools will learn to work with new sustainability models through the projects developed by this venture.

4.6.9 A PRACTICAL SYSTEMIC APPROACH: THE FOUR RETURNS MODEL

"If you want to build a ship, don't drum up people together to collect wood and don't assign them tasks and work, but rather teach them to long for the endless immensity of the sea."

Antoine de Saint-Exupéry

The time is ripe to scale up the restoration of degraded ecosystems with business involvement by matching companies, investors, and local communities and organizations in long-lasting ERPs, which employ existing experiences and know-how to optimize the value of the land and to mitigate risks. Long-lasting ERPs mean having a common long-term vision, sharing in common interests, and developing cohesion between goals and returns. The ROI that investors and companies can expect from ERPs depends on the nature of the partnership, the duration of the project, and the local ecosystem. Potential ROIs include an increase in agricultural output, carbon credits and new market development, a marked increase in local products and jobs, the development of sustainable resourcing, development of new business-to-business peer groups with attendant business opportunities, being an "early mover" on emerging issues (biofuels, loans, local agro-development, biodiversity offsetting) with attendant benefits (such as cornering a market niche), and enhancing CSR with positive implications for brand and reputation.

The approach is systemic: The landscape is understood to be an ecosystem, not a production unit for one crop or product, while the perspective is long term (intergenerational).

4.6.9.1 ECOSYSTEM DEGRADATION LEADS TO FOUR LOSSES; RESTORATION DELIVERS FOUR RETURNS

The maximization of ROI per hectare leads to ecosystem degradation, which creates four losses: job loss, economic loss, biodiversity loss, and meaningfulness loss. These losses increase over time if functional ecosystems degrade and cease to provide ecosystem services.

By reversing these degradation trends, successful ERPs should focus on maximizing four returns per hectare, based on an integrated three-zone approach:

- *Inspirational capital*: People engagement, innovation, awareness, and passion
- *Social capital*: Jobs, income, security, and social cohesion

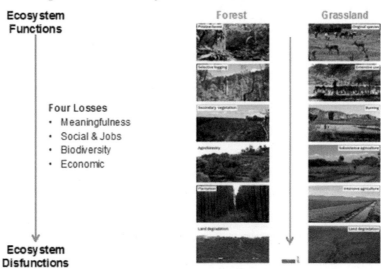

FIGURE 4.6.10

Four losses as a result of ecosystem degradation W. Ferwerda (lecture in Four Returns, Wageningen University, 2014). Photo credit: Ben ten Brink, Netherlands Environmental Assessment Agency.

- *Natural capital*: Fertile soils, hydrology, biodiversity, biomass, and carbon storage
- *Financial capital*: Financial performance (e.g., increases in agriculture, timber, water production) and demonstrable CSR

Four Returns from Restoration

4 Returns	Different Entities	Values measured
Return of Inspirational Capital	• Meaningfulness, spiritual/ holistic awareness, Gross National Happiness, re-sacralize nature • Local culture wisdom & outreach • Landscape leaders, commitment to local ownership, less corruption • Understanding meaning of long term commitment of companies, investors • Time for inner reflection, worship	• % of stakeholder group / yr/ ha: # local cultural, social, religion events • # 'defining moments' of people involved • % of stakeholder group /yr / ha • % of stakeholder group /yr / ha committed; % -/- corruption benchmark • % responding to long term commitment • # volunteers • % of free time to rest and think
Return of Social Capital	• Jobs • Security • Local social cohesion • Education & Social Services	• # of new jobs / project / municipality - ha • # various savings yr / project • # of social ventures / yr / project • # schools, trainings, services / project
Return of Natural Capital	• Biodiversity • Invasive species • Vegetation cover • Top soil • Water	• # of (native) species / yr / ha • % decrease / yr / ha • % coverage / yr / ha; % cloud formation • mm layer / yr / ha; % microbes; % C / ha • % humidity; # stream flow (m3 / yr / ha)
Return on Financial Capital	• Agriculture, Carbon, Timber • Leisure, hunting, bush harvesting • Real estate & other incomes • Water • Decrease erosion, increase topsoil	• Yield / yr / ha • Yield / yr / ha • Value / yr / ha • Production m3 / yr • Decrease costs input chemicals / ha/ yr

SOURCE: Nature Resilence: ecological restoration by partners in business for next generations. RSM, IUCN CEM (2012); McKinsey (2012), Ecosystem Return Foundation (2014)

FIGURE 4.6.11

Maximizing four returns per hectare.

4.6.9.2 THREE LANDSCAPING ZONES

In every project, three landscaping zones should be defined that will produce the previously discussed returns:

- A *natural zone* to restore the ecological foundation and biodiversity; this zone will have rich biodiversity, soil for ecosystem services, carbon sequestration, forest products, and opportunities for leisure and hunting.
- An *eco-agro-mix zone* to restore the topsoil and deliver low economic productivity; this zone will have partially restored biodiversity, soil in recovery, carbon sequestration, and timber supply, fruit trees, water supplies, and opportunities for leisure.
- An *economic zone* to deliver high economic productivity; this zone will have productive zones for sustainable agriculture and dedicated zones for real estate and infrastructure.

The restoration of such interconnected zones as parts of one plan creates landscapes in which an increase of biodiversity and vegetation cover will go hand in hand with newly developed agricultural lands. Within those mosaic landscapes, ecologically sustainable agricultural and economic zones will coexist.

The ERPs require a long time frame, combined with the flexibility to constantly develop creative solutions to combat complex stakeholder challenges. The approach is tailored to each location but has the underlying focus on optimization of the four returns per hectare (Four Returns video Commonland, 2015).

Three Landscape Zones

SOURCE: Living Lands, Baviaanskloof, South Africa. Photo: W. Ferwerda (2014)

FIGURE 4.6.12

Three landscape zones give four returns.

4.6.9.3 STAKEHOLDER MANAGEMENT: THEORY U

ERPs will succeed only if stakeholders collectively take accountability and understand that degraded landscapes are a negative by-product (externality) of economic success. To change the system, the economy must be updated to an ecosystem awareness that emphasizes the well-being of the whole. Local people should actively participate throughout the project cycle, thus gaining ownership over the process and continuing to take the necessary steps to maintain healthy lands even after the project has ended. Participation should be built on the strengths of people and on opportunities rather than focusing on their problems and needs, referring to the ability to sense and bring into the present one's highest future potential as an individual and as a group. Projects should promote micro-macro links; they should examine the influence of policies and institutions on livelihood options and require that policies be informed by local insights and by the priorities of those living in poverty. Partnerships should further draw on both the public and private sectors.

A theoretical perspective and a practical social technology to change people and groups can be found in a theory called *Theory U*. It offers a set of principles and practices for collectively creating stakeholders' desired future by following the movements of coinitiating, cosensing, coinspiring, cocreating, and coevolving (see Figure 4.6.6). In several projects, Theory U has been successfully implemented, leading to new and long-lasting local stakeholder partnerships. These partnerships can develop into restoration companies that can actively restore degraded ecosystems giving the four returns. Similar approaches are being promoted by the Future Earth program.[15]

4.6.9.4 LONG-TERM PARTNERSHIPS

Institutional investors are not yet investing in restoration projects due to the following barriers: (i) the unfamiliarity with restoration as a business opportunity; (ii) the lack of clarity about the risks; (iii) the long-term nature of restoration; and (iv) the lack of clarity regarding the exit strategy. The ROI time frame for businesses and investors is usually two to three years. Investors also want to have the possibility to easily end their participation when they experience difficulties. However, most of the financial returns occur over the long term. The same applies for other returns; it takes time, for example, for biodiversity to recover or to notice positive changes in local society. Donor-sponsored projects often have the same short-term commitment. Owing to the short running time of projects, partners are urged to spend much of their precious time on reporting from the outset. For these reasons, ERPs should secure a long term commitment of at least one generation: 20 years. While institutional investors like pension funds are not likely to commit, the game changers in a new restoration industry will be family-owned companies and impact investors that can build a track record and prepare the ground for institutional investors.

4.6.9.5 FOUR RETURNS DEAL MAKER AND BUSINESS DEVELOPER

The ERP can be organized by a matchmaker and business developer, a neutral agency, while working with the Four Returns model. This can be a small organization, with a team of professionals operating in a wider network of associates. The team will be connected to business investors, agriculture specialists, and ecology scientists. The deal making will be successful only if the Four Returns model is complementary to others and is endorsed by international experts, NGOs, scientists, business schools, private companies, and foundations.

Scaling up ecosystem restoration involves the following steps/activities, to be coordinated by a business developer:

- Selecting *Four Returns projects* by identifying and selecting existing small-scale restoration initiatives with the potential to be scaled up based on the Four Returns approach through a stage gate approach (Figure 4.6.14) and a set of criteria (Box 4.6.4 and Figure 4.6.15)
- Developing a *bottom-up stakeholder engagement process* to create the business model for each selected project based on cooperation with all partners and based on long-lasting commitments
- Actively creating a *Four Returns Development Company* through ERPs by matching companies, investors, people, and local organizations and develop these as operational restoration companies
- Establishing a *Four Returns Investment Fund* that invests in those operational restoration companies
- *Monitoring progress and communicating* by visually documenting and communicating projects and connecting a specific project to a wider local network and international partners

---

[15]Future Earth is the global research platform providing the knowledge and support to accelerate our transformations to a sustainable world.

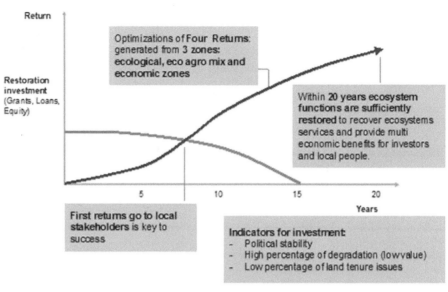

FIGURE 4.6.13

A business model for ecosystem restoration.

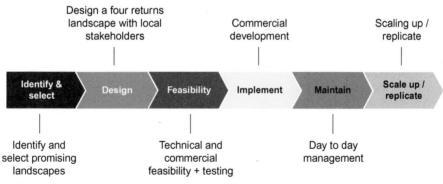

FIGURE 4.6.14

Stage gate approach from "restoration-ready" to "investor-ready" projects.

BOX 4.6.4 KEY ELEMENTS OF THE MULTICRITERIA BUSINESS CASE FOR ERPS

1. Identifying the business case (ROI over 20 years) for an ERP:
   * Leading stance on critical emerging issues: creating jobs, biofuels, sustainable agriculture, loans, local agrodevelopment, biodiversity offsetting, no net loss, etc.
   * New market development
   * Sustainable resourcing
   * Risks assessment and opportunities
   * Business-to-business peer group
   * Carbon sequestration
   * Meeting the demands of consumers/clients
   * Enhanced CSR
   * Ethics and future leadership in stakeholder management
   * Potential for stronger marketing, reputation, and goodwill
   * Increased innovation potential
   * An enhanced horizon scanning and awareness of new and forthcoming governmental policies
2. Identifying the right site:
   * Type of ecosystem and mosaic landscape and site size
   * Land tenure: local interests/conflicts/stability
   * Restoration potential in relation to agriculture, water, carbon, and jobs
   * Existence of local implementing organizations
   * Costs, benefits, risks, duration
   * Guaranties assessment: banks, development banks, and investors
   * ROI fairly distributed between locals, investors, other parties, and funders
   * Application of Criteria Ecosystem Restoration (IUCN Ecosystem Approach)
3. Identifying the right tools:
   * Participatory approach on the site and vocational training centers
   * Alternative local incomes
   * Ecosystem survey
   * Addressing local governance issues and legal issues
   * Using appropriate technical tools: analog/agroforestry, fencing, trenching, alluvial fans, land-life box/water box, permaculture, adding native mycorrhizae, microbiome treatments, and cultural and psychological tools
   * Establishing a finance and investment portfolio
4. Role of restoration business developer:
   * To ensure sustainable long-term management strategies
   * To support, problem-solve, and learn over the cycle of the partnership
   * To sign principles of cooperation for 20 years with investors, companies, and local organizations
   * To form project management teams on field sites and create local development companies
   * To guide the criteria and guidance monitoring process
   * To enable participation in business schools and to educate the various partners
   * To update the collective knowledge with new scientific findings
   * To maintain a project database for all partners

4.6.10 FROM RESTORATION-READY TO INVESTOR-READY: DEVELOPING A FOUR RETURNS/THREE ZONES/20 YEARS RESTORATION INDUSTRY

To find existing projects that can be scaled up, the local context must be understood. Many entities are already working in this field, such as REDD+ and governments (Working for Ecosystems in the Eastern Cape, South Africa, and the positive developments in El Salvador and Rwanda). Meanwhile, some

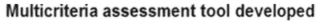

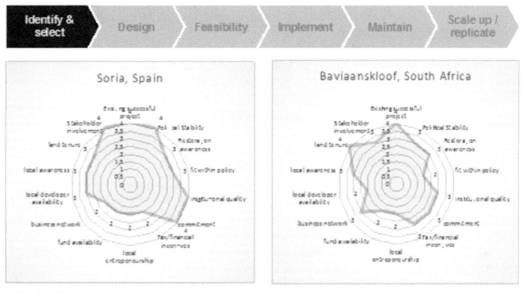

FIGURE 4.6.15

Examples of using a multicriteria assessment per project in Soria (Spain) and Baviaanskloof (South Africa).

business approaches already exist that are based on holistic livestock management. Companies like Unilever, Coca Cola, Nestlé, Heineken, and Mars are now participating in restoration activities. The proximity to the land means that companies in agriculture, mining, and water are more likely to be active in restoration activities than other businesses, although they operate on a balance between maximization of profit in the short term and optimization of commodity security in the long run, which is, of course, about establishing long-term landscape restoration and biodiversity conservation activities.

Scaling up means that a Four Returns business developer identifies existing and promising "scale up–ready" restoration projects, while keeping track of a group of potential investors and companies. This work should be based on a systemic approach, avoiding the maximization of ROI per hectare. Systemic business development should have the goal of making ecological restoration about maximizing the four returns per hectare.

A project qualifies if it meets several criteria assessment tools (Box 4.6.4). Spin diagrams per project provide an overview of the restoration readiness of an area and stakeholders, as well as the investment potential. An example is given in Figure 4.6.15.

4.6.10.1 THE TRANSITION FROM A DEGRADATION INDUSTRY TO A RESTORATION INDUSTRY

The aim of the Four Returns approach is to accelerate the transition process from a "degradation industry" to a "restoration industry." A pipeline of sorts must be created for promising projects with business models based on the Four Returns assessment tool. This pipeline forms the basis of an investment fund. As soon as projects are selected and presented, investors are asked to participate, particularly those with family and company capital. In the long run, as a track record is built, institutional funders are invited to participate by investing in the operational restoration companies, projects, or a landscape restoration business plan.

Measuring the progress of the Four Returns is very important: the progress serves as a means of justifying the interventions to the local stakeholders, investors, etc. In addition to the monitoring conducted by independent entities to safeguard the investors' interests, on-site monitoring, through the use of drones or local restoration training centers, is also beneficial.

4.6.11 CONCLUSION

"What we are doing to the forests of the world is but a mirror reflection of what we are doing to ourselves and to one another."

Chris Maser (2001) author and ecologist.

There is good news amidst the constant flow of distressing messages concerning environmental crises: ecosystems can be restored. The technology exists, the science is available, and the financial resources are ready. Such initiatives as the Chinese Loess Plateau Watershed Rehabilitation Project that was started by the World Bank and Chinese government in 1995, shows what is possible. Increasingly, governments are convinced to participate, as in Rwanda and Ethiopia, and after natural disasters (mangrove reforestation after the 2004 tsunami in Southeast Asia and the water management restoration after Hurricane Katrina in New Orleans, Louisiana, in the United States).

By promoting the business cases for restoration and involving institutional investors and companies, large-scale degraded landscapes of more than 100,000 ha could be restored, especially within the context of ERPs. Local stakeholder managers are critical to success in turning an ERP into an investable business opportunity. The Four Returns/Three Zones/20 Years approach includes all stakeholders and promotes cooperation to achieve the long-term goals of a four returns landscape development business plan. In 2013 Commonland was founded by the Rotterdam School of Management, IUCN CEM and the Common Foundation to implement the four returns approach at project level. Under the Commonland Foundation, an investment fund is setup as well as a 4 returns development company (Commonland, 2015).

Time and trust are required to produce these results. Governments, investors, companies, and other stakeholders that are interested in long-term, intergenerational projects are needed instead of those that are focused on the short-term activities that achieve no real sustainability impact. By widely adopting Four Returns, we can transform the degradation industry into a restoration industry.

ACKNOWLEDGMENTS

Many individuals have contributed to the Four Returns model. A full overview can be found in Ferwerda (2012). For this chapter, I am especially thankful to Ed Barrow, Dieter van den Broeck, John Loudon, Kosima Liu, John D. Liu,[16] Michiel de Man, Dominique Noome, Hans van Poelvoorde, Herman Rosa Chavéz, Wijnand Pon, Thekla Teunis, Astrid Vargas, Piet Wit, Herman Wijffels, Maurits Roodhuijzen, Sofia Marrone, and Petra van Veelen.

---

[16]John Liu is director of the Environmental Education Media Project: www.eempc.org.

REFERENCES

Anderson, R., 2009. The business logic of sustainability. TEDx, May. https://www.ted.com/talks/ray_anderson_on_the_business_logic_of_sustainability?language=nl.

Brundtland, G.H., 1987. Our Common Future. World Commission on Environment and Development. United Nations (UN). Geneva, Switzerland.

Clewell, A., Aronson, J., 2013. The SER primer and climate change. Ecol. Manage. Restor. 14 (3), 182–186.

Convention on Biological Diversity, 2011. The global biodiversity outlook 4. Convention on Biological Diversity (CBD), Montreal.

Corporation 2020, n.d. What is Corp 2020? http://www.corp2020.com/what-is-corp2020.html.

Costanza, R., et al., 2014. Changes in the global value of ecosystem services. Glob. Environ. Chang. 26, 152–158.

Crutzen, P.J., Stoermer, E.F., 2000. The "Anthropocene." IGBP Newsletter 41, 17–18.

De Groot, R.S., et al., 2013. Benefits of investing in ecosystem restoration. Conserv. Biol. 27 (6), 1286–1293.

Economics of Ecosystems and Biodiversity (TEEB), 2009. TEEB—The economics of ecosystems and biodiversity for national and international policy makers—summary: responding to the value of nature. TEEB, Geneva, Switzerland.

Elkington, J., 1997. Cannibals With Forks: The Triple. Bottom Line of 21st Century Business. Capstone. Oxford, 402 pp. ISBN 1-900961-27-X.

Ellen MacArthur Foundation, 2014. Towards the circular economy. Ellen MacArthur Foundation, Isle of Wight, UK.

Ferwerda, W., 2012. Nature Resilience: Ecological restoration by partners in business for next generations. 93 pp. Rotterdam School of Management, Erasmus University, IUCN Commission on Ecosystem Management. Rotterdam, The Netherlands.

Food and Agriculture Organization (FAO), 2005b. Helping forests take cover. RAP Publication, island. 2005/13. Rome, Italy.

Food and Agriculture Organization (FAO), 2005a. Habitat rehabilitation for inland fisheries: Global review of effectiveness and guidance for rehabilitation of freshwater ecosystems. FAO Fisheries Technical Paper 484. Rome, Italy.

Food and Agriculture Organization (FAO), 2011. The state of the world's land and water resources for food and agriculture (SOLAW)—managing systems at risk. FAO, Rome, and Earthscan, London.

Food and Agriculture Organization (FAO), 2013. Climate-Smart Agriculture Sourcebook. http://www.fao.org/docrep/018/i3325e/i3325e.pdf. Accessed on 18 Jan. 2015.

Four Returns video Commonland, 2015. https://www.youtube.com/watch?v=8zXj4dxMbeg. Accessed on 8 Aug. 2015. Commonland: http://www.commonland.com.

Global Partnership on Forest and Landscape Restoration (GPFLR), 2013a. The Bonn Challenge. http://www.forestlandscaperestoration.org/topic/bonn-challenge. Accessed on 18 Jan. 2015.

Global Partnership and on Forest Landscape Restoration (GPFLR), 2013b. The global partnership on forest landscape restoration (GPFLR). GPFLR, Toronto.

Groasis, 2014. Groasis Tech. http://www.groasis.com/en. Accessed on 15 Dec. 2014.

Hanson, C., Ranganathan, J., Iceland, C., Finisdore, J., 2008. The corporate ecosystem services review: guidelines for identifying business risks and opportunities arising from ecosystem change. World Resources Institute (WRI), Washington, DC.

Hanson, C., Ranganathan, J., Iceland, C., Finisdore, J., 2012. The corporate ecosystem services review: guidelines for identifying business risks and opportunities arising from ecosystem change—version 2.0. World Resources Institute, Washington, DC.

Index Mundi, 2014. World Economy Profile 2014. http://www.indexmundi.com/world/economy_profile.html. Accessed on 4 Jan. 2015.

International Biochar Initiative, 2014. "Biochar." www.biochar-international.org. Accessed on 27 Dec. 2014.

International Union for Conservation of Nature (IUCN), 2009. Operational guidelines for private sector engagement—version 2.0. IUCN, Gland, Switzerland.

International Union for Conservation of Nature (IUCN), 2012. Business Engagement Strategy. IUCN, Gland, Switzerland.

International Union for Conservation of Nature (IUCN), 2012. n.d. Enhancement of natural capital through forest and landscape restoration (FLR). IUCN Policy Brief, IUCN, Gland Switzerland. https://cmsdata.iucn.org/downloads/policy_brief_on_forest_restoration_2.pdf.

International Union for Conservation of Nature (IUCN), 2014. The IUCN Red List of Threatened Species. www.iucnredlist.org. Accessed on 16 Oct. 2014.

International Union for Conservation of Nature (IUCN) and Commission on Ecosystem Management (CEM), 2014. Red List of Ecosystems. www.iucn.org/about/union/commissions/cem/cem_work/tg_red_list. Accessed on 15 Jan. 2015.

Johansson, T.B., Patwardhan, A.P., Nakićenović, N., Gomez-Echeverri, L. (Eds.), 2012. Global Energy Assessment: Toward a Sustainable Future. Cambridge University Press, Cambridge UK and New York, NY, USA and the International Institute for Applied Systems Analysis, Laxenburg, Austria.

Lamb, D., Gilmour, D., 2003. Rehabilitation and restoration of degraded forests. IUCN, Gland, Switzerland, and Cambridge, UK, and World Wildlife Fund (WWF), Gland, Switzerland.

Land Life Company, 2014. Land Life Company. www.landlifecompany.com. Accessed on 6 Jan. 2015.

Maser, C., 2001. Forest primeval: The natural history of an ancient forest. Oregon State University Press, Oregon.

McKinsey & Company, 2011. Resource revolution: meeting the world's energy, materials, food, and water needs. McKinsey & Company, New York

Millennium Ecosystem Assessment, 2005. Ecosystems and human well-being. Island Press, Washington, DC.

Minnick, A., et al., 2014. Ethiopia commits to restore one-sixth of its land. World Resources Institute. October 21. http://www.wri.org/blog/2014/10/ethiopia-commits-restore-one-sixth-its-land.

Murcia, C., et al., 2014. A critique of the "novel ecosystem" concept. Trends Ecol. Evol. 29 (10), 548–553.

Nidulomu, R., Prahalad, C.K., Rangaswami, M.R., 2009. Why sustainability is now the key driver of innovation. Harv. Bus. Rev., 56–64. September.

Nkonya, E., Gerber, N., Von Braun, J., De Pinto, A., 2011. Economics of land degradation: the costs of action versus inaction. IFPRI Issue Brief 68. http://ebrary.ifpri.org/utils/getfile/collection/p15738coll2/id/124912/filename/124913.pdf.

Rockström, J., et al., 2009. Planetary boundaries: exploring the safe operating space for humanity. Ecol. Soc. 14 (2), 32.

Roosevelt, F.D., 1937. Letter to All State Governors on a Uniform Soil Conservation Law, February 26. http://www.presidency.ucsb.edu/ws/?pid=15373. Accessed on 16 Oct. 2012.

Sahan, E., and Mikhail, M., 2012. Private Investment in Agriculture: Why It's Essential and What's Needed. Oxfam Discussion Paper. https://www.oxfam.org/sites/. https://www.oxfam.org/files/dp-private-investment-in-agriculture-250912-en.pdf.

Scharmer, C., Kaufer, K., 2013. Leading from the emerging future: from ego-system to eco-system economies. Berrett-Koehler, San Francisco.

Senge, P., 1990. The fifth discipline, The art and practice of learning organization. Doubleday Currency, New York.

Senge, P., Scharmer, C., Jaworski, J., Flowers, B., 2004. Presence: An exploration of profound change in people, organisations, and society. Crown Publishing Group, New York.

Society for Ecological Restoration (SER), 2004. SER international primer on ecological restoration, version 2. Society for Ecological Restoration (SER), Washington, DC.

Spreckley, F., 1981. Social Audit: A management tool for co-operative working. Beechwood College, Leeds, U.K.

Steffen, W., Crutzen, P.J., McNeill, J.R., 2007. The Anthropocene: are humans now overwhelming the great forces of nature? Ambio: J. Human. Environ. 36 (8), 614–621.

Steffen, W., et al., 2015. Planetary boundaries: guiding human development on a changing planet. Science 347 (6223), 791–792.

Sukhdev, P., Wittmer, H., Miller, D., 2014. The economics of ecosystems and biodiversity challenges and responses. In: Helm, D., Hepburn, C. (Eds.), Nature in the Balance: The Economics of Biodiversity. Oxford University Press, Oxford, U.K., p. 448.

TEEB, 2012. The Economics of Ecosystems and Biodiversity in Business and Enterprise. Edited by Joshua Bishop. Earthscan, London and New York.

TEEB, 2010a. The Economics of Ecosystems and Biodiversity for Business – Executive Summary. www.teebweb.org.

TEEB, 2010b. The Economics of Ecosystems and Biodiversity Ecological and Economic Foundations. Edited by Pushpam Kumar. Earthscan, London and Washington.

United Nations, 2012. Resolution adopted by the General Assembly on 27 July 2012. 66/288. The future we want, September 11. UN General Assembly, Geneva, Switzerland.

United Nations Convention to Combat Desertification (UNCCD), 2012. Zero net land degradation: A sustainable development goal for Rio+20. United Nations Convention to Combat Desertification (UNCCD), Bonn, Germany.

United Nations Environmental Programme (UNEP), 2010. Dead planet, living planet: biodiversity and ecosystem restoration for sustainable development. A rapid response assessment. United Nations Environment Programme (UNEP), Global Resource Information Database (GRID), Arendal, Norway.

United Nations Environmental Programme (UNEP), 2011. Towards a green economy: pathways to sustainable development and poverty eradication. United Nations Environmental Programme (UNEP), Nairobi.

Van Andel, J., Aronson, J., 2012. Restoration Ecology: The new frontier. Wiley-Blackwell, Oxford, UK.

Wensing, D., 2012. Leaders for Nature: The Success Story of a Leading Business & Biodiversity Network. http://www.business-biodiversity.eu/default.asp?Menue=155&News=742.

World Business Council on Sustainable Development (WBCSD), 2013. Action 2020. http://www.wbcsd.org/action2020.aspx. Accessed on 14 Jun. 2014.

World Resources Institute, 2014. Atlas of Forest and Landscape Restoration Opportunities. www.wri.org/resources/maps/atlas-forest-and-landscape-restoration-opportunities. Accessed on 15 Apr. 2014.

RESTORING DEGRADED ECOSYSTEMS BY UNLOCKING ORGANIC MARKET POTENTIAL: CASE STUDY FROM MASHONALAND EAST PROVINCE, ZIMBABWE

Georgina McAllister

GardenAfrica, Burwash, East Sussex, UK

4.7.1 INTRODUCTION

In late 2009, GardenAfrica, a UK-based nongovernmental organization (NGO), joined forces with two Zimbabwean NGOs, the Fambidzanai Permaculture Centre and the Zimbabwe Organic Producers and Promoters Association, to explore the relative opportunities presented by organic certification and market development for communal smallholders in Zimbabwe. This collaboration resulted in a project called "Livelihood Security in a Changing Environment: Organic Conservation Agriculture."[1] The project was underpinned by social and market research that revealed a steadily growing domestic demand for organic goods—a demand so far met by imports from South Africa. At the same time, Zimbabwe's communal smallholders, who depend on highly vulnerable, subsistence-based agriculture, and increasingly degraded soils, remain net recipients of food aid. Organic agriculture (OA) in Zimbabwe is commonly viewed as backward and is only pursued by farmers without the means to invest in high-input farming. This initiative, therefore, sought to "square a difficult circle": facilitating both livelihood opportunities based on these emerging market demand realities, including meeting demand at the local "food-shed" level and to apply sound ecological management to restore ecosystem functions for sustained productivity and economic growth.

Among a series of other indicators gathered using different data collection methods and tools, the success of the project was measured through comparative increases in the productive diversity, yields, and income of the 591 participating farmers. Within the first 18 months of the project, agrobiodiversity increased by 122%, combined yields by 90%, and income by up to 78%. By the time the project neared

---

[1]Funded under Comic Relief's Trade, Enterprise, and Employment program, initially as an 18-month action research project and then extended another two years (2011–2015). See GardenAfrica (2013, pp. 5–6).

Land Restoration. http://dx.doi.org/10.1016/B978-0-12-801231-4.00026-4

the completion of its second phase, it had nearly doubled the number of participating farmers engaged in the target area. Furthermore, 3562 more members were incorporated into the national organic membership body, resulting in Zimbabwe's first 280 ha of locally certified land and farmers realizing multiple routes to market existed to enter the domestic supply chain.

Here, we explore the methods that have contributed to these outcomes and the social and ecological conditions that represent ongoing, but by no means insurmountable, challenges to the uptake of this project and its goals, as well as the realization of wider impacts.

4.7.2 THE WIDER CHALLENGE

Sub-Saharan Africa (SSA) has approximately 33 million small family farms, representing 80% of all farms on the subcontinent, which produce as much as 90% of SSA's agricultural output (Nagayets, 2005). Yet family farmers are continually excluded from decision making and are often unable to produce enough to feed their families throughout the year, primarily because they lack access to the appropriate advice, inputs, credit, and markets that would enable them to flourish. In addition, the average family farm size in SSA is decreasing such that about 60% of farms are smaller than 1 ha and control close to 20% of the farmland (Lowder et al., 2014). Under increasing pressure to produce sufficient output, both in terms of quantity (for household food, nutrition security, and trade), soils and other natural resources are thus depleted, increasing long-term vulnerability to climatic and economic shocks. Therefore, making the most sustainable use of their natural resources is essential to provide the best possible chance of building and sustaining viable farming livelihoods.

However, for many decades quantitative measurements have focused on fertilizer use and grain yield per hectare, paying little attention to the complex biophysical interactions that support ecosystem and social resilience. The sustainability of the agricultural ecosystem, or agrobiodivesity, and the subsequent resilience of the people who depend on it are centrally important to the global food system. Yet, the fact remains that what is considered measureable and, therefore valid, affects how smallholders benchmark their own success and how they subsequently value (or devalue) the resources upon which they themselves depend.

Today, 50% of Zimbabwe's smallholders are regular recipients of food aid and are unable to meet their daily food requirements (UN, 2013). The main cause of frequent food insecurity for many households is their reliance on subsistence-based agriculture, using a limited range of Green Revolution inputs, which are often poorly suited to the communal soils and social and environmental conditions in which they are applied. In addition to these socioeconomic realities, the emphasis placed on monocropping,[2] particularly of maize, degrades the very ecosystem upon which these farmers depend for health and livelihoods and ultimately increases their susceptibility to drought, pests, and pathogens, as well as being a high-risk strategy in the face of volatile grain markets.

While this case study highlights small-scale family farming in Zimbabwe, it is indicative of a wider challenge related to ecological realities and agricultural perceptions. For example, the average maize

---

[2]Monocropping is central to Green Revolution agriculture, introduced in the 1960s. It involves the scaled production of single, often hybrid crops that are heavily dependent on chemical inputs, which are ill suited to Africa's biophysical and social conditions. This form of production is highly susceptible to externalities. See also Elwell (1993).

yield in Zimbabwe is 734 kg/ha[3] (World Bank, 2015). Having started at below subsistence productivity, organic maize yields under this initiative have increased by 300%, with some farmers now achieving the equivalent[4] of 8 tons/ha. However, unless this maize is sold on the formal market, it is not calculated as part of national statistics; as such, it tends to be unmeasured and thus ignored.

So how much of what is being produced sustainably within these polycultures[5] is actually incorporated into mainstream development thinking, national poverty alleviation, or food security strategies? And what are we missing when we calculate grain for processing and not nutrients for people, livestock, and recycling?

4.7.3 ENGAGING DIFFERENT ACTORS TO STIMULATE CHANGE

The primary objective of the project has been to promote a shift toward agroecological farming to rebuild soil organic matter and protect it from further degradation, thus promoting a return to productive diversity through intercropping and rotation. By increasing biodiversity and habitats, a balance is restored between pests and their natural predators and attracts pollinators to improve yields. Such diversity is not only important in restoring the natural environment to minimize, or even mitigate, dependence on costly inputs, but also agrobiodiversity provides important outputs for farmers and their families, underpinning nutritional diversity and health, as well as reducing risks associated with monocropping.

With markets in mind, the second objective was to explore the relative opportunities presented by organic certification and market development for Zimbabwe's smallholder farming sector, with the aim of providing an outlet for participating farmers and their certified organic produce. The project hypothesis was that access to lucrative organic markets would stimulate the wider uptake of agroecological practices. Such change requires a systems view, with an emphasis on mapping and monitoring the behavioral aspects of value chain and other influential actors. The main elements are:

- Farming practice and natural resource use
- Community leadership and customary practice
- Consumer trends
- Private-sector purchasing practice and policy
- Public-sector policy and barriers to change

In the discussion that follows, we focus primarily on changing behaviors and the relationship between farming practices and local leadership, while providing some context in terms of the work undertaken toward organic market development to create a more conducive environment for long-term systemic change and impact.

---

[3]Whereas in countries such as the United States the average is 7,340 kg/ha.
[4]The word *equivalent* is used here, because, on their communal small holdings of between 1 and 1.5 ha each, farmers are encouraged to diversify production on their remaining land to include herbs, fruits, and vegetables, some for market and some for household (HH) consumption.
[5]Polycultures relate to the cropping of multiple species and varieties, including nitrogen-fixing crops in a single space to enhance soil health, reduce pests and pathogens, and increase diversity for ecosystems and human health.

4.7.4 ACTION RESEARCH AREA

The province of Mashonaland East (ME) in Zimbabwe was selected due to its proximity to the capital, Harare, where the majority of the organic demand is currently focused. ME's four agroecological zones, from semiarid to dry subhumid, provide a strong empirical basis for testing the project's permaculture[6] production methodology and the different strategies to be employed in response to any emerging resource challenges. The initial baseline survey, undertaken with participating farmers, revealed that all were producing at below subsistence level, with extremely low levels of agrobiodiversity. Those in areas with the most acute resource challenges were exposed to the highest levels of political insecurity, where land, food, and inputs were regularly used as a political tool. Correspondingly, levels of farmer coordination and cooperation (agency) in the identification of natural resource management challenges were low, which affected information sharing, increased transaction costs, and limited collective action to mitigate them. In addition, insecure land tenure represented a significant disincentive to the uptake of organic and other sustainable land use systems, which require longer-term investments for livelihood benefits to be fully realized.

4.7.5 OUR APPROACH

The project's approach has been to:

- Deliver a wide-ranging series of training courses
- Offer support and guidance for the establishment of peer networks for representation
- Provide farmers with field support for on-farm challenges
- Encourage and enable the engagement of and participation by women
- Engage influential actors (such as community leaders) who represent a potentially wider supportive environment but who can, sometimes unwittingly, represent barriers to change

All these approaches are part of a wider structured delivery, the components of which have been outlined in Figure 4.7.1. The bridging and enrichment of the two seemingly contradictory learning approaches—top-down for knowledge acquisition and horizontal for knowledge accumulation—have been enabled by the careful navigation between each, creating a reflexivity to accelerate practice-oriented change.

In partnership with the Agriculture Research and Extension services (Agritex),[7] the project was allocated two extension officers (AEWs) from each of the eight districts engaged in ME. The Agricultural Extension Worker (AEWs), renowned for being underresourced, demoralized, and unmotivated, were trained in organic farming, standards, participatory planning, and market development. Each AEW was then tasked with putting forward eight candidate farmers who they believed would adapt well to organic farming, had existing capacity, and could aid neighboring farmers to share skills and rebuild contiguous ecosystem corridors. Four farmers from each district that demonstrated an

---

[6]Permaculture is a social and agricultural system that mimics natural ecologies, working with nature to manage resources efficiently and equitably. As one of the approaches under the umbrella of agroecology, it is increasingly recognized as an appropriate system for production and resource management that resonates with traditional farming techniques and cultures.
[7]A department within the Ministry of Agriculture, Mechanization, and Irrigation.

Structured Delivery Components:

- Permaculture training and peer transfer;
- Change laboratory workshops;
- Local leadership exchanges for wider community-based natural resource management (CBNRM);
- Organic standards development and training;
- Market research, development, and facilitation;
- District Packhouse development and management;
- Market stimuli (i.e., organic media campaign);
- Institutional training and support to build farmer representation and agency; and
- Engagement with policy makers to address barriers to change.

FIGURE 4.7.1

Structured Delivery Components

aptitude and commitment to the ongoing training underwent a further training and selection process at Fambidzanai. Each of these 32 farmers then selected an average of 20 peer farmers to establish associations, creating more entry points for women. The decision was made early in the process to actively avoid the "lead farmer" terminology, to guard against information capture and emerging power relations, and to create the awareness among associations that all members have something to contribute. The intention was to ensure that members elected different individuals to attend each weeklong course, according to their aptitude and capacity to feed this information back to the group for wider accumulation.[8] Course attendance on a rotational basis would also enable the attendance of more women, who may have otherwise been occupied by other commitments. Here, it is important to note that, during the action research phase, the team made a decision not to apply any affirmative action in relation to gender[9] or to influence how associations planned their production (individually or communally) on the understanding that these decisions would need to be reached by associations themselves in order to gain traction, prior to any further actions being taken.

4.7.6 FACILITATING BEHAVIORAL CHANGE TO RESTORE ECOSYSTEM FUNCTIONS

The focus of the ongoing courses was to build confidence and skills in agroecological systems (Figure 4.7.2) and to enhance ecosystem functions to support the outcomes that farmers most wanted to achieve. By facilitating precise but adaptable training for production and market access (Figure 4.7.3), the project team was confident that farmers would soon see the desired food security and livelihood benefits. This knowledge acquisition was backed up with regular field support to assess the level and quality of feedback

---

[8]Weeklong residential courses were hosted at Fambidzanai Permaculture Centre: a 22-ha farm 30 km northwest of Harare.
[9]Here, it is interesting to note that, without affirmative action, 56% of participating farmers were women.

On-farm Resource Management:

Training for transformation;

Introduction to ecology;

Soil conservation and management;

Water management and tank building;

Companion planting;

Dryland cropping;

Integrated pest management;

Post harvest management; and

Livestock integration.

FIGURE 4.7.2

On- & Off-Farm Resource Management Training

Market-Focused Training:

Value addition and wild harvesting;

Bee-keeping and organic honey production;

Farming as a family business;

Agri-planning;

Participatory market systems;

Internal savings and lending; and

Association building and representation.

FIGURE 4.7.3

Market-Focused Training

to peers for knowledge exchange and accumulation at the association level, while enabling the training team to see where challenges lay and how improvements could be made regarding their own delivery.

Provided with training resources for distributing information to peers, associations were able to explore and adapt different techniques and technologies, according to site-specific challenges with soils or pests, wider resource degradation, and/or erosion. The quality and regularity of peer feedback was monitored during field visits to each of the 32 associations. Those most regularly reporting back experienced the most benefits, resulting in soil and natural pest management strategies, as well as effective management structures being established, all contributing to fast-changing production systems for incremental ecosystem restoration. This restoration was demonstrable through yield and income increases earlier than even the project team had forecast.

In addition, a series of field-based Change Laboratory Workshops (CLWs)[10] offered an opportunity for associations to come together at a central meeting point in each district to identify and prioritize the challenges they faced, while learning to address each challenge sequentially. These ranged from off-farm resource challenges due to a lack of erosion controls and watershed management to poor communal livestock management and dilapidated roads, which constrained market access. As part of the two-stage process, discussions were facilitated toward the development of achievable action plans. By the second round, farmers reported on their progress, by which time many had resolved these issues directly or engaged others to motivate the required change at the community level. This participatory learning enabled the farmers to discuss their priorities, needs, and opportunities, thereby creating an opportunity for interactive learning and knowledge sharing. Building confidence by facilitating participatory planning to affect change in their immediate natural and sociopolitical environment has proved pivotal in encouraging more proactive engagement in the wider market system.

4.7.7 ENGAGING LEADERSHIP FOR LAND TENURE SECURITY

It was noted fairly early in the process that that the association achieving the most significant changes and accessing markets had a supportive leadership. The recognition that they would need to comply with organic practices and standards for three years prior to acquiring full certification led this association to quickly apply to its leadership for virgin/reverted land (approximately 1.5 ha) on which no conversion period would be required and on which they would work communally to improve coordination and output. All other associations, while in full compliance and still making steady progress, were working disparately and were therefore struggling with coordination. The end of the action research phase and our subsequent evaluation enabled the team to assess the relative benefits of the different structures that had emerged independently and to feed this back to all the associations. This led to a leadership exchange, with headmen and chiefs visiting the top performing association to see what they had achieved and to discuss the choices and decisions made by their leadership, which had created the enabling environment for success.

The projectwide transformation from this point was rapid. Within three months, all but one association[11] had been granted secure access to virgin or reverted land, with all leaders stating that they would no longer prioritize high-input conventional agriculture, instead allocating land to "our organic farmers who are protecting the environment and bringing benefits to the community."[12] These associations are now fully certified, supplying supermarkets, wholesalers, and local schools, hospitals, and traders.[13] Both access to well-resourced land, on the basis of an ongoing duty of care, and the resulting access to markets have been a considerable incentive in motivating other farmers to convert to agroecological practices, driven by the sheer determination of these pioneering organic farmers.

[10]CLWs facilitated by Dr Mutizwa Mukute; see Mukute (2013).
[11]The remaining association is still waiting for land, due to restriction in terms of land availability.
[12]Community leader from Hwedza district (Community Exchange Report, April 2013).
[13]The project's sales analysis has demonstrated that considerably higher levels of fully certified goods are in demand and traded on informal markets, often with better prices and payment terms than those offered by wholesalers, representing market diversity and choice for organic farmers.

Perhaps most significantly, for ecosystems management and community-based natural resource management (CBNRM), community leaders have since become more aware and engaged in issues relating to the overexploitation of natural resources and the impact this has on farming livelihoods and the wider community. The dialogue that has been initiated has also opened the door for farmers to express concerns about declining biodiversity and groundwater levels, which have a direct impact on their livelihoods and well-being.

4.7.8 FARMER AGENCY: FACILITATING REPRESENTATION

Zimbabwe's national organic standards are formed around the internationally recognized Participatory Guarantee Scheme[14] by Zimbabwe Organic Producers & Promoters Association (ZOPPA) and have been formally incorporated into the Standards Association of Zimbabwe. As a membership organization, ZOPPA monitors the compliance of its organic membership and facilitates market access. All farmers and associations trained under this initiative are, therefore, incorporated as ZOPPA members for long-term support. While its membership has, until recently, been largely disparate, this initiative signifies Zimbabwe's first provincewide network of organic farmers coming under ZOPPA's compliance wing. As such, a provincial representation structure has been established to create a two-way flow of information: to ZOPPA about readiness of produce for market, and from ZOPPA to associations on market opportunities and how to schedule market production without compromising their natural resource base.

4.7.9 PROMISING ADVANCES

The team and farmers alike have been greatly encouraged by farmers' yields and income gains resulting from improved ecological practices, as well as the learning that continues to emerge from this initiative. After only 18 months, farmers had already increased the number of species cultivated on, or around, their land by 122%. With additional organic-compliant innovations to address field and postharvest losses,[15] yields of key crops saw a 90% increase, with corresponding income increases of up to 78%.

Significantly, the participatory approaches have led to an increase in confidence and self-esteem resulting in improved decision making at the HH and community level—such that 94.4% of participants reported that their social capital has been improved as a result of this intervention.[16] Although improved financial capital was the second most articulated change (88.9%), only 66.7% of the participants related the changes to improvements in natural capital. However, the reported changes in financial and social capital would have been unlikely had they not been underpinned by improvements in natural capital. But these changes may not have been as apparent to farmers, despite their commitment to growing their

---

[14]A low-cost participatory certification scheme recognized by the International Federation of Organic Agriculture Movements (IFOAM) under submitted standards, creating wider access to organic markets and specifically designed for smallholder farmers.

[15]Indigenous pesticidal plants were recommended by the Royal Botanic Gardens Kew (UK) and participating farmers. A total of 16 species were selected, each collected from three different plant populations for chemical analysis at Kew, with information returned to the project farmers in support of the safe biological management of field and postharvest pests.

[16]Results of the Most Significant Change (MSC) process undertaken with 90% of participating farmers, which assessed changes articulated in relation to subdomains clustered around natural capital, financial capital, and social capital.

natural resource-base in order to drive related benefits, such as income, health, social status and self esteem. This, therefore, represents important learning if the project is to understand and translate the relevance of ecosystem functions to that of creating sustainable livelihoods.

In relation to our initial hypothesis that organic market access and related income increases would lead to the wider uptake of agroecological practices, the initiative has seen the growth of existing associations with more members. A further eight organic associations have been founded and are benefiting from training, wider support, and representation. This has been entirely motivated by the farmers themselves, through farmer-based advocacy and demonstrations. With two farmers from each district now trained as standards trainers, this expansion is now entirely self-generating and has resulted in the near doubling of organic farmers under the project to almost 1000 smallholders. In addition, one farmer and standards trainer are in the process of training a further five associations without any support from the project. Increased awareness of the potential of organic markets has also led to other non-governmental organizations and institutions such as the United Nations Development Programme (UNDP) converting farmers under their conservation farming[17] programs to certified organic. Within the course of the project to date, ZOPPA's wider membership has increased from 300 members to 3862, accompanied by the expansion of Zimbabwe's locally certified organic land to 306 ha.

4.7.10 ONGOING AND EMERGING CHALLENGES

While many of the resource challenges are being addressed by better erosion control, soil management practices, and natural pest management strategies, not all associations have experienced such high levels of success in terms of diversity, yield, and income increases. The resource management successes undoubtedly validate the resource-use training, as witnessed in the different agroecological zones. However, the ongoing challenges are all together more difficult to quantify. These are thought to represent a combination of individual motivation, social organization, and institutionalized thinking around normative modalities of agricultural production—all of which are difficult to shift. This has been evident in the reluctance of some associations, despite training and exchanges, to diversify their production. Those associations cultivating the lowest level of diversity and varieties have the lowest confidence, yields, and income.

It is also important to note that there is a 100% correlation between those associations achieving the lowest gains to those teamed with the least motivated AEWs. This points to the central role played by well-motivated and supported AEWs in delivering food security and poverty alleviation, as well as wider environmental and climate change objectives.

The fact that not all associations have fared equally presents a challenge to the team. We may well be pleased with our maximum increases, but the associations achieving these represent the lowest-hanging fruit. It is clear that more work needs to be done to fully understand the barriers that persist for many producers. Perhaps, as already suspected, not all farmers are ready or able to enter markets. Yet for these farmers achieving food security also requires all the same conditions as those who meet with market success: fully functioning ecosystem services. Between 2013 and 2014, the top-

---

[17]Conservation farming has seen considerable growth in Zimbabwe, as is considered more appropriate for small-scale farmers in that it promotes the conservation and use of biomass, is zero tillage, yet does use direct microdosing of synthetic fertilizers.

performing district, which is also the most drought-prone, has earned $27,000 U.S. dollars (with a per capita income of $428), while the lowest performing district has earned just $6636 from their sales ($86 per capita). Although their food security has been improved by relative increases in agrobiodiversity, the target must always be to enable the lowest performing groups to reach the next level in order to generate disposable income for essential investments in health and education.

4.7.11 MARKET PRODUCTION VERSUS NATURAL RESOURCE USE

The project has had to navigate a delicate path. A pure market focus can have negative impacts on HH food security (relating to availability, access, and use) due to (i) sales being prioritized over consumption; (ii) dominance of men in commodified production; (iii) loss of agricultural biodiversity; (iv) production for home consumption is devalued; and (v) selection becomes focused on yields rather than other characteristics such as resilience, taste, etc.

This project has not found this to be the case, perhaps due to its focus on diversification for HH and market resilience, and has seen an increase in, and prominence of, women engaged in both production and marketing. And in cases where associations are focused on more targeted crops for market, the data demonstrate that their primary HH investment, after school fees and materials, is additional foodstuffs to augment food and nutrition security.

However, this initiative has found that market-led production, albeit ecological production, which is enhancing soil-moisture retention and protection, has had a negative impact on groundwater levels. The more successful associations are, the more groundwater they use. Despite water reuse and conservation practices, this is perhaps inevitable at scale. The solution is not a simple one: requiring CBNRM, which concentrates on watershed management to recharge groundwater supplies, thus reducing potential conflict between HH needs and irrigation. In the case of the two most successful associations in the most drought-prone district, members engaged their community to rebuild an old dam that had been damaged by heavy rains. There is now sinking water to recharge the water table while providing year-round water for irrigation and livestock. It has been interesting to note that associations in high-risk areas have improved capacity for engaging adaptive strategies, which has not been found in other districts. For more farmers to engage in these kinds of critical, though essentially off-farm activities requires time and foresight. Any project seeking to promote livelihood development through horticultural production and market development will need to address this resource conflict and manage it appropriately. Those farmers who navigate this delicate path successfully will thrive; those who do not will wither.

If such a proposition were possible, the role of NGOs engaged in promoting these types of activities as poverty alleviation strategies is to facilitate and enable foresight and to continue to encourage community leaders to lead on matters of environmental degradation and mitigation, as has emerged through this initiative.

4.7.12 CONCLUSION

From the outset of the action research phase, we were acutely aware that aligning the demands of the market with sound ecological practice would be a delicate balancing act. But it soon became clear when engaging a range of market actors that, in fact, the market was also demanding a diverse range of high-quality produce in volumes we could barely begin to meet. Monocropping, so it seemed, was advantageous to neither farmers nor the buyers they were supplying, calling in to question decades of advice from extension services and NGOs alike. And, while the increasing demands of an ever-more-centralized food system placed enormous pressure on our ecosystem services, this would require far too substantial an investigation for this case study. A central task for projects of the nature discussed here becomes one of encouraging and facilitating knowledge and skills to harness natural systems and processes and to coordinate and cooperate to bulk produce in order to scaled volumes and reduce transaction costs. This work is ongoing. We stress that while organic certification is not the only way to protect ecosystem services, ours and the farmers' experiences demonstrate that, where the conditions are conducive, organic certification can serve as a significant market-based mechanism to motivate and build confidence in farmer-led ecosystem restoration in order to realize wider benefits, through which viable farming communities can once again emerge.

REFERENCES

Elwell, H.A. 1993. Development and adoption of conservation tillage practices in Zimbabwe. In Soil tillage in Africa: needs and challenges. FAO Soils Bulletin 69. Rome, FAO.

GardenAfrica, 2013. Annual Report 2012–13. GardenAfrica, East Sussex, UK.

Lowder, S.K., Skoet, J., Singh, S., 2014. What do we really know about the number and distribution of farms and family farms worldwide? Background paper for The State of Food and Agriculture 2014. ESA Working Paper, 14 (02), Food and Agricultural Organization (FAO), Rome.

Mukute, M., 2013. Bridging and Enriching Top-down and Participatory Learning: The Case of Smallholder, Organic Conservation Agriculture Farmers in Zimbabwe. Southern Afr. J. of Env. Ed., 29–30.

Nagayets, O., 2005. Information Brief: Small Farms: Current Status and Key Trends. Prepared for the Future of Small Farms Research Workshop. Wye College, Wye, UK. June 26–29.

United Nations (UN), 2013. Zimbabwe Humanitarian Gaps 2013. Office for the Coordination of Humanitarian Affairs (OCHA), New York and Geneva, Switzerland.

World Bank Group, 2014. World Development Indicators. Available from; http://data.worldbank.org/indicator/AG.YLD.CREL.KG. Accessed 27.02.15.

A CONTINUING INQUIRY INTO ECOSYSTEM RESTORATION: EXAMPLES FROM CHINA'S LOESS PLATEAU AND LOCATIONS WORLDWIDE AND THEIR EMERGING IMPLICATIONS

4.8

John D. Liu[1], Bradley T. Hiller[2]

Director, Environmental Education Media Project; Visiting Fellow, Netherlands Institute of Ecology;
Ecosystems Ambassador, Commonland Foundation[1]
Freelance Consultant to the World Bank, Cambridge Institute for Sustainability Leadership,
University of Cambridge, University of Anglia Ruskin[2]

4.8.1 A JOURNEY BEGINS

In early September 1995, after spending 15 years working as an international news television producer and cameraman, John D. Liu embarked on a life-changing assignment that ultimately prompted this section (Figure 4.8.1). Liu was part of a documentary crew flying in a small Soviet-era copy of a Fokker Friendship aircraft (dubbed the "Friendshipsky"), which landed at a small, dusty airport in Yanan in Shaanxi Province, on China's Loess Plateau (see Box 4.8.1).

4.8.1.1 CONSIDERING THE IMPLICATIONS

The Loess Plateau has proven to be an excellent place to pursue a type of ecological forensics to witness and understand how human actions over time can destroy natural ecological function. The restoration process of the plateau is providing a living laboratory in which the potential of returning ecological function to long degraded landscapes can be studied. Essentially, what has been witnessed and documented on the Loess Plateau is that it is possible to rehabilitate large-scale damaged ecosystems. This realization has potentially enormous implications for human civilization. Witnessing firsthand many geopolitical events as a journalist, including the rise of China from poverty and isolation throughout the 1980s, the Tiananmen tragedy in 1989, the collapse of the Soviet Union, global terrorism, and much more, provided John D. Liu with references to compare to the restoration of the Loess Plateau. He quickly came to realize that, in terms of significance for a sustainable future for humanity, understanding what occurred on the Loess Plateau would be critical.

Land Restoration. http://dx.doi.org/10.1016/B978-0-12-801231-4.00027-6

FIGURE 4.8.1

John Liu filming on the Loess Plateau.

Yanan is perhaps most famous as the mountainous hideout where Mao Zedong led the Chinese communist Red Army to escape annihilation by the Nationalists (this retreat is often referred to as the "Long March"). The typical dwellings in Yanan at the time were man-made cave dwellings dug so deeply into the soil that they were almost invisible, which made Yanan a particularly good place for a revolutionary to disappear. By 1995, the Chinese communist revolution had moved on and China's socialist market economy was flourishing on the eastern coast and making waves worldwide. But Yanan remained virtually untouched by the agricultural changes, the industrial growth, and the increasing international influence that much of China was experiencing. Yet in this backwater, a new revolution was brewing.

The purpose of the assignment was a World Bank baseline study for an ambitious development project called the Loess Plateau Watershed Rehabilitation Project (see Box 4.8.2). The scene (as depicted in Figure 4.8.2) was of a completely barren landscape. This did not instill confidence in the possibility to restore the vegetation cover, biodiversity, or natural hydrological regulation. The

---

BOX 4.8.1 THE LOESS PLATEAU

The Loess Plateau in northwest China occupies approximately 640,000 km^2 and is the dominant geological feature in the middle reaches of the Yellow River basin. The plateau has been inhabited for more than 8,000 years (Peng and Coster, 2007; Wang et al., 2006) and while somewhat fragile for farming, historically the environment was rich (Liu and Ni 2002; Niu and Harris, 1996). Many studies have indicated that the plateau originally contained wide, large, flat surfaces and few gullies (Shi and Shao, 2000; Chen et al., 2001; Xu et al., 2004) and that the driving forces behind landscape, vegetation, and hydrological changes can be attributed to the dual effects of human land use and climatic changes (Ren and Zhu, 1994; Saito et al., 2001; Shi et al., 2002). The plateau's forest cover dropped to 7%–10%, down from historical estimates of 50% (Liu and Ni, 2002; Cai, 2002; Chinese Academy of Sciences, 1991). 70% of the plateau is affected by soil erosion, 58% of which is extremely severe (Chen et al., 2007; MWR, 2008), and the soil erosion rates are among the highest in the world (Fu, 1989). In addition to downstream sedimentation and eutrophication problems (Greer, 1979; Wang et al., 2006), dust storms (Luo et al., 2003), and landslides (Zhou et al., 2002) have also been problematic.

project sounded optimistic, though few seemed to actually believe that serious rehabilitation was possible. Within a few hours after arrival, all the team's equipment and clothes were covered with fine dust. The area was incredibly dry, and after the first day, it became apparent that regardless of how much water one consumed, it was rarely necessary to urinate. The overwhelming first impression of the Loess Plateau was of a virtual "moonscape."

FIGURE 4.8.2

Filming on the Loess Plateau.

BOX 4.8.2 A BRIEF SUMMARY OF THE LOESS PLATEAU WATERSHED REHABILITATION PROJECTS

From 1994 to 2005, two Loess Plateau Watershed Rehabilitation Projects were implemented in 48 counties in the Shanxi, Shaanxi, and Gansu provinces and the autonomous region of inner Mongolia. Over 35,000 km^2 of physical activities were performed and total investment amounted to US$550 million (Fock and Cao, 2005). The projects had objectives to achieve sustainable development in the Loess Plateau by increasing agricultural production and incomes, and improving ecological conditions in the tributary watersheds of the Yellow River (World Bank, 2003, 2005).

Each project contained about 1000 microwatersheds ranging from 1000–3000 ha (Darghouth et al., 2008). Counties (and priority microwatersheds) were selected based on specific criteria relating to project objectives (including severity of soil erosion, poverty level, experience with soil and water conservation works, leadership and commitment at the local government level, development potential and loan repayment capacity, and proximity to science and research organizations involved in SWC (World Bank, 2010; CPMO, 2005). The projects were built on existing government institutions (Darghouth et al., 2008) and implementation monitoring was focused on physical progress and outputs (World Bank, 1999).

The Loess Plateau projects are regarded as among the largest and most successful erosion control programs in the world (Fock and Cao, 2005; Liu, 2005), and one of the few large-scale working examples of the much promoted "win-win poverty-environment model" (Varley, 2005). While only active in the middle and upper watersheds, the scale of the projects' impact was impressive:

1. Socially, 2.5 million people were raised from poverty (Fock and Cao, 2005).
2. Ecologically, soil erosion was reduced over 920,000 ha of land, and vegetation cover increased from 17%–40% (MWR, 2008). The annual reduction of sediment inflow into Yellow River tributaries was estimated at 110 million tons (MWR, 2008) with correspondingly lower variability in downstream water flows. The grazing ban was adopted throughout much of the plateau, dramatically altering over 500,000 km^2 of landscape, more than 25 times larger than the original project areas (World Bank, 2005).

BOX 4.8.2 A BRIEF SUMMARY OF THE LOESS PLATEAU WATERSHED REHABILITATION PROJECTS—cont'd

3. Economically, the projects significantly contributed to the restructuring of the agricultural sector and adjustment to a more market-oriented economy (World Bank, 2005). The projects were considered more cost effective than infrastructure replacement and mitigation costs downstream (Liu, 2007).
4. Institutionally, the projects increased farmer participation and the effective project management structure was replicated for other national programs (World Bank, 2005).

Loess soils are wind-deposited glacial dusts (Peng and Coster, 2007) that are rich in minerals but highly-prone-to-wind-and-water-erosion. Imagine geological and atmospheric forces, including glacial movements high in the Himalayan Mountains, pulverizing rocks and the resultant dust being carried on the prevailing winds to the plains below in the China. Now imagine them as a continuous process occurring over hundreds of millions of years, building up deep sedimentary soil layers. While loess soils are found in many parts of the world, by far the largest loess deposits on Earth are found in China's Loess Plateau.

To understand the significance of the area, it is important to know that while in the 1930s the Loess Plateau was an ideal place for a revolutionary to hide, much earlier in its history it had been a very different kind of place. The plateau is the geographical birthplace of the Han Chinese, the most populous ethnic group on the planet. The mineral-rich soils are believed to be the second place on Earth where humans began to practice settled agriculture. This place was the center of power and affluence for the Han, Qin, and Tang dynasties, a long period during which China produced cultural, scientific, and artistic works that are some of the greatest achievements of humanity.

Due to the giant gullies that scarred the landscape, the Loess Plateau had been called the most eroded place on Earth. The gullies carry the names of the families that traditionally lived there. We became very familiar with the Ho Family Gully (Ho Jia Gou) near Ansai County in Shaanxi Province and returned several times during the succeeding years to document the results of the restoration project. As the restoration of the area progressed, a dramatic change appeared to take place. The once barren hills were covered with trees and vegetation. The relative humidity was completely different, with moisture in the air at all times, and dew glistening on the vegetation in the mornings. The soil moisture has also been positively affected: the area's vegetation is now better able to survive and thrive, even in prolonged periods of drought. The productivity of the agricultural lands has increased enormously, positively influenced by the natural vegetation returning to areas designated as ecological lands. As a counterintuitive outcome, the productivity of the agricultural lands has increased by reducing the area used in cultivation. This principle alone demands attention to the changes and their possible causes.

Biological diversity has returned naturally to areas that have been removed from agriculture. Vast amounts of carbon have been sequestered in the biomass and in the accumulated organic matter in the restored living soils.[1] Sedimentation loads in the rivers have been reduced (MWR, 2008; CPMO, 2005),[2] and along with this, the risk of flooding has diminished. The reduction of runoff and lower

---

[1] For example, Bai et al. (2005) demonstrate a 20-year trend of increasing biomass across the Loess Plateau despite decreasing rainfall during the same period. Given that the LP Projects directly established 7,500 km^2 of vegetation and led to vegetation recovery over 500,000 km^2 (World Bank, 2003, 2005), coupled with improved agricultural and soil management practices, regional biological carbon sequestration is expected to have increased and soil carbon emissions reduced.

[2] The LP Projects supposedly reduced annual sediment inflow into Yellow River tributaries by between 7.5 and 11 x 10^7 tons (MWR, 2008; CPMO, 2005).

frequency of flooding result in more water infiltrating in situ, reducing the incidence of drought and increasing the natural resilience of the region. The weather and microclimate are beginning to naturally regulate again, reducing the risk of extreme or erratic weather events. All these ecological improvements have ensured that even though they still face many challenges, the lives of the people of the plateau have been enormously improved. The importance of the principles for restoration of the plateau for local people, for China, and more broadly are significant. Putting such knowledge into practice could help to address certain components of human-induced climate change and help ensure greater food security for the global population, particularly marginalized communities.

The long-term inquiry that has been carried out required the consideration of geologic time, evolution, and human history leading to our present circumstances, as well as imagining and anticipating the significance for the future of humanity and the planet. The findings from the rehabilitation of the Loess Plateau point to and call for conscious decisions that humanity can make to avoid predicted catastrophic outcomes from climate change and ecosystem collapse. The experience on the plateau offers potential solutions to issues as broad ranging as unemployment, flooding, drought, food insecurity, and biodiversity loss. The lessons of the Loess Plateau may be able to help human civilization chart a more sustainable pathway.

4.8.1.2 HUMAN HISTORY IN THE LOESS PLATEAU

The Loess Plateau is in the upper and middle reaches of the Yellow River in northwestern China. The area is almost completely circumscribed in the south by a huge bend in Yellow River that flows east from the high northern Tibetan Plateau. The area of the plateau is 640,000 km^2, approximately the size of modern France. The plateau stretches across parts of seven different Chinese provinces, namely: Qinghai autonomous region, Gansu, Ning Xia autonomous region, inner Mongolia autonomous region, Shaanxi, Shanxi, and Henan. There is fossil evidence in the Shaanxi Natural History Museum suggesting that bands of humans or their ancestors were roaming in the area of the Loess Plateau approximately 1.5 million years ago. The area resembles in many ways the area in Ethiopia where the first fossil remains of humans were found. There is evidence that at that time, the forest cover was of climax height and enormous expanse. The grasslands stretching north to Siberia are still among the most magnificent on Earth. The majority of geographers believe that the civilization in the Loess Plateau was the second place on Earth after Mesopotamia where settled agriculture emerged. The Yellow River stemming from the northern Himalayas was once known as the "Mother river" because all the various tribes in the region developed along its banks. It was in this fertile plain where many early tribes vied for ascendancy. In the north, the plain's tribes included nomadic Mongols, Kazakhs, Kirgiz, and Xianbei. In the south, the Han built up a more elaborate sedentary society that eventually was able to surpass the migratory tribes to become the dominant ethnic group in the region. The Han continued to flourish; eventually building a great empire of many dynasties that shared the city of Xian, in the very heart of the Loess Plateau, as their capital. An example of the magnificence of their civilization can still be seen today at the excavated terracotta army site, the ceremonial burial guard for the first Emperor of the Qin dynasty. Situated on the trade routes to the ancient civilizations of Persia and Egypt, the Chinese civilization in this region thrived long before European cultures began to emerge. The descendants of the early civilization of this region are today's largest ethnic group.

In China, Shen Nong was the semimythical emperor said to be the originator of both agriculture and Chinese traditional medicine. Shen is credited with personally tasting all plants to see if they were edible or had healing properties. Mythical dynasties were replaced with real dynasties with elaborate palaces, great wealth, and power. It may have been unlikely that the rulers and the ruled of the time

FIGURE 4.8.3

1.6 billion tons of sediments eroded annually into the Yellow River from the denuded Loess Plateau.

could imagine that this powerful civilization could fail, but by 1,000 years ago the powerful and the privileged had left the region. The capital of later dynasties was moved to Beijing and the Loess Plateau became a place of legend, of poverty, and eventually became known as China's sorrow.[3]

While Chinese medicine and Chinese philosophy are very strong on conservation and acknowledge human existence emerging from natural systems based on the five elements—earth, water, wood, fire, and metal—the daily reality in China for thousands of years has been much less respectful to nature. Chinese historical and literary records are well maintained and numerous, and document evidence of deforestation to build palaces, to expand agriculture, and to reduce the risk of surprise attack from the nomadic warrior tribes in the north. In hindsight, it is possible to see what happened to this region: deforestation on a large scale. Without these forests, the powdery, loess geological soils were exposed to wind and water erosion, with clearly evident results (Figure 4.8.3).

Over millennia, as the plateau declined, the exploitation of nature continued until all the forests were cut down. At first, the soil was rich with organic material from generations of trees, plants, and perennial grasses, but much of the fertility was worn away quite rapidly. Surrounded by nature in a degraded state, the remaining population had to work hard to make a living. After the establishment of the People's Republic, the area was hit hard once again with an ill-conceived plan to settle semi-nomadic pastoralists in the area. This led to the ranging of large numbers of goats and sheep within a one day walk from their pens, and led the already devastated landscape to become essentially bare.

When a culture is ignorant of certain fundamental truths of how natural ecosystems function, then the cultural constructs they produce are inherently flawed. These practices can then be passed from generation to generation and can become enshrined as dogma. So the seeds of destruction are repeatedly reinforced and the cycle of poverty and ecological degradation continues until inevitably the ecosystem can no longer compensate and collapses, along with the fate of the civilization (see Diamond, 2005). On the other hand, if a civilization comes to realize that its survival and sustainability are dependent on functional ecosystems and aligns its behavior to what the Earth's ecological system needs to naturally regulate the atmosphere, hydrological cycle, soil fertility, biodiversity, weather, and climate, then that civilization will have reached a new level of collective consciousness. From this perspective, we can safeguard the survival of myriad forms of life, as well as protecting those parts of human civilization that we can be rightly proud of, such

---

[3]The Yellow River became known as China's sorrow due to almost 3000 major flooding and drought events over the past two millennia (Niu and Harris, 1996).

as our growing scientific knowledge, our tolerance of cultural diversity, and protection of all forms of life, including humanity. This study suggests that while we are facing our greatest challenge, we are also closer than we have ever been before to envisioning and creating a fair and functional society in harmony with the natural systems we depend on for life.

4.8.1.3 RESTORATION: THEORY AND PRACTICE

The fundamental lesson of the Loess Plateau rehabilitation is that it is possible to rehabilitate large-scale damaged ecosystems including those that have been degraded over the course of centuries or even millennia (Figure 4.8.4). This is of enormous importance given the huge areas of the Earth that have been degraded by humans since the advent of settled agriculture, and the emerging risk from human-induced climate change. Imagine the enormous degraded areas of the Middle East, the Mediterranean, large areas of central Asia, parts of North and South America, Australia, Europe, and north Africa that were once functional, biologically diverse, fertile, and productive. What if these areas could be restored? What would restoring the Earth to productivity over vast areas mean in terms of mitigation and adaptation to climate change, availability of food, economic security, social cohesion, and even military security?

FIGURE 4.8.4

Ho Family Gully (Ho Jia Gou) in late August 1995 and then again in late August 2009; these changes can be seen in several films produced by the Environmental Education Media Project (EEMP), including *The Loess Plateau Watershed Rehabilitation Project, The Lessons of the Loess Plateau, Hope in a Changing Climate, and Green Gold.*

4.8.1.4 INTERNALIZING EXTERNALITIES

Following the Chinese Revolution of 1949, the People's Republic of China was shunned and isolated from the Western world. In the early 1960s, the Chinese also broke relations with the Soviet Union and were essentially isolated until the death of Mao and other revolutionaries allowed for the opening to the outside world in the late 1970s. Since 1978, the Chinese have been implementing what they call socialist market economics. In theory, this comprises using market forces to spur productivity for the social well-being of everyone, not just those who produce and sell things. In terms of increasing productivity, the evidence is very clear in the nearly consistent double-digit growth of the Chinese gross domestic product (GDP) year upon year for decades. Yet, what we are seeing is that the GDP is not a holistic measure of societal progress. The way the GDP economy calculates growth is to simply exclude issues such as pollution, climate change, biodiversity loss, desertification, health consequences, poverty, and disparity, by calling these externalities. Without externalizing these effects it is impossible to see current economics as positive, and it highlights the deficiency of using such an imperfect measure that doesn't take into account issues central to society's well-being.

On the Loess Plateau in the 1990s, the Chinese scientific community had begun to realize that without a natural vegetative cover there was very little infiltration and retention of rainfall in situ during rainfall events. This massively increased evaporation rates, which led to very little of this moisture being available to plant life and other ecological functions like climate regulation, despite the area's average rainfall amount of approximately 500 mm annually. While this led to a localized cycle of ecological destruction and poverty, it also led to annual siltation of the Yellow River, because all the loess soils eroded into the river as runoff during even normal rainfall events, not to mention in extreme rainfall situations. When the Chinese calculated the cost of annually mitigating the sediments in comparison with the restoration of vegetation on the Plateau they realized that the cost of annually dredging the river and raising levees was vastly more expensive than restoring the vegetation on the plateau (World Bank, 1994).

The initial economic calculation led to the related realization that not only were the costs for restoring vegetation lower than those of annual sediment control, but also the exact value of the vegetation restoration was difficult to measure because along with the vegetation, myriad additional valuable benefits including improved soil moisture, relative humidity, carbon sequestration, biodiversity, and increased agricultural productivity emerged. Although it wass difficult to find absolute values, it was easy to see that the relative value of ecological function was vastly more than the value of production in the degraded landscapes.

4.8.1.5 CAPITAL INVESTMENT

When value was assigned to the perpetual functionality of the ecosystem and compared to the short-term value of the derivatives extracted from the system, making a capital investment in restoring the Loess Plateau was a straightforward choice. This was made possible by a US$500 million development loan (essentially a very complex revolving line of credit) provided by the World Bank, together with project design and strategic technical assistance. In the early 1990s in China, this was a significant investment, which also created the necessity for stringent management systems. National, provincial, and local project offices were created to develop the strategy, and to manage and report on the dispersal of the funds, as well as to oversee and document the implementation of the project.

4.8.1.6 INNOVATIONS IN IT TO IMPROVE STAKEHOLDER UNDERSTANDING AND MANAGEMENT

When the Chinese project management system was set up, it included enterprise software that could track investments and link them with geographic information system (GIS) satellite maps. Satellite maps were created for each watershed (even individual streams within each watershed were given addresses), which meant that every intervention and investment was connected to a unique address, allowing for very effective data collection in order to analyze the cost and benefit of every aspect of the project. This level of analysis provided insight for experts and locals alike. Farmers were presented with GIS images, together with a clear verbal explanation, which led to broad support for the project goals. Spatial analysis proved to be effective and was an integral part of the restoration effort.

4.8.1.7 DIFFERENTIATION AND DESIGNATION OF ECOLOGICAL AND ECONOMIC LAND

After several years of initial study and analysis of the economics of the project, it was possible to see that returning ecological function to the lands would be far more valuable than the meager harvests that had been extracted from the eroded gullies. This realization spurred the differentiation and designation of ecological and economic landscapes. This was first done theoretically on the satellite maps based on topography with every slope over 25% excluded from cultivation. Given that the predominant landscape was pitted by gullies, this meant removing quite significant amounts of land from cultivation. Both active measures (e.g., gully control measures, earth banks, plantings, infiltration improvement structures, etc.) and passive measures (e.g., removing grazing pressure, allowing natural colonization, and revegetation) were employed in the recovery of the ecological land, which returned vegetation to the denuded landscape within a few years. The return of natural vegetation ensured much greater infiltration and retention of moisture, which helped nurture agricultural lands. Microclimates below grass and tree canopies massively changed the relative humidity and even the temperature. Microbiological fauna as well as plant and animal biodiversity strongly rebounded (as confirmed by studies during project implementation [coordinated by Yangling Research Center (Hiller, 2012)], and expost [coordinated by the UK Department for International Development funded China Watershed Management Project (CWIECC, 2008)]. These positive results also illustrated that productivity is linked to ecological function and explains why it is possible to increase productivity by restricting the area in cultivation.[4] This particular finding is of extreme importance in many parts of the world where extensive agricultural systems leave no room for natural vegetation.

4.8.2 MOSAIC LANDSCAPE THEORY

There is a persistence of preconceived ideas about conservation or restoration: often the assumption is made that the point of these practices is to somehow return to an earlier time when nature was pristine. However, this is very unlikely given the huge current human population. Another and more accurate way to consider the ecological representation on the land is based on ecological function: ecosystem services are dependent on ecosystem function, and ecosystem services are fundamentally intertwined with productivity and well-being. Some systems may be more valuable if maintained primarily for ecological purposes, as the services they provide in that state (e.g., hydrological regulation, in situ retention of rainfall, microclimate regulation, etc.) are more valuable than their limited productivity as agricultural land. Other areas must continue to primarily produce food, fiber, or energy used by human society, and still others may be optimal as more mixed-use areas. Likewise, housing, urban areas, and industrial zones serve essential purposes. Regardless of the individual land use types, a mosaic view of the landscape provides a mix where the value of each land use is optimized and where ecological land uses are both set aside, and integrated into other land use types, particularly where there value is high. Whether the systems are protected for ecological conservation or used extensively for agricultural production or

---

[4]Improvements in yields were achieved through measures such as terracing of sloped areas (increasing water and soil retention), increased diversity of crop/horticulture/forestry production, focus on higher-value products, reduction of fallow land, etc. These improvements helped overall livelihoods increase despite some areas being set aside for ecological purposes.

other human uses, or whether the zones are of mixed use, they should, from the perspective of ecological function, always aim to reach their highest potential.

4.8.2.1 PARTICIPATORY ASSESSMENT MECHANISMS

It is often assumed that China is a tightly disciplined authoritarian country where leaders have the power to order people to restore their landscapes without facing any contestation. This is often not the case. In fact, Chinese people all have their own observations and opinions, and if they do not understand and agree with what the government tells them, it is very difficult to get them to act. At the beginning of the Loess Plateau watershed rehabilitation project, there was a long period of expert consultation as well as extensive use of participatory methods to engage the local people in the inquiry and in the execution of the project. By carefully explaining the relationship between vegetation cover, hydrological regulation, and fertility, local stakeholders began to see the benefit for themselves and more willingly participated. As soon as the results of restoration were visible, it was possible to increase the participation in some areas beyond the initial project sites.

4.8.2.2 PAYMENT FOR INCREASING ECOSYSTEM FUNCTION

In order to increase the uptake of project principles (both within and beyond initial project areas) the Chinese government also used another method to promote people's participation: they paid them. There were several iterations of remuneration schemes, including "return fields to forests," "return fields for grasslands," and the "grain for green" payments. The success of these efforts illustrates that it is possible to use public funds to improve the livelihoods of the people while also restoring ecological function. This demonstrates one potential way forward for people to be engaged in mitigation and adaptation to climate change if funding is made available directly to the people. This also suggests that rather than being considered expensive, conservation and restoration can be seen as cost effective.

Many will have heard or even used the term ecosystem services. In this chapter we have chosen to instead focus on the return of ecosystem function. Ecosystem services are actually derivatives of functional ecosystems, and calling ecosystem functions "services" speaks to the fundamental problem that can be seen historically. By valuing derivatives higher than ecological function, humanity created a perverse incentive to degrade the ecosystem. Instead, valuing ecosystem function higher than extraction, production, and consumption, reverses this and produces positive incentives to conserve, maintain, and restore ecological function. This is what is needed to transition human civilization onto a more sustainable pathway.

4.8.2.3 CONTINUING RESEARCH

By 2004, when the television film *Scaling Up Poverty Reduction in China* was being produced, it was apparent that the restoration of the Loess Plateau was extremely successful and massively important as an example for other parts of the world. If it is possible to restore long degraded, large-scale ecosystems in China, then can it be replicated elsewhere? If it is physically possible to reduce flooding, drought, mudslides, food insecurity, poverty, and disparity while simultaneously helping to mitigate and adapt to climate changes, then what is preventing us from doing this? The inference from the research suggests that humanity should shift its intention from extraction, manufacturing, trade, and consumption to putting restoration of ecological function firmly at the center of human goals, now and for future generations.

However, to properly understand and communicate the changes requires a profound understanding of a number of complex interacting systems. What was taking place on the Loess Plateau was virtually unknown to the rest of the world and what had been a development problem could be seen as a communications problem. Consequently, it became clear that communicating about such an enormous event was not going to be simple. Liu realized that it was necessary to continue to study about what he had witnessed. Hence, he pursued multiple avenues of inquiry, including pursuit of a Ph.D, and continued to utilize his wealth of experience as a journalist to share his learning with the public.

4.8.3 LESSONS

In order to observe with an understanding of the Earth's natural processes, it is necessary to grasp the context of what is being observed. Considering the broad sweep of geologic and evolutionary time, it is possible to imagine and understand that the Earth when it formed was a molten rock surrounded by (to humans) poisonous gases. Over billions of years, the Earth has been transformed into the planet on which humans now live. There are three long-term trends that are of special interest in understanding the formation of the planet's surface, atmosphere, hydrological cycle, and the natural regulation of the weather and climate. The first is the total colonization of the Earth by biological life. The second is differentiation and speciation leading to infinite potential variety in genetics. The third is the constant accumulation of organic materials as each generation of life dies and gives up its body to nurture the next generations of life.

From these trends emerge processes on which all life on Earth depends, from the tiniest living microbes to the giant mega-fauna on the land and in the sea. The engine of this growth is photosynthesis in plant cells, which by taking minerals from the geological materials, sunlight, and water, is able to produce living matter. Further evolution of life on Earth led to the emergence of conscious beings that can reflect on these processes and consider the broader implications of their actions.

Photosynthetic materials remove carbon dioxide and other gases from the atmosphere and continuously generate the fragile oxygenated atmosphere that allows more complex organisms including human beings to live and flourish. The absorption and respiration of plant biomass, together with soils combining geologic, decaying organic materials, and vast living microbiological communities, constantly filter and continuously renew the hydrological cycle, making fresh water available for life on Earth. These processes are also the basis of the fossil fuels, representing hundreds of millions of years of filtering carbon and other gases from the atmosphere, naturally regulating the climate, and storing fixed carbon in the Earth.

The Loess Plateau followed an evolutionary trajectory to reach a mixed landscape of climax forest, grassland, and wetland ecosystems. With normal levels of rainfall generally at or over 500 mm (except in the extreme north of the plateau), it is possible to imagine a quite wonderful landscape with high, closed canopy, temperate forests and vast grassland systems connected in the north to the steppe system in Mongolia and central Asia. The rich mineral soils combined with huge quantities of organic material would have made these some of the most fertile soils on Earth. The constant accumulation of organic materials would have acted as a sponge during extreme rainfall events and then slowly released the moisture during long periods without rainfall. The climax vegetative system supported fauna of all types including mammoths, top predators, large herds of migratory ungulates, small mammals, birds, reptiles, and insects in abundance.

Broadening this inquiry to other continents and synthesizing scientific opinion suggests that the percentages and total amounts of biomass and organic material in the soil are major determining factors in maintaining soil moisture, relative humidity, microbiological communities, fertility, and natural regulation of weather and climate. Continuous study and observation suggest that biodiversity is the representation of infinite potential variety in genetics and a higher order of functionality. It also has become clear that Earth's ecosystem emerges as an integrated process and if the accumulative trends of biomass, organic matter (necromass), or biodiversity are disrupted, the system can move from functional to dysfunctional. This can be seen to cause cascading consequences with the ultimate outcome of dysfunction being the collapse of the system. Extremely brittle systems will collapse quickly and extremely robust systems will take longer to collapse. However, it seems that unless the development trajectory is corrected, the ultimate outcome for even the most robust systems is collapse; it may simply be a matter of time.

In nature, exposed geological soils are often an indicator of catastrophic seismic events such as landslides, earthquakes, or volcanic eruptions. Human impact over time has changed this, and now many urban, industrial, extractive industries and even agricultural areas show indicators of extreme natural dysfunction. Humanity as a whole often does not react because we are acclimated to these situations. We are seeing and valuing more highly cultural systems and manufactured goods that only last a short time rather than the infinite natural systems that are the source of life. This is the fundamental problem: the logical underpinning of what we value flies in the face of all the evidence that shows that the functional systems are vastly more valuable than the derivatives from the current, flawed system.

Replacing destructive behaviors with practices that align with natural evolutionary trends will help place our civilization on a trajectory leading to the restoration of continuously accumulative and naturally regulating Earth systems. This suggests that the point of intervention is a paradigm shift that determines whether humanity continues toward destroying ecosystems or becomes sustainable. Hence, ecological restoration can be an important contribution to human civilization and ensure that functional and biologically diverse ecosystems survive into future generations. The lessons of the Loess Plateau illustrate that if humanity were to become conscious of the importance and value of ecological functions that emerge from natural evolutionary trends, it could trigger ecosystem restoration on a planetary scale, ushering in a new sustainable era of human civilization.

4.8.4 COMMUNICATING ABOUT THE CHINESE RESTORATION EXPERIENCES IN AFRICA

In 2006, Sir Ian Crute, then the director of the Rothamsted Research Institute, suggested that John D. Liu visit Africa and share the story of what he had learned on China's Loess Plateau with the governments and people of Africa. The Environmental Education Media Project (EEMP) initiated a comprehensive communication strategy to share lessons learned with the political elites, the scientific community, the development agencies, the press, and to broadcast television films aimed at the general public. This comprehensive communication approach was extremely effective in Rwanda where the recent experience of genocide, the topography, the population density, and food insecurity made this knowledge immediately relevant and clear.

The Rwandan government agreed that ecological function was necessary for the country to develop, and so they rewrote their land use policy laws mandating that their economic development should be linked to ecological functionality. The government has made a massive and continuous effort to educate the people about the relationship between ecological health, human well-being, and prosperity. In cooperation with international agencies, a series of agroecology, reforestation, housing, healthcare, connected communications, and distributed energy systems have been initiated. Preliminary results have been highly encouraging: Rwanda has had an average economic growth of 8.2% over the last five years during a global recession. Rwanda's hydrological resources are returning, which is of vital importance for all of Africa because Rwanda is at the headwaters of the White Nile and the Congo rivers. And perhaps most important of all, Rwanda has food security while there is famine in other parts of east Africa. (EEMP has made several films on restoration in Rwanda, including *Hope in a Changing Climate, "Forests of Hope, Emerging in a Changing Climate,* and *Choosing the Pathway to Sustainability*—all are available for viewing at http://www.whatifwechange.org.)

Beyond Rwanda, other African countries have become a focus of attention for related long-term studies. Through the interest and support of various agencies, further research, documentation and communications about restoration have been carried out in Ethiopia, Tanzania, Uganda, Mali, Ghana, Kenya, and South Africa. In all of these countries it is possible to recognize the need and the potential for ecological restoration. There have been various levels of accomplishment in each of these places and the knowledge of restoration and its potential continues to spread. Much of this work has been documented on film and these materials remain revealing and relevant.

4.8.4.1 FURTHER ACTIVITIES

It has become apparent that the historical development trajectory we found on the Loess Plateau was similar to historical developments on other continents. The degradation in some areas (e.g., in Australia, the Middle East, or the Mediterranean) began a very long time ago. It is still possible to see the remnants of human impacts in the Neolithic agricultural techniques that remain in use at the edges of some places.

In Jordan, it is possible to witness the potential of restoration at the Royal Botanical Gardens, which was founded and championed by Princess Basma Al-Ali. The knowledge gained by the scientific and field staff of the Botanical Gardens could form the basis of a widespread movement to restore ecological function in the Middle East. This and other similar initiatives are especially important because they simultaneously provide an outlet for people to actively participate in both increasing their own physical security and improving the Earth's natural systems. This may be what is needed now, in a time of widespread disaffection with current political and economic realities, to engage people in non-violent, meaningful, and productive activities. It may be through such means that the fundamental ability to ensure water security, food security, and peace can be achieved. An effort to create a collaborative research, training, and innovation center for ecological restoration is underway to take the thoughts described here and direct them at restoring ecological function in a specific site in Jordan, and to use this site to train as many people as possible in restoration methodologies, with the hope of growing peace through ecology throughout the region.

In Oman, there is a truly wonderful opportunity to restore a massively degraded area in the southern part of the country. The Dhofar region near Salalah receives moisture in the form of mist coming in from the sea. The highland region parallel to the coast of Dhofar was once heavily forested; however, through deforestation, overgrazing, and poor agricultural methods, this region has been seriously degraded. The scientific measurements of the amounts of moisture available have already been carried out and there is physical evidence that biophysical interventions to massively reintroduce native trees and grasses could restore the region. Efforts are ongoing by Oman's government and private initiatives to reforest the mountains, in order to capture all of the mist. This area has high potential and is especially interesting because, in its degraded state, it is fundamentally and visibly ruined, and if it were restored the contrast would be extremely dramatic.

During further studies in Australia and in South America, John D. Liu has documented several regions with high potential for restoration. In Australia, the Sustainable Land Management (SLM) Partners, other land managers, and cattle breeders are beginning to implement a system that has become known as "holistic pasture management." This methodology uses large numbers of domesticated cattle to mimic large migratory herds of ungulates on savannah or steppe land. The movements of the herd are tightly managed using satellite photography (i.e., GIS), ensuring that they continuously move, and while they heavily impact an area, they are only there for a short period of time and then they do not return to this area for several months. Following this method, the area is liberally fertilized and the surface, rather than being compacted, is actually aerated by the trampling of the herd. Results from SLM Partners and others suggest that, when practiced correctly, it is possible to generate larger quantities of high quality animal protein while simultaneously improving biodiversity and the infiltration and retention of moisture. Given the very large area of degraded grassland on Earth this method warrants further study.

There is also great potential for restoration in many equatorial countries. For example, El Salvador in Central America endured a long period of oppression, resulting in a highly stratified society and a highly degraded landscape. As the political struggles of the 20th century ended and the government is more keen to engage in ideas like restoration, there is now an opportunity to fundamentally change the intention of society and realize its productivity potential. If such systems near the equator were restored to the highest level of ecological function in natural landscapes, in agricultural landscapes, in mixed-use landscapes, and even in cities and industrial areas, it would be of enormous benefit in humanity's efforts to mitigate and adapt to human induced climate changes (particularly because of the high carbon sequestration potentials in these climates). The areas now seen as "poor" or "underdeveloped" are actually the areas of the highest potential because restoration could engage large numbers of people in meaningful work to restore ecological function needed to naturally regulate the hydrological cycle, weather, and climate.

4.8.5 WATER RETENTION LANDSCAPES

An important understanding emanating from the work on the Loess Plateau and elsewhere is the need for and usefulness of physical and biophysical water-harvesting methodologies. While the technologies to do this are fairly well developed, a more enlightened analysis has emerged in recent years to demonstrate that natural physical characteristics and biology are symbiotic parts of the same system.

In each part of the Earth, there are different amounts of rainfall and available moisture from mist, fog, and dewdrops caused by temperature differentials. One place where this is being actively studied is the Tamera community in Portugal. This is a good place to carry out this research, and the results are relevant for a large area of the Mediterranean and north Africa. Initial observations of the water retention landscape at Tamera confirm that the percentages and total amounts of organic matter and the percentages and total amounts of biomass are the criteria determining infiltration and retention of rainfall. The work in Tamera also confirms that communities collectively intending to increase ecological function can transform historical landscapes. This is of vital importance to a huge numbers of communities and suggests an alternative to the political gridlock that often delays responses without dealing with the physical levels of the problems of hydrological disruption and natural climate regulation.

For vast numbers of communities in both the developing and developed world, the work of local communities suggests an effective way forward that improves resilience, creates social cohesion, and employs community members in ways that create diversified wealth. This is not only beneficial for the local people and communities involved, but engages the efforts of all these people in activities that we know are effective ways to mitigate and adapt to climate change, providing a global benefit.

4.8.6 BIODIVERSITY

The issue of biodiversity may be one of the harder concepts to fully understand and communicate to a broad public. Through long-term inquiry into ecosystem function and dysfunction, Liu noted that the discussion has been on two different levels in this field. The first level is an environmental discussion that puts human needs and desires at the forefront. This perspective is necessarily limited and can never reach a full understanding of the implications of biodiversity. The best that can be hoped for in the environmental discussion is a "less bad" conclusion. The second level is an ecological discussion (of which humanity is a part) that leads to a realization and an understanding about biodiversity that can reorient our understanding of human history and possible future pathways.

Cultures often have cosmological narratives, which are taken by some to be religious truths. In some of these cosmologies, it is said that God creates human beings and puts them in paradise where all their needs are provided for. But then if human beings sin, they may be driven from the garden in shame, and required to toil to survive. Interestingly, when one studies evolution, one finds that by the time human beings emerge on the scene the Earth's systems have evolved until it is a wonderfully nurturing place— a paradise—but then human beings in their ignorance begin to damage the natural systems, cutting vast forests, devegetating vast areas, altering the water cycle, and paradise is lost. In this sense, our science and our religious cosmologies may tend to agree.

Thus, understanding species richness and distribution mapping for each biome presents an opportunity for collective study and enrichment. Species of course vary with latitude, hydrological regime, and other factors, but the basic methodology for mapping is the same. Essentially this is identifying, photographing, and describing the keystone species and the symbiotic relationships with the satellite species that grow together. This method of study can be done collectively and when displayed publicly and virtually on digital platforms can both engage and inform entire communities in understanding the natural Earth systems that they depend on.

4.8.7 LAND TENURE AND PRECEDENT?

In order for people to work toward improving their lives and their surroundings, it is almost always assumed that they must have tenure over the land or at the very least have a clear right to benefit from their own labor. This is a rational assumption but it can also be a simplification of a much larger discussion. Ensuring the rights of individuals is very important, but it is also necessary to look at the historical, cultural, and cosmological worldview of the group being analyzed to understand how the situation came to be as it is.

Researching ecosystem function from China's Loess Plateau to Africa and beyond has shown that human behavior can cause a cascading series of consequences that we have labeled in various ways. Climate change, food insecurity, desertification, poverty, disparity, biodiversity loss, can all be seen as symptoms of larger systematic dysfunction. It seems necessary to see how all these issues are connected and to address the root causes of the dysfunction or we will be forced to try to live with the consequences. In many cases, such as biodiversity loss and climate change, it is required that humanity evolves to a higher level of consciousness or knowingly chooses to cause future generations great harm.

Many human choices taken by society hundreds and even thousands of years ago can be traced to or near their beginnings. The motivations and logic with which the original decisions were made have been self-serving for certain groups or sometimes were simply wrong given the limitations of understanding at the time. We are currently using institutions that were created in earlier times, often with limited scientific understanding, illogical assumptions, or created with self-serving or outdated intentions. The question we need to ask now is: must we continue to live with the decisions made in the past or can we make new decisions based on what we now know?

At one time in the European world, the Catholic Church believed that the Earth was flat and the Sun revolved around the Earth. There was no evidence for this; in fact, there was a great deal of evidence that the Earth is round and revolves around the Sun. When scientists first began to tell the truth about this, they were persecuted (especially Gallileo), but eventually it became understandable to everyone that, in fact, the Earth is round. Once the collective consciousness of humanity was aware that the Earth was round it was impossible to espouse the "flat Earth" idea without ridicule. Slavery is another example where evolving societal conscience ended an era.

Now we are faced with the need to come to some rational decisions about human induced climate change and to act on them as a species. In order to successfully do this, we need to address many historical issues that underpin the earlier decisions that have created the problems that are coming to a head. This is the largest and most rapid consciousness shift and behavioral change that humanity has had to adapt to in its short history. We need to get this right.

4.8.8 THE PROMISE OF THE COMMONS

In 1968, Professor Garritt Hardin wrote *The Tragedy of the Commons*, discussing at length the dilemma that humanity faces, which does not have a technical solution (Hardin, 1968). Almost 50 years later, humanity collectively needs to have processed the knowledge and thoughts contained in this essay. We are required to move human consciousness to a new higher level in which the dilemma posed can be transformed into "The Promise of the Commons." In order to survive, we must ensure that human behavior shifts from selfish, oppressive, and unequal rights, in which minorities are rewarded for excessive extraction, manufacturing, and trade, to a new paradigm in which human actions help to restore Earth's fundamental ecological systems, contributing to sustainability, greater equality, and common wealth.

4.8.9 VALUING FUNDAMENTALS

Assessment of the value of revegetation compared to the cost of sediment control in the Loess Plateau led to a calculation of the value of ecosystem function in comparison with agricultural productivity (World Bank, 1994), and this revealed that ecosystem function could be vastly more valuable than agricultural production. In situations where this holds true and where is it fully understood, it has the potential to change the course of human history. Looking at the relative value of functional ecosystems and products extracted from them reveals a major flaw in the current human economic system: namely, that by valuing the derivatives higher than the source of life we have created a perverse incentive to degrade natural ecological systems. If we were to value the ecosystems more highly than the things that are extracted or manufactured, it would be virtually impossible to degrade or pollute because all incentives to do those things would have been removed.

This type of reasoning also makes it apparent that it is in humanity's interest to restore degraded landscapes across the planet. This would go some way to mitigate and adapt to climate change, promote employment opportunities, and greater food security. However, the current economic metrics are flawed and have elevated the value of derivatives extracted from the natural systems and manufactured goods above the natural processes that have created and constantly renewed life. We need to change this. Hence, the question that arises is "What is money, and what do we really value as a society?" It is in the answer to this question that the solution to the sustainability of our economy in relation to our environment may be found.

4.8.10 CONCLUSION

The fundamental implication of these observations is that degraded landscapes should be restored sooner rather than later. Currently, colleagues in many organizations are working to build a business case surrounding this. The moral, social, capital, policy, and technical requirements for the restoration of Earth systems on a planetary scale are now being defined. This positive development is beginning to accelerate and will endeavor to produce a transitional stage in human development that ushers in a new ecological and economic era for humanity.

ACKNOWLEDGMENT

John D. Liu is extremely grateful for the support, encouragement, and direction that has come from a very large number of individuals, development agencies, and educational institutions without which his study would not have gone on for so long or been so fruitful. These include the World Bank, United Nations Development Program (UNDP), The UK Department for International Development (DFID), The UK Department for Environment, Agriculture and Rural Affairs (DEFRA), The United Nations Environment Program (UNEP), The Global Environmental Facility, The Faculty of Natural Sciences and the Faculty of the Built Environment, University of the West of England (UWE), The Rothamsted Research Center, The Graduate Program at the University of Reading, The International Union for the Conservation of Nature (IUCN), The COMON Foundation, Critical Zone Hydrology Group at the Vrije University, Netherlands Institute of Ecology, Royal Netherlands Academy of Arts and Sciences (NIOO/KNAW), and Communications COMMONLAND Foundation.

Bradley T. Hiller expresses gratitude to the Center for Sustainable Development at the University of Cambridge, the (former) Agriculture and Rural Development Department at the World Bank, Pembroke College at the University of Cambridge, the Environmental Education Media Project, and all the research participants in China and Turkey who were willing to share their experiences and insights into project processes and impacts.

REFERENCES

Bai, Z.G., Dent, D.L., Schaepman, M.E., 2005. Quantitative global assessment of land degradation and improvement: pilot study in North China Report 2005/6. ISRIC—World Soil Information. Wageningen, Netherlands.

Cai, Q., 2002. The relationships between soil erosion and human activities on the Loess Plateau. In: Juren, J. (Ed.), Sustainable Utilisation of Global Soil and Water Resources. Technology and Method of Soil and Water Conservation. Proceedings of 12th International Soil Conservation Organization Conference, Beijing, May 26–31, 2002, vol. 3. Tsinghua University Press, Beijing, pp. 112–118.

Central Project Management Office (CPMO), August 2005. Implementation Completion Report. World Bank Loess Plateau Rehabilitation Project, Implementation Document. Xian, China.

Chen, L., et al., 2001. Land use change in a small catchment of northern Loess Plateau. China Agricultur. Ecosys. Env. 86, 163–172.

Chen, L., Wei, W., Fu, B., Lu, Y., 2007. Soil and water conservation on the loess plateau in china: review and perspective. Prog. Phys. Geo. 31 (4), 389–403.

Chinese Academy of Sciences (Comprehensive Survey Team on Loess Plateau of Chinese Academy of Sciences), 1991. The Natural Environmental Characteristics and Evolvement. China Science and Technology Press, Beijing.

China Water International Engineering Consulting Co., Ltd (CWIECC), June 2008. Effect of Soil and Water Conservation on Water Resources and Water Environment, Final Report. China Small Watershed Management Project (DFID Trust Fund Project).

Darghouth, S., et al., 2008. Watershed Management Approaches. Policies and Operations: Lessons for Scaling Up, Water Sector Board Discussion Paper Series. Paper No. 11, May 2008. World Bank, Washington DC.

Diamond, J., 2005. Collapse: How Societies Choose to Fail or Survive. Allen and Lane, an imprint of Penguin Books Ltd, London.

Fock, A., Cao, W., 2005. Small Watershed Rehabilitation and Management in a Changing Economic and Policy Environment, Exploration and Practice of Soil and Water Conservation in China. Proceedings of Seminar on Small Watershed Sustainable Development. Beijing.

Fu, B.J., 1989. Soil Erosion and its Control on the Loess Plateau of China. Soil Use Mgt. 5, 76–82.

Greer, C., 1979. Water Management in the Yellow River Basin of China. University of Texas Press, Austin.

Hardin, G., 1968. The tragedy of the commons. Science. 162 (3859), 1243–1248.

Hiller, B.T., 2012. Sustainability Dynamics of Large-Scale Integrated Ecosystem Rehabilitation and Poverty Reduction Projects. Unpublished Ph.D dissertationCenter for Sustainable Development. University of Cambridge, Cambridge, UK.

Liu, J.D., 2005. Environmental Challenges Facing China Rehabilitation of The Loess Plateau. Environment Education Media Project (EEMP).

Liu, J.D., 2007. Learning How to Communicate the Lessons of the Loess Plateau to Heal the Earth. Environment Education Media Project (EEMP). Earth's Hope.

Liu, G.Q., Ni, W.J., 2002. On Some Problems of Vegetation Rehabilitation in the Loess Plateau. In: Juren, J. (Ed.), Sustainable Utilisation of Global Soil and Water resources. Technology and Method of Soil and Water conservation. Proceedings of 12th International Soil Conservation Organisation Conf., Beijing, May 26–31, 2002. Vol. 3. Tsinghua University Press, Beijing, pp. 217–222.

Luo, J.N., et al., 2003. Information Comparable Method of Monitoring the Intensity of Dust Storm by Multisource Data of Remote Sensing. J. Nat. Disasters. 12 (2), 28–34.

Ministry of Water Resources (MWR), 2008. Research on the Soil and Water Conservation and the Rural Sustainable Development in China, Development Research Center of the Ministry of Water Resources. China Watershed Management Project. PR China.

Niu, W., Harris, W.M., 1996. China: The Forecast of its Environmental Situation in the 21st Century. J. of E. Mgt. 47, 101–114.

Peng, H., Coster, J., December 2007. The Loess Plateau: Finding a Place for Forests. J. Forest. 105, 409–413.

Ren, M.E., Zhu, X.M., 1994. Anthropogenic influences on changes in the sediment load of the Yellow River, China, during the Holocene, The Holocene 4, 314–320.

Saito, Y., Yang, Z., Hori, K., 2001. The Huanghe Yellow River and Changjiang Yangtze River Deltas: A Review on their Characteristics, Evolution and Sediment Discharge during the Holocene. Geomorphol. 41, 219–231.

Shi, H., Shao, M.A., 2000. Soil and Water Loss from the Loess Plateau in China. J. Arid Environ. 45, 9–20.

Shi, C., Dian, Z., Youa, L., 2002. Changes in Sediment Yield of the Yellow River Basin of China during the Holocene. Geomorphol. 46, 267–283.

Varley, R.C.G., 2005. The World Bank and China's Environment 1993–2003. Operations Evaluation Department, World Bank, Washington, D.C., USA.

Wang, L., et al., 2006. Historical Changes in the Environment of the Chinese Loess Plateau. Env. Sci. and Policy. 9, 675–684.

World Bank, 1994. Staff Appraisal Report. China, Loess Plateau Watershed Rehabilitation Project, Agriculture Operations Division. East Asia and Pacific Regional Office, Washington, D.C., USA.

World Bank, 1999. Project Appraisal Document for the Second Loess Plateau Watershed Rehabilitation Project, Report No. 18958, Rural Development and Natural Resources Sector Unit, East Asia and Pacific Region, Washington, D.C., USA.

World Bank, 2003. Implementation Completion Report for a Loess Plateau Watershed Rehabilitation Project, Rural Development and Natural Resources Sector Unit, East Asia and Pacific Region, Washington, D.C., USA.

World Bank, 2005. Implementation Completion Report for the Second Loess Plateau Watershed Rehabilitation Project, Report No: 34612, Rural Development and Natural Resources Sector Unit, East Asia and Pacific Region.

World Bank, 2010. Turkey National Watershed Management Strategy, Sector Note, Draft, World Bank, Europe and Central Asia Region, Sustainable Development Unit, Ankara, Turkey.

Xu, X.X., Zhang, H.W., Zhang, O.Y., 2004. Development of Check Dam Systems in Gullies on the Loess Plateau, China Env. Sci. Pol. 7, 79–86.

Zhou, J., et al., 2002. Landslide Disaster in the Loess Area of China. J. of Forestry Research. 13 (2), 157–161.

THE COMMUNITY AS A RESOURCE FOR LAND RESTORATION

POVERTIES AND WEALTH: PERCEPTIONS, EMPOWERMENT, AND AGENCY IN SUSTAINABLE LAND MANAGEMENT

5.1

Noel Oettle[1], Bettina Koelle[2]

Environmental Working Group, Nieuwoudtville, South Africa[1]
Indigo Development & Change, Nieuwoudtville, South Africa[2]

5.1.1 INTRODUCTION

This section critically explores the design and facilitation of interventions that support the development of greater agency in conservation of land-based resources while also extending social protection. A case study approach is taken to reflect upon the circumstances and endeavors of members of a small, isolated community of farming families in the Suid Bokkeveld area of the Northern Cape Province of South Africa to enhance their livelihoods and use their land-based resources in more sustainable ways within the dynamic context of climatic, political, social, and economic change. Members of this community were classified as "Coloured" by the apartheid regime and denied access to services provided to "white"[1] farming communities. In 1999, members of the community met to share their aspirations and to analyze what obstacles they faced in attaining them, as well as how best to overcome any impediments they encounter. Their journey since then is one that offers insights that may be of value to members of other communities, development practitioners, and policy makers.

5.1.2 HISTORY OF THE SUID BOKKEVELD

The Bokkeveld was home to San hunter-gatherers for millennia before the advent of pastoralism, which was introduced by Khoikhoi herders who migrated into the area, probably around the beginning of the first millennium AD. Following settlement of the Cape by the Dutch in 1652, the abrogation of the rights of indigenous peoples and appropriation of their resources led to conflict with the newly arrived settlers. The Suid Bokkeveld was settled by Europeans following the frontier war of 1739, leading to

---

[1]The term *white* denotes South African persons who claim descent from Caucasian ancestors and who were uniquely privileged during the colonial and apartheid eras.

Land Restoration. http://dx.doi.org/10.1016/B978-0-12-801231-4.00017-3

the eventual extermination or enslavement of the San hunter-gatherers and the virtual enslavement of the Khoikhoi pastoralists (Penn, 2005). The KhoiSan peoples of the Bokkeveld were entirely dispossessed of their land, indigenous culture, own language, and autonomy by a succession of events that resulted in their being vassals in a colony in which they had no political voice whatsoever and were regarded by those in power as of value only as labor.

Over the succeeding generations, the identity of the settlers came to be defined as "white," and that of all other inhabitants gradually coalesced into "colored." Generally speaking, property ownership was the privilege of the "whites."

Tradition, expedience, and legislation contributed to the development of chronic addiction to alcohol of agricultural workers in the Cape Colony (and subsequently in the Cape Province). This addiction was perpetrated by the provision of alcohol to farm workers as a daily ration, known as "dop" (London, 1999).

Following the abolition of slavery in the Cape Colony in 1834, and the granting of self-governance there in 1853, a constitution was adopted in 1854 that prohibited discrimination on the basis of race. Colored persons were able to gain access to Crown Land that was available for settlement. In the second half of the 19th century some colored farmers were able to gain a foothold on land in the Suid Bokkeveld, most of it the lower-altitude parts that were not desired by white farmers because of its low economic value and inaccessibility. By the early 20th century, a number of colored families held land under freehold title in the Suid Bokkeveld. Typically these farms were home to extended families and were used to produce largely subsistence crops of wheat and vegetables and to graze flocks of sheep and goats. Donkeys provided transport and traction for ploughing. Honey, rooibos, and other products of the veld were harvested for domestic use.

Following the formation of the Union of South Africa in 1910, colored men in the former Cape Province meeting the requirements regarding education or property did have access to limited democratic rights. However, the election of the Nationalist Party government in 1948 and the subsequent implementation of its policies of apartheid resulted in these rights being removed. The role of the rural colored person in the vision of the ideologues of apartheid was that of a provider of cheap labor to white farmers and associated local agricultural industries. The "dop" system was modernized and entrenched. Blatant political oppression by the apartheid state aligned with the economic interests of white farmers to enforce impoverished conditions of economic exploitation on colored people, ensuring the availability of abundant agricultural labor at low cost to the employer.

Posel (2001, p. 52) notes that "Apartheid's principal imaginary was of a society in which every 'race' knew and observed its proper place—economically, politically, and socially. Race was to be the critical and overriding faultline: the fundamental organizing principle for the allocation of all resources and opportunities, the basis of all spatial demarcation, planning and development, the boundary for all social interaction, as well as the primary category in terms of which this social and moral order was described and defended."

Although the Suid Bokkeveld was not declared a "white" Group Area,[2] the agreement of its neighbors and permission from the state were nevertheless required before land could legally be sold to another person who was, in terms of the apartheid laws, classified as being a member of another "race group." As a result, there was not a great deal of change in terms of land ownership in the apartheid era.

---

[2]The Group Areas Act of 1950 restricted residential and ownership rights to fixed property in large areas of the country according to "race," as defined in the Population Registration Act (Act No. 30 of 1950); see Posel (2001).

The small-scale farmers of the Suid Bokkeveld were able to retain their tenure to their farms, some of which also came to serve as refuges for the families of some farm laborers from the area.

The 1983 constitution, designed to create a platform for a new alliance of whites with previously disenfranchised colored and Indian minorities, had the effect of stripping the colored farmers of the Suid Bokkeveld of the last remaining services (to which they at least potentially had access) that had previously been provided by the Land Bank and the Department of Agriculture. Between 1984 and 1994, they lacked access to any sort of agricultural support services whatsoever.

As is revealed by this historical narrative, the colored members of the Suid Bokkeveld community have lived for centuries in the economic, social, and political shadow of the dominant political forces of the Cape Colony and subsequently those of the united, white-dominated South Africa. The result of the grinding oppression of the colonial and apartheid eras was the erosion of freedom and land tenure and the creation of a laboring class that could serve the economic interests of white landowners. Over many generations, the addiction of farm workers to alcohol was encouraged by white farmers (with the active collusion of the state; London, 1999), resulting in a pervasive culture of alcohol abuse and dependence throughout the Western and Northern Cape. The infamous dop system ensured that farm laborers were addicted to alcohol, access to which was controlled by white farmers in terms of privileges granted to them by the colonial and apartheid states (London, 1999).

The dop system and apartheid contributed to limiting the ability of colored people to assert a positive identity or participate fully in social, political, and economic life and thus to meet some of their basic human needs (Max-Neef, 1991). The stereotyping (by people of European descent) of people of indigenous descent as being inferior, and the denial of their land-based heritage by those holding political and/or economic power, further served to undermine their chances of achieving social and economic empowerment. Although the dop system was officially outlawed in the transition to democracy following the 1994 elections, alcohol abuse remains a powerful socially destructive force and is arguably the most prominent social pathology of rural colored communities.

The advent of democracy in 1994 found the small-scale farmers of the Suid Bokkeveld community lacking any formal institutions to represent their interests and a concomitant lack of collective experience in participation in democratic organizations. This was in marked contrast to the neighboring white community of large-scale farmers, whose participation in a number of different representative structures, including the elected government of the country, had been legitimized and supported for more than a century.

Isolation, poor infrastructure, low rainfall, intermittent drought, and very low nutrient soils limited the development of the local economy in the Suid Bokkeveld. The need to produce grain crops for subsistence on marginal land in the more arid and drought-prone southern and eastern areas of the Suid Bokkeveld led to erosion of topsoil and desertification of previously fertile pockets of land. However, physical and political marginalization also meant that the community was able to evolve in ways that were relatively free of outside interference and to find ways of meeting their human needs that were relatively successful. In this context, the church has played a significant role, creating a safe haven in which congregants' human needs could, to a limited extent, be satisfied.

Despite the advent of democracy in 1994, in 1999 the colored members of the Suid Bokkeveld community were still excluded from significant market access, influence in political decision-making processes, and access to information and knowledge crucial to the long-term sustainability of their enterprises and of their community. Material poverty was widespread.

Ploughing wheat lands in the Suid Bokkeveld.

5.1.3 GEOGRAPHY AND ECOLOGY OF THE SUID BOKKEVELD

The Suid Bokkeveld is a remote rural area situated on an incised sandstone plateau to the north of the Cederberg Mountains, in the Northern Cape Province of South Africa. The area lies south of the village of Nieuwoudtville (31° 23'S, 19° 07'E), covering approximately 1600 km². On its western margin, impressive cliffs delineate the western edge of the Great Karoo Plateau, and the adjacent sandstone highlands are covered in fynbos vegetation. Further east, shales and dolerites support the notably different vegetation of the Succulent Karoo. The area thus lies within the transition zone of the ecologically significant fynbos and Succulent Karroo biomes (Cowling et al., 1997).

Notable amongst the fynbos plant species is *Aspalathus linearis*, or rooibos, which in recent decades has become known worldwide as a health-giving beverage. Most other plants in this biome are of little commercial value, being of low nutritional value to livestock and producing no other commercially viable products. The adjoining shales sustain drought-resilient plants of the Succulent Karroo vegetation type, which, despite appearing sparse, scant, and unappetizing, nevertheless provide high-quality grazing for sheep.

Biodiversity richness (approximately 1350 plant species in the Nieuwoudtville area) and the occurrence of rare and endangered plant species have earned the Bokkeveld Plateau recognition as a global biodiversity hot spot, with high levels of endemism of geophytes in particular (Manning and Goldblatt, 2007). The area lies within the semiarid winter rainfall region of the western interior of South

Africa. The area receives most of its rainfall between April and September, with a peak in rainfall usually occurring during May and August. In the southernmost parts of the Suid Bokkeveld, a steep north–south and west–east rainfall gradient of 350–150 mm per annum reflects the altitude and rain shadow impacts of the landscape on precipitation from cold fronts arriving from the South Atlantic in the winter season. Periodic droughts impact most severely on the farmers in the lower rainfall areas. Direct impacts of prolonged drought conditions from 2003 to 2006 resulted in dramatic declines in agricultural production, mortality of crops and livestock, and the drying up of most water sources (Louw, 2006).

A Suid Bokkeveld landscape in the rainy season.

5.1.3.1 EXTREME EVENTS AND SOIL EROSION

The entire winter rainfall region of the Cape was affected by severe drought initiated by the failure of the winter rains of 2003 (Archer et al., 2008) and compounded by below-normal rains in the winters of 2004 and 2005. The drought was particularly severe in the northern reaches of the winter rainfall region, including the Suid Bokkeveld, where groundwater supplies were depleted. In the last summer of the drought (2005/2006), only one water source served the entire, dispersed community. The drought resulted in killing virtually all the cultivated rooibos plants in the Suid Bokkeveld (Louw, 2006). Rangelands were unproductive and unable to support the numbers of livestock usually kept by farmers. Drinking water for domestic use and livestock had to be transported for distances of up to 40 km. Whereas the drought was unprecedented in living memory, periodic droughts characterize the climate of the winter rainfall region of South Africa. It is anticipated that climate change will increase the likelihood of such droughts increasing in intensity and frequency in the future. Global climate models indicate that precipitation is likely to decrease in much of the winter rainfall region. This prediction is associated with the anticipated poleward displacement of midlatitude westerly winds and associated storm tracks (Archer et al., 2008).

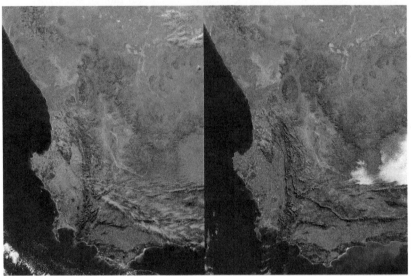

Comparative satellite images (http://eoimages.gsfc.nasa.gov/images/imagerecords/11000/11912/SouthAfrica_drought_lrg.jpg) showing the impact of the 2003 drought in the winter rainfall area (L.H. image: 21 July 2003. R.H. image: 21 July 2002). The location of the Suid Bokkeveld is indicated on both images.

Most rainfall in the Suid Bokkeveld occurs in the winter, when vegetation promotes absorption of rainwater and slows runoff. The anticyclonic winter rains tend to be gentle and soaking. However, the area is also exposed to summer rainfall events when moist tropical air occasionally causes extreme downpours of rain and hail, accompanied by violent winds. These events have caused extensive soil erosion on cultivated land, as well as damage to property. A recent example is a storm that wreaked havoc on some farms in January 2004. Torrential rains caused extensive soil erosion and the accompanying winds destroyed houses and water storage tanks; whereas significant soil erosion is inevitable when such rainfall events occur, injudicious land use exposes vulnerable soils to accelerated erosion in their course.

5.1.4 SOME KEY CONCEPTS FOR SUSTAINABLE DEVELOPMENT

5.1.4.1 PARTICIPATION

According to Max-Neef (1991), participation is a basic human need. Effective participation in the life of the community, including sharing and contribution to its aspirations, decision making, and activities, can thus be understood as vital human processes. The history of token participation in decision making in South

Erosion damage following the January 2000 Dobbelaarskop cloudburst.

Africa by people who had been disempowered by the more powerful (by either the state and its agencies, or by privileged citizens, or by both) has contributed to social dysfunction and given rise to widespread resistance to imposed measures. The typology of participation developed by Jules Pretty (1995) provides the development planner and practitioner with a valuable tool for selecting and assessing forms of participation appropriate to the developmental process that is being planned or implemented. The typology describes and ranks a range of types of participation, from manipulative participation through passive participation, participation by consultation, participation by material incentives, functional participation, and interactive participation to self-mobilization, which is seen as the ideal form of participation.

In the initial stages of an engagement with a community that has requested development support, interactive participation is both achievable and desirable. Pretty (1995) describes this form of participation as follows:

> [P]eople participate in joint analysis, development of action plans, and formation or strengthening of local institutions. Participation is seen as a right, not just the means to achieve project goals. The process involved interdisciplinary methodologies that seek multiple perspectives and make use of systemic and structured learning processes. As groups take control over local decisions and determine how available resources are used, so they have a stake in maintaining structures and practices. (p. 1252)

The process described here is one in which (by implication) members of the relevant community take over decision making and management of processes that have been initiated with input from development agents. This can be understood as a crucial step on the way to fully taking control of the development process. According to Pretty (1995), where self-mobilization takes place, people "participate by taking initiatives independently of external institutions to change systems" (p. 1252). As opposed to creating unequal dependencies, self-mobilization enables people and their organizations to develop contacts with external institutions for resources and technical advice they need, while retaining control over how resources are used, thus creating a wider net of interdependencies.

Pretty (1995) notes that "self-mobilisation can spread if governments and NGOs provide an enabling framework of support" (p. 1252). The unstated converse of this is that if the necessary enabling framework is absent, self-mobilization is not easily achieved. Interventions by outsiders that either intentionally or inadvertently impose external authority, undermine agency, deepen and exploit divisions in the community, create relationships of patronage, and deepen dependency can combine to undermine self-mobilization.

Pretty (1995) also states that "self-initiated mobilization may or may not challenge existing distributions of wealth and power" (p. 1252). Self-mobilization requires a common set of values or beliefs that people identify with and around which they feel motivated to organize and take action to achieve common goals. If these values and beliefs include notions of democratic equality, equity, and remedial support for the least privileged, it is to be expected that the more wealthy and powerful will feel that their status is under threat.

Government agencies tasked with rolling out extensive programs can frequently not effectively embrace more participatory processes and may resort to operating in the paradigms of manipulative participation and passive participation, as well as participation by consultation and by material incentives. This section explores the impact of such different dynamics and paradigms on the case study area.

5.1.4.2 AGENCY AND EMPOWERMENT

The ability to take appropriate action to improve one's situation or prevent negative impacts may be described as "agency." Agency requires both the ability and liberty to do what might be needed to achieve such results. In situations in which people's agency is constrained by oppression, a process of empowerment is needed to enable them to exercise agency in contested spheres of intervention, action, or expression. Empowerment can be understood as the ability to take action to change physical, economic, and social realities, and as such reflects a state of mind as well as liberty and utility of body. As such, it is an essential precondition for effective and sustainable development.

The World Bank Empowerment and Poverty Reduction Source Book (Narayan, 2002) notes that "empowerment refers broadly to the expansion of freedom of choice and action" (p. xviii). It further notes that "for poor people, that freedom is severely curtailed by their voicelessness and powerlessness" (p. xviii), and argues that "powerlessness is embedded in the nature of institutional relations" (p. 14).

Norbert Herriger (2006) proposes that empowering processes encourage people to discover their own strengths and to enhance self-determination and autonomy. He emphasizes that it is important to strengthen people's abilities to organize and direct their own lives and to provide resources for self-determination and independence. In contexts in which systematic disempowerment and concomitant racism have been prime mechanisms that enabled colonialists to control resources and exploit the labor of the disinherited, it is understandable that oppressed communities might aspire to empower themselves.

Whereas it is entirely possible for one person to disempower another (be it physically or psychologically), we will argue that conversely, it is not possible for one person to empower another. At best, a process that may lead toward a person becoming more empowered may be facilitated by kindling vision and enthusiasm, removing obvious obstacles, providing opportunities for education and other forms of learning, and creating the space within which people are able to grow, change, and stretch their wings.

The end of South Africa's civil war and the advent of democracy and adoption of a constitution that enshrines human rights and dignity contributed greatly to the creation of the necessary frame conditions for individual empowerment. However, many of the external constraints to empowerment remain (such as economic hardship, poor access to education, classism, and racism) and greater efforts are needed to create enabling conditions for members of historically disempowered communities to overcome these constraints.

5.1.4.3 PARTICIPATORY ACTION RESEARCH

Faced with the complexity and immediacy of the challenges faced by so many rural communities, development practitioners are tempted to intervene without an adequate understanding of the situation or an adequate plan of action. All too frequently, this may result in frustration, inappropriate action, waste of resources, and finally diminution of agency on the part of the community. In any situation that appears to require an intervention from the outside to bring about improvement, a sound "methodological basket" is essential to hold a selection of methods within an overall guiding framework. Participatory action research (PAR) provides a practical and appropriate overall methodological framework for advancing sustainable rural development.

Kemmis and McTaggart (2000) explore the field of PAR and observe that there is no unitary approach to PAR. The evolution of PAR owes most to the "press of contexts," in which it is practiced. The PAR approach allows the practitioner to be a researcher and argues that research conducted within, and not just on, practice can yield evidence and insights that assist in the critical transformation of practice.

According to Kemmis and McTaggart (2000), PAR is:

- An ethical approach that acknowledges coresponsibility for the outcomes of actions
- A social process that explores the relationship between the realms of the individual and the social
- Emancipatory, in that it aims to release people from the constraints of irrational, unproductive, unjust, and unsatisfying social structures that limit their self-development and self-determination

- Critical, in that "it is a process in which people deliberately set out to contest and to reconstitute irrational, unproductive, unjust and/or unsatisfying ways of interpreting the world, ways of working and ways of relating to others" (p. 598)
- Recursive or reflexive, in that it aims to help people investigate reality in order to change it
- An approach that aims to transform both theory and practice

They also note that although "the process of participatory action research is only poorly described as a mechanical sequence of steps, it is generally thought to involve a spiral of self-reflective cycles of:

1. planning a change,
2. acting and observing the process and consequences of the change,
3. reflecting on these processes and consequences, and then
4. re-planning,
5. acting and observing,
6. reflecting, and so on…" (p. 595).

This cyclical process provides an accessible and practical conceptualization of the PAR process in action.

5.1.5 THE PROCESS OF DEVELOPMENT IN THE SUID BOKKEVELD
5.1.5.1 OPENING THE DOOR TO CHANGE

In order to gain support from the newly elected African National Congress (ANC) government, colored landowners organized the Moedverloor Farmers' Association in 1996. The Northern Cape Department of Agriculture subsequently provided financial and technical support via the association to enable colored landowners. The aim was to improve the infrastructure on their farms, although this approach explicitly excluded those who did not own the land they farmed on. By 1998, the association had all but collapsed as a result of the capture of the government-provided resources by two of the largest-scale colored farmers, cessation of funding for the project by the Department of Agriculture (due to changes in national policy), and the resultant conflict within the community.

Faced with intractable disputes and deep dissatisfaction on the part of the vast majority of members of the community, the director of the Department of Agriculture invited the Environmental Monitoring Group (EMG) to work with the officials of the department to assist in developing its capacity to work with farmers in a more participatory and bottom-up manner. Following a request from the leaders of the two disputing factions, EMG in partnership with Indigo development and change agreed to design a developmental intervention.

The intention of the two NGOs was to support a community-based development process that would sustain both the livelihoods and natural resources of the community. Their point of departure was to work collaboratively with local people who were eager to engage in planning and taking action toward a positive outcome. A communitywide meeting was held on the Melkkraal farm on March 7, 1999, in the course of which participants defined common principles to guide future interactions. These principles and the values underlying them have shaped the development process since that time. Adey

(2007) notes that these principles illustrate the participatory and transparent nature of the process. The agreed principles were:

- Involvement in any project activity should include contributions and benefits.
- People's vision, enthusiasm, and contribution should be mobilized before benefits are achieved.
- The least advantaged should benefit the most.
- The project should benefit both the local community and the wider community.
- Everybody undertakes to work together in the spirit of mutual respect.
- There should be transparency regarding all project documentation.

The expectation of the nongovernmental organizations (NGOs) working in the Suid Bokkeveld was that interactive participation was a sound entry point in this community that lacked broad experience of collective self-organization and could best result in self-mobilization. They thus designed an intervention process to enable members of the community to "participate in joint analysis, development of action plans, and formation … of local institutions." (Pretty, 1995, p. 1252). The process was designed to enable experiential learning, with the intention that members of the community would gradually take greater control of decision-making processes and resource allocation and increasingly move in the direction of self-mobilization.

Following discussions with members of the community, the NGOs and members of the community agreed to engage in a PAR process that sought "to enhance people's ability to learn together in the course of taking action to improve their situation" (Environmental Monitoring Group, 1999) and was informed by the principle that change must be driven by those whose lives are to be improved and not by "outsiders."

The change process in the Suid Bokkeveld was shaped by an interpretation of the ethical aspect of coresponsibility for outcomes that has predicated that the process should be:

- Participatory: actively involving all relevant and willing players (especially those who are usually disregarded) as active researchers and change agents and not relegating some parties to the role of objects of the research of others or imposing external views of what change is desirable or not.
- Action-oriented: acting to improve the situation (and not just observing it).
- Research: a research process in which knowledge is developed, abilities to solve problems are enhanced, and theory is critically reviewed in an ongoing process of action and reflection.

5.1.5.2 CREATING A SHARED VISION AND MAPPING A PATHWAY TO A BETTER FUTURE

In the course of the workshop, participants created a vision for their own development and created a "problem tree" that reflected their analysis that poverty was their central challenge. They saw poverty as being derived from various root causes and in turn giving rise to a number of other challenges. At this point, poverty was implicitly understood to be of a material nature, associated with inadequate access to the resources needed for subsistence. Land degradation was identified as both a cause and a consequence of material poverty.

The NGO facilitation team gained deeper insight into the nature of the poverties experienced by community members in the course of interactions, conversations, observations, and shared reflections over the succeeding years and realized that the people of the Suid Bokkeveld had been prevented from meeting a number of their basic human needs, as defined by Max-Neef (1991). These basic needs include subsistence, protection, affection, understanding, participation, idleness, creation, identity, and freedom. This insight resulted in more focus on enabling people to meet these needs themselves.

It was clear at this stage that achieving the development vision that participants had created would involve using the material, cultural, and intellectual resources that they either had available or would be able to mobilize. The facilitating NGOs did not have ready access to funds, and the Department of Agriculture did not make any available. The NGOs, therefore, set out to make linkages with organizations that could provide stimulus, opportunity, and market access. However, it proved to be a slow process: farmers were not readily willing to enter into new modes of production or take the risks involved in selling their products to traders with whom they had no experience. Conversely, the NGOs were unwilling to advocate risk-taking on the part of the farmers. A year passed with little tangible progress toward realizing the vision.

In order to stimulate thinking about how best to achieve their goals, EMG and Indigo raised funding and organized an exchange visit of rooibos producers from the Suid Bokkeveld to small-scale producers in Wupperthal and to an exporter of organic rooibos (Oettlé et al., 2004). The stimulus and confidence building involved in exchanging experiences and ideas with other rooibos producers who were successfully selling their produce as organic and fair trade had a galvanizing effect. While still on their way home, the farmers decided to found a new business as a vehicle to achieve some of their aspirations. The NGOs subsequently facilitated feedback and follow-up workshops in the course of which a wider group from the community participated in endorsing the impulse, deciding what form the business should take and planning the practical steps that they would follow in the process (Oettlé et al., 2004).

5.1.5.3 THE HEIVELD COOPERATIVE AS A VEHICLE FOR LOCAL DEVELOPMENT

The initial impulse that led to the founding of the Heiveld Cooperative was the desire of farmers to be able to cut costs by processing their product in a collectively owned facility and to increase incomes by accessing markets more directly. The founding members decided that the most appropriate form for the business was a cooperative, and that it would "organise persons engaged in cultivating rooibos tea in the South Bokkeveld, especially those who have been disadvantaged because of their race, or because they are women, into an association that will be jointly owned and democratically controlled by all its members," (Constitution of the Heiveld Cooperative Limited, 2001) and would also "promote the interests of its members, by enabling its members to develop a sustainable economic activity, thus promoting the social and economic development of the South Bokkeveld community."

While in its formative stage, the Heiveld was approached by a European alternative trade organization with a view to purchasing fair trade and organic rooibos. Over the previous two years, the members had been exposed to, and had come to appreciate, the marketing opportunities for organic rooibos. Most were already de facto organic producers, so becoming certified organic producers was relatively easy. The concept of farming with nature was appealing to the ethos of the members, many of whom had had negative experiences of soil erosion and loss of productivity.

Furthermore, the Heiveld members had an immediate affinity with the core values of fair trade and were quick to embrace this approach. These are defined by FINE[3] as follows: "Fair trade is a trading partnership, based on dialogue, transparency, and respect, that seeks greater equity in international trade. It contributes to sustainable development by offering better trading conditions to, and securing the rights of, marginalized producers and workers—especially in the global South. Fair trade organizations, backed by consumers, are engaged actively in supporting producers, awareness raising and in campaigning for changes in the rules and practice of conventional international trade."[4]

The Heiveld initially rented a tea processing facility from a large-scale farmer in the area and entered into a three-year contract with the European company to supply organic rooibos produced under fair trade conditions. After three years the Heiveld had secured and paid for its own organic certification and was able to enter the market in its own right as an exporter. That same year, the Heiveld became the first rooibos producer organization to be certified by Fairtrade Labelling Organizations International (FLO), thus ensuring more secure access to premium funding. Under the standards set by FLO, the premium contribution of R5.00/kg is designated for development and is spent according to a Premium Plan endorsed by the membership at each annual general meeting. Over the years, these funds have been used for supporting the sustainable production of members (soil erosion control, training in organic production, provision of organic rooibos seed and plants, etc.), as well as for community development and for the development of the cooperative itself.

5.1.6 CONSERVING NATURAL RESOURCES

5.1.6.1 SUSTAINABLE USE OF ENDEMIC *ASPALATHUS LINEARIS*

The founding members of the Heiveld shared a desire to farm sustainably and raised a number of research questions on this subject. One set of questions had to do with the sustainable utilization of wild rooibos, which is harvested from a slower-growing subspecies of *Aspalathus linearis* that occurs in the Suid Bokkeveld. EMG was asked to identify a suitable researcher who would be in a position to work closely with the farmers to define the parameters of sustainable production from wild rooibos. Under the supervision of Professor Timm Hoffman of the University of Cape Town, master's student Rhoda Malgas (née Louw) was able to develop and test a relevant hypothesis with members of the Heiveld (Louw, 2006). Their work forms the basis for the Heiveld's policy on the sustainable harvesting of wild rooibos, which in turn has contributed to the industrywide standard on the wild harvesting of rooibos. Areas containing stands of wild rooibos are conserved by Heiveld members, who benefit from a secure market and the highest prices in the industry for their wild-harvested product. The environmental benefit of this accrues not only to the individual land users and their families; all of the biodiversity in these areas is conserved and carbon is retained in the soil, thus creating a global benefit as well.

---

[3]FINE is an informal association of the four main fair trade networks: Fairtrade Labelling Organizations International (FLO), the World Fair Trade Organization (WFTO), the Network of European Worldshops (NEWS!), and the European Fair Trade Association (EFTA).
[4]http://www.fairtrade-advocacy.org/component/content/archive?year=2009&month=12, accessed January 17, 2015.

Koos Koopman and Rhoda Malgas with a wild rooibos plant: farmer and student as coresearchers and cogenerators of knowledge.

5.1.6.2 CONSERVING SOIL

Soil erosion was another area of concern to the founding members, who in their lifetimes had witnessed the loss of fertile soils due to wind and water erosion, leading to land degradation, loss of productivity, and deepening poverty. As land becomes degraded, productivity drops and investments in plant material and fertility bring ever-decreasing returns.

The founding members of the cooperative shared their insights in analyzing the causes of the problem and devising solutions, which were subsequently implemented by the farmers, with supplementary support being provided by the premium funding, fair trade partners, and donor funds. A knowledge-exchange visit to the Kalahari enabled farmers to learn about how wind erosion had been controlled in the desert sand dunes in an area of similar average annual rainfall and soil texture to parts of the Suid Bokkeveld and to apply some of their learning back home.

In 2002, funding was secured from the National Department of Agriculture's LandCare Programme for these farmer-led interventions, and Heiveld members formed a local LandCare Committee to implement the project. LandCare is a national program whose objectives are to promote partnerships between communities, the private sector, and government in the management of natural resources; establish institutional arrangements to develop and implement policies, programs, and practices that will encourage the sustainable use of natural resources; and encourage skill development for sustainable livelihoods and opportunities for the development of business enterprises with a sustainable resource management focus while enhancing the long-term productivity of natural resources.

The Suid Bokkeveld LandCare Committee met regularly and planned project interventions in a democratic manner, with input from EMG and the Northern Cape Department of Agriculture. Land users were provided with planning and implementation support to manage soil erosion on their lands, and a number of innovative interventions were designed and implemented by the farmers to limit wind and water erosion and revegetate denuded areas. The design drew on their local knowledge and was stimulated by innovative ideas gained via knowledge exchange to rehabilitation work in the Mier area. The outcomes of the interventions were shared with other farmers and had a positive impact on rooibos production and incomes.

A crucial element of the success of these interventions was that the problem analysis and design of the intervention was undertaken by the farmers on the basis of their knowledge and assessment of what would be most likely to be affordable and practical in addressing the problem. Support was provided by advisors, and some of the additional resources necessary for implementation were provided from project funds. However, solutions were not imposed, leading to a strong sense of ownership for the interventions. Importantly, the farmers were committed to the maintenance and follow-up of these interventions. By implementing plans of their own design and subsequently monitoring the impacts, farmers were able to gain confidence and broaden their understanding of the dynamics of soil and water management, thus developing greater agency in land management.

The knowledge gained in this process was to subsequently be used by the Heiveld Cooperative to shape future support for sustainable land management interventions and to define best practices that are advocated in the Cooperative's Organic Management Plan.

Cultivated rooibos lands on Matarachope demonstrating best practices: organic rooibos grown from seed in lands made on the contour and protected by shelter belts of indigenous vegetation to control wind erosion and damage.

In the course of project implementation, officials of the Northern Cape Department of Agriculture decided to divert some of the resources for purposes not described in the project proposal and went ahead without discussing their plans with the members of the LandCare Committee. When the committee objected, they were informed that the decision of the officials was final. The committee subsequently asked for the intervention of an official of the National LandCare Programme and called a meeting to resolve the dispute.

Despite full support for the committee from the national official, the provincial officials refused to comply; they insisted that they would do as they felt was appropriate, even if it was neither in accordance with the project proposal nor the wishes of the committee. Officials then solicited letters from four members of the community complaining about the LandCare Committee in an effort to gain endorsement for the reallocation of the funds and to undermine the authority of the committee. This led to an intractable disagreement between them and the LandCare Committee, which had been democratically elected and was acting within its mandate.[5]

This incident soured the relationship between the farmers and the Northern Cape Department of Agriculture and marked the beginning of the end of the LandCare project. The LandCare Committee decided that once the project was completed, they no longer wanted to implement LandCare initiatives with the Northern Cape Department of Agriculture. Nevertheless, the Suid Bokkeveld LandCare project was voted one of the world's best at the International LandCare Conference held in Stellenbosch in 2004, and when an assessment of LandCare projects in the Northern Cape was undertaken, the Suid Bokkeveld project earned the Gold Award in the Premier Service Excellence Awards (Joemat-Pettersson, 2006). The reasons given for this recognition were related to the increases in productivity and improvement in livelihoods achieved through the project, recognition of the importance and potential of value-adding activities, the democratic and inclusive nature of the committee, the excellent financial and other records, the close relationship between the LandCare project and the Heiveld Cooperative, and the collaborative relationships with NGOs.

Building on the experience gained in the course of the LandCare project, in 2005, the Heiveld secured support from the Global Environmental Facility's Small Grant Programme (GEF SGP) for a project entitled "Combating land degradation and enhancing livelihoods in the Suid Bokkeveld." The goal of the project was that "farmers in the Suid Bokkeveld manage their soil, water and other natural resources in a manner that is sustainable despite the negative impacts of climate change, and achieve sustainable livelihoods."[6] This project enabled the cooperative to support its members with initiatives to improve management of soil and water on their farms and to employ two "Mentor Farmers." A formula was used whereby land users identified challenges relating to soil and water management and designed a set of interventions with the support of the Heiveld's Mentor Farmers and EMG staff. Land users undertook to provide at least 25% of the human and other resources needed for the intervention, with the bulk of the resources being provided by the project via the Heiveld.

The experience gained by the Heiveld in the course of the GEF SGP project, which was concluded in 2009, provided the basis for ongoing support to members to improve soil and water management and the

[5]Report of a second meeting held between role players in the Suid-Bokkeveld LandCare Project on the Landskloof farm. Meeting facilitated and report prepared by Lehman Lindeque, National Department of Agriculture. 3 July 2003.
[6]Proposal by the Heiveld Cooperative Limited to the Global Environmental Facility Small Grants Programme for South Africa (2005). Combating land degradation and enhancing livelihoods in the Suid Bokkeveld. Heiveld Cooperative, Nieuwoudtville.

framework for the standard of land use practice to which the members must adhere in order to retain their certification as organic producers. The Heiveld has retained the services of at least two experienced Mentor Farmers from the community who are well positioned to advise their fellow farmers on sustainable land management practices and who regularly inform the board of directors of the cooperative of their experiences and recommendations. Resources for farmer training and for improved soil and water management interventions are provided via the cooperative from a range of sources, including Fairtrade premium funds, donations by trading partners, and funding raised by EMG. The approach of enabling members to analyze challenges, design interventions, and implement these with supplementary resources is one that enables empowerment and greater agency on the part of the members of the cooperative.

Interventions designed by farmers and supported via the Heiveld include the establishment of mulched and vegetated buffer zones in cultivated rooibos lands, construction of drainage bunds on roads, use of cereals as nursery crops in cultivated rooibos lands, and construction of contour bunds in cultivated lands to enhance infiltration of rainwater and prevent soil erosion. In addition, areas of wild rooibos have been mapped and protected from cultivation, thus conserving biodiversity and sequestered carbon.

Hartweg Oktober demonstrates how a contour bund is controlling runoff from a rocky mountainside and enhancing the infiltration of water.

Buffer zones were established by farmers in old rooibos tea lands that had been seriously eroded by wind, building on traditional knowledge of building garden fences from restios (a tough endemic genus of reed). The restio barriers succeeded in breaking the winds and resulted in deposition of wind-blown sand, seed and organic material, and the reestablishment of vegetation, including rooibos tea. This technique was subsequently developed further and it has become standard practice for Heiveld members to create or retain vegetated buffer strips in their rooibos tea lands.

Some farm tracks in the Suid Bokkeveld were liable to become stormwater channels in severe rain events. Following the flooding of a homestead during heavy rains in 2002, the affected family designed and constructed a series of drainage bunds across their access road to prevent soil erosion and flooding. Constructed from shale and rock, the bunds have functioned well for more than 12 years, including during torrential rains in the 2012 cloudburst. Similar bunds have been constructed on tracks on other farms.

Heiveld farmers have developed techniques of sowing rooibos seed with a nurse crop of oats or rye that have enabled them to establish rooibos on sandy soils in more arid parts of the production area without having to resort to using seedlings, which tend to be more susceptible to fungal diseases. The nurse crop not only protects the young rooibos plants from damage from windborne sand but also increases the organic content of the soils and improves soil structure. All Heiveld members use this technique.

More than 5 km of contour bunds have been constructed to enhance infiltration of rainwater and prevent soil erosion on sloping, sandy rooibos tea lands, particularly where runoff of rainwater from rocky areas above the lands had caused severe erosion in the past.

The Heiveld has created a market for wild rooibos that has incentivized producers to conserve production areas on the farms, thus conserving biodiversity and sequestering carbon in soil and plant biomass. Four thousand ha of land are conserved according to regulations outlined in the Heiveld's internal Organic Management Plan, which also define how frequently and intensely the plants may be harvested. Producers obtain premium prices for the wild rooibos harvested in these areas, which is marketed by the Heiveld as sustainably harvested wild rooibos.

5.1.7 AGENCY AND DEVELOPMENT IN THE SUID BOKKEVELD

5.1.7.1 PERCEPTIONS AND AGENCY

At the start of the development process, people's experiences of self-organization were generally limited, and they lacked the self-confidence to organize themselves and collectively address common problems. There was at least one positive and inspiring example: an informal burial society had successfully provided an insurance service to members and given support to indigent members who had fallen into arrears with their payments. This served as a source of pride for those who had been involved in it.

On the other hand, the experiences of the Moedverloor Farmers' Association were of frustration, disappointment, and disempowerment. The most influential members of the community had succeeded in enriching themselves, and the somewhat less powerful landowners had failed to use the organization or its links with government to their advantage.

Most members of the community (in particular the landless and women) had no experience of collective organization beyond the somewhat hierarchical structures of the church. This dearth of positive experiences on the part of most members of the community created a negative perception about the ability of the community to self-organize.

In the initial March 1999 workshop, participants were asked to list the local resources that people had, or had access to, that could be used within a development process. One of the more confident women participants responded by vocally expressing the view that "we have nothing, nothing, nothing!" It was only with some encouragement that other women in the group were able to compile an impressive list of physical, biological, intellectual, and knowledge- and skills-based resources. The response of "We have nothing!" appears to reflect a perception that only those who are truly impoverished deserve to be supported, which does not serve as a sound platform for self-mobilization. On the other hand, the list of resources compiled by the group became a source of pride and hope from which aspirations for improvement could grow.

A knowledge-exchange visit to Wupperthal in 2000 enabled the visitors to perceive that their peers in this neighboring community had succeeded in collectively organizing and entering the rooibos market. This experience had so shifted the perceptions of the visitors that they had decided to form a collective business before returning to their homes.

The early successes of the Heiveld in entering the market, paying higher prices to members, building up capital, distributing profits to members, and providing services all benefited from, and in turn contributed to, a "can-do" attitude on the part of members and staff. Members volunteered their labor to build the first Heiveld tea court, thus contributing an important chapter to the Heiveld narrative—indeed, one that is frequently repeated by older members to those who subsequently joined the cooperative.

In the course of the past 14 years, members of the community have evolved more sustainable farming practices and developed the Heiveld Cooperative as a democratically governed collective business that directly markets the rooibos tea of its members to traders from more than 10 countries. By 2014, the cooperative had built up net assets of US$212,600, and in the same year, it exported rooibos to a value of more than US$450,000. It has consistently paid its members the highest price in the trade for their tea. The figures reflect the ability of the cooperative and its members to work to advance the interests of their community despite a range of challenges.

Perhaps of more importance in developing a positive attitude to dealing with adversity and complexity is how the organization and its members responded to crises and disappointments. The Heiveld board, membership, and staff were faced with major challenges on a number of occasions.

On more than one occasion, traders did not pay for the rooibos that they had purchased (and the cooperative lacked the financial reserves to be able to pay its members for this very rooibos, which they had produced and sold to the Heiveld).

A significant investment was made by the Heiveld in a packaging company that subsequently did not pay for the tea that it packaged and marketed and proved to be unprofitable; finding the way out of this dilemma was a process that involved all the members in passionate debate and problem solving.

A government initiative to process and market rooibos sought to undermine the Heiveld in a number of ways, including recruiting key members of staff and forcing the members to produce nonorganic rooibos for delivery to the government-owned factory. Despite incentives and threats offered by government officials, the Heiveld found ways to support its members to retain their organic certification and assert their independence.

Over the years, the Heiveld has had to deal with a number of disputes with staff and members and with suppliers, service providers, and certifiers. In some years, sales have been poor, resulting in serious cash-flow crises. Through all of these experiences, the Heiveld and its members and staff have been able to broaden their experience and skills, to build their reputation, and to gain self-confidence. The positive attitude and self-confidence of the organization and its members draw on these experiences and contribute to a growing belief that they can overcome any obstacles that are put in their way.

5.1.7.2 LEARNING AND CHANGE

The PAR approach to identifying problems and finding solutions in a transparent, ethically sound, and participative way has proved to be highly effective. The work undertaken primarily between Heiveld member Koos Koopman and student Rhoda Malgas to explore the parameters of sustainable harvesting proved to be of great value to the members, the organization, and the wider community. Partnerships with climate, animal, soil and vegetation scientists, social researchers, hydrologists, and entomologists have been shaped by the same methodology and have contributed to the broadening of knowledge and perceptions on the part of all.

Collective reflection has become part of the culture of the Heiveld and of the Suid Bokkeveld community. Following interactive events and business-related events and processes (trade fairs or tea-harvesting seasons), the members and staff welcome the opportunity to share their thoughts and emotions and reflect on their experiences and thus engage as active learners.

5.1.7.3 EVOLVING AGENCY: SELF-MOBILIZATION AND THE POVERTY TRAP

Having a source of funding available for development and sound democratic decision-making procedures has enabled the members of the cooperative to play a proactive role in developing not only their individual and collective enterprises but also their community. Decisions taken by the members to invest their joint funds in sustainable production both reflected and broadened the uptake of sustainable land management approaches.

In their initial 1999 analysis of the challenges facing them, members of the Suid Bokkeveld community defined poverty as their core challenge, and indeed they were seeking to escape the poverty trap that had come about as a result of historical, cultural, economic, and political forces.

Carpenter and Brock (2008) describe a "poverty trap" as a system where organization has not coalesced to drive forward solutions to problems, even though good ideas for solving the problems may exist and sufficient raw materials may be available. The Heiveld Cooperative was founded in order to forge an important key to unlocking the poverty trap, and it has succeeded in coalescing the many ideas and inspirations of its members and staff into a dynamic system that is able to respond proactively to opportunities and challenges.

Carpenter and Brock (2008) argue that fluctuations of internal demand or external shocks can generate pulses of adaptive capacity, which may collectively serve to generate the necessary traction to liberate the system from the trap that it is caught in. The successes that arise from these actions reflect movement that the community itself will increasingly be able to sustain. The narrative of this entire section has described a growing and more focused internal demand from members of the community to improve their situation, linked to the ability to respond positively to the sorts of external shocks that galvanize movement when people are able to respond proactively and overcome challenges and difficulties.

In terms of the usual measures of material poverty, the members of the Heiveld currently earn better returns on their agricultural enterprises and for their labor than in the past and thus enjoy a better standard of living. Possibly more significantly, the process that they have engaged in has contributed to their being able to meet other human needs as well. Participation and understanding have been broadened and deepened through interaction in workshops and other learning events and in the democratic governance of the cooperative itself. Identity and pride have found increasing expression through the cooperative and their product, enabling them to strengthen what Ives (2014, p. 709) describes as the "supposedly deficient, placeless identities" ascribed to colored people in the rooibos-growing areas of South Africa. More sustainable land use practices and access to water resources have provided greater protection to people.

5.1.8 CONCLUSION

In this section, we have sought to describe the endeavors of a small group of farmers in a remote corner of South Africa to improve their situation and to reflect on how this process has contributed to more sustainable use of the land. The perceptions of people that their lives have been enriched through their own individual and collective agency reflects and contributes to the processes of empowerment and learning that have taken place in the course of this journey.

Central to the story is an indigenous plant: rooibos and especially wild rooibos, which are so valued by members of the Heiveld Cooperative that some enthusiastically express their love for the plant (*Adapting on the Wild Side*, 2008). Writing about colored residents of the rooibos growing area, Ives (2014) notes that they "imagined their own redemption not through an indigenous culture, but through their identification with an indigenous plant. They described an idea of heritage that was directed toward a future in which heritage was not a burden linked to a violent and oppressive past, but a resource that they could use if and when they wanted" (p. 710).

The development of the Heiveld Cooperative reflects and parallels the journeys of individual members and employees. The remarkable development of this small business as a vehicle for sustainable development and as a proactive supporter of sustainable land use has depended on, and simultaneously stimulated, a number of changes in the community. The most significant changes include:

- Enthusiastic engagement in processes of learning and reflection
- Growing confidence of members in their own business and the integrity and rightness of its objectives and modus operandi
- A deepening sense of identity, self, and place, linked to a deeper appreciation of rooibos
- Diminution of dependence and growing interdependence
- Increased agency to act to address challenges and bring about improvement
- Interventions by farmers to conserve soil and biodiversity on their land
- Increased incomes and growing food security
- Greater sense of ownership of the resources on which people depend

The journey toward sustainability and empowerment must follow an uncharted and rough path, and the travelers themselves are best positioned to decide what to do to best keep on track and avoid painful obstacles. In the course of their journey, members of the Suid Bokkeveld community have deepened their knowledge and ability to manage their natural resources optimally so as to increase their well-being now and in the future, within a process rooted in their shared values and aspirations.

REFERENCES

Adey, S., 2007. A Journey without Maps: Towards Sustainable Subsistence Agriculture in South Africa. Ph.D. thesis. Wageningen University, Wageningen, the Netherlands.

Archer, E.R.M., Oettlé, N.M., Louw, R., Tadross, M., 2008. "Farming on the Edge" in arid western South Africa: Adapting to climate change in marginal environments. Geography, 98–107. June 2008.

Carpenter, S.R., Brock, W.A., 2008. Adaptive capacity and traps. Ecol. Soc. 13 (2), 40 (online). http://www.ecologyandsociety.org/vol13/iss2/art40/.

Constitution of the Heiveld Cooperative Limited, 2001. Heiveld Cooperative Limited, Nieuwoudtville.

Cowling, R.M., et al., (Ed.), 1997. Vegetation of Southern Africa. Cambridge University Press, Cambridge, UK.

Environmental Monitoring Group, 1999. Report of the Suid Bokkeveld Community Workshop, 27–28 March 1999, Cape Town.

Herriger, N., 2006. Empowerment in der Sozialen Arbeit. Kohlhammer, Stuttgart.

Ives, S., 2014. Farming the South African "bush": Ecologies of belonging and exclusion in rooibos tea. Amer. Ethnol. 41 (4), 698–713.

Joemat-Pettersson, T., 2006. Budget Vote Speech 12 of 2006/2007, Presented to the Northern Cape Legislature by MEC for Agriculture and Land Reform, June 15. Accessed January 17, 2015, from, http://www.gov.za/t-joemat-pettersson-northern-cape-agriculture-and-land-reform-prov-budget-vote-200607.

Kemmis, S., McTaggart, R., 2000. Participatory Action Research. In: Denzin, N., Lincoln, Y. (Eds.), Handbook of Qualitative Research. Sage Publications, London, Thousand Oaks, CA, New Delhi, pp. 567–605.

London, L., 1999. The dop system, alcohol abuse and social control amongst farm workers in South Africa: A public health challenge. Soc. Sci. Med. 48, 1414.

Louw, R.R., 2006. Sustainable Harvesting of Wild Rooibos (*Aspalathus linearis*) in the Suid Bokkeveld, Northern Cape. M.Sc. thesis, University of Cape Town, Cape Town, South Africa.

Manning, J., Goldblatt, P., 2007. Nieuwoudtville–Bokkeveld Plateau and Hantam. Botanical Society of South Africa, Cape Town, South Africa.

Max-Neef, M.A., 1991. Human Scale Development: Conception, Application, and Further Reflections. Apex Press, New York.

Narayan, D. (Ed.), 2002. Empowerment and Poverty Reduction: A Sourcebook. World Bank, Washington, DC.

Oettlé, N., Arendse, A., Koelle, B., Van Der Poll, A., 2004. Community Exchange and Training in the Suid Bokkeveld: A UNCCD Pilot Project to Enhance Livelihoods and Natural Resource Management. In: Wiesma, G.B. (Ed.), Environmental Monitoring and Assessment. Kluwer Academic Publishers, Dordrecht, the Netherlands.

Penn, N., 2005. The Forgotten Frontier: Colonist and Khoisan on the Cape's Northern Frontier in the 18th Century. Double Storey Books, Cape Town, South Africa.

Posel, D., 2001. What's in a Name? Racial Categorisations under Apartheid and Their Afterlife. Transformation 47 ISSN 0258–7696. University of KwaZuluNatal, Durban, South Africa.

Pretty, J.N., 1995. Participatory learning for sustainable agriculture. World Dev. 23 (8), 11–17.

Wege, T., 2008. Adapting to the Wild Side, Plexus Films. Environmental Monitoring Group, Cape Town.

ALL VOICES HEARD: A CONFLICT PREVENTION APPROACH TO LAND AND NATURAL RESOURCES*

5.2

Lynn Finnegan
Quaker United Nations Office, Geneva

5.2.1 INTRODUCTION

Land, livelihoods, and peace are closely interconnected. Much work is being done on the local, national, and international levels to better understand these interconnections, particularly in places where land and natural resources shape peoples' livelihoods, food security, well-being, and identity. Unsustainable, exclusive, top-down land management can lead to destructive conflict, exacerbating underlying tensions, and inequalities already present in the structures of a society. To prevent destructive conflict and build peace, land management and policy must take into account the needs and aspirations of all the people and groups involved.

The key terms used throughout this chapter are provided in Box 1.

Climate change is making it increasingly important to understand the links between land, lives, and peace. What impact does climate change have on social, political, and economic relations that are centered around land? Without significant mitigation of greenhouse gas (GHG) emissions, increases in global temperatures will increasingly place a greater strain on people's ability to respond to changes and resolve conflicts cooperatively. Rising temperatures result in uncertainty and variability in natural resources via a number of mechanisms, such as changes in precipitation, changing growing seasons, rising sea levels, and more extreme weather events. Intensified climate changes could result in greater inequality, polarization, segregation, and even violence. But this is not inevitable. Conflict in itself is not negative; it is a part of life that can drive change and development in a society if handled constructively. Conflict can also provide an opportunity for cooperative action and can actually bring groups together. Pastoralists and farmers in the Horn of Africa, for example, are cooperating to strengthen livelihoods in the face of changing climates and longer drought periods through sharing crop residues and grazing areas (Lind et al., 2010).

To understand how this is occurring, it is necessary to look at the structures, practices, and processes in place to manage conflict: how are decisions made around land? Who is involved in such decisions,

*Thanks go to Diane Hendrick for her help in the preparation of this chapter.

Land Restoration. http://dx.doi.org/10.1016/B978-0-12-801231-4.00018-5

BOX 1

Destructive Conflict

Conflict in itself is not negative. It is an inevitable part of life and can function as a catalyst for change and development in society if handled constructively. Conflict becomes destructive when it leads to a breakdown of communication among groups, damaging social relations and exacerbating tensions that can lead to violence.

Peace Building and Conflict Prevention

Peace building is both the development of human and institutional capacity for resolving conflicts without violence and the transformation of the conditions that generate destructive conflict. In this sense, it is closely allied to the prevention of destructive conflict and is not just relevant in postconflict settings.

International Legal Frameworks

There are numerous international legal frameworks that are relevant to the prevention of destructive conflict around land and natural resources. This section focuses on legal instruments that are part of the United Nations (UN). Some of the instruments are "hard" law, which are legally binding for national governments that have signed up to them and that have supervisory and oversight mechanisms at the international level. Other instruments are "soft" law, which are not legally binding but provide guidelines or minimum standards for national governments. Several soft-law instruments have come to be widely accepted as containing recognized principles within international law. Further, soft-law instruments often lead to the development of binding hard-law in the future, and thus are good indicators of the direction that international legal frameworks may take.

Community Empowerment

Community empowerment seeks to enable local people to play an active role in the decisions that affect their communities. It occurs when communities have access to relevant knowledge and develop the appropriate skills to analyze their situation, organize in an inclusive way, and manage their different interests cooperatively. Thus, the community becomes a confident and competent partner in dialogue and negotiations, whether with local or national authorities, outside investors, or both.

Land-Related Projects

The category of "land-related projects" includes any projects or activities that are centered around land and have important impacts on how land is owned, managed, used, and accessed by local people. These can include land restoration projects, agriculture, forestry, mining activities, etc.

and who is left out of dialogue among groups? Are marginalized and vulnerable groups part of the discussion and adequately listened to? Are women actively involved in decision-making processes? Are there safe spaces for open and honest dialogue?

Land-related projects often involve multiple stakeholder groups, government officials, nongovernmental organizations (NGOs), grass-root practitioners, and external investors such as foreign governments or private companies. The decision-making processes involved in land-related projects are shaped by social, political, and power relations within and between stakeholder groups. External stakeholders often have a greater financial, legal, and institutional capacity and, as such, may disproportionately influence the direction and nature of decision making in ways that can be unaligned with the needs and aspirations of local groups. For example, this phenomenon can be seen in large-scale land and forest concessions, agriculture plantations, or mining projects where local groups have difficulty understanding and asserting their rights. Without a voice in decisions that affect them, top-down land governance can bring profound changes to local people's ownership of, and access to, land and natural resources that they have traditionally inhabited, used, and managed.

In light of such power disparities, it is important to consider the role that government officials, policy makers, private actors, civil society organizations (CSOs), and local communities can play in supporting a conflict prevention approach, enabling all voices and groups to be heard during the planning and implementation of land management projects.

This section of this chapter focuses on the supporting role of law and policy in preventing destructive conflict around land. Law and policy that shapes land management practices are predominantly implemented by state institutions, but they are also influenced by regional and international legal frameworks and supported by a wide range of actors. Three key areas for policy and practice will be considered:

- Supporting participatory decision making
- Empowering local communities
- Building resilience to environmental change

This section will highlight some of the relevant international legal frameworks that can help shape national policy and provide support for all stakeholders mentioned previously to work toward peaceful management of natural resources.

5.2.2 ROLE OF LAW AND POLICY: PARTICIPATORY DECISION MAKING
5.2.2.1 PUBLIC PARTICIPATION IN DECISIONS AROUND LAND

One of the key roles of law and policy is to support and ensure participatory decision making. A national framework can contribute to enabling environments that involve all stakeholders in joint decisions, planning, and implementation around land and natural resource projects, as well as respecting and protecting freedom of speech and freedom of association. There is a large volume of research that highlights the importance of public consultation, joint decision making, and free, prior, and informed consent (FPIC) around environmental projects that will affect local communities (UN-REDD Programme, 2013). "Consultation" with local communities means more than consulting local stakeholders about predetermined analyses and plans for a certain project, and in effect communicating a decision already made. It means facilitating communication and dialogue between stakeholders, working toward mutual understanding and being aware of the potential need to manage conflict. Participation of all concerned citizens in environmental projects is provided for in the following legal instruments:

The Rio Declaration on Environment and Development
The Rio Declaration on Environment and Development, which came out of the 1992 UN Conference on Environment and Development in Rio, is seen as the foundation for many principles in current international environmental law. Article 10 of the Rio Declaration recognizes that environmental issues are best handled with the participation of all concerned citizens. It also explicitly highlights the vital role of women, youth, and Indigenous peoples in effective environmental management.

The Aarhus Convention
The Aarhus Convention on Access to Information, Public Participation in Decision Making, and Access to Justice in Environmental Matters was developed in 1998 within the UN Economic Commission on Europe (UNECE). It has 46 European parties at present, but it is open to all countries, not just UNECE states. The convention clearly defines three pillars of public participation relating to the environment, as its name reflects: access to information, participation in decision making, and access to justice. It recognizes that sustainable development can be achieved only through the involvement of all stakeholders and focuses on the relationship and interactions between people and government authorities. It takes a rights-based approach, linking environmental rights and human rights, recognizing the right of everyone to have a voice in environmental matters.

Indigenous Peoples' Rights and International Law
Indigenous peoples are recognized as "peoples" in international law, bringing a fundamental right to self-determination and sovereignty over their lands and natural resources (Carmen, 2010). International legal instruments relating to Indigenous peoples tend to be soft law, providing nonbinding guidelines for national governments. The UN Declaration on the Rights

Continued

CONT'D

of Indigenous Peoples (UNDRIP), however, sets out minimum standards negotiated among all countries at an international level and is listed by the Office of the UN High Commissioner for Human Rights (OHCHR) as a universal human rights instrument. Adopted by the UN General Assembly in 2007, it affirms the human rights of the 370 million Indigenous peoples worldwide. One of the few legally binding treaties specifically addressing Indigenous peoples is the ILO Convention No. 169 on Indigenous and Tribal Peoples (known as "ILO Convention 169"), passed in 1989. It states that Indigenous peoples have "the right to participate in the use, management, and conservation of their resources." The Convention, however, only has 22 signatories, the majority of which are Latin American. Both UNDRIP and ILO Convention 169 oblige governments to meaningfully consult Indigenous communities and uphold the FPIC principle, requiring that no decision be made without communities' informed consent given freely, without coercion, intimidation, or manipulation, where a proposed project or measure directly affects Indigenous peoples and their natural resources. Further, human rights treaty monitoring bodies, such as the Committee on Economic Social and Cultural Rights and the Committee on the Elimination of Racial Discrimination, have also issued general comments on Indigenous peoples, and so are being used to monitor Indigenous people's rights.

5.2.2.2 IMPROVING DIALOGUE CHANNELS AMONG ACTORS

Improving opportunities for dialogue among groups is crucial for building bridges among stakeholders who often have very different values, perceptions, and priorities around environmental issues, land, and livelihoods. Open and honest dialogue among local communities, civil society, government officials, and private actors can help groups garner a better understanding of their own and each other's positions. This is especially true if it is sustained throughout all stages of land management, from early planning to implementation and ongoing evaluation of impacts. It is important that the people who are most affected by change in terms of access to land, biodiversity, livelihood activities, and areas of cultural importance have ongoing opportunities to voice their concerns and assert their rights. Multistakeholder platforms, when part of an ongoing dialogue rather than a one-time event, can help build trust among user groups and support collective and adaptive learning that responds to local needs.

Conversatorios of Citizen Action in Colombia, for example, made a long-term commitment to build capacity among vulnerable groups involved in managing coastal resources. The project brought together government, institutional, and private-sector representatives with members of the public to make joint commitments around land and natural resources. The process, which takes place over a period of several years, helped change attitudes between the communities and institutions involved. Community members said they had more confidence in dealing with institutions, while also perceiving them to be more transparent. Private- and public-sector institutions reported seeing community groups as constructive partners where they had previously perceived them as "hostile and uninformed" (Candelo et al., 2008).

Working within a group with diverse backgrounds and perceptions can be very challenging, but multistakeholder platforms can also provide long-term channels to help change attitudes and address power imbalances among groups (Roberts and Finnegan, 2013). Similarly, bringing local voices to national and international fora can help change attitudes among policy makers and help states become more aware of the impacts that national frameworks are having on local communities.

5.2.2.3 INCLUDING VULNERABLE AND MARGINALIZED GROUPS

Governance and management of land that does not include vulnerable and marginalized groups can become ineffective, resulting in the fuelling of local tensions and underlying conflicts. Poorly planned or uncoordinated public participation can worsen social marginalization by excluding traditionally vulnerable groups who may not be readily accepted as legitimate participants in public decision making. One such example is noted by the LandNet West Africa Network, which acknowledges regional similarities in the challenges faced by women in land tenure arrangements. Local governance structures are often dominated by men, and land rules are often interpreted in ways that discriminate against women. This feeds into environmental decision making, which is accentuated by women's lower literacy levels and lack of knowledge and understanding of the legal system involved. These are serious barriers to women becoming equal and effective partners in environmental planning and land management. This contributes to women experiencing less secure land tenure arrangements (FAO, 2013).

Government officials, CSOs, and project planners can take an active role in improving representation and involvement of traditionally excluded groups. In Mozambique, the Centre for Legal and Judicial Training (CFJJ) works with local communities and district-level government officers to improve understanding of the 1997 National Land Law and the land-related conflicts and challenges that local communities face. This has helped link local communities to agencies that provide legal support, aiding them in having a greater voice when interacting with potential outside investors. It has also contributed to changing attitudes between district officials and rural farming communities who would traditionally be marginalized from decision making and land policy (Tanner and Baleira, 2006; Cotula and Mathieu, 2008).

Questions need to be asked in order to make participation effective: Do the times, places, and formats of consultations inhibit a particular group from being fully present? For example, do they coincide with the commitment of young people, women, or other marginalized groups with school, child care or tending and harvesting in the fields? Conversatorios of Citizen Action in Colombia, for example, work with local communities to improve community management of natural resources, and the workshops are carefully tailored to minimize written materials and focus on verbal communication given the low literacy rates among participants (Roldan, 2008; Candelo et al., 2008).

A natural resource management program in Nepal works with local government structures using a decentralized and participatory approach to rural water management. The project involves public consultations to identify local priorities, and specific meetings are held with women or other marginalized groups if their voices are not being heard in the main meetings (van Koppen et al., 2014; Rautanen et al., 2014).

5.2.2.4 INTERNATIONAL HUMAN RIGHTS LAW

International human rights law emphasizes the universality of human rights and the responsibility of states to actively identify, acknowledge, and protect vulnerable groups. Human rights include the right to water, food, and health of all people. UN human rights conventions include protection of the most vulnerable people, including small-scale farmers, landless workers, fishers, and Indigenous peoples. The International Covenant on Economic, Social and Cultural Rights (ICESCR) deals with many of the substantive rights relating to the environment, and the Convention on the Elimination of All Forms of Discrimination Against Women (CEDAW) is the only treaty to deal explicitly with rural women (Article 14). CEDAW recognizes the particular challenges that rural women face and the significant roles they play in the economic survival of their families, calling on states to eliminate discrimination against rural women and ensure that they participate in and benefit from all rural development and development planning.

All the human rights conventions are hard law: they are legally binding for national governments that have ratified them, and most have supervisory and oversight mechanisms at the international level. Each convention has a periodic review mechanism and a supervisory body, where national governments, national human rights institutions, and CSOs are invited to comment on the implementation of the convention in question. The shadow, or alternative reports provided by stakeholders are considered by the treaty's committee of experts during country reviews (ESCR-net and International Women's Rights Action Watch Asia Pacific, 2010). Alongside the reporting processes, "special procedure" mandates

Continued

CONT'D

allow independent experts and special rapporteurs to research and report progress by thematic topics. Current mandates include the Independent Expert on Human Rights and the Environment, along with the Special Rapporteur, on the human right to safe drinking water and sanitation[1].

5.2.2.5 ACCESS TO JUSTICE

Inclusive decision making and rights relating to the environment mean little without accountability and access to justice if people's rights are violated. Human rights activists, as well as environment and development practitioners, continue to call for more effort by states to provide effective avenues for accountability and settlement of disputes. Rural and Indigenous communities are among the most vulnerable groups who face serious obstacles to accessing formal justice systems. Long distances to courts, high travel costs, differences in language, and cultural misunderstandings all represent serious barriers to accessing the national legal system[2].

One way of achieving access to justice for rural and vulnerable communities is through the promotion of customary justice systems (Ubink, 2011). Customary justice systems, such as community courts, are often local peoples' preferred method for resolving disputes around land, are seen as more legitimate by local communities, and have a deeper understanding of local needs and context. National frameworks that recognize customary justice processes and make efforts to bring community courts into the national legal system can help improve access to justice, as well as decrease the backlog and burdens in formal court systems. Nonetheless, there are risks in using customary processes; without ongoing evaluation and active efforts to improve inclusiveness, participation of women and stigmatized groups can remain low and power relations can be slow to change. Overall, however, dispute settlement methods that are grounded in local institutions (and broader than court-based adjudication) can empower communities to resolve land and natural resource disputes without violence and other forms of destructive conflict.

The Aarhus Convention

The Aarhus Convention's third pillar is access to justice, calling on governments to provide an independent grievance mechanism to bring accountability and redress if the convention's procedural rights have been violated. Aarhus provides a mechanism for complaints that allows civil society to communicate directly to the convention through the Compliance Committee[3]. This mechanism has been used by a wide range of stakeholders—from international NGOs like Friends of the Earth, to national environmental law organizations, small community organizations, and concerned individuals—in more than 20 countries that have signed on to the convention.

---

[1]These mandate holders are held by John Knox and Léo Heller, respectively.
[2]The UN Expert Mechanism on Access to Justice (EMRIP) is doing an ongoing study on access to justice. More information is available at www.ohchr.org/EN/Issues/IPeoples/EMRIP/Pages/AccessToJustice.aspx.
[3]For more information, see www.unece.org/env/pp/cc.

5.2.3 ROLE OF LAW AND POLICY: EMPOWERING LOCAL COMMUNITIES

A legal framework that respects public participation in decision making is not necessarily enough to prevent destructive conflict around issues related to land. Even if states provide a supportive and participatory environment and human rights policy, it can be difficult to achieve equitable and sustainable land management. Even if external actors are willing to engage and negotiate with local groups, it can be difficult for them to immediately and quickly find a way into a local community, as well as to tell whether the people they are engaging with are truly representative of all the community's interests. Thus, a necessary element of a conflict prevention approach to land is empowerment of local communities to become competent partners in land and natural resource management. To play an active role in decisions that affect them, communities need access to the relevant knowledge and skills (namely, the ability to analyze their situation, organize in an inclusive way, and manage their different interests cooperatively). Communities can then become confident partners in dialogue and negotiations with other stakeholders.

5.2.3.1 CAPACITY BUILDING AT ALL LEVELS

For a community empowerment approach, capacity building is necessary across all levels, from local communities and CSOs to government officials and policy makers (UN-EU, 2012). In contexts where land, water, forests, and other natural resources are central to people's lives, local social structures, and institutions need sufficient capacity to respond to land-related conflicts in a constructive way.

It is not necessarily feasible or desirable that government officials provide capacity building to local communities, as an independent body or local institution may be better positioned to facilitate community empowerment work. If opposing interests are perceived, or there is suspicion about government officials' objectives, it can be difficult to build the trust needed to allow for effective capacity building. States can finance capacity-building initiatives, but independent groups often carry them out better. Such work includes helping communities build the necessary skills to identify, analyze, and manage conflicts, as well as help improve public speaking and communication skills; many of these methods have been used by peace-building practitioners for decades.

Capacity building is also required within formal government in order for government officials to take on a more facilitative role in supporting collective and adaptive decision making that responds to local needs rather than implementing predetermined directives. Climate change adds another level of urgency to this need. Rising GHG emissions will lead to greater uncertainty in precipitation levels, growing seasons, extreme weather events, and significant seasonal temperature increases.

5.2.3.2 ACCESS TO INFORMATION

Access to adequate information on environmental matters is essential if CSOs and local communities are to understand the options available to them and make informed decisions. This means communicating information about potential land projects in an accessible way and in appropriate languages, taking into consideration literacy rates and groups that may need information in different formats. The information needs to be understandable by local communities for them to make effective use of it. Superficial information in inappropriate formats can exacerbate unequal access to land and natural resources by disadvantaging groups with less sophisticated language and communication skills.

Impartial and factual information is needed for national-level proposals or projects that could have potential economic, social, and environmental impacts on local communities. This information is needed at an early enough stage that civil society can still influence the plans. Across eastern and central Europe, there are several Aarhus centers that help facilitate such access to timely and accurate information around proposed plans, in accordance with the Aarhus Convention (as discussed previously). The centers help disseminate environmental information to civil society and empower groups to address their concerns and take appropriate action[4]. On a global scale, The Access Initiative (TAI) is an international network that aims to link civil society organizations working on environmental rights and access to information. TAI was established in 1999 and now has more than 150 partner organizations working in more than 45 countries[5].

CSOs and local communities also need to understand their legal rights over land and natural resources. There is often a real lack of awareness of national laws relating to procedural rights, land, water, biodiversity, forests, fisheries, and other natural resources at the local level. In such cases, legal empowerment work is needed, including education programs that work with rural workers, pastoralists, farmers, youth clubs, and other rural groups about national land laws and the rights and procedures available to them. When the rights and responsibilities of different groups are known and understood by all parties, local communities can be more assertive in claiming and exercising their rights over land that they have traditionally occupied and managed. In Uganda, for example, the 1998 Land Act provides an opportunity for communities to register their communally held lands. The process is complicated and time consuming, however, with very few communities having managed to complete it. The Land and Equity Movement in Uganda (LEMU) and Namati are two organizations that work with local communities to help them understand and engage with the Land Act and the legal process involved, bridging the gap between the land law in writing and in practice (Knight et al., 2012; LEMU and Norwegian Refugee Council, 2009).

5.2.3.3 RECOGNIZING LOCAL TENURE RIGHTS

Tenure systems regulate how individuals and groups gain access to land and other natural resources and determine the rights and duties associated with land use and ownership. Individual and collective tenure rights shape local ownership of, and access to, land. Effective governance of tenure is, therefore, an integral element of sustainable land use, food security, and social cohesion. In many parts of the world, small-scale farmers, landless workers, fishers, and Indigenous peoples have owned, managed, and used land for many generations without their tenure rights being recognized in formal national law. If local communities have insecure tenure rights, there is little incentive to invest long term in sustainable land-use practices. An absence of clear tenure rights is also often a barrier to community participation in decision making when external actors are involved. For example, in many countries, the state officially owns all land and many of the natural resources, like forests and subsoil minerals. This can have profoundly negative impacts on local communities' tenure security if their access to land is subject to top-down policies and agreements being signed between government officials and external investors. This has become a major issue as external actors invest in natural resources without recognizing the tenure rights of local communities[6].

---

[4]For more information, see www.osce.org/secretariat/89067.

[5]For more information, see www.accessinitiative.org.

[6]For a range of resources and statistics, see the work of the Land Matrix, available at www.landmatrix.org.

Another tension between customary rights and formal rights can be found if customary tenure systems discriminate against particular groups, especially those with rights that would be more strongly protected in formal law. Women and girls, for instance, face serious obstacles to inheriting land in many regions (FAO, 2013). There have been many efforts over the last five years to document access to land and advocate for more equitable governance of tenure, such as the work coordinated by the International Land Coalition[7].

5.2.3.4 THE CFS VOLUNTARY GUIDELINES ON TENURE

In 2012, the UN Committee on World Food Security (CFS) adopted guidelines relating to the responsible governance of tenure of lands, forests, and fisheries (VGGT, 2012). The guidelines outline rights and responsibilities relating to tenure, urging governments to recognize, respect, and protect the customary land, forest, and fishery rights of Indigenous peoples and other communities, even if they are not currently protected by law (Articles 7 and 8). A framework should also be put in place to provide a competent and impartial body to resolve disputes, providing access to justice when people believe that their legitimate tenure rights have not been recognized. The UN Food and Agricultural Organization (FAO) provides introductory documents and online training on the guidelines, and capacity building is available for countries on request[8].

5.2.3.5 INDIGENOUS PEOPLE'S RIGHTS

Indigenous people's rights in international law lay out provisions to improve relationships between states and Indigenous communities, creating minimum standards for their survival, dignity, and well-being. ILO Convention 169 states that governments shall recognize Indigenous ownership of lands they traditionally occupy, including land used for subsistence, nomadic, and traditional activities. UNDRIP similarly states that Indigenous peoples have the right to their traditional lands, territories, and resources, and these shall be given legal recognition and protection by the state (Article 26). Governments must not forcibly remove Indigenous peoples from their lands or territories (Article 10) and shall provide effective mechanism for redress if FPIC is not respected.

5.2.4 ROLE OF LAW AND POLICY: BUILDING RESILIENCE

The term *resilience* is used in many disciplines, including ecology, engineering, and disaster management, to consider the capacity of various systems to cope with, respond to, and recover from shocks or disturbances. The international development community has become increasingly interested in resilience as a useful way of thinking about societies' capacities to cope with change, learn to live with uncertainties, and transform systems when necessary. With the overlapping and interrelated challenges of climate uncertainty, economic instability, population change, food and water insecurity, and natural disasters, resilience thinking looks at three elements of capacity: absorptive capacity, adaptive capacity, and transformative capacity (Béné et al., 2013). For instance, if we consider climate change, a resilience approach would seek to develop the ability to *absorb* shocks such as changes to crop yields, to *adapt* to precipitation change, and to *transform* a fundamental part of the system when it is no longer viable, by switching land use or livelihood activities.

---

[7]For more information, see www.landcoalition.org.
[8]For more information, see www.fao.org/nr/tenure.

Resilience recognizes that development efforts happen in complex environments; a change in one variable, such as growing seasons, can cause changes in other variables that are difficult to predict or extrapolate. It also recognizes that changes at a certain level or sector has impacts on others, and thus refers to all scales from the household level to the community and then to the societal level. This encourages a holistic approach to project planning that encompasses both state and nonstate institutions.

The concept of resilience can lend useful insights when we consider a conflict prevention approach to land and natural resource management. The concepts of adaptive and transformative capacity in particular are closely linked to a community's decision-making abilities—how the community makes decisions and who is involved in the decision-making process (de Weijer, 2013). Resilient communities, supported by policy, will be more able to manage changing climates, natural disasters, tensions, and conflicts around land. Resilience is shaped largely by social relations and individual characteristics, but policy frameworks can still create and support an enabling environment in which resilience can emerge. An important element of this is supporting local institutions.

5.2.4.1 SUPPORTING LOCAL INSTITUTIONS AROUND LAND AND NATURAL RESOURCES

Both resilience and peace-building approaches emphasize the importance of supporting local ownership of social institutions (de Carvalho et al., 2014). Local institutions help shape the resilience of a society to absorb, adapt, and respond to shocks. Building resilience is best done with a thorough understanding of local vulnerabilities and the power relations involved, so as not to inadvertently increase the resilience of some groups at the expense of others. This actually reinforces poverty and exclusion rather than building peaceful relations around land (Béné et al., 2013).

By focusing on the capacity of local communities to change and adapt, resilience thinking encourages a move away from more interventionist, top-down approaches to development efforts. Supporting local organization and ownership requires a long-term approach, which can seem incompatible with the time scales involved in policy making, especially at the national level. These differences can easily create tensions between policy makers and local practitioners. Law and policy frameworks, however, can work to strengthen local institutions that support adaptive learning, collective action, and conflict resolution.

In Nepal, for example, Community Forest User Groups (CFUGs) are the main institution used to implement community-based forest management. They successfully link across scales of organization and different social groups using the forest, and have a policy that at least one third of representatives are women and 10% from disadvantaged groups (Adhikari and Adhikari, 2010). During the decade of violent conflict in Nepal from 1996–2006, many broader institutions of governance failed in the region, but the CFUGs remained effective. They adapted to changing circumstances and continued to be a mechanism for forest management. Not only that, but the CFUGs became a presence that helped moderate the broader impact of the violent conflict on local communities (Ratner et al., 2013).

Land and natural resource management should consider the vulnerabilities of each stakeholder group, whether they are pastoralists, small farmers, youth, women, seasonal rural workers, ethnic minorities, Indigenous communities, or people with disabilities. Which groups need to be resilient (and to what)? How can land practices and decision making help build resilience?

5.2.4.2 HYOGO FRAMEWORK FOR ACTION: BUILDING THE RESILIENCE OF NATIONS AND COMMUNITIES TO DISASTERS

The 2005 World Conference on Disaster Reduction developed the Hyogo Framework, a 10-year action plan for building resilience to disasters that could be brought on by climate change and variability. The framework called on national and local governments to prioritize disaster risk reduction and build a strong institutional base to implement strategies, including identifying, assessing, and monitoring disaster risks and enhancing early warning. The 2015 review of the Framework resulted in the Sendai Framework for Disaster Risk Reduction 2015-2030, which was adopted by UN Member States on 18 March 2015. The Framework is the first major agreement of the post-2015 development agenda. The new framework calls for "disaster risk governance" and stressed the need for coordination within and across sectors, as well as participation of all relevant stakeholders. It acknowledged that the "knowledge, experience, and resources" of non-State stakeholders will be necessary in this. The hope is that strengthening disaster risk governance for prevention, mitigation, preparedness, response, recovery, and rehabilitation will foster collaboration and partnership across mechanisms and institutions for the implementation of agreements relevant to both disaster risk reduction and sustainable development (resolution adopted by the General Assembly, 2015).

5.2.5 CONCLUSIONS AND RECOMMENDATIONS

This section has considered some of the conditions necessary to achieve peaceful, equitable, and sustainable land and natural resource management with the focus on the role of law and policy. Government officials, policy makers, the private sector, CSOs, and local communities are all important and relevant actors that can work to support constructive handling of land-related conflicts.

Important elements of a conflict prevention approach to land management involve supporting participatory decision making, improving the dialogue between government officials and other stakeholders (including marginalized groups), respecting customary land tenure rights, and empowering communities to understand and exercise their legal rights.

The international legal frameworks outlined throughout this text show some of the instruments relevant to preventing destructive conflict around land and natural resources. Some of them are hard law (i.e., legally binding) while others are soft law (i.e., nonlegally binding). It is important to remember, however, that, according to Madiodio Niasse, director of the International Land Coalition, speaking at the 2013 Policies Against Hunger Conference in Berlin, "[I]t is not whether or not an international instrument is mandatory or voluntary, what really matters in the end is the extent to which it is known, referred, and actually used". Much work is needed to raise awareness of these laws and guidelines among national governments, CSOs, and local communities. Many of these can be part of the empowerment process that allows local communities to assert and protect their rights and become effective partners in land and natural resource management. Several of the legal instruments provide opportunities for states to share information and recommendations, and receive capacity building and technical assistance from UN agencies. These opportunities are available for states and stakeholder groups and should be used more fully.

Diagram showing the relevant international legal 'landscape'

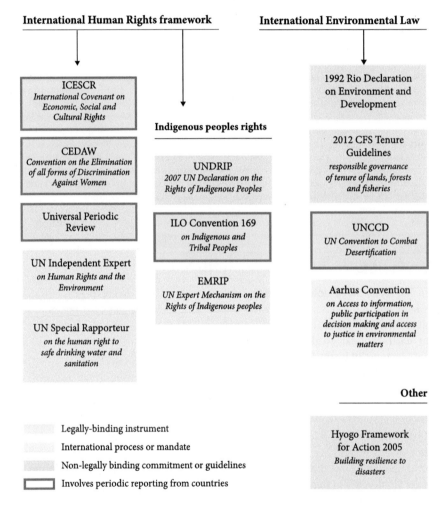

The international legal 'landscape' relevant to
building peace around land and natural resources

International Human Rights framework

ICESCR
International Covenant on Economic, Social and Cultural Rights

CEDAW
Convention on the Elimination of all forms of Discrimination Against Women

Universal Periodic Review

UN Independent Expert
on Human Rights and the Environment

UN Special Rapporteur
on the human right to safe drinking water and sanitation

Indigenous peoples rights

UNDRIP
2007 UN Declaration on the Rights of Indigenous Peoples

ILO Convention 169
on Indigenous and Tribal Peoples

EMRIP
UN Expert Mechanism on the Rights of Indigenous peoples

International Environmental Law

1992 Rio Declaration on Environment and Development

2012 CFS Tenure Guidelines
responsible governance of tenure of lands, forests and fisheries

UNCCD
UN Convention to Combat Desertification

Aarhus Convention
on Access to information, public participation in decision making and access to justice in environmental matters

Other

Hyogo Framework for Action 2005
Building resilience to disasters

Legally-binding instrument

International process or mandate

Non-legally binding commitment or guidelines

Involves periodic reporting from countries

* We aim to highlight some of the instruments relevant to a conflict prevention approach to land and natural resources. It is not intended to be an exhaustive list.

5.2.5.1 KEY SUMMARY POINTS

The following are the major issues discussed in this section:

- Improving dialogue among groups and actors involved in land-related projects is essential, supporting participatory decision making that gives all stakeholders a voice in the process.
- Excluding vulnerable and marginalized groups undermines a conflict prevention approach to land management.

- Access to justice is essential for providing accountability to all parties and groups involved, providing a way of settling disputes and providing redress where rights have been violated.
- Capacity building is required at all levels for communities to become competent partners in dialogue and negotiations, and government officials to facilitate decision making that responds to local needs.
- CSOs and local communities need access to information that allows them to understand the choices available and make informed decisions that are in line with their development and livelihood aspirations.
- Local livelihoods and food security are better protected when customary land tenure rights (both individual and collective) are recognized in formal law and policy.
- Strengthening local ownership and control of institutions helps build resilient communities that are more able to cope with social and environmental change and uncertainty; improving the resilience of all stakeholders involved in land-related projects can help them absorb environmental shocks, adapt to changes and uncertainties and transform to fundamentally new ways of doing things when necessary.

REFERENCES

Adhikari, J.R., Adhikari, B., 2010. Political conflicts and community forestry: understanding the impact of the decade-long armed conflicts on environment and livelihood securities in rural Nepal. International workshop on collective action, property rights, and conflict in natural resources management. Siem Reap, Cambodia. June 28–July 1, 2010.

Béné, C., Newsham, A., Davies, M., 2013. Making the most of resilience. IDS In Focus Policy Briefing 32. Institute of Development Studies.

Candelo, C., et al., 2008. Empowering communities to co-manage natural resources: impacts of the Conversatorio de Acción Ciudadana. In: Fighting poverty through sustainable water use, vols. I, II, III, IV, CGIAR challenge programme on water and food. p. 204.

Carmen, A., 2010. The right to free, prior, and informed consent: a framework for harmonious relations and new processes for redress. In: Heartley, J., Joffe, P., Preston, J. (Eds.), Realizing the UN declaration on the rights of indigenous peoples: triumph, hope, and action. First Nations Summit Society, Purich Publishing Ltd.

Cotula, L., Mathieu, P. (Eds.), 2008. Legal empowerment in practice: using legal tools to secure local land rights in africa. UN food and agriculture organization (FAO) and International institute for environment and development (IIED). Highlights from the international workshop "Legal empowerment for securing land rights" Accra, 13-14 March 2008.

de Carvalho, G., de Coning, C., Connolly, L., 2014. Creating an enabling peace-building environment: how can external actors contribute to resilience? Background paper produced for: *ACCORD Conference on "Drawing Coherence between Peacebuilding Frameworks"*, 19-20 February 2014, Johannesburg, South Africa.

de Weijer, F., 2013. Resilience: a trojan horse for a new way of thinking? (ECDPM Discussion Paper 139), ECDPM, Maastricht.

ESCR-net and International Women's Rights Action Watch Asia Pacific, 2010. Participation in ICESCR and CEDAW reporting processes: guidelines for writing on women's economic, social, and cultural rights in shadow/alternative reports. Available at https://docs.escr-net.org/usr_doc/CEDAW_CESCR_reporting_guidelines_FINAL_Oct_6_2010.pdf.

Food and Agricultural Organization (FAO), 2013. Governing land for women and men: a technical guide to support the achievement of responsible, gender-equitable governance of tenure, governance of tenure technical guide no. 1.

General Assembly on 3 June 2015,69/283. Sendai Framework for Disaster Risk Reduction 2015–2030 A/RES/69/283 23 June 2015.

Knight, R., et al., 2012. Protecting Community Lands and Resources: Evidence from Liberia, Mozambique and Uganda, International Development Law Organization (IDLO) and Namati.

Land and Equity Movement in Uganda (LEMU) and Norwegian Refugee Council, 2009. How can women's land rights be best protected in the National Land Policy?. Briefing paper, accessed from, http://land-in-uganda.org/lemu/wp-content/uploads/2013/11/SecuringWomensLandRightsinSouthernandEasternAfrica.pdf.

Lind, J., Ibrahim, M., Harris, K., 2010. Climate change and conflict: moving beyond the impasse. IDS In Focus Policy Briefing 15. Institute of Development Studies, p. 3.

Ratner, B.D., et al., 2013. Resource conflict, collective action, and resilience: an analytical framework. Intl J Commons 7 (1), 189–208.

Rautanen, S.-L., van Koppen, B., Wagle, N., 2014. Community-driven multiple-use water services: lessons learned by the rural village water resources management project in Nepal. Water Alternatives 7 (1), 160–177.

Roberts, E., Finnegan, L., 2013. Building peace around water, land, and food: policy and practice for preventing conflict. Quaker United Nations Office, Geneva, Switzerland. Accessed from http://www.quno.org/resource/2013/9/building-peace-around-water-land-and-food-policy-and-practice-preventing-conflict.

Roldan, A.M., 2008. A collective action to recognize commons and to adopt policies at multiple government levels. Paper submitted to the International Association for the Study of the Commons (IASC), UK, July 2008.

Tanner, T., Baleira, S., 2006. Food and Agriculture Organization of the United Nations (FAO), Livelihood Support Programme (LSP) Working Paper 28, Access to Natural Resources Sub-Programme, "Mozambique's legal framework for access to natural resources: The impact of new legal rights and community consultations on local livelihoods". Available in ftp://ftp.fao.org/docrep/fao/009/ah249e/ah249e00.pdf.

Ubink, J. (Ed.), 2011. Customary Justice: Perspectives on Legal Empowerment, Legal and Governance Reform: Lessons Learned. No. 3, International Development Law Organization (IDLO).

United Nations–European Union (UN-EU), 2012. Strengthening capacity for conflict-sensitive natural resource management, toolkit and guidance for preventing and managing land and natural resource conflict. UN inter-agency framework team for preventive action, UN-EU partnership. accessed at http://postconflict.unep.ch/publications/GN_Capacity_Consultation.pdf.

United Nations–Reducing Emissions from Deforestation and Forest Degradation in Developing Countries (UN-REDD) Programme, 2013. Legal companion to the UN-REDD programme guidelines on free, prior, and informed consent (FPIC): international law and jurisprudence affirming the requirement of FPIC. UN-REDD Programme, Geneva, Switzerland.

van Koppen, B., et al., 2014. Scaling up multiple-use water services: accountability in the water sector. Practical Action Publishing, Rugby, UK.

Voluntary Guidelines on Responsible Governance of Tenure of Land, Fisheries and Forests in the Context of National Food Security (VGGT), Food and Agriculture Organization of the United Nations, Rome, 2012.

GENDER IN THE CONTEXT OF LAND RESTORATION

LAND RESTORATION, AGRICULTURE, AND CLIMATE CHANGE: ENRICHING GENDER PROGRAMMING THROUGH STRENGTHENING INTERSECTIONAL PERSPECTIVES

6.1

Mary Thompson-Hall

Basque Centre for Climate Change, Bilbao, Spain

6.1.1 INTRODUCTION

The restoration of degraded agricultural ecosystems represents vital opportunities for both mitigation of and adaptation to global climate change (Mbow et al., 2014; Beddington et al., 2011; Lal, 2004; Albrecht and Kandji, 2003). Regarding mitigation, it is estimated that with restoration, improved use, and better management, global agricultural soils have the potential to offset global carbon emissions by 1.2–3.1 billion tons of carbon (C) per year (Lal, 2011, p. 1). Further, it is now accepted that enhancing degraded agricultural landscapes through integrating agroforestry with conventional cropping and livestock systems presents significant opportunities for increasing carbon sequestration through both enhancing soils and adding woody biomass, though the exact levels are still debated (Ramachandran Nair and Nair, 2014; Vermeulen et al., 2012). From an adaptation standpoint, studies show that increasing the soil carbon of degraded lands by 1 ton of C/ha/year could increase production of cereals and legumes by 32 ± 11 million tons/year and roots and tubers by 9 ± 2 million tons/year in developing countries (Lal, 2011, p. 6). The use of agroforestry as a land restoration strategy is also viewed as offering promising benefits for bolstering livelihoods and food security strategies among those vulnerable to climate change. This is achieved through increasing nutritional and market opportunities, therefore increasing their adaptive potential (Verchot et al., 2007).

At the heart of the success of land restoration initiatives are the uptake and maintenance by local land users. Therefore, designing and implementing programs that are relevant, affordable, and attainable for farmers working under varying site-specific conditions is essential. This type of programming necessitates increasingly nuanced understandings of the experiences of farmers that transcend simplified portrayals of land users that risk reducing difference into unrealistic categories that inhibit rather than support mitigation and adaptation efforts. In this section, consideration is given to how increasing levels of complexity are being incorporated into land restoration scholarship, with a focus on the role of gender analysis and climate change. Drawing on examples from broader agriculture, development, and climate change literature, opportunities are illuminated for deeper engagement with cutting-edge approaches for understanding multidimensional social contexts. We argue that such approaches are key to achieving the most tailored and effective initiatives for restoring agrarian landscapes, addressing negative impacts of climate change, and increasing the ability of vulnerable people to adapt to change.

6.1.2 INCORPORATING SOCIAL DIFFERENCE INTO LAND RESTORATION RESEARCH AND PROGRAMMING

The necessity to design and implement restoration programs that not only take into account the biophysical factors of restoring degraded agricultural lands but that also acknowledge site-specific socioeconomic contexts of project areas and that engage local communities is a notion that has been acknowledged and built upon for the past several decades (Dreber et al., 2014; Palmer and Bennett, 2013; Botha et al., 2008). Research from this perspective goes beyond one-sided narratives that frame land users solely as destructive agents; it provides alternative approaches that highlight the potential for farmers to be proactive participants in land restoration that benefits both the farmers and the biophysical environment (Garen et al., 2011). Often, these works explore how community members in agrarian settings participate in the design and implementation of restoration projects, and whether the community benefits, and if so, how. For example, Garen et al. (2011) examines contemporary and historical differences in community-level tree planting practices, needs, and preferences across a number of different communities in Panama. This level of analysis is helpful for portraying site-specific differences in land use practices and farmer tendencies, though it may not always sufficiently capture the potential opportunities for designing land restoration initiatives that bring the greatest benefit to the majority of people. Dreber et al. (2014), for instance, investigates the benefits and drawbacks of participatory approaches used for understanding decision-making factors of local livestock farmers regarding restoring and managing degraded drylands in southern Africa. Their research illustrates that helpful information can be gained from community-level participatory analysis but that certain social and economic differences, including farming practices, available funding, and awareness levels, can prove problematic for representative evaluations of the participatory processes involved and for the identification of best practices (Dreber et al., 2014).

One approach that land restoration researchers and practitioners have used to integrate heterogeneity of community members into their work is gender analysis. Their efforts augment an extensive body of scholarship and practice that have been taking place since the 1970s and 1980s, which explore gender-related topics within development, agriculture, and environmental conservation. This research, as evidenced within academic and gray literature, highlights important differences between the experiences of men and women farmers, including that in many agrarian communities in the Global South, women farmers often take on a disproportionate workload on the farm and within the household, while at the same time lacking equal levels of decision-making power and land tenure rights as men (Agrawal, 2003; Brody et al., 2008; Djoudi and Brockhaus, 2011; FAO, 2011; Quisumbing and Pandolfelli, 2008; Udry and Goldstein, 2008). Examples can be found within the land restoration literature where gender has been utilized as a relevant category of social difference that incorporates a greater degree of complexity into experiences of farmers. For example, Tefera et al. (2007) included differing opinions of men and women on the enclosures of degraded lands in Ethiopia in their research on the Grain for Green project in China's Shaanxi Province, Cao et al. (2009) discuss women's perceptions of negative livelihood impacts from the program, and Liu et al. (2011) point out potential ways that women could be disadvantaged by programs to convert marginal lands into the production of energy crops.

Apart from academia, such approaches are rapidly gaining traction within international agriculture, conservation, and development circles and are being circulated across a broader audience outside traditional literature sources by means of social media outlets. For instance, the International Union for the Conservation of Nature (ICUN) has initiated a blog series featuring scientific case studies on the topic of gender and restoration, including a study from Burkina Faso that discusses the benefits of utilizing shea trees for landscape restoration due to their multiple beneficial uses for women and agricultural lands (Aguilar, 2014; Elias, 2014). While gender and restoration work remains rare overall, to date, it has made important inroads in highlighting the need for the mainstreaming of gender analysis throughout all levels of restoration programming, from local participation in activities to more gender balance at top institutional planning levels (Gurung et al., 2006; Bossio et al., 2010).

6.1.3 CLIMATE CHANGE, GENDER, AND LAND RESTORATION

Today, concerns over negative implications of global climate change are pervasive across the scientific literature. For researchers working on vulnerability and adaptation of small farmers in agrarian settings, the occurrences and threats of negative impacts such as drought, floods, higher temperatures, and extreme weather events are spurring efforts to incorporate clearer understandings of the unique experiences of different social groups so that more effective adaptation strategies can be developed. Within this literature, gender is a central focal point upon which many understandings of difference are based. Much work has been devoted to illustrating the links between increased vulnerability of women and persistent inequality between men and women in the Global South regarding control over planting decisions on farms, marketing decision making, land use and ownership rights, educational opportunities, along with household and farm labor. Disparities between men's and women's participation and input in international, national, and regional climate agreements are also highlighted as problematic for effectively mainstreaming the concerns and needs of women in policy (Skinner, 2011; Boyd, 2002). Further, programming for decreasing the vulnerability and increasing the adaptive capacity of small farmers in many areas has also been criticized for relying on outdated notions that only men are farmers and therefore ignoring the activities of half the land users affected by the programs (see also Ahmed and Fajber, 2009; Buvenic, 1986; Saito and Weidemann, 1990). While much of the focus within the gender and climate change literature has tended to center on the victimization of women or the disproportionate hardships they face within agrarian settings compared with men, attention has also been paid to the unique roles that women play in agrarian landscapes. In particular, focus on their abilities to actively change their behaviors and practices in ways that increase their adaptive capacity and that of their families and communities (Swai et al., 2012; Djoudi and Brockhaus, 2011; Babugura et al., 2010; Mitchell et al., 2007; Demetriades and Esplen, 2008).

Integration of the agrarian land restoration, gender, and climate change literature is sparse, but those works that are available show similar evidence of how women can be disproportionately affected by land restoration and management solutions. At the same time, women can also act as proactive agents engaging in meaningful restoration activities that increase the adaptive capacity of themselves and their communities (Kiptot and Franzel, 2012; Gurung, 2006). Examples from within programming literature are increasingly available; for instance, a paper for the International Fund for Agricultural Development (IFAD; Gurung, 2006), gives an in-depth look at the unique experiences that women in dryland areas face when dealing with desertification. It emphasizes the importance of acknowledging and learning from the traditional agricultural and land management knowledge of women, as well as strengthening collective groups and organizations of women. These works have affirmed the fundamental role that gender analysis plays in land restoration research and practice, as well as how it has enriched the understandings of agricultural landscapes in vital ways.

6.1.4 DRAWBACKS OF CONVENTIONAL BINARY GENDER ANALYSIS

Across the bodies of work on land restoration, gender, and climate change, gender analysis largely tends to rely on conventional blanket comparisons of the experiences of men versus women, with no deeper exploration of the nuanced differences within these groups. While these traditional comparisons have played a crucial role in bringing the interests and concerns of women farmers to the forefront of research and policy, an emergent literature is now questioning whether binary comparisons of men and women farmers are the most effective means of understanding social difference in agrarian settings. Scholars within this line of thinking are building on approaches that originate in critical feminist theories that adopt a more nuanced gender lens when analyzing the differential situations of small farmers (Goheen, 1991; Grigsby, 2004; Pearson and Jackson, 1998; Wangari, Thomas-Slayter, and Rocheleau, 1996; Bigombe Logo and Bikie, 2003; Carr, 2008; Lawson, 1995).

These approaches center on the idea that gender is a socially constructed aspect of identity that gains meaning through the confluence of a plethora of other social markers and that this confluence is dynamic and flexible. Such an idea is encompassed within the concept of intersectionality, defined by Davis (2008, p. 68) as "the interaction between gender, race, and other categories of difference in individual lives, social practices, institutional arrangements, and cultural ideologies and the outcomes of these interactions in terms of power." Intersectional approaches have gained traction within broader agriculture and development literatures for several decades but are now emerging with increasing frequency within bodies of work focusing specifically on small farmers, decision making, and livelihoods among other subjects (Carr, 2013, 2014). These more nuanced explorations of gender are still rare within the climate change adaptation literature (Carr and Thompson, 2014; Kaijser and Krosnell, 2014); however, a growing number of researchers are raising important questions regarding how legitimate the empirical links are within studies that use gender as a primary indicator of an individual or groups being more or less vulnerable (Arora-Jonsson, 2011; Harris, 2006).

For example, Arora-Jonsson (2011) explores how current discourses and available literature on the issue of gender and climate change most often portray women summarily as either "virtuous" or "vulnerable" in their relationships with their environment (i.e., either imbued with an inherent sense of environmentalism and stewardship that is lacking in men or victimized by the actions of men that leave them with little control within their changing environments). These portrayals and the underlying assumptions that accompany them, she explains, are obstructing a more realistic understanding of the complexities of vulnerabilities in agrarian settings because, according to Arora-Jonsson (2011, p. 750),

"The specificity of vulnerability may differ. A generalized belief in women's vulnerability silences contextual differences." Therefore, we see that it has become common for simplistic categorizations of women and men as less vulnerable and more vulnerable to take the place of more meaningful investigations of context-specific differences that might offer more realistic insights into how to design more effective strategies for addressing the impacts of climate change within agricultural communities.

Land restoration researchers are only beginning to scratch the surface of discovering potential opportunities for utilizing intersectional approaches in their work. There are increasing examples though of more detailed explorations of social difference across land users experiencing climate change impacts in degraded areas. The following section first introduces a conversation of what kinds of beneficial insights might be gained by applying ideas of intersectional gender and livelihoods analysis to the topic of land restoration and climate change and then highlights the few examples of research that are at the forefront of this type of research today.

6.1.5 EXPANDING INTERSECTIONAL GENDER ANALYSIS WITHIN LAND RESTORATION AND CLIMATE CHANGE RESEARCH

Degraded agrarian areas, though currently operating at low or negligible productivity levels, are situated within underlying sets of social contexts and histories associated with use and ownership (Foley et al., 2005; Bromley, 2009). These contexts feature complex webs of power relationships in which numerous social factors including gender, ethnicity, caste, class, age, marital status, and migratory status intersect in dynamic ways to produce unique experiences and realities for different land users. Within a given agrarian area, these intersections can result in higher levels of privilege and security for some users and in marginalization and increased vulnerability for others. When a group from within this web or from an outside linked context decides to undertake actions to rehabilitate or restore the degraded lands, working to understand the social landscape can be as important to the project's success as understanding the biophysical factors at play.

Regarding improved fallows for soil restoration in Zambia, Phiri et al. (2004) looked at the intersection of gender and household head status, as well as marital status, for the purpose of examining differences in the likelihood of land users to utilize improved fallows. Torri (2013) outlines the differential impacts of community-based conservation initiatives involving the restoration of degraded agricultural soils on different women within local farming communities in the Sariska region in Rajasthan, India.

Useful examples of these complex social relations can also be found within works regarding reforestation and agroforestry, where a number of authors have raised important points concerning the complexities involved with trees, tenure, and gender (Rocheleau and Edmunds, 1997; Sjaastad and Bromley, 1997). For instance, Unruh (2008) warns of potentially unforeseen outcomes of afforestation and reforestation projects in Africa associated with land restoration and climate change mitigation in the form of carbon sequestration. Highlighted are the complex intersections between tree planting and land tenure in some areas of the African continent and how different forms of socially constructed identity such as gender, owner or tenant, migrant, or other "outsider" status, combined with local customs and traditions, can dictate the options many people have for choosing to plant trees or not (Unruh, 2008,

p. 703). He emphasizes that providing trees to community members for planting as part of tree-planting programs without taking customary tenure norms into account can not only lead to a stagnant or failed restoration/sequestration project but also risk aggravating preexisting tensions and conflicts over ownership of contested lands (Unruh, 2008). Also relating to tree-planting programs, trees represent complexes of interconnected use rights for different people at different times in many situations—differences that may fall along gendered lines but that often intersect with other aspects of identity as well (Saunders et al., 2002; Fortmann, 1985).

6.1.6 CONCLUSIONS: LOOKING TOWARD INTEGRATING LAND RESTORATION, CLIMATE CHANGE, AND INTERSECTIONAL GENDER RESEARCH

For many organizations operating at the nexus of agriculture, climate change, and restoration of degraded land, gender has become an important theme. Among many others, this includes the Food and Agriculture Organization (FAO) of the United Nations and the research centers and programs of the Consultative Group on International Agricultural Research (CGIAR). Intersectional approaches are progressing rapidly within the programmatic strategies of these groups, as can be seen in the "Strategy for Research and Action" report of the CGIAR Research Program on Forests, Trees, and Agroforestry, which stresses that "gender-based disadvantages may not always be the most urgent in all settings and that substantial differentiation can exist among men and women and not only between them" (CIFOR, 2013, p. 1). Similarly, in "Practical Tips for Gender-Responsive Research" for Bioversity International, Elias (2013, p. 1) advises that "since 'women' and 'men' are not homogenous categories, we have to dig deeper and examine how gender is cross-cut by many other forms of social difference: wealth status, age, ethnicity, caste, and migrant or indigenous status, among others."

Coalescing and advancing knowledge of these interconnected spheres of agrarian land restoration, gender, and climate change are intricate tasks. The expansive body of work falling under the heading of climate-smart agriculture (CSA) is one domain that may hold potential as a space for innovative collaborations across the specialized disciplines involved. As defined by FAO, CSA "promotes production systems that sustainably increase productivity, resilience (adaptation), reduces/removes GHGs (mitigation), and enhances achievement of national food security and development goals" (FAO, 2015, p. 2). CSA approaches may encompass practical actions, such as reducing grazing pressures on degraded grasslands, soil restoration, improvement and better management of pastures, restoration of wetland areas drained for agricultural production, initiating agroforestry practices in degraded areas, as well as policy and institutional techniques like new finance mechanisms for large- and small-scale land restoration (McCarthy et al., 2011; FAO, 2013).

In a recent publication on gender and climate change adaptation actions in Africa relating to CSA approaches, Twyman et al. (2014, p. 10) acknowledge that "other social factors as race, class, ethnicity, religion, age, etc., also influence a person's position in society, as well as the power dynamics that these imply." In this example, as in certain of the programmatic publications mentioned in this section, although gender is recognized in principle as a multidimensional social construction that intersects with other social markers to produce unique and dynamic experiences of power, vulnerability, and adaptive capacity for different land users, the translation of these ideas into nonbinary gender analysis in rigorous, peer-reviewed case studies is exceedingly rare. Neufeldt et al. (2013) have criticized CSA as being too generalized to effect meaningful changes in social-ecological systems that will leave them better equipped to face impacts from global climate change. In their critique, the authors point to a lack of sufficient consideration within CSA programming for differences between the needs and incentives of different actors (i.e., land users and stakeholders) and social, political, and cultural dynamics (Neufeldt et al., 2013, p. 2). Supporting such a view is the persistent focus on dualistic comparisons of the experiences of men and women within and across most CSA and land restoration research. This persistence can partly be attributed to a scarcity of methodological approaches available for exploring intersectional social complexity in agrarian settings worldwide. However, these shortcomings also represent an opportunity for land restoration researchers to build on cutting-edge methodological and conceptual advances in other related areas of expertise (Carr, 2013, 2014; Bezner Kerr, 2014) in order to develop new strategies for restoration programs that are more tailored, flexible, and legitimate for farmers facing the negative impacts of climate change.

REFERENCES

Agrawal, B., 2003. Gender and land rights revisited: exploring new prospects via the state, family, and market. J. Agr. Change 3 (1–2), 184–224.

Aguilar, L., 2014. Introducing On Gender and Restoration: A case study series. IUCN Blog. http://www.iucn.org/about/work/programmes/forest/?17147/On-Gender-and-Restoration.

Ahmed, S., Fajber, E., 2009. Engendering adaptation to climate variability in Gujarat, India. Gender and Devel. 17 (1), 33–50.

Albrecht, A., Kandji, S.T., 2003. Carbon sequestration in tropical agroforestry systems. Agr. Ecosys. Environ. 99 (1), 15–27.

Arora-Jonsson, S., 2011. Virtue and vulnerability: Discourses on women, gender, and climate change. Global Environ. Change 21 (2), 744–751.

Babugura, A., Mtshali, N., Mtshali, M., 2010. Gender and Climate Change: South Africa Case Study. Accessed February 7, 2013, from, https://www.donorplatform.org/resources/63/file/2010_Heinrich-Boll-Stiftung_Gender-and-Climate-Change_South-Africa.pdf.

Beddington, J., et al., 2011. Achieving Food Security in the Face of Climate Change: Summary for policy makers from the Commission on Sustainable Agriculture and Climate Change. Publication for the Climate Change, Agriculture, and Food Security Program of the CGIAR. https://ccafs.cgiar.org/publications/achieving-food-security-face-climate-change-summary-policy-makers-commission#.VZqRthuqpBc.

Bezner Kerr, R., 2014. Lost and found crops: Agrobiodiversity, indigenous knowledge, and a feminist political ecology of sorghum and finger millet in northern Malawi. Annals Assoc. Amer. Geog. 104 (3), 577–593.

Bigombe Logo, P., Bikie, E.-H., 2003. Women and Land in Cameroon: Questioning Women's Land Status and Claims for Change. In: Wanyeki, L.M. (Ed.), Women and Land in Africa: Culture, Religion, and Realizing Women's Rights. Zed Books, London, pp. 31–66.

Bossio, D., Geheb, K., Critchley, W., 2010. Managing water by managing land: Addressing land degradation to improve water productivity and rural livelihoods. Agri. Water Manage. 97 (4), 536–542.

Botha, M.S., Carrick, P.J., Allsopp, N., 2008. Capturing lessons from land-users to aid the development of ecological restoration guidelines for lowland Namaqualand. Bio. Conserv. 141 (4), 885–895.

Boyd, E., 2002. The Noel Kempff project in Bolivia: Gender, power, and decision-making in climate mitigation. Gender and Develop. 10 (2), 70–77.

Brody, A., Demetriades, J., Esplen, E., 2008. Gender and Climate Change: Mapping the Linkages. Brighton, UK. http://siteresources.worldbank.org/EXTSOCIALDEVELOPMENT/Resources/DFID_Gender_Climate_ Change.pdf.

Bromley, D.W., 2009. Formalising property relations in the developing world: The wrong prescription for the wrong malady. Land Use Policy 26 (1), 20–27.

Buvenic, M., 1986. Projects for women in the third world: Explaining their misbehavior. World Develop. 14 (5), 653–664.

Cao, S., et al., 2009. Attitudes of farmers in China's northern Shaanxi Province towards the land-use changes required under the Grain for Green Project, and implications for the project's success. Land Use Policy 26 (4), 1182–1194.

Carr, E.R., 2008. Men's crops and women's crops: The importance of gender to the understanding of agricultural and development outcomes in Ghana's central region. World Develop. 36 (5), 900–915.

Carr, E.R., 2013. Livelihoods as intimate government: Reframing the logic of livelihoods for development. Third World Quarterly 34 (1), 77–108.

Carr, E.R., 2014. From description to explanation: Using the livelihoods as intimate government (LIG) approach. Appl. Geog. 52, 110–122.

Carr, E.R., Thompson, M.C., 2014. Gender and climate change adaptation in agrarian settings: Current thinking, new directions, and research frontiers. Geog. Comp. 8 (3), 182–197.

Center for International Forestry Research (CIFOR), 2013. Gender in the CGIAR Research Program on Forests, Trees and Agroforestry: A Strategy for Research and Action. CIFOR, Bogor, Indonesia.

Davis, K., 2008. Intersectionality as buzzword: A sociology of science perspective on what makes a feminist theory successful. Feminist Theory 9 (1), 67–85.

Demetriades, J., Esplen, E., 2008. The gender dimensions of poverty and climate change adaptation. IDS Bull. 39 (4), 24–31.

Djoudi, H., Brockhaus, M., 2011. Is adaptation to climate change gender neutral? Lessons from communities dependent on livestock and forests in northern Mali. Intl. Forest. Rev. 13 (2), 123–135.

Dreber, N., et al., 2014. Towards improved decision-making in degraded drylands of Southern Africa: An indicator-based assessment for integrated evaluation of restoration and management actions in the Kalahari Rangelands. Planet@Risk 2, 1.

Elias, M., 2013. Practical Tips for Conducting Gender-Responsive Data Collection. Bioversity Intl, Rome 1–2. http://www.bioversityinternational.org/index.php?id=244&tx_news_pi1%5Bnews% 5D=2575&cHash=ab962a09cf1c9b4097010bbcdf587fdb.

Elias, M., 2014. The Mighty Shea: How Women and Men Sculpt Landscapes—and why this Matters for Restoration. IUCN Blog. http://www.iucn.org/about/work/programmes/forest/?18248/The-Mighty-Shea-How-women-and-men-sculpt-landscapes—and-why-it-matters-for-restoration.

Foley, J.A., et al., 2005. Global consequences of land use. Science 309 (5734), 570–574.

Food and Agricultural Organization (FAO), 2011. The State of Food and Agriculture 2010–2011: Women in Agriculture Closing the Gender Gap for Development. Available at: Food and Agricultural Organization of the United Nations, Rome. http://www.fao.org/docrep/013/i2050e/i2050e00.htm. (accessed 07.02.13.).

Food and Agricultural Organization (FAO), 2013. FAO Success Stories on Climate-Smart Agriculture. Available at: Food and Agriculture Organization of the United Nations, Rome. http://www.fao.org/3/ a-i3817e.pdf.

Food and Agricultural Organization (FAO), 2015. FAO Climate Smart Agriculture Website. Available at: Food and Agriculture Organization of the United Nations, Rome. http://www.fao.org/climate-smart-agriculture/72610/en/.

Fortmann, L., 1985. The tree tenure factor in agroforestry with particular reference to Africa. Agroforest. Sys. 2 (4), 229–251.

Garen, E.J., et al., 2011. The tree planting and protecting culture of cattle ranchers and small-scale agriculturalists in rural Panama: Opportunities for reforestation and land restoration. Forest Ecol. Manage. 261 (10), 1684–1695.

Goheen, M., 1991. The Ideology and Political Economy of Gender: Women and Land in Nso, Cameroon. In: Gladwin, C.H. (Ed.), Structural Adjustment and African Women Farmers. University of Florida Press, Gainesville, FL, pp. 239–256.

Grigsby, W.J., 2004. The Gendered Nature of Subsistence and Its Effect on Customary Land Tenure. Society and Natural Resources 17 (3), 207–222.

Gurung, et al., 2006. Gender and Desertification—Expanding Roles for Women to Restore Dryland Areas. Publication for the International Fund for Agricultural Development (IFAD), Rome.

Harris, L.M., 2006. Irrigation, gender, and social geographies of the changing waterscapes of southeastern Anatolia. Environ. Plan. D—Soc. Space 24 (2), 187–213.

Kaijser, A., Kronsell, A., 2014. Climate change through the lens of intersectionality. Environ. Politics 23 (3), 417–433.

Kiptot, E., Franzel, S., 2012. Gender and agroforestry in Africa: A review of women's participation. Agroforestry Systems 84 (1), 35–58.

Lal, R., 2004. Soil carbon sequestration impacts on global climate change and food security. Science 304 (5677), 1623–1627.

Lal, R., 2011. Sequestering carbon in soils of agro-ecosystems. Food Policy 36, S33–S39.

Lawson, V., 1995. The politics of difference: examining the quantitative/qualitative dualism in post-structuralist feminist research. Prof. Geog. 47 (November 1994), 449–457.

Liu, T.T., et al., 2011. Strengths, weaknessness, opportunities, and threats analysis of bioenergy production on marginal land. Energy Proc. 5, 2378–2386.

Mbow, C., et al., 2014. Achieving mitigation and adaptation to climate change through sustainable agroforestry practices in Africa. Curr. Opin. Environ. Sustain. 6, 8–14.

McCarthy, N., Lipper, L., Branca, G., 2011. Climate-smart agriculture: Smallholder adoption and implications for climate change adaptation and mitigation. Mitig. Clim. Change Agri. Working Paper, 3.

Mitchell, T., Tanner, T., Lussier, K., 2007. We know what we need! South Asian Women Speak Out on Climate Change Adaptation. ActionAid International, London. http://www.actionaid.org/publications/we-know-what-we-need-south-asian-women-speak-out-climate-change-adaptation.

Neufeldt, H., et al., 2013. Beyond climate-smart agriculture: Toward safe operating spaces for global food systems. Agri. Food Sec. 2 (1), 12.

Palmer, A.R., Bennett, J.E., 2013. Degradation of communal rangelands in South Africa: Towards an improved understanding to inform policy. Afr. J. Range Forage Sci. 30 (1–2), 57–63.

Pearson, R., Jackson, C., 1998. Introduction: Interrogating Development: Feminism, Gender, and Policy. In: Jackson, C., Pearson, R. (Eds.), Feminist Visions of Development: Gender Analysis and Policy. Routledge, London, pp. 1–16.

Phiri, D., et al., 2004. Who is using the new technology? The association of wealth status and gender with the planting of improved tree fallows in Eastern Province. Zambia. Agri. Sys. 79 (2), 131–144.

Quisumbing, A., Pandolfelli, L., 2008. Promising Approach to Address the Needs of Poor Female Farmers. IFPRI Note. 13. International Food Policy Research Insitute (IFPRI).

Ramachandran Nair, P.K., Nair, V.D., 2014. "Solid–fluid–gas": The state of knowledge on carbon-sequestration potential of agroforestry systems in Africa. Curr. Opin. Environ. Sustain. 6, 22–27.

Rocheleau, D., Edmunds, D., 1997. Women, men, and trees: Gender, power, and property in forest and agrarian landscapes. World Develop. 25 (8), 1351–1371.

Saito, K.A., Weidemann, C.J., 1990. Agricultural extension for women farmers in Africa. Policy, Research, and External Affairs working papers; no. WPS 398. Washington, DC: World Bank. http://documents.worldbank.org/curated/en/1990/04/700289/agricultural-extension-women-farmers-africa.

Saunders, L.S., Hanbury-Tenison, R., Swingland, I.R., 2002. Social capital from carbon property: Creating equity for indigenous people. Philos. Trans. Royal Soc. London 360, 1763–1775.

Sjaastad, E., Bromley, D.W., 1997. Indigenous land rights in sub-Saharan Africa: Appropriation, security, and investment demand. World Develop. 25 (4), 549–562.

Skinner, E., 2011. Gender and Climate Change Overview Report. Institute of Development Studies. Brighton, UK, October 2011, pp. 91.

Swai, O., Mbwambo, J., Magayane, F., 2012. Gender and perception on climate change in Bahi and Kondoa districts, Dodoma region. Tanzania. J. Afr. Stud. Develop. 4 (9), 218–231.

Tefera, S., Snyman, H.A., Smit, G.N., 2007. Rangeland dynamics of southern Ethiopia: (2). Assessment of woody vegetation structure in relation to land use and distance from water in semiarid Borana rangelands. J. Environ. Manage. 85 (2), 443–452.

Torri, M.C., 2013. Power, structure, gender relations, and community-based conservation: The case study of the Sariska region, Rajasthan, India. J. Intl. Women's Stud. 11 (4), 1–18.

Twyman, J., et al., 2014. Adaptation Actions in Africa: Evidence that Gender Matters. CGIAR Research Program on Climate Change, Agriculture, and Food Security (CCAFS), Copenhagen. CCAFS Working Paper No. 83.

Udry, C., Goldstein, M., 2008. The profits of power: Land rights and agricultural investment in Ghana. J. Pol. Econ. 116, 981–1022.

Unruh, J.D., 2008. Carbon sequestration in Africa: The land tenure problem. Glob. Environ. Change 18 (4), 700–707.

Verchot, L.V., et al., 2007. Climate change: Linking adaptation and mitigation through agroforestry. Miti. Adapt. Strat. Global Change 12 (5), 901–918.

Vermeulen, S.J., Campbell, B.M., Ingram, J.S.I., 2012. Climate change and food systems. Ann. Rev. Environ. Res. 37 (1), 195.

Wangari, E., Thomas-Slayter, B., Rocheleau, D., 1996. Gendered Visions for Survival: Semi-Arid Regions in Kenya. In: Rocheleau, D., Thomas-Slayter, B., Wangari, E. (Eds.), Feminist Political Ecology: Global Issues and Local Experiences. Routledge, New York, pp. 127–154.

GENDER ROLES AND LAND USE PREFERENCES— IMPLICATIONS TO LANDSCAPE RESTORATION IN SOUTHEAST ASIA

6.2

Delia C. Catacutan[1], Grace B. Villamor[2]

World Agroforestry Centre (ICRAF-Vietnam), Hanoi, Vietnam[1]
Center for Development Research, University of Bonn, Bonn, Germany[2]

6.2.1 INTRODUCTION

The Global Environment Fund (GEF) defines land degradation as the process by which the land is no longer able to sustain its economic and natural ecological functions due to natural processes or human activity. Land degradation is caused by various processes and interactions between two interlocking and complex systems: the natural ecosystem and the human social system (Shrestha, 2011). In southeast Asia, more than 50% of the land has experienced dramatic transformation, mainly from forest to agriculture (Zhao et al., 2008). This phenomenon has resulted in land degradation, primarily due to deforestation and consequently soil erosion resulting from the region's high rainfall. The issue of land degradation is important to the region as the human population continues to increase along with rapid economic development.

Land restoration has been proposed to reverse degradation. However, to make land restoration efforts effective, social factors including gender appreciation of land uses must be considered (Villamor et al., 2014a; Ban et al., 2013; Bernard et al., 2014). We argue in this section that the pathway to land restoration is influenced by the way that land is appreciated by multiple direct stakeholders. The diversity of these stakes is an important consideration in efforts to reclaim or restore land. Through three case studies conducted in Indonesia, Vietnam, and the Philippines, we highlight the roles that women play in agriculture and decision making, as well as their choices over land use. We found that context-specific gender roles and an appreciation of land uses play a role in predicting future changes in land-use and land restoration efforts. Our case study findings are instructive for future land restoration efforts in southeast Asia.

Land Restoration. http://dx.doi.org/10.1016/B978-0-12-801231-4.00029-X

6.2.2 GENDER AND LAND MANAGEMENT NEXUS

The past few years have witnessed a dramatic increase in global attention to gender, its role in development, and inequalities that exist between men and women (Kiptot et al., 2014). Yet, there remains a paucity in tangible actions that lead to greater impacts on the ground. There are serious discrepancies in the gender and land management nexus. While women in Central and West Africa, for example, produce three-quarters of food crops, they have almost no rights to land. With regard to agroforestry, women are the principal holders of knowledge and managers of traditional home gardens; about 60% of the practitioners of innovative practices, such as domestication of indigenous fruit trees and production of dairy fodder, are also women (World Bank et al., 2009). In eastern Zambia, female-headed households are more likely to adopt improved fallows than male-headed households, regardless of household size, previous experience with natural fallows, age, and membership in farmer groups (Gladwin et al., 2001).

Among other factors, the discrepancy lies in land tenure and gender-biased decision-making roles. Some researchers argue that it is about culturally embedded asymmetries in power relations between men and women. In forested and agricultural landscapes of developing countries and subsistence economies, rural women depend critically on natural resources but tend to be excluded from governance and decision-making processes (Catacutan et al., 2015; Agarwal, 2010; Benjamin, 2010; Buffum et al., 2010; Colfer, 2005; German et al., 2010; Mwangi and Mai, 2011). As a result, their contributions to sustainable resource management are often limited by their insecure rights to forests and trees, which constrain their incentives for undertaking sustainable management practices and the range of actions that they can take with regard to forest management (Meinzen-Dick et al., 1997; Rocheleau and Edmunds, 1997; Yadama et al., 1997).

Furthermore, the difference in adoption of agricultural and natural resource management technologies between men and women in male-headed and female-headed households is often linked to their variations in accessing and controlling related resources. Women are often confined to the lower end of the value chain of agroforestry products (retailing), which limits their control over and returns from the productive process (Kiptot and Franzel, 2012). Women also have less access to extension information such as on new technologies and crop varieties, due to the differing education levels between men and women. In many cases, when women are given the chance to join extension activities, lower education and literacy levels limit their appreciation of the potential benefit of adopting innovations (Kiptot and Franzel, 2012).

In sum, disparities between men and women affect land-use and land management decisions. These may be in the form of gender-assigned roles, perceptions, or resource endowments. However, very few empirical studies on gender-specific land-use decision making are available (Kiptot and Franzel, 2012; Villamor et al., 2014c). In the next section, we present three case studies to demonstrate the need for further examination of gender-specific roles and land-use decision making, with respect to land restoration.

6.2.3 CASE STUDIES

We used three case studies to highlight the gender dimensions of land restoration (Figure 6.2.1). These cases were selected primarily because their geographical areas are experiencing dramatic land use transitions. They are also sites of research by the Consultative Group on International Agricultural Research (CGIAR) on gender dimensions of forest, trees, and agroforests (FTA). The case studies

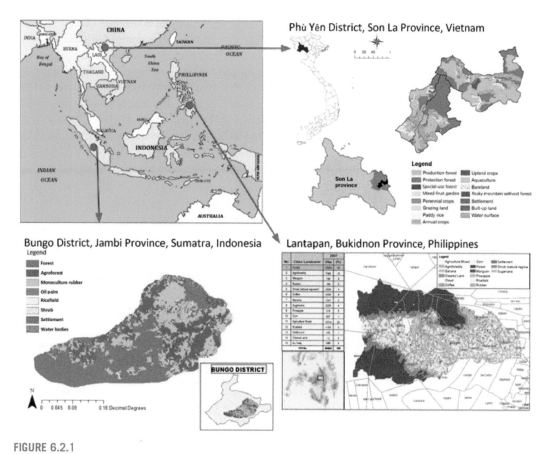

FIGURE 6.2.1

Location of case study sites in southeast Asia.

were guided by an overarching hypothesis that "appreciation of tree cover and its associated ecosystem services varies according to gender and ecological knowledge" (van Noordwijk and Villamor, 2014).

The first site is Bungo District, in Jambi Province, West Sumatra, Indonesia. The district is composed of three villages, spanning a total land area of 15,000 ha. The area is dominated by rubber agroforests, which are a primary source of livelihood in the region, while the province is the third-largest rubber producer in Indonesia. The second site is the municipality of Lantapan, Bukidnon Province, in the southern Philippines. Lantapan has a total land area of 35,465 ha and is partially located within the biodiversity-rich Mt. Kitanglad Range Natural Park. Due to its favorable climatic conditions, the municipality is dominated by large-scale banana, pineapple, and sugarcane plantations. The third site is Phù Yên, a rural district in the northwest region of Vietnam, which is located 135 km northeast of Son La Province (around 174 km of Hanoi). The total land area of the district is around 123,655 ha, of which 6496 ha is under agriculture and about 59,493 ha is forest.

A mixed-method approach was used across all case studies, including role-play games (RPGs), surveys, focus group discussions (FGDs), and land-use change analysis. However, in this text, we report only on the results of the land-use RPG for the Bungo case study and the household survey of gender roles and decision making for the Lantapan and Phù Yên case studies. The household surveys in Lantapan and Phù Yên involved a total of 605 respondents, of which almost half were women. In the RPG, the villagers were assigned the roles of farmers/villagers and outside agents [e.g., oil palm company, logging concession, nongovernmental organization (NGO) for biodiversity conservation, green rubber company, and local government watershed protection board] in three separate male-only and female-only configurations over six rounds or years per game, for the purpose of observing how men and women respond to outside agents who come to promote either the conversion or conservation of land (Villamor et al., 2014c).

6.2.3.1 WOMEN'S PREFERENCE OVER HIGHLY PROFITABLE LAND USES IN INDONESIA

In Bungo district, our findings indicated that women were more inclined to shift to highly profitable land uses such as oil palm plantation from traditional rubber agroforests and rice fields, contrary to men who preferred tree-based systems such as agroforests and forests (Villamor et al., 2014c). Furthermore, in the RPG played exclusively by women, we observed that they enthusiastically interacted with outside agents and accepted offers to shift from current land use to oil palm plantation. The women also accepted the offer of the logging company to log their forest patches for additional income while the men (in the men-only configuration) refused to accept the same offer (Figure 6.2.1).

During the game, both genders were exposed to various stressors, such as forest fires, population growth, and decreased price of rubber, to stimulate a change in their decision over the use of low-income-generating land (e.g., forests). However, the men's group did not change their decision and kept their forest patches, while the women's group did. In the end, the women's group accumulated higher income than the men's group (Figure 6.2.2). This contrasts with the situation in Africa, where the by-products of agroforestry systems with the highest economic value are often subjugated by men (Kiptot and Franzel, 2012).

Several reasons can be used to explain women's preference for oil palm plantations. First, the economic returns of oil palm production are enticing for women, who are often in charge of household budgeting—so the financial difficulty that is mostly felt by women may be connected to this behavior. Second, we found that women in this study area were actively involved in productive activities that exposed them to the cash-oriented local economy; in fact, many women are involved as wage workers in large oil palm plantations (Villamor et al., 2015). Third, an underlying factor influencing women's proactive behavior toward highly profitable land uses can be the matrilineal nature of the Minangkabau ethnic group (Villamor et al., 2015; Blackwood, 2008). Some communities in Sumatra, such as the Minangkabau and Jambi Melayu ethnic groups, practice a matrilineal kinship system, whereby land (particularly rice farms) is bequeathed by the mother to her daughters or nieces (Martial et al., 2012). Within this matrilineal system, women have stronger land rights and experience a relative absence of gender discrimination (Otsuka et al., 2001; Quisumbing and Otsuka, 2001). In this case, the matrilineal system could be the source of power for women to make decisions independently from their husbands. Moreover, it was found that the majority of the women in the study area have higher educational attainment than men, and thus, they have a clear understanding

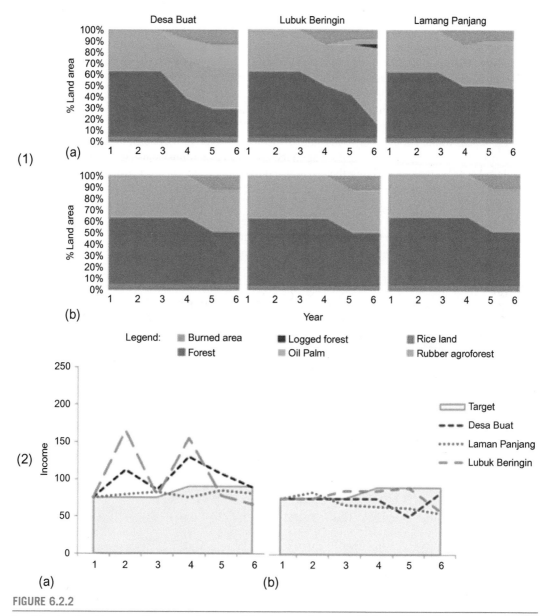

FIGURE 6.2.2

Land-use trend and income of women-only groups (a) and men-only groups (b) during RPGs (Villamor et al., 2014c).

of the long-term implications of their choices. However, there were concerns that women's positive behavior toward profitable land uses may run counter to land restoration efforts. Reclaiming degraded lands may impose restrictions on certain land uses or practices (e.g., oil palm plantations) that lowers yield and income and, hence, may simply not be wanted by women.

6.2.3.2 INCREASING FEMALE-HEADED HOUSEHOLDS IN THE PHILIPPINES

According to the Philippine Commission on Women,[1] as the number of men migrating into urban centers for seasonal or long-term employment increases, rural areas of the Philippines find a growing number of women-headed households. Although our study did not look at the impacts of outmigration, the high divorce rate in the study area contributed to this peculiarity. A total of 30% of surveyed households ($n = 302$) in the Lantapan municipality were female-headed, compared with 5% and 13% in the Indonesian and Vietnamese case studies, respectively.

In Lantapan, women household heads singlehandedly tend the farm, decide what to plant, market the produce, and support children's schooling in the absence of their husbands. This situation imposed multiple burdens on women, making them interested in land-use options that provide quick and high returns on investment, such as commercial vegetables. However, commercial vegetable production in Lantapan has already been pushed to the forest frontiers due to the growing number of large-scale, export-oriented agroindustrial companies (e.g., pineapple and banana) that took possession of prime lands in the municipality, and farming in the forest frontiers requires the women to work in a harsher environment. In addition, vegetable production often involved unsustainable practices that trigger soil erosion and the use of chemicals and pesticides that can be harmful to women's health. Moreover, our survey revealed that both men and women favor maize and high-value commercial vegetables but differ in their choice of perennial crops, particularly between coffee and banana (Figure 6.2.3).

Conversion from maize and vegetables to large-scale banana plantations began in the late 1990s, when multinational companies began to rent land from smallholder farmers on a 25-year contract. On one hand, women prefer to grow coffee over renting land to multinational banana plantation companies. Men, on the other hand, prefer converting to banana plantations, perhaps to initiate the possibility of employment at the plantation companies. In fact, we found that many male farmers had already left farming to work in banana plantations, contributing to a shift in the role of women and spurring changes in the local economy. The survey suggested that decisions on land use were based on profitability (40%), marketability of produce (22%), land suitability (12%), and household consumption (8%).

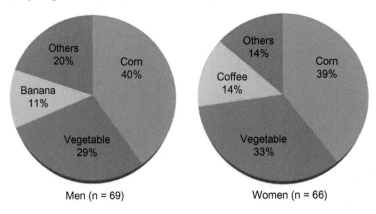

FIGURE 6.2.3

Preferred land-use options by men and women who agreed to change their current land uses ($N = 302$; Source: 2013 household survey).

---

[1]Based on statistics about 2009 accessed on January 12, 2015, from http://www.pcw.gov.ph.

As female-headed households increase and women are more involved in household decision making, it becomes necessary to adapt extension services and land restoration initiatives to accommodate the changing social/gender structure. Initiatives should also be mindful of women's perspectives regarding land-use approaches that maximize output and income.

6.2.3.3 SHARED ROLES AND JOINT DECISION MAKING IN VIETNAM

The situation in Vietnam is somewhat different from that in the Philippines, as decisions were shared between men and women. Using the "gender roles framework" (Razavi and Miller, 1995), we explored the question of who does what for all productive, reproductive, and community roles and tasks. Table 6.2.1 shows that almost half the roles and activities identified were jointly performed by men and women. Productive roles, such as planting trees and crops and harvesting, were undertaken jointly. Similarly, reproductive roles such as taking care of the children, doing the laundry, and cleaning the house were also done jointly, unlike the Minangkabau tribe in Bungo District, where all reproductive roles were performed mainly by women. In Vietnam, men and women apparently share the burden of managing farms and households and presumably share the benefits too. It is thus important to involve both genders in planning for land restoration interventions; otherwise, initiatives that undermine either men or women may distort the gender norm that is already observed in Phù Yên District. Similar to the Lantapan case, the share of female-headed households in Phù Yên, which is 13% of the sample, could not be ignored since they also perform the roles formerly played by their spouses. This calls for land restoration interventions targeting female-headed households. To restore degraded lands, strategies must align to women's specific needs and limitations.

6.2.4 GENDER IMPLICATIONS WITH LAND RESTORATION

What can future land restoration efforts learn from our case studies? The context in the three case studies varies, but elsewhere, men and women may be confronted with similar challenges when pursuing land restoration efforts. As land is central to the economic activities of men and women, it is a key determinant to the sustainability of livelihoods in rural southeast Asia. It is thus important to highlight the gender dimensions of land restoration. Finally, our case studies highlight three important points for consideration by any land restoration effort in the region and elsewhere.

6.2.4.1 LAND RESTORATION EFFORTS MAY BE UNWANTED BY WOMEN WHO SEEK GREENER PASTURES BY ENGAGING IN HIGHLY PROFITABLE LAND USES THAT MAY NOT ALIGN WITH RESTORATION GOALS

In the Indonesian case, traditional gender stereotypes whereby women are expected to focus only on subsistence crop production have been demystified. This is linked not only to cultural factors but also to global developments in the agricultural sector leading to changes in the choice of crops. Within our broader study of the gender dimensions of FTA in Bungo District, Villamor et al. (2014b) argued that gender-specific factors may contribute explicitly or implicitly to the pattern of multifunctionality or provisioning of goods and ecosystem services at the landscape level. They also concluded that

Table 6.2.1 Perspectives of men and women in terms of joint roles (N = 303), 2013 survey.

Activity	Men said... (%)		Women said... (%)		Fisher exact test (*p*)
	Joint	Individual	Joint	Individual	
Productive					
1. Planting crops	94	6	86	14	0.089
2. Planting trees	98	2	88	12	**0.017**
3. Feeding animals	61	39	64	36	0.392
4. Weeding the field	86	14	81	19	0.296
5. Fertilizing	82	18	80	20	0.468
6. Spraying	80	20	75	25	0.305
7. Watering/Irrigating	34	66	23	77	**0.040**
8. Producing seedlings	17	83	13	86	0.284
9. Pruning trees	72	28	69	31	0.284
10. Harvesting crops	10	90	16	84	0.117
11. Harvesting trees	96	4	93	7	0.315
12. Hauling crops	63	37	60	40	0.421
13. Marketing/selling crops	27	73	36	64	**0.092**
14. Farm budgeting	50	50	48	52	0.424
15. Buying farm inputs	20	80	24	76	0.255
16. Keeping farm records	22	78	18	82	0.284
Reproductive					
1. Collecting firewood	1	99	1	99	0.686
2. Fetching water	78	22	82	18	0.411
3. Preparing meals	7	93	6	94	0.535
4. Taking care of children	89	11	88	12	0.556
5. Washing clothes	98	2	92	8	**0.048**
6. Cleaning house	97	3	95	5.4	0.579
7. Attending children's needs	84	16	91	9	0.236
8. Household budgeting	80	20	70	30	**0.048**
Community					
1. Attending community meetings	2	98	1	99	0.311
2. Village planting	50	50	42	58	0.145
3. Church/School meetings	22	78	26	74	0.318
4. Building schools/bridges etc	40	60	39	61	0.491
5. Cleaning public places	32	68	39	61	0.162

Note: Numbers in bold text represent significant results.

involving the women in Bungo District in reducing emissions from degradation and deforestation (REDD) programs may be counterproductive, as they may simply change their land use when the reward for changing is high. As land restoration relates to enhancing the multifunctionality of landscapes, therefore, women's perspectives and land-use choices cannot be taken for granted.

6.2.4.2 GENDER-SPECIFIC ROLES AND PREFERENCES ARE NOT STATIC

The Lantapan case study also showed the importance of considering gender-specific contexts and women's choices related to land restoration efforts. Although the context in Lantapan is totally different from Bungo District in Indonesia, the eminent roles of women in productive activities have similarly indicated the importance of understanding their changing position in the economic sphere and how these changes are likely to influence future land investment decisions. Women have emerged from their traditional reproductive roles and are now engaged in the market economy. While this can be seen as an opportunity for self-empowerment for women, this may also have created household disarray and a gender rebalance in decision making. The economic and social structure of Lantapan is indicated by the fact that more women are becoming heads of households due to male outmigration, more livelihood options, or changes in marital status.

The Vietnamese situation is somewhat different from those in Indonesia and the Philippines, but it confirms the evidence that joint gender decision making leads to better economic and environmental outcomes (Villamor et al., 2014b). However, further study is needed to better understand the enabling conditions that maintain this type of decision making.

6.2.4.3 BUILDING ON GENDER-SPECIFIC CONTEXTS TO DESIGN GENDER-RESPONSIVE AND GENDER-FOCUSED LAND RESTORATION APPROACHES

Regardless of context-specific differences, the argument running through the case studies is that knowledge about changing gender roles, decision making, and land use preferences can be better used to inform future investments in land. With increased roles in productive tasks and decision making and integration into the mainstream economy, women can become agents of both land degradation and restoration. Understanding gender-specific differences over land use and land management practices can and should be the entry point, followed by nuanced design and implementation of gender-responsive and gender-focused land restoration approaches and investments. Undermining gender-specific appreciation of land uses may lead to unsuccessful land restoration efforts in southeast Asia and elsewhere.

REFERENCES

Agarwal, B., 2010. Does women's proportional strength affect their participation? Governing local forests in South Asia. World Devel. 38 (1), 98–112.

Ban, N.C., et al., 2013. A social-ecological approach to conservation planning: Embedding social considerations. Front. Ecol. Environ. 11 (4), 194–202.

Benjamin, A.E., 2010. Women in community forestry organizations: An empirical study in Thailand. Scand. J. Forest Res. 25, 62–68.

Bernard, F., et al., 2014. Social actors and unsustainability of agriculture. Curr. Opin. Environ. Sustain. 6, 155–161.

Blackwood, E., 2008. Not Your Average Housewife: Minangkabau Women Rice Farmers in West Sumatra. In: Ford, M., Parker, L. (Eds.), Women and Work in Indonesia. Taylor & Francis, Oxon, UK, pp. 17–40.

Buffum, B., Lawrence, A., Temphel, K.J., 2010. Equity in community forests in Bhutan. Intl. Forest. Rev. 12 (3), 187–199.

Catacutan, D., Naz, F., Nguyen, H.T., 2015. The Gender Dimensions of Agroforestry Adoption in Northwest Vietnam. International Forestry Review (in press).

Colfer, C.J.P., 2005. The equitable forest: Diversity and community in sustainable resource management. Resources for the Future and CIFOR Publication, Washington, DC.

German, L.A., Karsenty, A., Tiani, A.M. (Eds.), 2010. Governing Africa's Forests in a Globalized World. Earthscan Forest Library, Earthscan and CIFOR, London.

Gladwin, C., et al., 2001. Addressing food security in Africa via multiple strategies of women farmers. Food Policy 26, 177–207.

Kiptot, E., Franzel, S., 2012. Gender and Agroforestry in Africa: Who benefits? The African Perspective. In: Nair, P.K.R., Garrity, D. (Eds.), Agroforestry—The Future of Global Land Use. Springer, Dordrecht, the Netherlands, pp. 463–497.

Kiptot, E., Franzel, S., Degrande, A., 2014. Gender, agroforestry and food security in Africa. Curr. Opin. Environ. Sustain. 6, 104–109.

Martial, T., Helmi, N.E., Martius, E., 2012. Land and tree tenure rights on agroforestry (parak) system at communal land in West Sumatra, Indonesia. J. Agri. Ext. Rural Develop. 4 (19), 486–494.

Meinzen-Dick, R., et al., 1997. Gender, property rights and natural resources. World Devel. 25 (8), 1305–1315.

Mwangi, E., Mai, Y.H., 2011. Introduction to the special issue on forests and gender. Intl. Forest. Rev. 13 (2), 119–122.

Otsuka, K., et al., 2001. Evolution of land tenure institutions and development of agroforestry: Evidence from customary land areas of Sumatra. Agri. Econ. 25 (1), 85–101. http://dx.doi.org/10.1016/S0169-5150(00)00098-0.

Quisumbing, A.R., Otsuka, K., 2001. Land inheritance and schooling in matrilineal societies: Evidence from Sumatra. World Devel. 29 (12), 2093–2110.

Razavi, S., Miller, C., 1995. From WID to GAD: Conceptual Shifts in the Women and Development Discourse. United Nations Research Institute for Social Development, Geneva, Switzerland.

Rocheleau, D., Edmunds, D., 1997. Women, men, and trees: Gender, power, and property in forest and agrarian landscapes. World Devel. 25 (8), 1351–1371.

Shrestha, R.P., 2011. Land Degradation and Biodiversity Loss in Southeast Asia. In: Trisurat, Y., Shrestha, R.P., Alkemade, R. (Eds.), Land Use, Climate Change and Biodiversity Modeling: Perspective and Applications. IGI Global, Hershey. http://dx.doi.org/10.4018/978-1-60960-619-0.ch015.

van Noordwijk, M., Villamor, G.B., 2014. Tree cover transition in tropical landscapes: Hypotheses and cross-continental synthesis. GLPnews 10, 33–37.

Villamor, G.B., et al., 2014a. Assessing stakeholders' perceptions and values towards socio-ecological systems using participatory methods. Eco. Proc. 3 (1), 1–12.

Villamor, G.B., et al., 2014b. Gender differences in land-use decisions: Shaping multifunctional landscapes? Curr. Opin. Environ. Sustain. 6, 128–133.

Villamor, G.B., et al., 2014c. Gender influences decisions to change land use practices in the tropical forest margins of Jambi, Indonesia. Mitig. Adapt. Strat. Glob. Change 19, 733–755. http://dx.doi.org/10.1007/s11027-013-9478-7.

Villamor, G.B., Akiefnawati, R., van Noordwijk, M., Desrianti, F., Pradhan, U., 2015. Land-use change and shifts in gender roles in central Sumatra, Indonesia. International Forestry Review 17 (1), 1–15.

World Bank, Food and Agricultural Organization (FAO), International Fund for Agricultural Development (IFAD), 2009. Gender in Agricultural: Sourcebook. Agroforestry Landscape: Gendered Space, Knowledge, and Practice. Module 15. Washington, DC.

Yadama, G.N., Pragada, B.R., Pragada, R.R., 1997. Forest Dependent Survival Strategies of Tribal Women: Implications for Joint Forest Management in Andhra Pradesh, India. RAP Publication 1997/24. Regional Office for Asia and the Pacific. FAO, Washington, DC.

Zhao, S., et al., 2008. The theoretical model for quantifying the value of ecosystem services. J. Biotech. 136 (Supplement 1), S767.

COMMUNITIES, RESTORATION, RESILIENCE

DROUGHT-MANAGEMENT POLICIES AND PREPAREDNESS PLANS: CHANGING THE PARADIGM FROM CRISIS TO RISK MANAGEMENT

7.1

Donald A. Wilhite

University of Nebraska, Lincoln, NE, USA

7.1.1 INTRODUCTION

The implementation of a drought policy based on the philosophy of risk reduction can alter a nation's approach to drought-management by reducing the associated impacts (risk). This concept helped motivate the World Meteorological Organization's (WMO) Congress, at its sixteenth session held in Geneva in 2011, to recommend the organization of a High-level Meeting on National Drought Policy (HMNDP). Accordingly, the WMO, the Secretariat of the United Nations Convention to Combat Desertification (UNCCD), and the Food and Agriculture Organization of the United Nations (FAO), in collaboration with a number of UN agencies, international and regional organizations, and key national agencies, organized and held the HMNDP in Geneva from March 11 to 15, 2013. The theme of the HMNDP was "Reducing Societal Vulnerability—Helping Society (Communities and Sectors)."

Concerns about the spiraling impacts of drought on a growing number of sectors, the current and projected increase in drought frequency and severity, and the outcomes and recommendations from the HMNDP are drawing increased attention from governments, international and regional organizations, and nongovernmental organizations (NGOs). These impacts, regardless of the setting, can only be partially attributed to deficient or erratic precipitation. Drought is a complex natural hazard, and the impacts associated with it are the result of numerous climatic factors and a wide range of societal factors that together define the level of societal resilience. Population growth, redistribution, and changing consumption and production patterns are some of the factors that define the vulnerability of a region, economic sector, or population group. Other factors include poverty and rural vulnerability; increasing water demand due to urbanization; poor soil- and water-management practices; climate variability and change; changes in land use; environmental degradation; and greater awareness of the need to preserve the integrity of ecosystems. Although the development of drought policies and preparedness plans can be a challenging undertaking, the outcome of this process can significantly increase societal resilience to these events.

Land Restoration. http://dx.doi.org/10.1016/B978-0-12-801231-4.00007-0

7.1.2 NATIONAL DROUGHT POLICY: BACKGROUND

Simply stated, a national drought policy should establish a clear set of principles or operating guidelines to govern the management of drought and its impacts. The overriding principle of drought policy should be an emphasis on risk management through the application of preparedness and mitigation measures (HMNDP, 2013). This policy should be directed toward reducing risk by developing a better awareness and understanding of the drought hazard and the underlying causes of societal vulnerability to such events. It should also develop a greater understanding of how a proactive approach and adopting a wide range of preparedness measures can increase societal resilience. Risk management can be promoted by encouraging the improvement and application of seasonal and shorter-term forecasts; developing integrated monitoring and drought early-warning systems and associated information delivery systems; developing preparedness plans at various levels of government; adopting mitigation actions and programs; creating a safety net of emergency-response programs that ensure timely and targeted relief; and providing an organizational structure that enhances coordination within and between levels of government and with stakeholders. The policy should be consistent and equitable for all regions, population groups, and economic sectors, and consistent with the goals of sustainable development.

The incidence of drought and our vulnerability to it has increased globally. With this increase, greater attention has been directed toward reducing drought-related risks by introducing plans to improve operational capabilities (i.e., climate- and water-supply monitoring, building institutional capacity) and mitigation measures that are aimed at reducing drought impacts. This change in emphasis is long overdue. Mitigating the effects of drought requires the use of all components of the cycle of disaster management (Figure 7.1.1) rather than only the crisis-management portion of this cycle. Typically, when drought occurs, governments and donors have responded with impact assessment, response, recovery, and reconstruction activities to return the region or locality to a predisaster state.

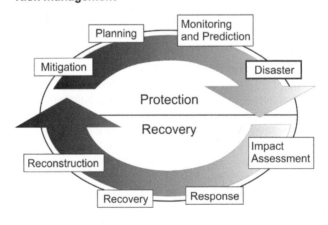

FIGURE 7.1.1

Cycle of disaster management.

Source: National Drought Mitigation Center.

Historically, little attention has been given to preparedness, mitigation, prediction/early-warning actions (i.e., risk management), and the development of risk-based national drought-management policies that could reduce impacts and lessen the need for government and donor interventions in the future. Crisis management only addresses the symptoms of drought, as they manifest themselves in the impacts that occur as a direct or indirect consequence of drought. Risk management, on the other hand, focuses on identifying where vulnerabilities exist (e.g., particular sectors, regions, communities, or population groups) and addresses these risks through systematically implementing mitigation and adaptation measures that will lessen the risk to future drought events. Countries have generally moved from one drought event to another with little, if any, reduction in risk because societies have consistently emphasized crisis management over risk management. In addition, in many drought-prone regions, another drought event is likely to occur before the region fully recovers from the previous event. If the frequency of drought increases in the future, as projected for many regions, there will be less recovery time between these events.

Progress on drought preparedness and policy development has been slow for a number of reasons, including the slow-onset characteristics of drought and the lack of a universal definition. Drought and climate change share the distinction of being creeping phenomena whose impacts occur slowly over a long period of time, which makes it difficult for people to recognize the implications. This difficulty of public recognition for drought makes early warning, impact assessment, and response difficult for scientists, natural resource managers, and policy makers. The lack of a universal definition often leads to confusion and inaction on the part of decision makers since scientists may disagree on the existence and severity of drought conditions (i.e., the onset and differences in recovery time among meteorological, agricultural, and hydrological drought). Severity is also difficult to characterize since it is best evaluated on the basis of multiple indicators and indices, rather than on the basis of a single variable. The impacts of drought are also largely nonstructural and spatially pervasive. These features make it difficult to assess the effects of drought and to respond in a timely and effective manner. Drought impacts are not as glaring as the impacts of other natural hazards, which makes it difficult for the media to communicate the significance of the event and its impacts to the public. Public sentiment in response is often lacking in comparison to other natural hazards that result in loss of life and property, particularly in more developed countries.

Associated with the crisis-management approach is the lack of recognition that drought is a normal part of the climate. Climate change, and associated projected changes in climate variability, will likely increase the frequency and severity of drought and other extreme climatic events. In the case of drought, the duration of these events may also increase. Therefore, it is imperative for all drought-prone nations to adopt a more risk-based approach to drought-management that would increase resilience to future droughts.

It is important to note that each occurrence of drought provides a window of opportunity to move toward a more proactive risk-management policy. Immediately following a severe drought episode, policy makers, resource managers, and all affected sectors are aware of the impacts that occurred and the deficiencies in the government's response. This is the appropriate time to approach policy makers with the concept of developing a national drought policy and preparedness plans in order to increase societal resilience. At the peak of the severe and widespread 2012 drought in the United States, Charles Fishman, author of *The Big Thirst: The Secret Life and Turbulent Future of Water*, wrote an op-ed contribution to *The New York Times* entitled "Don't Waste the Drought." Fishman (2012) correctly asserts: "The pain of drought, a slow-motion disaster, is very real. Drought can lead to paralysis and pessimism—or it can inspire us to fundamentally change how we use water. Water doesn't respond to wishful thinking. If it did, prayer services and rain dances would be all we need."

7.1.3 DROUGHT POLICY DEVELOPMENT: A TEMPLATE FOR ACTION

To provide guidance on the preparation of national drought policies and planning techniques, it is important to define the key components of drought policy, its objectives, and steps in the implementation process. An important component of national drought policy is increased attention to drought preparedness in order to build institutional capacity to deal more effectively with this pervasive natural hazard.

A constraint to drought preparedness has been the dearth of methodologies available to policy makers and planners to guide them through the planning process. Drought differs in its characteristics between climate regimes, and impacts are locally defined by unique economic, social, and environmental characteristics. A methodology developed by Wilhite (1991) and revised to incorporate greater emphasis on risk management (Wilhite et al., 2000; Wilhite et al., 2005) has provided a set of generic steps that can be adapted to any level of government (i.e., local, state or provincial, or national) or geographical setting to develop a drought-mitigation plan.

The Integrated Drought-Management Program (IDMP) recognizes the urgent need to provide nations with guidelines for the development of national drought-management policies.[1] To achieve this goal the drought-preparedness planning methodology, referred to previously, has been modified to define a generic process by which governments can develop a national drought policy along with drought-preparedness plans at various levels of government that support the principles of that policy. This process is described below to provide a template that governments or organizations can follow to reduce societal vulnerability to drought. The preferred approach is to create a national drought policy that is a stand-alone policy because droughts have much different characteristics than floods and other natural hazards. In some cases, however, countries may prefer to create a national drought policy as a subset of an existing policy such as a natural-disaster risk reduction, sustainable development, integrated water resources, or climate-change adaptation plan. These policies or strategies already exist in many countries.

7.1.3.1 DROUGHT POLICY: CHARACTERISTICS AND THE WAY FORWARD

As a starting point in the discussion of drought policy, it is important to identify the various types of drought policies that are available and have been utilized for drought-management. The most common approach followed by both developing and developed nations is postimpact government (or nongovernment) interventions. These interventions are normally relief measures in the form of emergency assistance programs that provide money or other specific types of assistance (e.g., livestock feed, water, food) to the victims (or those experiencing the most severe impacts) of the drought. This reactive approach, characterized by the hydro-illogical cycle (Figure 7.1.2), is seriously flawed from the perspective of vulnerability reduction since the recipients of this assistance are not expected to change behaviors or resource-management practices as a condition of the assistance. Although drought

---

[1]http://droughtinformation.org (November 2014).

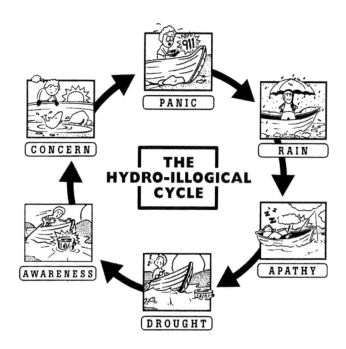

FIGURE 7.1.2

The hydro-illogical cycle depicts the crisis management or reactive approach to drought management normally used by governments.

Source: National Drought Mitigation Center.

assistance provided through emergency-response interventions may address a short-term need, it may, in the longer term, actually decrease the coping capacity of individuals and communities by fostering greater reliance on these interventions rather than increasing self-reliance. For example, livestock producers that do not maintain adequate on-farm storage of feed as a drought-management strategy will be those that first experience the impacts of extended precipitation shortfalls. These producers will be the first to turn to the government or other organizations for assistance in order to maintain herds until the drought is over and forage supplies return to adequate levels. This reliance on the government for relief is contrary to the philosophy of encouraging self-reliance through an investment by producers to improve their drought-coping capacity. Government assistance or incentives that encourage these investments, along with a change in how governments respond, would alter the expectations of livestock producers as to the role of government in these response efforts. The more traditional approach of providing relief is also flawed in terms of the timing of the assistance provided. It often takes weeks or months for assistance to be received, at times well beyond the window of when the relief would be of greatest value in addressing the impacts of drought. In addition, those livestock producers who employed appropriate risk-reduction techniques are usually ineligible for assistance because they mitigated the impacts that would have occurred and likely do not meet the eligibility requirements. This approach rewards those that have not adopted appropriate resource-management practices.

Although there is at times a need to provide emergency response to various sectors (i.e., postimpact assessment interventions), for the purpose of moving toward a more proactive risk-management approach, it is critical that the two drought-policy approaches described below become the cornerstone of the policy process.

The second type of drought-policy approach is the development and implementation of policies and preparedness plans. This would include organizational frameworks and operational arrangements developed in advance of drought and maintained between drought episodes by government or other entities. This approach represents an attempt to create greater institutional capacity focused on improved coordination and collaboration within and between levels of government, with stakeholders in the primary-impact sectors, and with the plethora of private organizations with a vested interest in drought-management (i.e., communities, natural resource or irrigation districts or managers, utilities, agribusiness, farm organizations, and others).

The third type of policy approach emphasizes the development of preimpact government programs or measures that are intended to reduce vulnerability and impacts. This approach could be considered a subset of the approach listed previously. In the natural-hazards field, these types of programs or measures are commonly referred to as mitigation measures. Mitigation in the context of natural hazards differs from mitigation in the context of climate change, where the term describes efforts to reduce greenhouse gas (GHG) emissions. Mitigation in the context of natural hazards, however, refers to actions taken in advance of drought to reduce impacts in the future. Drought-mitigation measures are numerous, but they may be more confusing to the general public in comparison to mitigation measures for earthquakes, floods, and other natural hazards where the impacts are largely structural. Impacts associated with drought are generally nonstructural, and the impacts are less visible, more difficult to assess in a timely fashion (e.g., reductions in crop yield), and do not require reconstruction as part of the recovery process. Drought mitigation measures would include establishing comprehensive early-warning and delivery systems; improved seasonal forecasts; increased emphasis on water conservation (demand reduction); increased or augmented water supplies through greater utilization of groundwater resources, water reutilization, and recycling; construction of reservoirs; interconnecting water supplies between neighboring communities; drought-preparedness planning to build greater institutional capacity; and awareness building and education. In some cases, such water-resource augmentation measures are best developed jointly with a neighboring state (or country), or at least such measures should be coordinated if they might have an impact on the other riparian state (or downstream use in general). Insurance programs, currently available in many countries, would also fall into this category of policy types.

7.1.3.1.1 Principal Elements of a Drought Risk-Reduction Policy Framework

Drought policy should emphasize four principle components during the development process: (1) risk and early warning, including vulnerability analysis, impact assessment, and communication; (2) mitigation and preparedness, including the application of effective and affordable practices; (3) awareness and education, including a well-informed public and a participatory process; and (4) good governance and an effective policy framework, including political commitment and responsibilities (UNISDR, 2009). Another important component of this framework is the inclusion of policy options for emergency response and relief. In all cases, when severe drought occurs, governments and other organizations must provide some form of emergency relief to those sectors most affected. It is crucial as a part of a drought risk-reduction policy, however, that this assistance to be provided in a form that does not run counter to the goals and objectives of the national drought policy, which would include a strong emphasis on the sustainability of the natural resource base.

7.1.4 NATIONAL DROUGHT-MANAGEMENT POLICY: A PROCESS

The challenge that nations face in the development of a risk-based national drought-management policy is complex. It requires political will at the highest level possible and a coordinated approach within and between levels of government and with the diversity of stakeholders that must be engaged in the policy development process. A national drought policy could be a stand-alone policy. Alternatively, as noted previously, it could contribute to or form a part of a national policy for disaster risk-reduction with holistic and multihazard approaches that focus on the principles of risk management.

The policy would provide a framework for shifting the paradigm: from one traditionally focused on a reactive crisis management to one that is focused on a proactive risk-based approach. The latter is intended to increase the coping capacity of the country, thereby creating greater resilience to future episodes of drought.

The formulation of a national drought policy, while providing the framework for a paradigm shift, is only the first step in vulnerability reduction. The development of a national drought policy must be intrinsically linked to the development and implementation of preparedness and mitigation plans at the provincial, state, and local levels. These plans will be instrumental in executing a national drought policy.

The process that is briefly outlined here is intended to provide a template or roadmap that countries can follow in the development of a national drought-management policy and drought preparedness or mitigation plans at the sub-national level. In other words, the process defined below is not intended to be prescriptive but rather to be adapted by countries based on their institutional infrastructure, legal framework, and so forth. This process has been modified from a 10-step drought planning process or methodology developed in the United States for application at the state level. Currently, 47 of the 50 U.S. states have developed drought plans, and the vast majority of these states have followed the guidelines provided by the 10-step process in their preparation or revision of these plans.[2] This drought-planning methodology has also been followed in other countries for the development of national drought strategies. The process, originally developed in the early 1990s, has been revised numerous times and places greater emphasis on mitigation planning with each revision. Now, this original methodology has been modified once again to reflect an emphasis on capacity building for a national drought-management policy, including the development of drought-preparedness plans that are necessary to support a national policy.

The steps listed here provide an outline of the process for policy and preparedness planning. The process is intended to be generic—i.e., applying this methodology in each country setting would require adapting it to the current institutional capacity, political infrastructure, legal system, and technical capacity. The reader is referred to a more complete description of this policy-development process published by the IDMP (2014).

The 10 steps in the drought policy and preparedness process are:

Step 1: *Appoint* a national drought-management policy commission.
Step 2: *State* or *define* the goals and objectives of a risk-based national drought-management policy.
Step 3: *Seek* stakeholder participation; *define* and *resolve* conflicts among key water-usage sectors, also considering transboundary implications.
Step 4: *Inventory* data and financial resources available and *identify* groups at risk.

---

[2]http://drought.unl.edu/Planning/PlanningInfobyState.aspx (November 2014).

Step 5: *Prepare/write* the key tenets of a national drought-management policy and preparedness plans, which would include the following elements:
- Monitoring, early warning, and prediction
- Risk and impact assessment
- Mitigation and response

Step 6: *Identify* research needs and *fill* institutional gaps.

Step 7: *Integrate* science and policy aspects of drought-management.

Step 8: *Publicize* the national drought-management policy and preparedness plans and *build* public awareness.

Step 9: *Develop* educational programs for all age and stakeholder groups.

Step 10: *Evaluate* and *revise* national drought-management policy and supporting preparedness plans.

Step 1: *Appoint* a national drought-management policy commission

The process for creating a national drought-management policy should begin with the establishment of a national commission to oversee and facilitate policy development. Given the complexities of drought as a hazard and the cross-cutting nature of managing all aspects (including monitoring, early warning, impact assessment, response, mitigation, planning) there is a critical need to coordinate the activities of many agencies/ministries of government at various levels and the private sector, including key stakeholder groups, and civil society. To ensure a coordinated process, the president/prime minister or other key political leader must take the lead in establishing a national drought-policy commission. Otherwise, it may not garner the full support and participation of all relevant parties.

The purpose of the commission is twofold. First, the commission will supervise and coordinate the policy development process. This includes gathering all of the necessary resources of the national government. By pooling the government's resources, this initial phase will likely require only minimal new resources coupled with a redirection of existing resources (e.g., financial, data, human) in support of the process. Second, once the policy is developed, the commission will become the authority responsible for the implementation of the policy at all levels of government. The principles of this policy will be the basis for the development and implementation of preparedness or mitigation-based plans at the sub-national level. In addition, the commission will be tasked with the activation of the various elements of the policy during times of drought. The commission will coordinate actions, implement mitigation and response programs or delegate this action to local or provincial/state government, and either initiate policy recommendations to the president (or other appropriate political leader and/or the appropriate legislature body) or implement specific recommendations within the authority of the commission and the ministries represented.

The membership of national drought commissions that have engaged in the policy development process in specific countries may provide useful insight regarding composition. An example comes from Mexico, where the president, Enrique Peña Nieto, announced a national drought program on January 10, 2013. The composition of this commission is provided in the previously cited IDMP report (2014). The goals of this program are early warning and early action to identify preventive actions that would lead to timely decisions to prevent and/or mitigate the effects of drought.

Step 2: *State* or *define* the goals and objectives of a risk-based national drought-management policy

Drought is a normal part of climate and there is considerable evidence and growing concern that the frequency, severity, and duration of droughts are increasing in many parts of the world—or will increase in the future as a result of anthropogenic climate change. The HMNDP was convened largely in response to this concern, as well as to the ineffectiveness of the traditional crisis-management approach or response to the occurrence of drought. The ultimate goal of HMNDP was to provide a forum and launch initiatives to create more drought-resilient societies. The IDMP is an outgrowth of the HMNDP.

The essential elements of a national drought-management policy, as identified through the HMNDP process, are:

- Developing proactive mitigation and planning measures, risk-management approaches, and public outreach and resource stewardship.
- Enhancing collaboration among national, regional, and global observation networks and developing information delivery systems that improve public understanding of, and preparedness for, drought.
- Creating comprehensive governmental- and private-insurance and financial strategies.
- Recognizing the need for a safety net of emergency relief based on sound stewardship of natural resources and self-help at diverse governance levels.
- Coordinating drought programs and response efforts in an effective, efficient, and customer-oriented manner.

Following the commission's formation, its first official action should be to establish specific and achievable goals for the national drought policy and a timeline for implementing the various aspects of the policy, as well as a timeline for achieving these goals. Several guiding principles should be considered as the commission formulates a strategy to move from crisis management to a drought risk-reduction approach. First, assistance measures, if employed, should not discourage agricultural producers, municipalities, and other sectors or groups from the adoption of appropriate and efficient management practices that help to alleviate the effects of drought (i.e., assistance measures should reinforce the goal of increasing resilience or coping capacity to drought events). Those assistance measures employed should help build self-reliance to future drought episodes. Second, assistance should be provided in an equitable (i.e., to those most affected), consistent, and predictable manner to all without regard to economic circumstances, sector, or geographic region. It is important to emphasize that the assistance provided is not counterproductive or a deterrent for self-reliance. Third, the protection of the natural and agricultural resource base is paramount, so any assistance or mitigation measures adopted must not run counter to the goals and objectives of the national drought policy and long-term sustainable development goals.

As the commission begins its work, it is important to inventory all emergency response and mitigation programs that are available through the various ministries at the national level. It is also important to assess the effectiveness of these programs and the past disbursement of funds through them. A similar exercise should be implemented at the state or provincial level in association with the development of drought preparedness and mitigation plans.

To provide guidance in the preparation of national drought policies and planning techniques, it is important to define the key components of drought policy, its objectives, and steps in the implementation process. Commission members, supporting experts, and stakeholders should consider many questions as they define the goals of the policy:

- What are the purpose and role of government in drought mitigation and response efforts?
- What is the scope of the policy?
- What are the country's most vulnerable economic and social sectors and regions?

- Historically, what have been the most notable impacts of drought?
- Historically, what has been the government's response to drought and what has been its level of effectiveness?
- What is the role of the policy in addressing and resolving conflict between water users and other vulnerable groups during periods of shortage?
- What current trends (e.g., climate, drought incidence, land and water use, population growth) may increase/decrease vulnerability and conflicts in the future?
- What resources (human and financial) is the government able to commit to the planning process?
- What other human and financial resources are available to the government (e.g., climate-change adaptation funds)?
- What are the legal and social implications of the plan at various jurisdictional levels, including those extending beyond the state borders?
- What principal environmental concerns are exacerbated by drought?

A generic statement of purpose for the drought policy and preparedness plans is to reduce the impacts of drought by identifying principal activities, groups, or regions most at risk and developing mitigation actions and programs that reduce these vulnerabilities. The policy should be directed at providing government with an effective and systematic means of assessing drought conditions, developing mitigation actions and programs to reduce risk in advance of drought, and developing response options that minimize economic stress, environmental losses, and social hardships during drought.

Step 3: *Seek* stakeholder participation; *define* and *resolve* conflicts among key water-use sectors, considering also transboundary implications

As noted in Step 1, a public participation specialist is an important contributor in the policy development process due to the complexities of drought as it intersects with society's social, economic, and environmental sectors, as well as the dependence of these sectors on access to adequate supplies of water to support an area's diverse livelihoods. As drought conditions intensify, competition for scarce water resources increases and conflicts often arise. These conflicts cannot be addressed during a crisis, and thus it is imperative for potential conflicts to be addressed during nondrought periods when tension among these groups is minimal. As a part of the policy-development process, it is essential to identify all citizen groups (i.e., stakeholders), including the private sector, that have a stake in the process and their interests. These groups must be involved early and continuously for fair representation to ensure an effective drought-policy development process at the national and local (provincial) levels. In the case of transboundary rivers, international obligations under agreements that the state is a party to should also be taken into account. Discussing concerns early in the process gives participants a chance to develop an understanding of one another's various viewpoints, needs, and concerns, and leads to collaborative solutions. Although the level of involvement of these groups will vary notably from country to country and even within countries, the power of public-interest groups in policy making is considerable across many settings. In fact, these groups are likely to impede progress in the policy-development process if they are excluded from the process. The commission should also protect the interests of stakeholders who may lack the financial resources to serve as their own advocates. One way to facilitate public participation is to establish a citizens' advisory council (as noted in Step 1) as a permanent feature of the commission's organizational structure in order to keep information flowing and address/resolve conflicts among stakeholders.

A national drought-policy development process must be multilevel and multidimensional in its approach. Thus, the goals of state or basin plans should mirror or reflect national policy goals. State or provincial governments need to consider if district or regional advisory councils should be established and what their composition might be. These councils could bring stakeholder groups together to discuss their water-use issues and problems and seek collaborative solutions in advance of the next drought.

Step 4: *Inventory* resources and *identify* groups at risk

An inventory of natural, biological, human, and financial resources—including the identification of constraints that may impede policy development—may need to be initiated by the commission. In many cases, much information already exists about natural and biological resources through various provincial and national agencies/ministries. It is important to determine the vulnerability of these resources to water shortages that result from drought. The most obvious *natural* resource of importance is water (i.e., location, accessibility, quantity, quality), but a clear understanding of other natural resources such as climate and soils is also important. *Biological/ecological resources* refer to the quantity and quality of grasslands/rangelands, forests, wildlife, wetlands, and so forth. *Human resources* include the labor needed to develop water resources, lay pipeline, haul water and livestock feed, process and respond to citizen complaints, provide technical assistance, provide counseling, and direct citizens to available services.

It is also imperative to identify constraints to the policy-development process and to the activation of the various elements of the policy and preparedness plans as drought conditions develop. These constraints may be institutional, financial, legal, or political. The costs associated with policy development must be weighed against the losses that will likely result if no plan is in place (i.e., the cost of inaction). As stated previously, the goal of a national drought policy is to reduce the risk associated with drought and its economic, social, and environmental impacts. Legal constraints can include water rights, existing public trust laws, requirements for public water suppliers, transboundary agreements (e.g., specifying that a certain volume or share of river flow across the border has to be guaranteed), and liability issues.

The transition from crisis to risk management is difficult because, historically, little has been done to understand and address the risks associated with drought. To solve this problem, areas of high risk should be identified, as should actions that can be taken before a drought occurs to reduce those risks. Risk is defined by both the exposure of a location to the drought hazard and the vulnerability of that location to periods of drought-induced water shortages (Blaikie et al., 1994). Drought is a natural event; it is important to define the exposure (i.e., frequency of drought of various intensities and durations) of various parts of the country, province, or watershed to the drought hazard. Some areas are likely to be more at risk than others because of greater exposure to the hazard, which inhibits or shortens the recovery time between successive droughts. As a result of current and projected changes in climate and the frequency or occurrence of extreme climatic events, including droughts, it is important to assess historical as well as projected future exposure to droughts. Vulnerability, on the other hand, is affected by social factors such as population growth and migration trends, urbanization, changes in land use, government policies, water-use trends, diversity of economic base, and cultural composition. The commission can address these issues early on in the policy-development process, but the more detailed work associated with this risk or vulnerability process will need to be directed to specific working groups at the state or provincial level as they embark on the process of drought-preparedness planning.

These groups will have more precise local knowledge and will be better able to garner input from local stakeholder groups.

Step 5: *Prepare/write* the key tenets of a national drought-management policy and preparedness plans

Drought-preparedness/mitigation plans are the instruments through which a national drought policy is carried out. It is essential for these plans to reflect the principles of the risk-based national drought policy that centers on the concept of risk reduction. What is defined below is the creation of institutional capacity that should be replicated within each state or province within a country, with formal communication and reporting to a national drought commission.

At the outset, it is important to point out that preparedness planning can take two forms. The first form, response planning, is directed toward the creation of a plan that is activated only during drought events and usually for the purpose of responding to impacts. This type of planning is reactive and the responses that are forthcoming, whether from national or state government or donor organizations, are intended to address specific impacts on sectors, population groups, and communities and, therefore, reflect the key areas of societal vulnerability. In essence, responding to impacts through emergency measures addresses only the symptoms of drought (impacts), and these responses are usually untimely, poorly coordinated, and, often, poorly targeted to those most affected. As noted earlier, this largely reactive approach actually leads to an increase in societal vulnerability since the recipients of drought relief or assistance programs become dependent on government and other programs because of the assistance provided to survive the crisis. This approach discourages the development of self-reliance and implementation of improved resource-management practices that will reduce risk in the longer term. Stated another way, why should the potential recipients of emergency assistance institute more proactive mitigation measures if government or others are likely to bail them out of a crisis situation? Emergency measures are appropriate in some cases, particularly with regard to providing humanitarian assistance, but they need to be used sparingly and be compatible with the longer-term goals of a national drought policy that focuses on improving resilience to future events.

The second form of preparedness planning is mitigation planning. With this approach, the vulnerabilities to drought are identified as part of the planning process through the analysis of both historical and more recent impacts of droughts. These impacts represent those sectors, regions, and population groups that are most at risk. The planning process can then focus on identifying actions and governmental or nongovernmental authorities that can assist in providing the necessary resources to reduce the vulnerability. In support of a risk-based national drought policy, mitigation planning is the best choice if risk reduction is the goal of the planning process. The discussion below shows how states/provinces might go about creating a plan that emphasizes mitigation.

Each state/provincial drought task force should identify the specific objectives that support the goals of the national drought policy. The objectives that should be considered include the following:

- Collect and analyze drought-related information in a timely and systematic manner.
- Establish criteria for declaring drought emergencies and triggering various mitigation and response activities.
- Provide an organizational structure and delivery system that ensures information flow between and within levels of government and to decision makers at all levels.
- Define the duties and responsibilities of all agencies in the event of drought.

- Maintain a current inventory of government programs used in assessing and responding to drought emergencies and in mitigating impacts in the longer term, if available.
- Identify drought-prone areas of the state and vulnerable economic sectors, individuals, or environments.
- Identify mitigation actions that can be taken to address vulnerabilities and reduce drought impacts.
- Provide a mechanism to ensure timely and accurate assessment of drought's impacts on agriculture, industry, municipalities, wildlife, tourism and recreation, health, and other sectors.
- Keep the public informed of current conditions and response actions by providing accurate and timely information to media in print and electronic form (e.g., via television, radio, and the Internet).
- Establish and pursue a strategy to remove obstacles to the equitable allocation of water during shortages, and establish requirements or provide incentives to encourage water conservation.
- Establish a set of procedures to continually evaluate and exercise the plan and periodically revise the plan so it will remain responsive to local needs and reinforce national drought policy.

The development of a drought-mitigation plan begins with the establishment of a series of committees to oversee development of institutional capacity necessary for the plan, as well as its implementation and application during times of drought when the various elements of the plan are in use. At the heart of the mitigation plan is the formation of a state- or provincial-level drought task force that mirrors to a large extent the makeup of the national drought commission (i.e., representatives from multiple agencies/ministries, key stakeholder groups). The organizational structure for the drought plan (Figure 7.1.3) reflects the three primary pillars of the plan: monitoring, early warning, and information

FIGURE 7.1.3

The organizational structure of the drought mitigation plan includes a monitoring committee and a risk assessment and mitigation committee. Each committee must work collaboratively and in support of the Drought Task Force.

Source: Wilhite et al. (2005).

delivery; risk and impact assessment; and mitigation, preparedness, and response. It is recommended that a committee be established to focus on the first two of these requirements; the drought task force can, in most instances, carry out the mitigation and response functions since these are heavily policy oriented.

These committees will have their own tasks and goals, but well-established communication and information flow between committees and the task force is necessary to ensure effective planning.

Monitoring, Early Warning, and Information-Delivery Committee. A reliable assessment of water availability—its outlook for the near and long term—is valuable in both dry and wet periods. During drought, the value of this information increases markedly. A monitoring committee should be a part of each state or provincial committee because it is important to interpret local conditions and impacts, and communicate this information to the national drought-policy commission and its representative from the national meteorological service. In some instances, a monitoring committee may be set up for certain regions with similar climatic conditions and exposure to drought, rather than for each state or province. However, the makeup of this committee should include representatives from all agencies with responsibilities for monitoring climate and water supply. It is recommended that data and information on each of the applicable indicators (e.g., precipitation, temperature, evapotranspiration, seasonal climate forecasts, soil moisture, streamflow, groundwater levels, reservoir and lake levels, and snowpack) be considered in the committee's evaluation of the water situation and outlook. The agencies responsible for collecting, analyzing, and disseminating data and information will vary considerably from country to country and province to province. Also, the data included in systematic assessments of water availability and future outlooks will need to be adjusted for each setting to include those variables of greatest importance for local drought monitoring.

The monitoring committee should meet regularly, especially in advance of the peak-demand season and/or the beginning of the rainy season(s). Following each meeting, reports should be prepared and disseminated to the provincial-level drought task force, the national drought-policy commission, and the media. The chairperson of the monitoring committee should be a permanent member of the provincial drought task force. In many countries, this person would be the representative from the national meteorological service. If conditions warrant, the task-force leadership should brief the provincial governor or appropriate government official about the contents of the report, including any recommendations for specific actions. A public information specialist should screen public dissemination of information to avoid confusing or conflicting reports on the current status of conditions.

The primary objectives of the monitoring committee are to:

- Adopt a workable definition of drought that could be used to phase in and phase out levels of state and national mitigation actions and emergency measures associated with drought conditions.
- Establish drought-management areas (i.e., subdivide the province or region into more conveniently sized districts by political boundaries, shared hydrological characteristics, climatological characteristics, or other means such as drought probability or risk). These subdivisions may be useful in drought-management since they may allow drought stages and mitigation and response options to be regionalized as the severity of drought changes over time.
- Develop a drought-monitoring system. The quality of meteorological and hydrological networks is highly variable from country to country and region to region within countries (e.g., number of stations, length of record, amount of missing data). Responsibility for collecting, analyzing, and disseminating data is divided among many government authorities. The monitoring committee's

challenge is to coordinate and integrate the analysis so decision makers and the public receive early warning of emerging drought conditions.

- Inventory data quantity and quality from current observation networks. Many networks monitor key elements of the hydrologic system. Most of these networks are operated by national or provincial agencies, but other networks may also exist and could provide critical information for a portion of a province or region. Meteorological data are important but represent only one part of a comprehensive monitoring system. These other physical indicators (soil moisture, streamflow, reservoir and groundwater levels, etc.) must be monitored to reflect impacts of drought on agriculture, households, industry, energy production, transportation, recreation and tourism, and other water-use sectors.
- It is also imperative to establish a network of observers to gather impact information from all of the key sectors affected by drought and to create an archive of this information. Both quantitative and qualitative information are important.
- Determine the data needs of primary users for information and decision-support tools. Developing new or modifying existing data-collection systems is most effective when the people who will be using the data are consulted early and often to determine their specific needs or preferences and the timing for critical decision points.
- Develop and/or modify current data and information-delivery systems. People need to be warned of drought as soon as it is detected, but often they are not. Information must reach people in time for them to use it to make decisions. In establishing information delivery channels, the monitoring committee needs to consider at what point and what kind of information people need to receive and in what form (i.e., maps, graphs, text). Knowledge of these decision points will make a difference as to whether the information provided is used or ignored.

Risk-Assessment Committee. Risk is the product of exposure to the drought hazard (i.e., probability of occurrence) and societal vulnerability, represented by a combination of economic, environmental, and social factors. Therefore, in order to reduce vulnerability to drought, it is essential to identify the most significant impacts and assess their underlying causes. Drought impacts span many sectors and normal divisions of government authority.

Membership of the risk-assessment committee should include representatives or technical experts from economic sectors, social groups, and ecosystems most at risk from drought. The committee's chairperson should be a member of the drought task force to ensure seamless reporting. Experience has demonstrated that the most effective approach to follow in determining vulnerability to and impacts of drought is to create a series of working groups under the aegis of the risk-assessment committee. The responsibility of the committee and working groups is to assess sectors, population groups, communities, and ecosystems most at risk and identify appropriate and reasonable mitigation measures to address these risks. Working groups would be composed of technical specialists representing those areas referred to previously. The chair of each working group, as a member of the risk-assessment committee, would report directly to the committee. Following this model, the responsibility of the risk-assessment committee is to direct the activities of each of the working groups. These working groups will then make recommendations to the drought task force on mitigation actions to consider adding to the mitigation plan. Mitigation actions are identified in advance and implemented in order to reduce the impacts of drought when it occurs. Some of these actions represent programs that are long term in nature while others may be actions that are activated when drought occurs. The activation of these measures at appropriate times is determined by the triggers (i.e., indicators and indices) identified by the monitoring committee in association with the risk-assessment committee in relation to the key impacts (i.e., vulnerabilities) associated with drought.

The number of working groups will vary considerably among provinces, reflecting the principal impact sectors of importance to various regions within a country and their respective vulnerabilities to drought because of differences in the exposure to drought (frequency and severity) and the most important economic, social, and environmental sectors. More complex economies and societies will require a larger number of working groups to reflect these sectors. It is common for the working groups to focus on some combination of the following sectors: agriculture, recreation and tourism, industry, commerce, drinking water supplies, energy, environment and ecosystem health, wildfire protection, and health.

To assist in the drought-mitigation planning process, a methodology is proposed to identify and rank (prioritize) drought impacts through an examination of the underlying environmental, economic, and social causes of these impacts, followed by the selection of actions that will address these underlying causes. What makes this methodology different and more helpful than previous methodologies is that it addresses the causes behind drought impacts. Previously, responses to drought have been reactive in nature and focused on addressing a specific impact, which is a symptom of the vulnerability that exists. Understanding why specific impacts occur provides the opportunity to lessen these impacts in the future by addressing these vulnerabilities through the identification and adoption of specific mitigation actions. A more complete description is included in the IDMP report (2014).

Mitigation and Response Committee. It is recommended that mitigation and response actions be under the purview of the drought task force. The task force, working in cooperation with the monitoring- and risk-assessment committees, has the knowledge and experience to understand drought-mitigation techniques, risk analysis (economic, environmental, and social aspects), and drought-related decision-making processes. The task force, as originally defined, is composed of senior policy makers from various government agencies and, possibly, key stakeholder groups. They are, therefore, in an excellent position to recommend and/or implement mitigation actions, request assistance through various national programs, or make policy recommendations to a legislative body or political leader.

As a part of the drought planning process, the national drought-policy commission should inventory all assistance programs available from national sources to mitigate or respond to drought events. Each provincial drought task force should review this inventory of programs available from governmental and nongovernmental authorities for completeness and provide feedback to the commission for the improvement of these programs to address short-term emergency situations as well as suggest long-term mitigation programs that may be useful in addressing risk reduction. In some cases, additional programs might be available from the provinces or states that have supplemented programs available at the national level. Assistance should be defined in a very broad way to include all forms of technical, mitigation, and relief programs available. As stated previously, the national drought commission should undertake a similar exercise with national programs and evaluate their effectiveness in responding to and mitigating the effects of previous droughts.

Writing the Preparedness/Mitigation Plan. With input from each of the committees and working groups and the assistance of professional writing specialists, the drought task force will draft the drought-mitigation plan. After completion of a working draft, it is recommended that public meetings or hearings be held at several locations to explain the purpose, scope, and operational characteristics of the plan and how it will function in relation to the objectives of the national drought policy. Discussion must also be presented on the specific mitigation actions and response measures recommended in the plan. A public information specialist for the drought task force can facilitate planning for the hearings, prepare media announcements for the meetings, and provide an overview of the plan.

After the draft plan has been vetted at the state, provincial, or basin level, it should be submitted to the national drought commission for review to determine if the plan meets the requirements mandated by the commission. Although each state-level plan will contain different elements and procedures, the basic structure should conform to policy standards provided to the states at the outset of the planning process by the national drought commission.

Step 6: *Identify* research needs and *fill* institutional gaps

The national drought-policy commission should identify specific research needs that would contribute to a better understanding of drought—its impacts, mitigation alternatives, and needed policy instruments—and would lead to a reduction of risk. These needs are likely to originate from the state-level drought task forces that are implemented to develop mitigation plans. It will be the task of the commission to collect these needs into a set of priorities for future action and funding.

Step 7: *Integrate* science and policy

An essential aspect of the policy and planning process is integrating the science and policy aspects of drought-management. The policy makers' understanding of the scientific issues and technical constraints involved in addressing problems associated with drought is often limited. Likewise, scientists and managers may have a poor understanding of existing policy constraints for responding to the impacts of drought. In many cases, communication and understanding between the science and policy communities must be enhanced if the planning process is to be successful. This is a critical step in the development of a national drought policy. Members of the National Drought-Policy Commission have a good understanding of the policy-development process, and the political and financial constraints, associated with proposed changes in public policy. They are also aware of the difficulties inherent in a change in the paradigm for the recipients of drought emergency assistance to a new approach focused on drought risk reduction. However, those persons at the state or community level who are embedded in the preparedness-planning process are less aware of these constraints but have an excellent understanding of drought-management actions, local conditions, and the key sectors affected and their operational needs. Linking the policy process with critical needs requires an excellent communication channel from state-based drought task forces and the commission.

In essence, this communication channel is necessary to distinguish what is feasible from what is desirable for a broad range of science and policy options. Integration of science and policy during the planning process will also be useful in setting research priorities and synthesizing current understanding. The drought task force should consider a wide range of options for drought risk reduction and evaluate the pros and cons of each in terms of their feasibility and potential outcomes.

Step 8: *Publicize* the drought policy and plans, *build* public awareness and consensus

If there has been good communication with the public throughout the process of establishing a drought policy and plan, there may already be an improved awareness of goals of the drought policy, the rationale for policy implementation, and the drought planning process by the time the policy is ready to be implemented. The public information specialists engaged in this process at the commission level and at the state level are vital in this regard. Throughout the policy and planning development process, it is imperative for local and national media to be used effectively in the dissemination of information about the process.

Step 9: *Develop* education programs

A broad-based education program focused on all age groups is necessary to raise awareness of the new strategy for drought-management, the importance of preparedness and risk reduction, short- and long-term water-supply issues, and other crucial prerequisites for public acceptance and implementation of drought policy and preparedness goals. This education program will help ensure that people know how to manage drought when it occurs and that drought preparedness will not lose ground during non-drought years. It would be useful to tailor information to the needs of specific groups (e.g., elementary and secondary education, small business, industry, water managers, agricultural producers, home-owners, utilities). The drought task force in each state or province and participating agencies should consider developing presentations and educational materials for events such as a water-awareness week, community observations of Earth Day and other events focused on environmental awareness, relevant trade shows, specialized workshops, and other gatherings that focus on natural-resource stewardship or management.

Step 10: *Evaluate* and *revise* drought policy and mitigation plans

The tenets of a national drought policy and each of the preparedness or mitigation plans that serve as the implementation instruments of the policy require periodic evaluation and revision. This is in order to incorporate new technologies, lessons learned from recent drought events, changes in vulnerability, and so forth. The final step in the policy development and preparedness process is to create a detailed set of procedures to ensure an adequate evaluation of the successes and failures of the policy and the preparedness plans at all levels. Oversight of the evaluation process would be provided by the National Drought Policy Commission, but the specific actions taken and outcomes exercised in the drought-affected states or provinces would require the active involvement of those specific drought task forces. The policy and preparedness process must be dynamic or the policies and plans will quickly become outdated. Periodic testing, evaluation, and updating of the drought policy are necessary to keep the plan responsive to the needs of the country, states, and key sectors. To maximize the effectiveness of the system, two modes of evaluation must be in place.

Ongoing Evaluation. An ongoing or operational evaluation keeps track of how societal changes such as new technology, new research, new laws, and changes in political leadership may affect drought risk and the operational aspects of the drought policy and supporting preparedness plans. The risk associated with drought in various sectors (economic, social, and environmental) should be evaluated frequently, while the overall drought policy and preparedness plans may be evaluated less often. An evaluation under simulated drought conditions (i.e., computer-based drought exercise) is recommended before the drought policy and state-level plans are implemented and periodically thereafter. It is important to remember that the drought policy and preparedness planning process is dynamic.

Another important aspect of the evaluation process, and the concept of drought exercises, is linked to changes in government personnel, which generally occurs frequently. If the goals and elements of the national drought policy are not reviewed periodically and the responsibilities of all agencies revisited, whether at the national or state level, governmental authorities will not be fully aware of their roles and responsibilities when drought recurs. Developing and maintaining institutional memory is an important aspect of the drought policy and preparedness process.

Postdrought Evaluation. A postdrought evaluation or audit documents and analyzes the assessment and response actions of government, NGOs, and others and provides for a mechanism to implement recommendations for improving the system. Without postdrought evaluations of both the drought policy and the preparedness plans at the local level, it is difficult to learn from past successes and mistakes, as institutional memory fades.

Postdrought evaluations should include an analysis of the climatic, social, and environmental aspects of the drought—i.e., its economic, social, and environmental consequences; the extent to which predrought planning was useful in mitigating impacts, in facilitating relief or assistance to stricken areas, and in postdrought recovery; and any other weaknesses or problems caused or not covered by the policy and the state-based plans. Attention must also be directed to situations in which drought-coping mechanisms worked and where societies exhibited resilience; evaluations should not focus only on those situations in which coping mechanisms failed. Evaluations of previous responses to severe drought are also a good planning aid. These evaluations establish a baseline for later comparisons so trends in resiliency can be documented.

To ensure an unbiased appraisal, governments may wish to place the responsibility for evaluating the effectiveness of the drought policy and each of the preparedness plans in the hands of NGOs such as universities and/or specialized research institutes.

7.1.5 SUMMARY AND CONCLUSION

For the most part, previous responses to drought in all parts of the world have been reactive, reflecting what is commonly referred to as the *crisis-management approach*. This approach has been ineffective (i.e., assistance poorly targeted to specific impacts or population groups), poorly coordinated, and untimely; more important, it has done little to reduce the risks associated with drought. In fact, the economic, social, and environmental impacts of drought have increased significantly in recent decades. A similar trend exists for all natural hazards.

The intent of the policy development and planning process described in this report is to provide a set of generic steps or guidelines that nations can use to develop the overarching principles of a national drought policy aimed at risk reduction through a national drought-policy commission. This policy would be implemented at the sub-national (i.e., provincial or state) level through the development and implementation of drought mitigation and preparedness plans that follow the framework or principles of the national drought policy. Following these guidelines, a nation can significantly change the way they prepare for and respond to drought by placing greater emphasis on proactively addressing the risks associated with drought through the adoption of appropriate mitigation actions. These guidelines are generic in order to enable governments to choose those steps and components that are most applicable to their situation. The risk-assessment methodology embedded in this process is designed to guide governments through the process of evaluating and prioritizing impacts and identifying mitigation actions and tools that can be used to reduce the impacts of future drought episodes. Both the policy-development process and the planning process must be viewed as an ongoing, continuous evaluation of the nation's changing exposure and vulnerabilities and how governments and stakeholders can work in partnership to lessen risk.

REFERENCES

Blaikie, P., Cannon, T., Davis, I., Wisner, B., 1994. At risk: Natural hazards, people's vulnerability, and disasters. Routledge Publishers, London.

Fishman, C., 2012. Don't waste the drought. The New York Times. August 16.

HMNDP. 2013. Final declaration from the high-level meeting on national drought policy. Available from: http://hmndp.org.

Integrated Drought Management Programme (IDMP), 2014. National drought management policy guidelines: A template for action. Prepared by D.A. Wilhite, Global Water Partnership and the World Meteorological Organization.

United Nations International Strategy for Disaster Reduction (UNISDR), 2009. Drought risk reduction framework and actions. Geneva, Switzerland.

Wilhite, D.A., 1991. Drought planning: A process for state government. Water Res. Bull. 27 (1), 29–38.

Wilhite, D.A., Hayes, M.J., Knutson, C., Smith, K.H., 2000. Planning for drought: Moving from crisis to risk management. J. Am. Water Resour. Assoc. 36, 697–710.

Wilhite, D.A., Hayes, M.J., Knutson, C.L., 2005. Drought preparedness planning: Building institutional capacity (Chapter 5). In: Wilhite, D.A. (Ed.), Drought and Water Crises: Science, Technology, and Management Issues. CRC Press, Boca Raton, FL, pp. 93–136.

NOT THE USUAL SUSPECTS: ENVIRONMENTAL IMPACTS OF MIGRATION IN GHANA'S FOREST-SAVANNA TRANSITION ZONE*

7.2

Kees van der Geest[1], Kees Burger[2], Augustine Yelfaanibe[3], Ton Dietz[4]

United Nations University Institute for Environment and Human Security, Bonn, Germany[1]
Development Economics Group, Wageningen University and Research Centre, Wageningen, The Netherlands[2]
Faculty of Integrated Development Studies (Wa Campus), University for Development Studies, Tamale, Ghana[3]
African Studies Centre, Leiden, The Netherlands[4]

7.2.1 INTRODUCTION

In 2008, the United Nations Environment Program (UNEP) released the *Atlas of Our Changing Environment: Africa*, which is a collection of satellite images and ground photographs that aims to expose major environmental changes on the African continent. The atlas contains two LANDSAT satellite images depicting large-scale land degradation in Ghana's forest-savanna transition zone (see Figure 7.2.1). On the first image,[1] captured in 1973, the area is densely vegetated with only small patches of more barren land in the north and around major settlements. The dark green areas in the image are forest reserves. The white dots in the south and midwest of the first image are clouds. In the second image,[2] taken in 2002 and 2003, much of the vegetation in the northern part has disappeared (UNEP, 2008, p. 185).

In the atlas, UNEP primarily blames farmers for environmental degradation: "about one-third of the land area [of Ghana] is threatened by desertification, caused mainly by slash-and-burn agriculture and over-cultivation of cleared land, resulting in widespread soil erosion and degradation." Other factors that contribute to land degradation, such as surface mining and logging, are also mentioned (UNEP, 2008, pp. 182–185), but the villains in UNEP's discourse are small-scale farmers who use unsustainable farming methods that cause land degradation.

---

*Fieldwork for this paper was conducted in the context of a PhD project at the University of Amsterdam, funded by the Dutch Council for Scientific Research (NWO).
[1]Landsat-1 MSS, November 25, 1973, bands 2, 4, and 1 (UNEP, 2008: 360).
[2]Landsat-7 ETM+, December 24, 2002, and February 19, 2003, bands 7, 4, and 1 (UNEP, 2008, p. 360).

Land Restoration. http://dx.doi.org/10.1016/B978-0-12-801231-4.00030-6

463

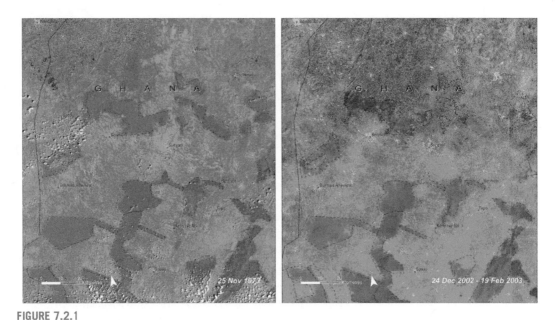

FIGURE 7.2.1

Land degradation in the Brong Ahafo Region (LANDSAT, 1973–2003) (UNEP, 2008, p. 185).

The area covered by the LANDSAT images is primarily located in Ghana's Brong Ahafo Region (see Figure 7.2.2). The northern part of the LANDSAT images, which experienced the most degradation, is a prime destination area for migrants from Ghana's Upper West Region, who belong to the Dagaba ethnic group. Most of the degraded area lies in the Wenchi District.[3] The vast majority of migrants from the Upper West Region who settle in the Brong Ahafo Region are small-scale farmers who migrate to rural areas in search of better agroecological conditions (Abdul-Korah, 2007; Van der Geest, 2011a, 2011b; Rademacher-Schultz et al., 2014). Their native home in the north of Ghana has only one rainy season, the soils are less fertile—especially in the more densely populated parts of the Upper West Region—and agroecological conditions have deteriorated over the past decades (Van der Geest, 2004, 2009; Amanor and Pabi, 2007; Dietz et al., 2013). This section investigates what role settler farmers from Northwest Ghana have played in the alleged "savannization" of the Northern forest-savanna transition zone.

Theoretically, the environmental impact of immigration can follow two lines. First, migration alters the population size in the areas of origin and destination. As Hugo (1996, p. 121) states, "Other things being equal (which of course they rarely are), emigration will reduce environmental pressures at the origin and increase them at the destination." Second, if other things are not equal, the impact of migrants on the natural environment differs from that of non-migrants. Hugo uses Ehrlich and Ehrlich's (1990, p. 58) IPAT equation—environmental impact (I) is a function of population size (P), affluence (A), and technology (T)—to indicate that the differential environmental impacts of migrants and

---

[3]In this section, we use the boundaries of the Wenchi District as it existed until 2004. In that year, the district was split into the Wenchi Municipal District in the West and the Tain District in the East.

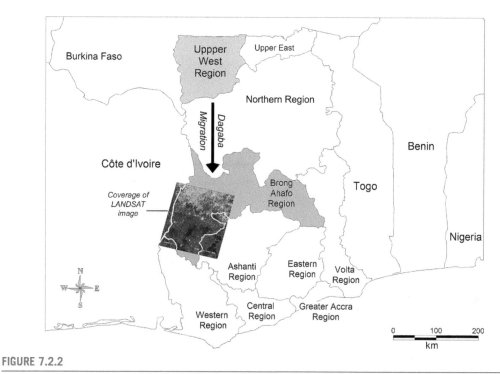

FIGURE 7.2.2

Ghana, Dagaba migration, and coverage of the LANDSAT image.

natives can be caused by differences in the productive and consumptive domains. This section focuses on the productive domain.

Migrant farmers are often thought to have less sustainable cultivation practices than native farmers. An important explanation is that they usually do not own the land they farm and often regard their stay on this land as temporary, which might reduce the incentive to apply environmentally sound farming methods. Moreover, they often have to rent land, which might encourage them to "mine the soil" (Afikorah-Danquah, 1997; Codjoe, 2006). Another common explanation is that migrant farmers do not have the same knowledge about local environmental conditions as native farmers, which can also lead to unsustainable farming practices (Lambin et al., 2001).

In an analysis of narrative structures in environmental discourses, Adger et al. (2001) distinguish three archetypal actors: heroes, villains, and victims. These archetypes are clearly discernible in three existing studies by Southern Ghanaian scholars about the differential impact of settlers' and natives' farming practices on vegetation cover (Adjei-Nsiah, 2006; Afikorah-Danquah, 1997; Codjoe, 2006). The authors of these three studies portray native farmers as the "heroes" who preserve soil fertility and vegetation cover and even convert savanna to forest. Meanwhile, settlers play the role of "villains" who cause deforestation and mine the soil. To complete the narrative of heroes, villains, and victims, Codjoe (2006, p. 103) argues that settlers have invaded areas "to the detriment of indigenous people," thereby assigning native farmers the victim role. Below are some characteristic quotes from these studies:

"... [migrants] tend to be more aggressive in their farming practices compared with indigenous populations mainly because of insecurity of tenure."

(Codjoe, 2006, p. 103)

"... practices which hasten the conversion of the forest to savannah are intentionally employed by the migrants ... so that the environment will resemble that of their home of origin...."

(Adjei-Nsiah, 2006, p. 58)

"[Native] landowners ... generally use a minimum tillage system of cultivation based on cutlass technology, and show a preference for long fallows.... such practices ... generally allow the regeneration of forest fallows...."

(Afikorah-Danquah, 1997, p. 42)

"... immigrants' practices ... can be associated with the savannisation of forest and fallow, reduction of tree cover in savannas and in some circumstances soil degradation."

(Afikorah-Danquah, 1997, p. 43)

These studies conclude that migrants have less sustainable farm practices because of their use of hoe technology (with mounds and ridges), the land tenure system that does not allow them to plant trees, their larger farm sizes, crop rotations,[4] and their alleged preference for savanna conditions. All studies focused on insecure land tenure conditions as an important underlying cause of migrants' short-term exploitation of the land.

The remainder of this chapter is structured as follows. We first use census data to study migration from the Upper West Region to the Brong Ahafo Region and to identify periods and areas of increased population growth in the region (1960–2000). The following sections are structured around six lines of evidence showing that migrant farmers from the Upper West Region have been falsely accused of causing environmental degradation through unsustainable farm practices. First, we show that the extent of deforestation is much less than UNEP's alarming picture suggests. Second, remotely sensed vegetation data are used to show that most environmental degradation took place before the large-scale arrival of migrants from the northwest. Third, we use the key findings of studies on local land use/cover change (LUCC) to assess the most detrimental and beneficial types of land use systems for vegetation cover; this review shows that those researchers who blame migrants for environmental degradation fail to consider the most crucial factors. Fourth, local discourses of environmental change and its causes in the study area show that the immigration of farmers from northwest Ghana hardly plays a role in this degradation. Fifth, an analysis of perceptions of land use by "the other" confirms that native farmers do not think that migrants' farm practices cause environmental degradation. Sixth, socioeconomic and land-use survey data are used to show that migrants' farm practices differ from those of native farmers, but no evidence suggests their practices are less sustainable. Further, there are strong indications that migrants' low external input agriculture in bush fallow systems allows for land regeneration, while the capital-intensive agricultural practices of native farmers lead to more permanent land cover change. The findings challenge earlier studies that blamed settler farmers for environmental degradation.

---

[4]Adjei-Nsiah (2006) considers crop rotations with pigeon pea (common among native farmers) to be a positive strategy and crop rotations with cowpea or groundnuts (common among migrant farmers) to be negative.

7.2.2 DAGABA MIGRATION

The Dagaba people hail from the savanna of northwest Ghana, but many have migrated southward. At the time of the population census in 2000, 51% of the Dagaba were living outside the Upper West Region and 36% had migrated to southern Ghana. Within this group, 51% had migrated to the Brong Ahafo Region, where they constituted 6.8% of the total population (Ghana Statistical Service, 2002). Apart from the movement of the more permanent Dagaba settlers, there is also a seasonal inflow of Dagaba migrants who utilize the off-season in the north to earn some money as farm laborers in the South.

Within the Brong Ahafo Region, the most popular destination districts of the Dagaba migrants are situated in the north of the region (see Figures 7.2.3a and 7.2.3b), with the Wenchi District having the largest number and proportion of Dagaba people, followed by the neighboring Techiman District. The districts in the southwest of the Brong Ahafo Region are not important destination areas for these migrants. Figure 7.2.3 also shows the location of the LANDSAT image that reveals large-scale environmental degradation. The districts that have received more Dagaba immigrants have experienced much more degradation than the other districts. The question is whether there is causality in this association and, if so, whether degradation is just a consequence of increased population pressure or also due to differences in land use sustainability between the migrants and native farmers.

Between 1970 and 2000—roughly the time between the first and second LANDSAT images shown in the introduction of this section—population growth in the northern districts of the Brong Ahafo Region was much higher (over 3% per year) than in the southern districts (approximately 2.5% per year). In the 1960–1970 intercensal period, when Brong Ahafo South was an important cocoa frontier, population growth was higher (4.3% per year).[5] In those periods, more than half the population in Brong Ahafo South was composed of immigrants from outside the region or outside of Ghana (60.9% in 1960 and 51.4% in 1970).[6] Brong Ahafo North had a much smaller proportion of immigrant population (12.3% in 1960 and 20.2% in 1970). In the 1970s and early 1980s, when the cocoa sector was in crisis and some hitherto poorly accessible and uninviting areas in the north of the region had been "opened up" through government interventions (Amanor and Pabi, 2007), the agricultural frontier shifted from south to north and from cocoa cultivation to food crop farming.

The rest of this section is structured around six lines of evidence that migrant farmers are erroneously blamed for environmental degradation in the forest-savanna transition zone.

7.2.3 FIRST LINE OF EVIDENCE: ENVIRONMENTAL DEGRADATION OVERSTATED

LANDSAT images, published in UNEP's *Atlas of Our Changing Environment*, show large-scale land degradation in Ghana's forest-savanna transition zone (1973–2003). However, UNEP does not state that the first image was taken at the end of the rainy season and the second image at the peak of the dry season. The 1973 LANDSAT image, in which the environment looks very green, was taken on November 25, while UNEP used a combination of two dates[7] for the 2002/2003 image, in which

---

[5]Census Office (1964); Central Bureau of Statistics (1984); Ghana Statistical Service (2005). See van der Geest (2011b, p. 72).

[6]Census Office (1964: 27; 1973, p. xxxviii).

[7]UNEP does not provide any information about the motivation behind this choice and the procedure followed.

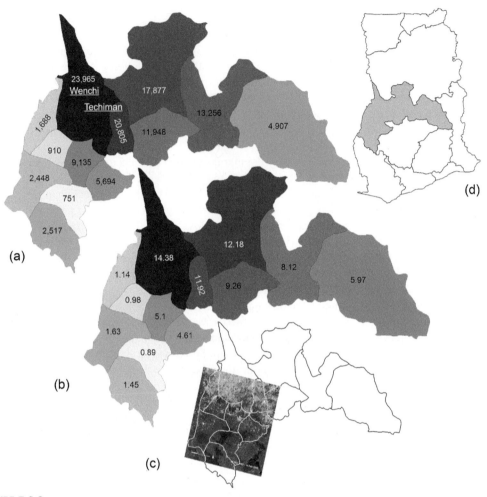

FIGURE 7.2.3

Dagaba immigration per district of the Brong Ahafo Region. (a) Total number of Dagaba immigrants;
(b) percentage of Dagaba population by district; (c) districts covered by LANDSAT image; (d) location of the
Brong Ahafo Region in Ghana.

Source: Maps drawn by authors; data from Ghana Population and Housing Census (2000). The district data on Dagaba immigration were acquired through a special data request at the Ghana Statistical Services.

the northern part of the area looks barren: December 24, 2002, and February 19, 2003. Figure 7.2.4 depicts the seasonality and long-term trend of vegetation cover in the Wenchi District, located in the center of the allegedly degraded area. The figure is based on remotely sensed Normalized Difference Vegetation Index (NDVI) data over a period of 25 years.[8] November is the last month of the rainy

---

[8]The NDVI database is described in more detail in van der Geest et al. (2010)

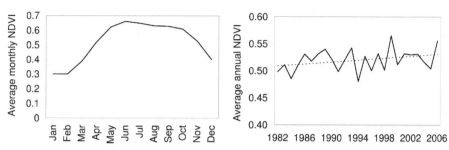

FIGURE 7.2.4

Average monthly NDVI and NDVI trend in the Wenchi District (1982–2006). Note: On the NDVI scale, a value of zero indicates bare soil and a value of 1 indicates dense forest.

Source: Figures by authors; the NDVI database was described in more detail in van der Geest et al. (2010).

season in the northern part of the LANDSAT image. January and February are the driest months of the year. In the NDVI scale, a value of zero indicates bare soil and a value of one indicates dense forest. The average NDVI value in the Wenchi District (1982–2006) is 0.300 for January and February, while in November, the average NDVI value is 0.528. If the latest LANDSAT image had been recorded in November 2002, the contrast with the 1973 image would have been much smaller and the extent of environmental degradation would have appeared much less alarming.

This is not to say that the area has not experienced land degradation. Most studies and the inhabitants (discussed next) confirm a negative long-term trend in vegetation cover. However, the timing of the changes is important and will be discussed in the second line of evidence.

7.2.4 SECOND LINE OF EVIDENCE: MOST ENVIRONMENTAL DEGRADATION, IF ANY, OCCURRED BEFORE THE LARGE-SCALE IMMIGRATION OF SETTLER FARMERS FROM THE NORTH

In the Wenchi District, located in the allegedly degraded area, the population grew slowly, from 98,091 to 105,115 inhabitants in the 1970–1984 intercensal period (0.5% per year).[9] In 1988, the administrative boundaries changed (Ghana Statistical Service, 2005, p. 18). A part of the old Wenchi District was lost to the neighboring Kintampo and Techiman districts. Despite this reduction in territory, the district population increased to 166,641 in 2000 (2.9% per year).[10] Thus, the population within the old district boundaries must have grown by well over 3% annually. The low growth figures for 1970–1984 and the high growth figures for 1984–2000 are a strong indication that most migrants came to the Wenchi District in the late 1980s and 1990s. This is confirmed by household survey data.[11]

The second graph in Figure 7.2.4 shows the vegetation trend for the 1982–2006 period. In this period of 25 years, NDVI values *increased* moderately, especially after 1984 (see Figure 7.2.4).[12] This

[9]Central Bureau of Statistics (1984).
[10]Ghana Statistical Service (2005). The population density in the Wenchi District was 33.3 people/km^2 in 2000.
[11]Of the 203 migrants we interviewed, only 16.4% arrived before 1984.
[12]See also van der Geest et al. (2010).

would mean that most degradation took place between 1973 and 1984, which was a period of wide-spread drought in West Africa (Hulme, 2001; Mortimore and Adams, 2001; Batterbury and Warren, 2001). In Ghana's forest-savanna transition zone, drought and bush fires decimated cocoa plantations (Dei, 1992; Ruf, 2007). In focus group discussions, covered next, farmers explained that in the affected areas, cocoa was usually not replanted because of increased risk of fire invasion, resulting in a southern shift of cocoa cultivation. Cocoa was usually replaced by field crops, such as maize and cassava, with significant effects on green cover—especially in the dry season.

If large-scale immigration of settler farmers from northwest Ghana started in the 1984–2000 inter-censal period and if longitudinal vegetation data suggest that most environmental degradation took place before 1984, then clearly immigration has not been a primary cause of environmental degradation in the district.

7.2.5 THIRD LINE OF EVIDENCE: THE STUDIES THAT BLAME MIGRANTS FOR ENVIRONMENTAL DEGRADATION NEGLECT THE MOST CRUCIAL CAUSES OF LAND DEGRADATION

Several scholars have studied LUCC in Ghana's forest-savanna transition zone, most of which used LANDSAT images combined with socioeconomic and agricultural surveys. Pabi and Attua (2005) analyzed land cover change in six locations in the Wenchi District based on 1984 and 2000/2001 LANDSAT images. They found that "dense woodland" (forest) had reduced in five out of the six sites. However, they emphasize that land conversion in the forest-savanna transition zone is not a unilinear process toward degradation but rather a complex, dynamic, and multidirectional interplay of human and natural factors, such as farming systems, land tenure, market access, and policy environment. The sites with mechanized and high-input agriculture showed the least potential for regeneration, because all vegetation, as well as the stumps and roots, were removed. In areas where bush fallow systems dominated, there was not only much conversion of forest to farmland but also many changes in the opposite direction. Based on the same data, Amanor and Pabi (2007, p. 61) conclude: "In contrast with the main narratives of modern environmentalism, there is considerable evidence that the activities of farmers in the transition zone do not lead to a downward spiral of degradation. Localized farming practices often encourage regeneration of the root and coppice mat in the soil and promote rapid regeneration of many species." The exceptions are fields that are permanently cultivated and on which farmers use tractors to plow the land and inorganic fertilizer to replenish the soil. In the mid-1980s, when Ghana implemented structural adjustment policies, subsidies on inorganic fertilizer were removed and the output prices of food crops declined because of cheap imports. The fields that were used for more permanent cultivation could no longer yield adequately and were largely abandoned. Even after more than 10 years, these areas showed very few signs of regeneration. Areas under bush fallow, on the contrary, showed considerable conversion of fallow land to open or dense woodland (Amanor and Pabi, 2007).

The authors of the studies that conclude that migrants have more detrimental farm practices (Adjei-Nsiah, 2006; Afikorah-Danquah, 1997; Codjoe, 2006) did not look at factors that turned out to be crucial in analyses of land cover change under different farming systems. Rather, they use some questionable criteria for pronouncing migrants' land use to be less environmentally sustainable. For example, in Adjei-Nsiah's (2006) work, crop rotations with pigeon pea—common among native farmers—are considered positive, while those with cowpea or groundnuts—common among migrant farmers—are not. Similarly, migrants' habit of using hoes to make ridges—usually considered a sustainable soil and water conservation technique—is negatively assessed by these studies, while "zero tillage" is assessed positively, despite the fact that the pesticide (Roundup, made by Monsanto, Ltd.) used in this farming method has a very bad record with regard to environmental sustainability and has been banned from European markets. Another questionable argument is migrants' alleged and inadequately documented preference for savanna conditions (Adjei-Nsiah, 2006, p. 58). The negative assessment of migrants' farming practices could be due to some degree of scapegoating. As Hugo (1996, p. 123) warned: "there are considerable dangers that the migrants involved can become scapegoats for a general failure to adopt sustainable policies of land and other resource use in the destination areas." Interestingly, as discussed in the fourth and fifth lines of evidence, the native farmers we interviewed do not share the view of these academics. They do not think that migrants have more degrading farm practices and point to other drivers of environmental change.

7.2.6 FOURTH LINE OF EVIDENCE: IMMIGRATION OF FARMERS FROM NORTHWEST GHANA HARDLY PLAYS A ROLE IN LOCAL DISCOURSES OF ENVIRONMENTAL DEGRADATION

Focus group discussions were held in eight study sites in the forest-savanna transition zone with the aim of capturing people's—possibly conflicting—perceptions of environmental change and its causes. The discussions were structured by a community questionnaire, which included several questions about trends in soil fertility, vegetation, and rainfall. In the case of negative trends, the respondents were asked about the causes of these trends and whether, and if so how, the arrival of settlers from northern Ghana had contributed to negative trends in soil fertility and vegetation cover.

The group discussions yielded an interesting overview of local perceptions of environmental change and its causes. The participants perceive the trends to be mostly negative: they thought that soils had become less fertile and that the vegetation cover and rainfall amounts had reduced. They attributed these negative trends to a combination of timber logging, increased pressure on farmland, charcoal burning, and uncontrolled bush fires caused by hunters, palm wine tappers, and charcoal burners. Reduced fallows were seen as an important cause of soil fertility decline in the more densely populated study locations, while, in the sparsely populated areas, uncontrolled bush fires were perceived to be the most important cause of fertility decline and of deforestation.

When asked specifically about the contribution of migrants from northern Ghana to soil fertility decline, most groups could not think of any, but one group mentioned that the land tenure system (fixed rent for a number of years) contributed to soil fertility decline. One participant said: "If the land is given out for four years, the settler will make sure he gets the maximum out of it and won't leave the land to rest." The same group also thought that settler farmers contributed more to loss of forest cover. They said: "Migrants try to make their farms as large as possible so they have to cut more trees."

In sum, only one out of the eight group discussions cited immigrant farmers from the North as drivers of environmental degradation. Moreover, they were not mentioned initially as such; it was only after specifically asking about possible contributions of immigrants to soil fertility decline and deforestation that the comments surfaced.

Table 7.2.1 Perception of Differences in Farming Methods by Settlers and Natives

Description	Settlers (75)	Natives (72)	Total (147)
There is no difference.	22	7	29
Settlers have bigger farms; they farm on a larger scale.	12	16	28
Settlers use their own strength; natives hire labourers or tractors.	11	12	21
Settlers use hoes and natives use cutlasses.	6	11	17
Settlers achieve better yields or larger produce.	6	8	14
Settlers have to rent land or do sharecropping.	7	7	14
Settlers go to the farm every day.	0	10	10
Settlers depend on farming more than natives.	3	6	9
Settlers organise labor parties.	1	6	7
Settlers cultivate more food crops, natives more cash and tree crops.	4	3	7
Settlers work harder.	3	3	6
Settlers do more intercropping; natives do more monocropping.	3	2	5
Natives don't weed properly.	5	0	5

Note: Only differences mentioned by five respondents or more are included in the table. The total number of households surveyed was 155, but eight respondents failed to answer this question.

7.2.7 FIFTH LINE OF EVIDENCE: NATIVE FARMERS SEE DIFFERENCES IN FARMING TECHNIQUES BETWEEN THEMSELVES AND SETTLER FARMERS, BUT THEY DON'T THINK THAT SETTLERS' METHODS ARE MORE DESTRUCTIVE

A questionnaire was administered among 155 farm households in four study locations[13]; about half the respondents were settler farmers and half were native farmers. As part of the questionnaire, we asked respondents to identify differences in the way settlers and natives farmed. This was an open question without set answers to choose from. The answers of both settlers and natives were coded and clustered. Table 7.2.1 shows the most frequently mentioned differences identified by both settlers and natives in descending order of frequency.

About a quarter of the settler farmers and about a tenth of the native farmers said they could not think of any differences in farming methods between the two groups. The other perceptions listed in Table 7.2.1 give a rough idea about some differences in farming behaviour. First, settlers are perceived to have bigger farms and larger harvests than natives. Second, farming is more often the main occupation of settlers. Third, native farms are more capital-intensive, while settler farms are more labor-intensive. Fourth, native farmers are more likely to farm on their own land, whereas settlers have to rent land. Fifth, settlers and natives grow different crops: natives are more likely to grow cash crops, while settlers tend to grow food crops. Sixth, settler farmers tend to use hoes to prepare the land and weed, while natives prefer a cutlass.

---

[13]The methods and study sites are described in more detail in van der Geest (2011b, pp. 76–79).

Natives did not explicitly accuse settler farmers of contributing to land degradation through environmentally unsustainable farm practices. Only one native respondent said that the migrants' use of hoes "spoils the soil." The use of mounds and ridges, which Adjei-Nsiah (2006) laments was not linked to negative effects on the land at all. In general, natives' perception of settlers' farming skills was quite positive. They seem to admire the Dagaba's strength and dedication. Settler farmers did have some negative statements about natives' farm practices and behavior: five of them mentioned that natives do not weed properly and two settlers also complained that natives keep the best lands for themselves and only allow settler farmers to cultivate the infertile portions.[14]

In the sixth and final line of evidence, discussed next, the results of a socioeconomic and agricultural survey among native and settler farmers are analyzed. Some of the findings on perceptions are confirmed, some are refuted, and some additional differences are identified.

7.2.8 SIXTH LINE OF EVIDENCE: A SURVEY AMONG SETTLER FARMERS AND NATIVE FARMERS SHOWS DIFFERENCES IN FARMING TECHNIQUES BUT NO EVIDENCE THAT SETTLERS' METHODS ARE MORE DEGRADING

This section compares the environmental sustainability of migrants' and natives' farm practices. Several aspects of their farming systems are discussed, such as land tenure, farm size, crop mix, tools used, tillage methods, capital inputs, and tree cutting and planting. The analysis shows that there are differences in farming techniques between migrant farmers and native farmers, but there is no evidence that settlers' methods are more degrading to the land.

Before this comparison is made, however, some basic socioeconomic information of the respondents and their households is provided (see Table 7.2.2). The main difference between settlers and natives in the four localities is that the native farmers had higher education levels and more nonfarm income.

Table 7.2.2 Socioeconomic Profile by Migrant Status		
	Native Farmers	**Settler Farmers**
Households	73	82
Average household size	5.5	5.8
Female-headed households	11%	8%
Average age of household head	49	42
Education years of household head	5.6	2.1
Households with nonfarm income	84%	45%
Average nonfarm income (€)	369	168
Settlement year (average)	N/A	1995

---

[14]See also Leach and Fairhead (2000, p. 34).

Table 7.2.3 Land Tenure (% of Fields)

Land Tenure System	Native Farmers	Settler Farmers
Borrowed	44	6
Owned	43	0
Fixed rent	10	30
Sharecropping	1	33
Stool land	0	27
Taungya	2	1
Other	0	2
Total	100	100

Note: Taungya = Reforestation land of the government (see Owusu, 2007, p. 46).

Table 7.2.4 Farm Size, Crop Mix, and Sales (€)

	Native Farmers		Settler Farmers	
Farm size (acres)	6.0		8.0	
Total crop sales (€)	422		419	
Crop Mix	**Households**	**Sales (€)**	**Households**	**Sales (€)**
Maize	92%	234	96%	183
Yam	86%	23	91%	74
Cassava	97%	25	78%	16
Sorghum	26%	5	37%	27
Legumes*	56%	20	83%	71
Vegetables**	92%	79	79%	31
Cashew	58%	17	10%	0

Notes: Legumes included groundnuts, beans, bambara beans, and cowpeas (in order of frequency). Vegetables included okra, pepper, tomato, garden eggs, and peppers (in order of frequency).

As noted in other studies dealing with farm practices of migrants and natives, the land tenure situation of these groups varies, which may give rise to differences in decision making at the farm level. In Adjei-Nsiah et al. (2004, p. 343), a migrant farmer explains why he does not invest in long-term soil fertility management strategies, illustrating the role of land tenure conditions in land-use decisions:

"I will never plant pigeon pea again because when I planted pigeon pea to improve the fertility of my farmland, the landlord asked me to quit the land because one of his sons was coming to farm on the land when he observed that the fertility of the land had improved."

In our study areas, land tenure conditions for settler farmers are also quite insecure. Table 7.2.3 (Owusu, 2007, p. 46) confirms that the majority of natives farmed on their own land or on borrowed land (free of charge), but that most settler farmers had to rent land, engage in sharecropping, or pay a fixed amount per year to traditional authorities to use communal land. In most arrangements, tenant farmers are not sure whether they can rent the land next year. However, as we will see next, this does not necessarily mean that these farmers apply environmentally destructive farming methods.

The qualitative analysis of differences in farming styles showed that settlers are perceived to have larger farms than native farmers. The perceptions analysis also gave some clues as to why their farms are larger: settlers have to farm more and work harder to survive, because—in comparison with most native farmers—they have to rent land or sharecrop, and they have less non-farm income. In addition, they are expected to send remittances to their relatives in the Upper West region. The perception that settlers have larger farm sizes than natives is confirmed in Table 7.2.4. The fact that settlers' farms tend to be larger than natives' farms could be an indication that migrants contribute more to the conversion of forest to farm and fallow land than native farmers do. However, as we will see next, settler farmers usually farm the old fallows of native farmers, and there are indications that their style of farming, which is less capital-intensive and more labor-intensive than that of native farmers, allows for faster regeneration of soil and vegetation after farms have been abandoned.

In the literature on land use and land cover change in Ghana's forest-savanna transition zone, an important distinction is made between indigenous and modern farm practices: the environmental impact of the traditional system of bush fallowing differs from that of modern systems that are more sedentary and rely more on external inputs in that, for modern, intensive, and sedentary cultivation, less land has to be cleared, but recovery once the fields are abandoned is slow. Amanor and Pabi's (2007) analysis of farming systems and land cover change showed that high-input farming systems are more detrimental than cultivation under bush fallow systems because they cause a more permanent conversion from forest to savanna. Conversion in the opposite direction, from fallow to woody vegetation, was common in areas where bush fallowing was dominant.

The most typical crop to be cultivated under a modern, mechanized and high-input regime is maize, especially when monocropped. As shown in Table 7.2.4, maize is an important cash crop for both natives and settlers, though native farmers record slightly higher sales of maize. The most typical crop in the bush fallow system is yam. On yam farms, most trees are left standing, because yam does well under shady conditions. Similar proportions of settlers and natives cultivate yam, but for settler farmers, yam is much more important as a cash crop: they recorded more than three times the yam sales of native farmers, indicating that they plant a larger proportion of their fields with yam (see Table 7.2.4).

Besides maize, other crops that are cultivated under modern regimes are cashew and vegetables (when cultivated commercially). Table 7.2.4 shows that both these crops are more popular among native farmers, for whom vegetables were the main cash crop after maize. Settler farmers also engage in vegetable cultivation, but this is mostly done at a very small scale, for home consumption. Cashew cultivation is relatively new in the area: more than half the native respondents had planted cashew trees, but most trees are still young and have only recently started to bear fruits. Although cashew plantations add to green cover in the area, this new crop may also have some environmental drawbacks (see

Table 7.2.5 Tools Used for Land Preparation and Weeding (% of Farmers)		
	Native Farmers	**Settler Farmers**
Fire	100	95
Cutlass	100	100
Hoe	97	99
Tractor	10	5
Chemicals	18	7

Amanor and Pabi, 2007). Interestingly, as Table 7.2.4 shows, settler farmers are also going into cashew cultivation, especially in the savanna zone, where tenure arrangements are less exclusive. Apparently, the rules that prevent migrants from planting trees are more flexible than often assumed.

Integrating legumes in the crop mix is an important strategy to protect the soil from erosion and depletion (Amanor and Pabi, 2007). Table 7.2.4 shows that legume cultivation is much more common among settler farmers who are used to this practice from their home areas.

Despite the fact that settlers have larger farms, their crop sales were similar to those of native farmers. This is an indication that native farmers practice a capital-intensive style of agriculture, but it should be noted that part of the produce from sharecroppers is not included in tenants' crop sales.

An important aspect of land use sustainability is the type of tool used to clear the land and to weed. Most studies that conclude that settler farmers from northern Ghana have less sustainable farm practices lament their use of hoes and praise the natives' use of cutlasses. Table 7.2.5 shows that this conclusion may be primarily based on stereotyping, as no significant differences were found in the three most common tools used for land preparation and weeding—fire, cutlass, and hoe—by migrant and native farmers. More pronounced differences between settlers and natives exist in the use of tractors for plowing and chemicals for weeding, which are both more commonly used by native farmers. The use of tractors is associated with a more permanent conversion from tree cover to grassland and the most common chemical used for weeding is Monsanto's Roundup. At the time of the survey, this herbicide was, somewhat contraintuitively, promoted by Ghana's Ministry of Food and Agriculture as an environmentally sound alternative to manual weeding. Over the past decade, however, evidence has increasingly shown the very negative consequences of applying Roundup herbicide to soil fertility, biodiversity, and human health (Relyea, 2005; Samsel and Seneff, 2013).

Besides the tools used for land preparation and weeding, the questionnaire also inquired about the specific farm practices and tillage methods used by settlers and natives. Regarding intercropping, we specifically referred to sowing different crops intermixed in the same field. Thus, intercropping does not refer to a field having several portions, each with a different crop. Regarding crop rotation, we inquired specifically about the application of crop sequences that aim to restore soil fertility. Soil and water conservation (SWC) measures involved anti-erosion measures, such as ridges along the contours. Table 7.2.6 shows that a higher proportion of settlers employ methods that are meant to improve

Table 7.2.6 Farming Techniques (% of Farmers)

	Native Farmers	Settler Farmers
Practice intercropping	40	43
Practice crop rotation	27	46
Application of inorganic fertilizer	21	14
Application of animal dung	0	8
Application of compost	0	1
Planting cover crops	58	64
Physical SWC	3	32

Note: Physical SWC are physical soil and water conservation methods.

Table 7.2.7 Farm Expenditure (€)

	Native Farmers	Settler Farmers
Hired labor	53	70
Fertilizer	12	2
Chemicals	6	2

Table 7.2.8 Tree Cutting and Planting

	Native Farmers	Settler Farmers
Cut trees?	56%	41%
How many?	162	87
Planted trees?	72%	14%
Tree acreage	2.14	0.49

the fertility of the soil. The only method that was more common among native farmers was the application of inorganic fertilizer, which is associated with soil impoverishment in the long term.

These findings go against the "received wisdom" (Leach and Mearns, 1996) that migrant farmers do not invest in soil fertility management strategies. In fact, Dagaba farmers in the Wenchi District apply some of the techniques that they are accustomed to at home—where soils are less fertile and rain is scarcer—to maintain the fertility of the soil. This phenomenon has been noted by Lambin et al (2001, p. 263) who write: "In some cases, these 'shifted' agriculturalists exacerbate deforestation because of unfamiliarity with their new environment; in other cases, they may bring new skills and understandings that have the opposite impact."

The picture that emerges is that native farmers have a more capital-intensive style of farming, while Dagaba migrants farm in a more labor-intensive way. In the section about perceptions of farming methods of "the other," an important difference noted by both groups was that natives depend more on hired labor, though the survey results do not confirm this (see Table 7.2.7). Dagaba farmers in the sample spent more money on hired labor than native farmers did. Possibly, the perception is fed by the fact that most laborers, both on settlers' and natives' farms, are Dagaba seasonal migrants. Other capital inputs obtained in the questionnaire are the purchase of inorganic fertilizer and chemicals, which are much more common among native farmers, indicating that they tend to farm in a more "modern" way.

The last aspect of land-use sustainability analyzed here concerns farmers' direct impact on tree cover. To establish a field, one would expect farmers to remove trees. Table 7.2.8 shows that this is not necessarily the case: almost 60% of settler farmers and 44% of native farmers indicated that the tree cover on the lands they farmed was sufficiently open to start farming without removing any trees. On average, settler farmers estimated that they removed 87 trees on the fields they presently cultivated (11 trees per acre). Native farmers removed an average of 162 trees on an average farm size of 6 acres, amounting to 27 trees per acre. From a mainstream environmentalist point of view, it could be argued that asking farmers how many trees they cut is similar to asking a thief how many wallets he has stolen. In the local setup, however, this was quite a matter-of-fact question, as removing trees to farm is just one of many activities that need to be carried out for a good harvest. The fact that migrant farmers removed fewer trees from their fields than native farmers did is not all that surprising: clearing trees is a demanding job and—perhaps somewhat paradoxically for a Western audience—is seen as an investment in the farm, at least when tree densities are high. With low tenure security, a settler farmer would prefer to farm on land that is already more open. Moreover, native landowners usually do not give out their more virgin and woody lands, because they prefer to cultivate these lands themselves. Settler farmers tend to cultivate the old fallows of native landowners.

Native farmers tend to cut more trees on their farmlands, but they also plant more trees. Almost three quarters of the native respondents had planted trees, mostly cashew, with 90% of these tree plantations established in the 10 years prior to the survey. The average size of native farmers' tree plantations was more than 2 acres. It would be interesting to find out whether a positive effect of cashew plantations on vegetation cover will be visible in the next round of LANDSAT images.

The survey revealed that settler farmers—though they plant fewer trees than native farmers—are much more involved in tree planting than previously thought: about a third of the settler farmers in the savanna and forest zones have planted trees on their farms. In the forest zone, tree planting was usually part of the land tenure arrangement, but settlers in the savanna zone have started to establish their own cashew plantations.

7.2.9 EVALUATION OF SURVEY FINDINGS ON LAND-USE SUSTAINABILITY

The land-use survey conducted among settlers and native farmers revealed a number of differences in farm practices, both between settlers and natives. However, the findings do not confirm that migrants have more detrimental farm practices, as is commonly believed. Although they have larger farms and plant fewer trees than natives, migrants' performance on most other aspects of land-use sustainability was better. Settlers cut fewer trees; they make less use of tractors and other capital inputs that are associated with more permanent land cover change; they cultivate more yam, which is associated with less tree removal and faster regeneration of vegetation cover; they integrate legumes in the crop mix, which is an effective soil fertility management strategy; and some also maintain their use of physical soil and water conservation measures, which they brought from their home areas in the north, on their farms in the Brong Ahafo Region, a practice which is not common among native farmers.

7.2.10 CONCLUSION

This chapter aims to investigate the role of settler farmers from northwest Ghana in the alleged "savannization" of the forest-savanna transition zone. Several sources suggest that this role has been considerably negative. First, LANDSAT images from 1973 and 2003, published by UNEP (2008), reveal large-scale land degradation in a prime destination area for Dagaba migrants from Northwest Ghana, identifying population growth and primitive agricultural practices as major causes of deforestation in this region. Second, several studies that compare the land use of migrant and native farmers in Ghana's forest-savanna transition zone conclude that migrants' farm practices are less sustainable, contributing to deforestation and soil fertility decline. The picture that emerges from these studies is that immigration contributes to environmental degradation, both by increasing the human pressure on natural resources and because immigrants allegedly use less sustainable farm practices. This text has challenged, with a variety of data, the notion that migration must be an important cause of land degradation—there can be no smoke without fire.

First, UNEP's land degradation narrative is challenged by the seasonality of vegetation cover. The 1973 LANDSAT image, in which the vegetation looks lush, was taken in November, at the end of the rainy season, while the 2003 image, in which the area looks more degraded, was taken at the peak of the dry season. NDVI data suggest this makes a large difference and further shows that between 1982 and 2006, the trend in vegetation has not been negative. This could mean that most land degradation took place in the 1970s and early 1980s, that not much degradation has occurred at all, or that it is a combination of these two possibilities.

Second, census data show that most population growth in the study area took place in the 1984–2000 intercensal period (annual growth rates of more than 3%), while in the 1970–1984 intercensal, annual growth rates were less than 1%. The census data for 1984 do not provide district-level immigration data, but the settlement history of Dagaba migrants in the area suggests that most arrived between 1984 and 2000. If most of the degradation took place before the mid-1980s and population growth and immigration increased sharply after the mid-1980s, other factors must have been at play. Possibly, the widespread drought and bush fires that caused havoc in the early 1980s have had a lasting impact.

Third, in group discussions with both settlers and natives, we inquired after participants' perceptions of environmental change. In most areas, people perceived more negative than positive environmental trends (less rainfall, more deforestation, and less fertile soils). In the local discourse of environmental change, a variety of causes is given for negative trends: uncontrolled bush fires, caused mainly by hunters, charcoal burners, and palm wine tappers, played an important role, as well as increased pressure on natural resources because of population growth and timber logging. When asked specifically about the role of migrant farmers, all but one group said that migration does not play a role, other than altering population size. Besides the group discussions, native farmers were also asked individually whether they perceived any differences in farming methods between themselves and migrant farmers; many differences were mentioned, but native farmers did not perceive migrants as having less sustainable farm practices.

Fourth, a land use survey was conducted among settlers and native farmers to assess differences in farming methods. Again, substantial differences were found, but migrants' methods were not found to be less sustainable. While they have larger farms and the land tenure system limits their rights to plant trees, they do perform better on most other indicators. Migrant farmers tend to have more labor-intensive practices that are associated with faster regeneration of vegetation cover, while native farmers tend to have more capital-intensive practices that cause a more permanent conversion from forest to grassland.

The findings of this study challenge the validity of earlier studies, which blame migrants for land degradation in the forest-savanna transition zone. The arrival of migrants from northwest Ghana may have increased pressure on farmland and vegetation cover, but no evidence was found that their farm practices have a more negative environmental impact than the practices of native farmers. Earlier studies comparing the farm practices of settlers and native farmers were all carried out by academics from southern Ghana who may be more familiar with land-use practices of southern Ghanaians than of farmers from northern Ghana. The results of these studies can lead to the scapegoating of migrant farmers from northern Ghana, which is a known danger in studies dealing with the environmental impact of immigration (Hugo, 1996).

REFERENCES

Abdul-Korah, G.B., 2007. Where is not home? Dagara migrants in the Brong Ahafo Region, 1980 to present. Afr. Affairs 106 (422), 71–94.

Adger, W.N., Benjaminsen, T.A., Brown, K., Svarstad, H., 2001. Advancing a political ecology of global environmental discourses. Develop. Change 32, 681–715.

Adjei-Nsiah, S., 2006. Cropping systems, land tenure and social diversity in Wenchi, Ghana: Implications for soil fertility management. Ph.D. thesis, Agricultural University of Wageningen, the Netherlands.

Adjei-Nsiah, S., et al., 2004. Land tenure and differential soil fertility management practices among native and migrant farmers in Wenchi, Ghana: Implications for interdisciplinary action research. NJAS Wageningen J. Life Sci. 52 (3–4), 331–348.

Afikorah-Danquah, S., 1997. Local resource management in the forest-savanna transition zone: The case of Wenchi District, Ghana. IDS Bull. 28 (4), 36–46.

Amanor, K.S., Pabi, O., 2007. Space, time, rhetoric and agricultural change in the transition zone of Ghana. Hum. Ecol. 35, 51–67.

Batterbury, S., Warren, A., 2001. The African Sahel 25 years after the great drought: Assessing progress and moving toward new agendas and approaches. Glob. Environ. Chang. 11 (1), 1–8.

Census Office, 1964. 1960 Population Census of Ghana. Volume III: Demographic Characteristics. Government of Ghana, Accra.

Census Office, 1973. 1970 Population Census of Ghana, Vol. 3: Detailed Demographic Characteristics. Government of Ghana, Accra.

Central Bureau of Statistics, 1984. Preliminary Report of the 1984 Population Census of Ghana. Government of Ghana, Accra.

Codjoe, S.N.A., 2006. Migrant versus indigenous farmers. An analysis of factors affecting agricultural land use in the transitional agro-ecological zone of Ghana, 1984–2000. Geografisk Tidsskrift–Danish J. Geog. 106 (1), 103–113.

Dei, G.J., 1992. A forest beyond the trees: Tree cutting in rural Ghana. Hum. Ecol. 20 (1), 57–88.

Dietz, T., van der Geest, K., Obeng, F., 2013. Local Perceptions of Development and Change in Northern Ghana. In: Yaro, J. (Ed.), Rural Development in Northern Ghana. Nova Science Publishers, New York, pp. 17–36.

Ehrlich, P.R., Ehrlich, A.H., 1990. The Population Explosion. Simon & Schuster, New York.

Ghana Statistical Service, 2002. 2000 Population and Housing Census: Summary Report of Final Results. Government of Ghana, Accra.

Ghana Statistical Service, 2005. 2000 Population and Housing Census. Analysis of District Data and Implications for Planning: Brong Ahafo Region. Government of Ghana, Accra.

Hugo, G., 1996. Environmental concerns and international migration. Int. Migr. Rev. 30 (1), 105–131.

Hulme, M., 2001. Climatic perspectives on Sahelian desiccation: 1973–1998. Glob. Environ. Chang. 11 (1), 19–29.

Lambin, E.F., et al., 2001. The causes of land-use and land-cover change: Moving beyond the myths. Glob. Environ. Chang. 11, 261–269.

Leach, M., Fairhead, J., 2000. Challenging neo-Malthusian deforestation analyses in West Africa's dynamic forest landscapes. Popul. Dev. Rev. 26 (1), 17–43.

Leach, M., Mearns, R., 1996. The Lie of the Land: Challenging Received Wisdom on the African Environment. James Currey Ltd., Oxford, UK.

Mortimore, M.J., Adams, W.M., 2001. Farmer adaptation, change and "crisis" in the Sahel. Glob. Environ. Chang. 11 (1), 49–57.

Owusu, V., 2007. Ph.D. thesis, Migrants, Income, and Environment: The Case of Rural Ghana. Free University of Amsterdam.

Pabi, O., Attua, E.M., 2005. Spatio-temporal differentiation of land use/cover changes and natural resource management. Bull. Ghana Geogr. Assoc. 24, 89–103.

Rademacher-Schulz, C., Schraven, B., Mahama, E.S., 2014. Time matters: Shifting seasonal migration in Northern Ghana in response to rainfall variability and food insecurity. Clim. Dev. 6 (1), 46–52.

Relyea, R.A., 2005. The lethal impact of Roundup on aquatic and terrestrial amphibians. Ecolog. Appl. 15 (4), 1118–1124.

Ruf, F., 2007. The Cocoa Sector. Expansion, or Green and Double Green Revolutions. Background Note. Overseas Development Institute, London.

Samsel, A., Seneff, S., 2013. Glyphosate's suppression of cytochrome P450 enzymes and amino acid biosynthesis by the gut microbiome: Pathways to modern diseases. Entropy 15 (4), 1416–1463.

United Nations Environmental Programme (UNEP), 2008. Africa: Atlas of Our Changing Environment. UNEP, Nairobi, Kenya.

Van der Geest, K., 2004. We're Managing! Climate Change and Livelihood Vulnerability in Northwest Ghana. African Studies Centre, Leiden, the Netherlands.

Van der Geest, K., 2009. Migration and Natural Resources Scarcity in Ghana. EACH-FOR Case-Study Report. Case Study Report for the Environmental Change and Forced Migration Scenarios Project (EACH-FOR).

Van der Geest, K., 2011a. North-South migration in Ghana: What role for the environment? Int. Migr. 49 (s1), e69–e94.

Van der Geest, K., 2011b. The Dagara Farmer at Home and Away: Migration, Environment, and Development in Ghana. African Studies Centre, Leiden, the Netherlands.

Van der Geest, K., Vrieling, A., Dietz, T., 2010. Migration and environment in Ghana: A cross-district analysis of human mobility and vegetation dynamics. Environ. Urban. 22 (1), 107–123.

THE GLOBAL RESTORATION INITIATIVE

7.3

Kathleen Buckingham, Sean DeWitt, Lars Laestadius
World Resources Institute, Washington, DC

7.3.1 INTRODUCTION

Restoring degraded lands and landscapes are essential for human livelihoods and well-being, long-term food security, climate stability, and biodiversity conservation. In order to counteract land degradation and to improve livelihoods, we need to restore forests and increase the productivity of existing agricultural lands at the same time. Only then can we create landscapes that are diverse, productive, and resilient. Combining existing principles and techniques of development, conservation, and natural resource management, landscape restoration utilizes a "landscape approach." The landscape approach aims to tackle agriculture and forestry issues together rather than in isolation, seeking solutions which can benefit local communities. The landscape approach seeks to address the widespread environmental, social, and political challenges that transcend traditional management boundaries (CGIAR, 2013).

The vast expanses of degraded landscapes present an immense opportunity to achieve multiple objectives not afforded to conventional conservation practices, which focus primarily on biodiversity protection. Large-scale landscape restoration is not at odds with conservation; conservation of land is one part of the matrix of options identified in the landscape approach. Restoration, therefore, affords both primary and secondary support to biodiversity objectives, through protection, as well as in many cases, reducing pressure on conservation areas by expanding opportunities for human well-being through agriculture, forestry, and sustainable resource extraction. This is the aim of the Global Restoration Initiative at the World Resources Institute (WRI).

In the absence of human impact, forests would cover nearly half the Earth's landmass and be the dominant land-based ecosystem . The actual situation differs significantly from the potential, however. About 28% of the potential forest land has been cleared; making way primarily for agricultural crops and grazing land for livestock, and another 19% has been degraded. A significant share, 37%, is now secondary, fragmented forest, and only 15% is primary, intact forest—with vast stretches unperturbed by roads or other clear signs of recent human impact (Laestadius et al., forthcoming) (see Figure 7.3.1).

The Food and Agriculture Organization of the United Nations (FAO) conducted a study to quantify the extent of degraded land globally and found that a third of all land is considered to be moderately or severely degraded (Nachtergaele et al., 2011) (Figure 7.3.2).

With global population projected to reach 9.6 billion people by 2050 and consumption projected to increase by 69%, we need to effectively utilize our lands and restore vitality to a major portion of these degraded lands (Searchinger et al., 2013).

Land Restoration. http://dx.doi.org/10.1016/B978-0-12-801231-4.00031-8

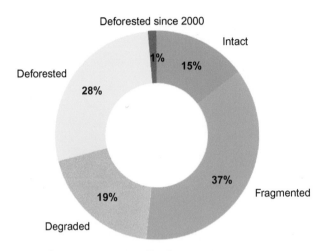

Land area: 100% = 7.5 billion ha

Source: Laestadius et al. (forthcoming). *The Carbon Potential of Forest Landscape Restoration.* Washington DC: World Resources Institute

FIGURE 7.3.1

Current status of lands where forests can grow.

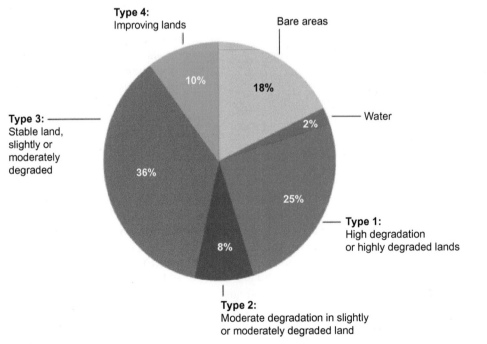

FIGURE 7.3.2

Extent of degraded lands globally (Nachtergaele et al., 2011).

7.3.2 OPPORTUNITY

A global analysis conducted by the WRI, the International Union for the Conservation of Nature (IUCN), and partners on behalf of the Global Partnership on Forest and Landscape Restoration indicates that more than 2 billion ha of cleared and degraded forest lands—an area twice the size of China—are not used for productive agriculture or human habitation. In this context, "degraded forest land" refers to areas that have had their natural forest or woodland cover cleared or significantly diminished, and now contain low levels of biodiversity, low stocks of carbon (less than about 40 metric tons per hectare), and are currently not used intensively as croplands or settlements.

A substantial portion of these vast areas has the potential to be restored. Some could be restored to natural forests, and others could be restored to agroforestry systems, where trees are integrated into farmland. Still others could be restored to productive agricultural land that, when combined with other measures, can relieve pressure to clear more forests to feed the planet. In short, the opportunity is for "restoration," where landscapes that once were forests or woodlands are restored into something productive again.

Restoration is essential for achieving sustainable, climate-smart landscapes (Figure 7.3.3). Today, the world's stock of degraded land is growing due to forest clearing and unsustainable land management practices by loggers, farmers, and ranchers (item 1). Likewise, forests continue to be converted, primarily into croplands and grazing lands (item 2). For the sake of human well-being, climate stability, and the environment, the world needs to break this pattern. Instead, we need a world in which the amount of forest cover grows, while the productivity of existing agricultural land increases.

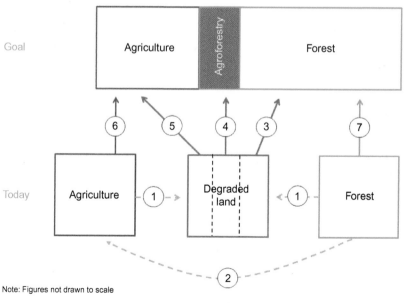

Note: Figures not drawn to scale

FIGURE 7.3.3

Illustrative dynamics of change in land use.

This goal can be achieved in part by restoring degraded land. The following describes how this process may take place:

- Some degraded lands could be restored to natural forests (especially on slopes, in riparian areas, in areas of high biodiversity, etc.) (item 3).
- Some degraded lands could be restored to agroforestry or mixed forest-agriculture landscapes (especially in areas where food security is a major concern) (item 4).
- Some degraded lands could be restored to highly productive agricultural land following principles of climate-smart agriculture or sustainable intensification (item 5).

Other concurrent and complementary strategies are also needed if sustainable climate-smart landscapes are to be achieved. In particular:

- Efforts to improve the productivity of croplands and grazing lands in a manner that mitigates and adapts to climate change, through, for example, "conservation agriculture," which implements practices such as reduced tillage, maintenance of soil cover to increase soil organic matter, crop rotation and rain water harvesting, need to be up-scaled (item 6).
- Efforts to avoid deforestation of the remaining natural forests of the world need to be catalyzed through activities such as improved law enforcement, better monitoring and transparency, alternative livelihood development, and payments for ecosystem services, amongst others (item 7).

WE'VE SEEN SOME PROGRESS . . .

Restoration success stories do exist. For instance, country examples present a myriad of benefits that contribute to climate change mitigation, provide water benefits, enhance economic livelihoods, facilitate conflict resolution, and empower women through greater gender awareness. Such examples are identified below.

7.3.3 CLIMATE CHANGE MITIGATION

In Ethiopia, soil degradation has contributed to economic losses of USD 1 billion to USD 2 billion annually, in a country where farming provides approximately 85% of the total employment and 47% of GDP (Tedla 2007). The Humbo District of Ethiopia, located 420 km southwest of the Ethiopian capital Addis Ababa, forests were largely cleared by the late 1960s (Brown et al., 2011), and an estimated 85% of Humbo residents lived in poverty in the early 2000s (World Bank, n.d.).

In the early 2000s, the nonprofit development organization World Vision, with financial and technical support from the World Bank and its BioCarbon Fund, developed a program to restore native vegetation to approximately 2700 ha in the Humbo region. The Humbo project encourages farmers to apply a practice called *Farmer Managed Natural Regeneration (FMNR),* where farmers allow native trees and shrubs to regrow underground root systems from live stumps. Seven community cooperatives with legal land titles managed the project and established a system to monitor restoration and associated carbon stock improvements (Rinaudo et al., 2008). Benefits, both expected and already emerging from this system, include (Brown et al., 2011):

- Projected sequestration of approximately 880,000 tons of carbon dioxide equivalent for an operating lifetime of 60 years
- Projected sale of more than 338,000 tons of carbon credits by 2017, with guaranteed purchase by the BioCarbon Fund
- Less flooding, erosion, and siltation, which community members have already observed
- Increase in domestic firewood availability, which community members have already noted
- Emerging increase in presence of wildlife.

7.3.4 WATER BENEFITS

Starting in the 16th century, the tropical rainforests surrounding Rio de Janeiro were gradually cleared to make way for sugarcane fields, coffee plantations, and pastureland. By the 19th century, reduced vegetation combined with lower-than-average rainfall levels resulted in a crisis for the city's water supply. By the mid-1800s, these water crises threatened Rio de Janeiro's growth (Drummond, 1996). Efforts to restore tropical forests to the area first began in the early 1860s. The following 15 years saw 68,000 trees planted, with another 20,000 during the subsequent 10 years (Rodrigues et al., 2007). Efforts to reintroduce additional native animal species took place in the 20th century. And in 1961, the Brazilian government designated the area as a national park (Freitas et al., 2006). Today, located in the middle of Rio de Janeiro, Tijuca National Park is currently one of the largest urban forests and replanted tropical forests in the world (Drummond, 1996).

Restoration provided environmental benefits that, in turn, addressed the water crises and improved human well-being. In particular, restoration of vegetation modulated water flows, slowing runoff during rainy seasons, and retaining water during drier seasons. It also increased soil moisture, reduced erosion, and created recreational opportunities for citizens. The restoration of Tijuca has had a number of positive impacts, including (Freitas et al., 2006):

- Improved water supplies for Rio.
- Improved air quality for Rio.
- Recreational opportunities for Rio's citizens and visitors.
- Reestablishment of 25 bird species, 7 mammal species, and 1 reptile species.
- Habitats for 49 species of mammals, of which 11 are on regional threatened species red lists and 4 are on the IUCN Red List of Threatened Species.

7.3.5 ECONOMIC LIVELIHOODS

The Maradi and Zinder regions lie in southern Niger. From the late 1960s through the 1980s, these regions, along with the wider Sahel, suffered periodic droughts and crop failures that led to widespread hunger and exacerbated rural poverty. The landscape was threatened by severe desertification. Yet, this outlook started to change by the mid-1980s. Many areas in Maradi and Zinder started undergoing restoration to productive agroforestry landscapes. Driving this restoration was a locally directed practice of FMNR. Since 1985, more than 1 million rural households in Niger have protected and managed trees in agroforestry landscapes across approximately 5 million ha in Maradi and Zinder (Winterbottom et al., 2013). These practices have generated a number of benefits in the region, including:

- *Increased household income and diversity of income.* The restored landscapes can generate grain, edible leaves, fodder for livestock, and honey. Likewise, farmers can use tree cuttings for construction poles and fuel wood. As a result, many farmers have doubled or tripled their incomes through the sale of these products (WRI, 2008). Gross income in the region has grown by $17–21 million (Haglund et al., 2011), which translates to around $1000 per household each year, according to the World Agroforestry Center (Pye-Smith, 2013). Pye-Smith (2013) found that gross income per capita was $167 for FMNR adopters compared to $122 for nonadopters. Extrapolating this $1000 per household per year of added income from FMNR to the entire 5 million ha implies aggregate income benefits of $900 million per year (Sendzimir et al., 2011).[1]
- *Increased resilience.* Farmers with more trees on their farms were better able to cope with the impacts of the 2004–2005 droughts because they were able to sell tree products such as firewood and fodder, which provided them with income to buy grain (Yamba, 2006).
- *Increased land values.* Land values have risen by 75-140% through the creation of specialized local markets in buying, rehabilitating, and reselling of degraded lands (Abdoulaye and Ibro, 2006).

7.3.6 CONFLICT REDUCTION

Increases in population, agricultural expansion, and the demand for wood products caused large-scale deforestation of Nepal's forests (LFP, 2013). In response, Nepal established a system of community forestry, whereby community forest users protect and manage state forests in partnership with the government. At the national scale, 14,500 community forest user groups (CFUGs) and 1.6 million households (33% of the population) were involved in forest and landscape restoration projects (SADC, 2009). Community forestry has helped regenerate substantial areas of degraded forests and has contributed to the improvement of livelihoods and the empowerment of communities and individuals (SADC, 2009). Sustained donor commitment from several countries has contributed to this success. For example, the UK-funded Livelihoods and Forestry Program (LFP) alone covered over 600,000 ha of forest land affecting an estimated 660,000 households. It introduced subsidized improved wood-burning stoves and biogas stoves and increased growing stock in community forests by 2 m^3/ha annually (LFP, 2008).

Community Forest User Groups play an important role in peace building. During the Nepalese Civil War (1996-2006), forest user groups were among the only local institutions that continued to operate. Given that the ecological conditions, property rights systems and social conditions were broadly similar across all sites, the patterns of increasing forest density in both conflict and non- conflict sites may be the combined outcome of long-enduring institutions along with underlying institutional embeddedness owing to shared values, trust and reciprocity (Karna et al., 2010). Community forestry supports inclusive democracy by aiming for equal representation of women and men in groups, and reserving 33 percent of leadership positions for women. Groups also provide land to poor families and scholarships for children from disadvantaged families. The improvements in forest conditions have had a significant impact on the time management of the rural women and girls who require less time to collect firewood, fodder, etc. because of greater availability of resources. This leaves more time for other activities such as attending school and childcare (SADC 2009).

---

[1]Approximately 900,000 households (4.5 million people) live in the areas restored by FMNR in Maradi and Zinder.

7.3.7 GENDER

According to the government of India (2011), Out of India's 329 million ha of land, 146 million are degraded (35 million of which were once forested lands). Causes of degradation and deforestation include high population pressure, low government investment in water use efficiency measures, conversion of land for agriculture to help meet food security needs, and unsustainable agricultural, livestock, and forestry practices (Government of India, 2011). The majority of degraded areas are rainfed, which are often characterized by erratic, deficient, and delayed rainfall patterns. Beginning in the 1970s, the Indian government began turning its attention to rainfed regions to help address national food security concerns. While rainfed areas are, on average, a third less productive in terms of crop yield than the national average, these regions represent the majority (65%) of arable area (Sharma et al., 2006). As of 2005, more than 45 million ha of rainfed and degraded lands have been treated through watershed development programs (Palanisami et al. 2009). Government investment continues to grow, with an estimated spending of $4 billion per year (Gray and Srinidhi, 2013).

Participation by local communities is widely promoted in watershed development and has been cited as a key factor for success in national watershed development guidelines. For example, guiding principles of the national strategy include gender and equity sensitivity, decentralization of project management, community participation, monitoring and evaluation, and capacity building (Government of India, 2011). Watershed development interventions include ecosystem-based interventions (e.g., afforestation, agroforestry), technical interventions (e.g., human-built interventions for soil and water conservation and drought mitigation), and social interventions (e.g., women's self-help group development, capacity building for community resource groups). Restored watersheds can reduce the burden on women for water and fuelwood gathering and the promotion of female self-help groups have improved women's participation in decision-making (Gray and Srinidhi, 2013).

7.3.8 . . . BUT HURDLES REMAIN

While a handful of countries have successfully restored at scale, too few countries have embarked on this path. Why? Our analysis points to three important reasons, discussed next.

7.3.8.1 LACK OF INSPIRATION OR MOTIVATION

Too few decision makers are inspired to pursue restoration or are not yet convinced that the benefits outweigh the costs. Part of the reason is that restoration has been given much less attention to date than its complementary issues, including avoided deforestation and agriculture yield growth via increased use of mineral fertilizers, irrigation, and hybrid seeds. But another aspect is that they don't know the answer to one or more questions: What is restoration? Where is it possible? Why do it? How big is the opportunity?

7.3.8.2 MISSING ENABLING CONDITIONS

Too often, countries or regions lack one or more critical "enabling conditions" that are necessary to support the spread of restoration across large areas. While bright spots of local pilot restoration projects may exist, conditions needed for scaling up from small areas to large transformed landscapes are often missing. In particular:

- Policies encouraging restoration may be absent (e.g., insecure or unclear land tenure, lack of rights to benefit from trees on one's land, or poor policy enforcement).
- Economic incentives may fall short of stimulating restoration (e.g., opportunity costs outweigh restoration, lack of positive financial incentives for restoration, financial incentives not readily accessible, negative financial incentives in place that encourage practices that keep trees off landscapes).
- Social engagement may be insufficient to secure local buy-in (e.g., local communities not being engaged in decisions regarding restoration, little local benefit sharing).
- Ecological conditions for restoration may be too tough (e.g., unsuitable soil or water conditions, large amount of invasive species, insufficient seed or root stocks for forest regeneration).

7.3.8.3 INSUFFICIENT ON-THE-GROUND IMPLEMENTATION

Even if a country or region is inspired to pursue restoration and has the adequate enabling conditions, on-the-ground restoration may still not occur for a number of reasons. In particular, many catalysts or aids are missing, such as the following:

- Local champions or leaders who persistently push the restoration agenda
- Local capacity and knowledge of how to restore landscapes (whether via human-assisted planting or via natural autoregeneration)
- Resources to finance and sustain restoration efforts
- Monitoring of restoration to track progress, enable adaptive management, and maintain momentum

7.3.9 SOLUTIONS?

The Global Restoration Initiative aims to catalyze a political and social movement to restore vitality to degraded lands. These degraded lands can be restored into healthy mosaics of sustainable agriculture, agroforestry, and forest systems— generating economic, ecological, and social benefits to people and to the planet.

The strategy to build this global restoration movement has three components:

- *Inspire:* Map opportunities, quantify benefits, conduct awareness campaigns, secure restoration commitments, build restoration champions at the global, national, subnational, and community levels
- *Enable:* Address gaps in governance systems to support restoration; e.g., access to information, appropriate policies, participation, processes, and institutional capacities and coherence
- *Mobilize:* Ensure sufficient technical knowledge, financing, and monitoring systems are in place to replicate and scale restoration activities at large scale

This vision requires longer-term thinking from governments, businesses, civil society, and local communities. The realization of this vision will result in sustainable land use across the globe. To achieve this vision, the drivers of degradation must be effectively managed, and land must be restored in a cost-effective way to achieve sustainability.

To advance this vision, it is important to recognize three key principles, discussed next.

7.3.9.1 RETURNING AN ECOSYSTEM TO ITS FORMER LANDSCAPE MAY NOT BE POSSIBLE OR DESIRABLE IN SOME PLACES

Landscapes have always changed over time, and human-caused climate change is exacerbating that changeability. The Intergovernmental Panel on Climate Change (IPCC) highlighted in its 2014 report that the globe has seen a significant warming trend since 1850. The world is facing unprecedented warming, sea level rise, and more extreme weather events (IPCC, 2014). Rather than bringing back previous ecosystems, landscape restoration must build landscapes that are capable of coping with the future—landscapes that are resilient to climate change and increasing landscape pressures.

Returning the landscape to the former vegetation is not always preferable. Puerto Rico has seen new ecosystems emerge that did not exist before the arrival of Europeans in the late 1400s. These so-called novel ecosystems offer resilience in a time of changing climate. Nonnative tree species are often criticized as being invasive, but in Puerto Rico, they have played an important role in reversing forest fragmentation and deforestation without causing the extinction of local animals or plants (Lugo and Helmer, 2004).

With increasing global populations, pressure on land use for agriculture and ecology will rise. Restoring the forests that once covered most of the United Kingdom, the Netherlands, and Rwanda would not work for today's populations. In Rwanda, for example, 85% of the population makes a living from subsistence farming of degraded, formerly forested lands (NISR, 2012). Restoring productivity to these lands by adding trees is essential; restoring forests to them is unrealistic. Restoration and livelihoods must go hand in hand.

7.3.9.2 RESTORATION IS ABOUT CREATING MULTIPLE SOCIETAL BENEFITS, AND PLANTING TREES MAY NOT BE A PART OF IT

Planting trees can be an effective way to restore landscapes, but not always—far from it. For example, afforestation has been accepted as an important strategy for preventing soil erosion on the Loess Plateau in north-central China, but scientists are increasingly questioning the long-term sustainability of afforestation in such an arid environment. Moreover, tree planting is biologically risky. The total survival rate of trees in the Loess Plateau has been low in some areas, and only 25% of the 400,000 Chinese pine trees planted in northern Shaanxi as part of China's "Grain for Green" policy survived. Natural regeneration, sometimes with the assistance of fire suppression and reduced grazing, can be a better approach than tree planting (Cao et al., 2011).

Nor are trees always the answer. Other plants have characteristics that make them more useful in some cases. Bamboo, for example, can help protect sloping land, regenerate denuded areas, or quickly restore riparian zones. In India, the International Network for Bamboo and Rattan (INBAR) developed a bamboo restoration project on a denuded brick-mining site that regenerated the land quickly. Following the initial benefit yielded from bamboo as a pioneer species, the site has served as productive bamboo agroforestry, helping the environment and fueling job growth (INBAR, 2013).

Tree planting is often expensive. Restoring all the trees that the planet needs by planting is neither practically feasible nor financially possible. Tree planting should be seen as one of a suite of restoration activities. A promising alternative is Farmer Managed Natural Regeneration, which in countries such as Ethiopia and Niger has allowed native trees and shrubs to regrow from remnant underground root systems into healthy agroforestry systems.

7.3.9.3 INACTION COSTS MORE THAN PEOPLE THINK

Every year that goes by without restoration comes with a cost—a cost that is often paid by poor people who live off degraded land. Deforestation in Ethiopia has left less than 3% of the country's native forests standing, with disastrous results (UNFCCC, 2009). Early action, on the other hand, can bring unforeseen benefits as demonstrated in the Rio example. A water crisis in the mid-1800s triggered the restoration of the rainforests that once surrounded Rio de Janeiro. The restored forests of Tijuca have resulted in one of the largest urban national parks in the world, allowing the city to grow and benefit in ways unforeseen 150 years ago (Drummond, 1996).

7.3.10 CONCLUSION

History tells us that large-scale restoration is possible. Not only can restoration be done, but it can also deliver significant benefits to people and the planet. First, countries with successful restoration programs were motivated by a wide variety of benefits, including improved water quality, soil retention, increased wood supplies, and job creation. Second, the desired benefits can change over time. More recently, the focus has expanded to recreation, wildlife and biodiversity conservation, and climate change mitigation benefits. Third, there are three common themes to successful restoration:

- *A clear motivation.* Decision makers, landowners, and citizens were inspired or motivated to restore food crops, forests, and trees on the landscape.
- *Enabling conditions in place.* These included ecological, market, policy, social, and institutional conditions.
- *Implementation capacity and resources.* Capacity and resources were in place and mobilized to implement restoration on a sustained basis.

The case studies indicate that it is essential to assess the benefits and awareness of restoration with the availability of leadership, knowledge, finance, and incentives. Different factors were important in different case studies, suggesting that context is important. The truly essential element, however, is swift and substantial action. There is much learning to be done, but we know that we need to start vast global efforts at landscape restoration for the benefit of both people and the planet.

REFERENCES

Abdoulaye, T., Ibro, G., 2006. Analyse des impacts socio-économiques des investissements dans la gestion des ressources naturelles: Étude de cas dans les régions de maradi, tahoua et tillabéry au niger. Etude Sahélienne, CRESA, Niamey.

Brown, D.R., Dettmann, P., Rinaudo, T., Tefera, H., Tofu, A., 2011. Poverty alleviation and environmental restoration using the clean development mechanism: A case study from Humbo. Ethiopia. Environ. Manage. 48, 322–333.

Cao, S., Chen, L., Shankman, D., Wang, C., Wang, C., Wang, X., Zhang, H., 2011. Excessive reliance on afforestation in China's arid and semi-arid regions: Lessons in ecological restoration. Earth-Sci. Rev. 104, 240–245.

Consultative Group on International Agricultural Research (CGIAR), 2013. Landscapes approach for sustainable development. Available at: http://www.cgiar.org/consortium-news/landscapes-approach-for-sustainable-development/. (accessed 25.11.14.).

Drummond, J., 1996. The garden in the machine: An environmental history of Brazil's Tijuca Forest. Environ. Hist. 1 (1), 83–105.

Freitas, S.R., Neves, C.L., Chernicharo, P., 2006. Tijuca National Park: two pioneering restorationist initiatives in Atlantic Forest in Southeastern Brazil. Brazil J. Biol. 66 (4), 975–982.

Government of India, 2011. Common guidelines for watershed development projects–2008 (Revised edition 2011). National Rainfed Area Authority, Planning Commission, New Delhi, India.

Gray, E., Srinidhi, A., 2013. Watershed development in India: economic valuation and adaptation considerations. World Resources Institute, Washington, DC. Working paper.

Haglund, E., Ndjeunga, J., Snook, L., Pasternak, D., 2011. Dry land tree management for improved household livelihoods: farmer managed natural regeneration in Niger. J. Environ. Manage. 92, 1696–1705.

IPCC, 2014: Climate change 2014: Impacts, adaptation, and vulnerability. Part A: global and sectoral aspects. contribution of working group II to the fifth assessment report of the intergovernmental panel on climate change. In: Field, C.B., Barros, V.R., Dokken, D.J., Mach, K.J., Mastrandrea, M.D., Bilir, T.E., Chatterjee, M., Ebi, K.L., Estrada, Y.O., Genova, R.C., Girma, B., Kissel, E.S., Levy, A.N., MacCracken, S., Mastrandrea, P.R., White, L.L., (Eds.). Cambridge University Press, Cambridge, United Kingdom and New York, NY, USA.

INBR: International Network for Bamboo and Rattan. 2013. Greening Red Earth: Restoring landscapes, rebuilding lives; INBAR Working Paper No. 76; IDRC. CRDI, INBAR; Beijing, China.

Karna, B., Shivakoti, G.P., Webb, E.L., 2010. Resilience of community forestry under conditions of armed conflict in Nepal. Environmental Conservation 37 (2), 201–209.

Laestadius, L., et al., forthcoming. The Carbon potential of forest landscape restoration. World Resources Institute. Washington, DC.

Livelihoods and Forestry Programme, 2008. Seven years of the Enhancing rural livelihoods through forestry in Nepal Contributions and Achievements. Available from http://www.msfp.org.np/uploads/publications/file/LFP_7_yrs_Achieveemnt_[2]_20120711122850.pdf. Accessed 17th August 2015.

Livelihoods and Forestry Programme, 2013. A decade of the livelihoods and forestry programme. In: Livelihoods and Forestry Programme.

Lugo, A.E., Helmer, E., 2004. Emerging forests on abandoned land: Puerto rico's new forests. Forest Ecol. Manage. 190, 145–161.

Nachtergaele, F., Biancalani, R., Petri, M., 2011. Land degradation SOLAW background thematic report 3. Food and Agriculture Organisation (FAO), Rome.

National Institute of Statistics of Rwanda (NISR), 2012. Food security improves in Rwanda, despite challenges. National Institute of Rwanda. Available from http://statistics.gov.rw/publications/article/food-security-improves-rwanda-despite-challenges.

Palanisami, K., Kumar, D.S., Wani, S.P., Giordano, M., 2009. Evaluation of watershed development programmes in India using economic surplus method. Agri. Econ. Res Rev. 22, 197–207 (July–December).

Pye-Smith, C., 2013. The Quiet Revolution: How Niger's farmers are re-greening the parklands of the Sahel. ICRAF Trees for Change: 12, World Agroforestry Centre, Nairobi, Kenya.

Rinaudo, T., Dettman, P., Tofu, A., 2008. Carbon Trading, Community Forestry and Development: Potential, challenges and the way forward in Ethiopia. World Vision Annual Review 2008. World Vision. Available from http://my.worldvision.com.au/Libraries/3_3_Responses_to_Poverty_2008_case_studies/Carbon_Trading_Community_Forestry_and_Development.pdf.

Rodrigues, R.R., Martins, S.V., Gandolfi, S., 2007. High Diversity Forest Restoration in Degraded Areas: Methods and Projects in Brazil. Nova Science Publishers Inc, Brazil.

SADC: Swiss Agency for Development and Cooperation. 2009. Asia Brief – Partnership Results Community Forestry in Nepal. SDC – South Asia Division. Available from http://www.intercooperation.ch/offers/news/Asia%20Brief%20Community%20Forestry%20final.pdf/view?searchterm=nepal%20forestry.

Searchinger, S., et al., 2013. Creating a Sustainable Food Future: Interim Findings: A Menu of Solutions to Sustainably Feed More than 9 Billion People by 2050. World Resources Institute, Washington, DC.

Sendzimir, J., Reij, C.P., Magnuszewski, P., 2011. Rebuilding resilience in the Sahel. Regreening in the Maradi and Zinder regions of Niger. Ecol. Soc. 16 (3), 1. Available from http://www.ecologyandsociety.org/vol16/iss3/art1/.

Sharma, B.R., Rao, K.V., Vittal, K.P.R., Amarasinghe, U.A., 2006. Realizing the potential of rainfed agriculture in India. International Water Management Institute. Available at, http://nrlp.iwmi.org/PDocs/DReports/Phase_01/11.%20Potential%20of%20Rained%20Agriculture%20-%20Sharma%20et%20al.pdf.

Swiss Agency for Development and Cooperation and Department for International Development (SDC and DFID), 2012. Development Assistance in Action: Lessons from Swiss and UK Funded Forestry Programmes in Nepal. Available from https://www.gov.uk/government/uploads/system/uploads/attachment_data/file/214309/dev-asst-action-lessons-swiss-uk-forestry-progs-np.pdf.

Tedla, S., 2007. Environment and Natural Resources as a Core Asset in Wealth Creation, Poverty Reduction, and Sustainable Development in Ethiopia. International Union for the Conservation of Nature (IUCN). October. Available from https://cmsdata.iucn.org/downloads/igad_ethiopia_report_2007.pdf.

United Nations Framework Convention on Climate Change (UNFCCC), 2009. Project design document form for afforestation and reforestation project activities (CDM-AR-PDD) Version 04. Available from http://cdm.unfccc.int/Projects/DB/JACO1245724331.7/view.

Winterbottom, R., Reij, C., Garrity, D., Glover, J., Hellums, D., McGahuey, M., Scherr, S., 2013. Improving land and water management. Installment 4 of "Creating a Sustainable Food Future". World Resources Institute Working Paper. Available from http://www.wri.org/sites/default/files/improving_land_and_water_management_0.pdf.

World Bank, n.d. Humbo Reforestation Project: Delivering multiple benefits. Available from http://wbcarbonfinance.org/docs/FINAL_STORY_green-growth-humbo.pdf.

World Resources Institute (WRI), 2008. World resources 2008: roots of resilience—growing the wealth of the poor. WRI, Washington DC.

Yamba, B., Larwanou, M., Hassane, A., and Reij, C., 2005. "Niger study: Sahel pilot study report." U.S. Agency for International Development and International Resources Group, Washington DC.

SELECTED CASE STUDIES

INDIGENUITY: RECLAIMING OUR RELATIONSHIP WITH THE LAND

8.1

Lili Hernandez Boesen, Stephen Hinton, Stephen Freeman

The Humanitarian Water and Food Award

8.1.1 INTRODUCTION

The Humanitarian Water and Food Award (WAF) identifies and promotes initiatives that have, in a sustainable and innovative way, brought water and food security to areas in distress. For many of the nominated initiatives, the foundation of their strategy for success has been reclaiming land, and indeed, reclaiming their relationship with the land, among the people involved, its climate, geography, plants, and animals. WAF staff members have found that people in the industrialized world generally only have a vague grasp of the quantity of land available for food production, its use, and its condition.

Arable land is defined by the Food and Agriculture Organization of the United Nations (FAO) as land under temporary crops (double-cropped areas are counted once), temporary meadows for mowing or for pasture, land under market or kitchen gardens, and land that is temporarily fallow; land that is abandoned as a result of shifting cultivation is excluded. It is a matter of simple mathematics that the greater the human population, the more productive land is needed. The reality is, of course, that the amount of productive land available for agriculture is not increasing. As soil erodes, the available arable land per person declines at a faster rate than population growth (Nattrass, 2013; for a detailed discussion of available land per person, see Wackernagel et al., 1999). Unfortunately, many have turned to cutting down forests to create more arable land. This is not a solution, however; it only perpetuates many problems, including the climate crisis. Importantly—and somewhat perversely—many developed countries require the agricultural development of land beyond their borders.

This section examines efforts that have been made to increase the productivity of agriculture (particularly local agriculture) in order to feed the world's populations in a more sustainable way.

8.1.2 RECLAIMING OUR RELATIONSHIP WITH THE LAND

8.1.2.1 GHOST ACRES AND CLIMATE CHANGE

The WAF gleaned several insights into food security from the UK Transition Towns Initiative (Hopkins, Thurstain-Goodwin, and Fairlie, 2009), which raises awareness about the fact that much of Europe's food comes from beyond its borders (Figure 8.1.1). Some of this food comes from so-called

Land Restoration. http://dx.doi.org/10.1016/B978-0-12-801231-4.00032-X

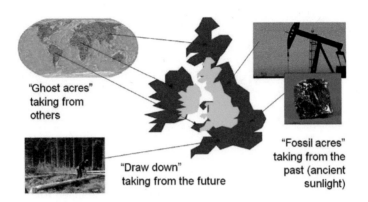

FIGURE 8.1.1

The UK: Living beyond our means. Transition Network (2007).

ghost acres (trading arable lands), which are defined as vast areas of land used to cultivate animal feed and processed food ingredients, several for consumption in distant places. This is one of several problems related to industrialized food and farming systems. Products normally are delivered in zones far from where they were produced, which therefore has increased the levels of fossil fuel dependency as a result of complicated food supply chains, including major greenhouse gas (GHG) emissions, unsustainable levels of irrigation, nonrecyclable food package pollution, and increased food freight transportation (Jones et al., 2011). Furthermore, according to some estimates, 2% of all fossil fuel energy is used to produce artificial fertilizers as the food demand grows. The indiscriminate use of fertilizers also have a major impact on soil deterioration and water contamination, affecting human health and the natural cycle of important nutrients in ecosystems (Galloway et al., 2008).

An important challenge, thus, is to find means of supplying modern cities with local food in order to reduce the dependence on fossil fuels (FAO, 2011) and to reduce the impact of climate change, which aggravates land degradation, reduces food security, and fuels unrest in many parts of the world.

8.1.2.2 LAND AND BUSINESS

Industrial activities can have a negative impact in different ways, such as air emissions, accidental spills into the environment, overuse of resources like energy, water, or raw materials, and the general land deterioration as collateral damage related to their core businesses. This may not be apparent without an evaluation of the impact on the ecosystem of the extended supply chain of food and water (Reisch et al., 2013). In the narrowest sense, this makes sense, since the manufacture of a screwdriver has little ostensible connection with food and water security and stability. However, the economy is a web of interlinked supply chains that have vastly varied interests. As the economy is, at its roots, based on natural resources, the environmental impacts of economic development affect every single company and interest.

According to Porter and Kramer (2011), doing good for society is likely to become the biggest brand differentiator in the coming decades. Corporations ought, therefore, to start supporting food and water security, as land restoration, land reclamation, and pollution prevention are real investments in prosperity. If all businesses contribute to create a food-secure world, entrepreneurship, prosperity, and ultimately peace will flourish.

8.1.2.3 FOOD AND WATER SECURITY, PEACE, AND PROSPERITY

Corporate social responsibility (CSR), the notion that corporations have a responsibility to the society and environment from which they draw their resources and whom they serve, is increasingly carried out in a way that supports businesses. For example, Lush, a UK-based cosmetics company, has been sending agricultural experts to the villages from which it obtains raw materials. These experts have helped villagers improve food production and have also taught them to refine the raw materials to the exact standard required by the cosmetics industry, thus bringing value-added production to these villages and increasing their returns (for instance, rather than selling crude natural resources to a refining plant, where much of the value of a product is added, they are able to secure the increased revenue from their products; also see FAO, 2008).

8.1.2.4 INDIGENUITY

The concept of indigenuity (a combination of the words *indigenous* and *ingenuity*) has been used to describe the relation between indigenous American tribes facing the climate change effects on their communities in respect with their environment by using traditional techniques to provide water and food for their people (Wildcat, 2013). WAF adopts this concept for a global reach and a general holistic and sustainable relationship between people and the land on which they live and also procures its fauna and flora to assure water and food security and community empowerment worldwide. "Indigenuity is developing of new relationships with each other and the land we rely on" (T. Lindgreen, personal communication, October 2010).

These efforts include diverse activities in the communities to help them reclaim their ecosystems, including developing an intimate knowledge of the ecosystems (soil, rainfall, etc.); establishing new roles for the corporations located on the arable lands so they can take care of the environmental impacts and support efforts to help everyone in the community participate in collaborations (CSR); redesigning agricultural practices to lower the impacts and offer integral and healthy solutions; collecting rainwater for procuring clean and fast access to this fundamental element; land stewardship with grazing animals for land recovery and providing food and economic support to the communities; and sharing land stewardship knowledge through schools and other educational resources (Figure 8.1.2).

This can be achieved as a global team effort, which WAF impulse by collaborating with different organizations such as The Prem Rawat Foundation and the WAF Initiatives.

The Prem Rawat Foundation created a program called "Food for People" which provides the necessary tools for communities to have food and water using their own resources (Prem Rawat Foundation, 2015). Prem Rawat is also WAF Patron, providing the support and exposure to communicate WAF's message of hope and better solutions for water and food secuirty challenges. (WAF, 2015b).

WAF Initiatives are selected every year by a WAF professional comittee. The best Initiatives are innovative, sustainable, replicable and deliver positive impacts and empowerment to their communities. This Initiatives are celebrated, promoted and supported by WAF for establishment of strategic collaborations and for obtain the necessary resources to achieve their goals. (WAF Award, 2015c).

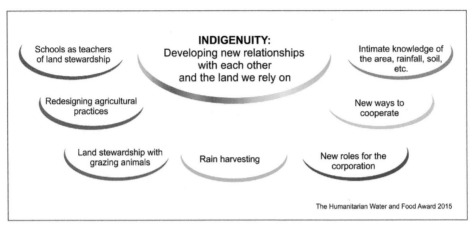

FIGURE 8.1.2

Indigenuity and some actions related to it. WAF (2015).

8.1.3 WAF INITIATIVES SHOW THE WAY FORWARD

8.1.3.1 RETHINKING LIVESTOCK: CATTLE ARE ANTIDESERTIFICATION ALLIES

8.1.3.1.1 Case Study: Africa Centre for Holistic Management (WAF, 2014)

The more deserts spread, the less inhabitable the Earth becomes. Desertification, exacerbated by climate change, is accompanied by soil erosion, which is mainly caused by the inability of the soil to retain water. And when the scarce rain finally does come, it washes even more soil away.

The Africa Centre for Holistic Management in Victoria Falls, Zimbabwe uses cattle in a way that mimics the behavior of grazing animals that created the savannas (Savory, 1999) thousands of years ago. Many believe that overgrazing caused desertification in the first place, leading to banning grazing in some areas to help restore the landscape. The Africa Centre for Holistic Management, however, believes that livestock can be part of the solution. By keeping a herd together in a small space for a short time, the animals' hooves break up the hard ground, allowing air and water to penetrate the soil. They trample down old grass, fertilize the area with their dung and urine, and keep perennial grasses healthy with their grazing. As the herd moves on, the land, thus suitably prepared, can restore itself.

The results the center achieved around the Dimbaganone River illustrate the effectiveness of this approach. The number of cattle and goats increased by 400%, as did the water flow in the river. This increased flow led to less competition between humans, livestock, and wildlife for water resources.

The center, thus, organized herds and planned grazing in a way that has increased the number of animals supported by the land, increased the fertility of the fields, and driven desertification back, creating a livestock watershed/wildlife habitat management tool for managing cattle that can be copied and reused on a wider scale (Figure 8.1.4a,b). These results can be achieved without expensive and complicated technology or massive use of chemicals, increasing the likelihood of reproducing these efforts elsewhere.

FRAMEWORK FOR HOLISTIC MANAGEMENT

© 2012 Savory Institute

Whole Under Management	Decision Makers	Resource Base	Money

Holistic Context		Forms of Production	
	Statement of Purpose		
	Quality of Life		
	Future Resource Base		

Ecosystem Processes	Community Dynamics	Water Cycle	Mineral Cycle	Energy Flow

Conventional Decision Making	Objectives	Goals	Vision	Mission

Tools	Human Creativity	Technology	Fire	Rest	Grazing	Animal Impact	Living Organisms	Money and Labor

One or More Factors	Past Experience	Expert Opinion	Research Results	Expediency	Compromise	Cultural Norms	Cost, Etc.

Testing Questions Objectives and Actions	Cause and Effect	Weak Link • Social • Biological • Financial	Marginal Reaction	Gross Profit Analysis	Energy/Money Source Pattern of Use	Sustainability	Society and Culture

Management Guidelines	Learning and Practice	Organization and Leadership	Marketing	Time	Stock Density and Herd Effect	Cropping	Burning	Population Management

Planning Procedures (Unique to Holistic Management)	Holistic Financial Planning	Holistic Land/ Livestock Planning	Holistic Planned Grazing

Feedback Loop	Replan	Plan (Assume Wrong with Environment) Control	Monitor

Areas depicted in blue are components of conventional decision making still integrated into the holistic framework.

FIGURE 8.1.3

Holistic management framework used by the Africa Centre for Holistic Management. Savory Institute (2012).

(a)   (b)

FIGURE 8.1.4

The Africa Centre for Holistic Management in action. Savory Institute (2015).

8.1.3.2 RETHINKING FOOD PROVISION: GROW FOOD EVERYWHERE

The solution to ghost acres is to grow food locally, which can increase community cohesion and education, while developing new techniques to change the way food is grown.

8.1.3.2.1 Case Study: Todmorden, the Incredible Edible Initiative (WAF Award Nominee 2010)

The "Incredible Edible" initiative (Figure 8.1.5a,b), taking place in Todmorden, a town in Northern England, focuses on growing food everywhere for people to take as they wish. Incredibly, it is working, showing that kindness can indeed help. It is also an example of working in public spaces that has encountered very little vandalism. It was created to address the lack of high-quality, nutritious, and accessible food, as the food being brought into local supermarkets was found to be either too expensive or of poor quality. The founders of the initiative acted on the basis of the vision of everyone working together to achieve a world where all human beings share responsibility for the future well-being of our planet and our collective society. Inspired by its success, other enthusiasts have started projects to add to the initial concept, such as the Incredible Aqua Garden and the Incredible Farm.

FIGURE 8.1.5

Incredible Edible UK initiative (2015).

Incredible Edible aims to provide access to good local food for all by promoting cooperation, providing learning opportunities from field to classroom to kitchen, and supporting local business. Its approach is continuously developing, adding such elements as rainwater harvesting, local chicken rearing, food box delivery, and food enterprises springing up in the town as the need arises.

The initiative's network is an umbrella group to support and inspire members who believe that providing public access to healthy, local food can enrich communities. The UK network alone has more than 50 independent groups and, as a worldwide movement, it stretches from Canada to New Zealand, with 700 groups in 25 countries. Each of these global groups is different, but typically, their work involves setting up community growing plots, reaching out to schools, and backing local food suppliers. The initiators demonstrate how pressure can be taken off ghost acres, how nutrition can be improved, and thus how a much closer-knit community can be created. And they have fun doing it, which cannot be discounted as a motivational factor for change.

8.1.3.3 RETHINKING FOOD PROVISION: GROW FOOD IN FOOD DESERTS

The term *food deserts* is defined by the WAF as food-insecure urban areas where, for economic, logistical, or other reasons, fresh, nutritious food is unavailable to local residents. Fast food is often the only alternative, bringing with it obesity and other health problems.

8.1.3.3.1 Case Study: Growing Power (WAF Award Finalist 2014)

Growing Power, a nongovernmental organization (NGO) based in Milwaukee, Wisconsin, aims to help people find a pathway from malnutrition to healthy eating. It works to transform deserted urban properties into highly productive community food provision centers (Figure 8.1.6). The project has demonstrated effective ways to address the growing problem of food deserts in developed Western countries where young people and children too often have no or very limited access to fresh food. Growing Power's approach uses abandoned city plots and ingeniously integrates multilevel greenhouses, vermiculture, composting, aquaponics, beekeeping, and smaller livestock with waste recycling, solar energy, and heat recycling. Will Allen, the chief executive officer (CEO) of Growing Power, believes that a healthy local food system is the prerequisite for a healthy community: "If people can grow safe, healthy, affordable food, if they have access to land and clean water, this is transformative on every level in a community." (Growing Power, 2014).

In Growing Power's concept, everything works in an integrated way. For example, dirty water from fish farming is recycled by growing vegetables over the tanks, converting the waste into usable nutrients; meanwhile, compost provides a source of heat for the greenhouses.

Presently, Growing Power feeds a variety of community members, including schools and formerly poorly served neighborhoods, with organic production. Moreover, the initiative offers a wide array of education and demonstration opportunities, with a focus on youth.

FIGURE 8.1.6

Growing Power (2015).

Their message is clear, attractive, and spreading. Over the course of 2014, the organization installed 30 gardens at licensed day-care centers in Milwaukee, trained more than 400 people in community food systems, and distributed 3000 food bags of fresh produce in the "Farm to City Market Basket" program. A total of 30,000 visitors toured the Community Food Center and over 1450 people attended an Urban Agriculture and Small Farms Conference organized by the NGO to share ideas on the topic of "Building a Fair Food Economy to Grow Healthy People."

By setting up intensive urban agriculture to serve and involve the community, Growing Power demonstrates that urban food insecurity can be overcome through the power of the community to self-organize and a simple combination of available solutions.

8.1.3.4 RETHINKING WATER TECHNOLOGY: THE SUN AS PURIFIER AND HEATER

8.1.3.4.1 Case Study: SOLVATTEN—Solar-Powered Water Purification (WAF Award Finalist 2010)

The Swedish company Solvatten, based in Stockholm, has developed solar-powered technology to purify drinking water, eliminating the need to cross long distances to find uncontaminated sources and collect firewood to boil, and thus purify, the water, as many families in sub-Saharan Africa typically do. Not only the practice of boiling water using wood is time consuming and can expose households to toxic levels and black carbon particles if it is not controlled properly, but it also razes forests for the wood, leading to a host of environmental problems, including increasing CO_2 emissions, erosion, and decreased local precipitation in already parched areas.

The technology is a water purifier that uses sunlight and requires no fuel, which means that it can be used to purify local water sources (Figure 8.1.7a). This mechanism heats the water to 75 °C, which is hot enough to kill most germs. The device looks like an ordinary 10-L container filled with water. Inside, the can is split in half by two transparent walls that can be folded out like the covers of a book. The raised container is placed in the sun and the translucent walls are permeable to ultraviolet (UV) light, killing harmful microorganisms. The treatment takes about 3–6 h, with an indicator showing when the water is drinkable. Macroorganisms that survived the UV treatment are removed with a built-in filter, resulting in purified water that meets World Health Organization (WHO) guidelines for safety.

The system units are extremely portable (Figure 8.1.7b), making them easy to handle for both children and adults. They are safe, reliable, sustainable, and socially and culturally acceptable. The process

(a) (b)

FIGURE 8.1.7

Solvatten (2015).

requires no chemicals or energy sources other than the sun, and the process can be repeated two or three times a day, yielding 20–30 L of clean drinking water from a single 10-L container. Correctly managed, a single Solvatten container can provide clean drinking water for up to 10 people every day for a period of 7–10 years.

This invention is used around the world to improve health, empower women, and help the environment. The risks associated with gathering firewood in forests (including the exhaustion that people feel after traveling long distances from their villages, not to mention the potential rapes and assaults they may suffer on the road) are reduced. Additionally, the decreased need for less firewood corresponds to a higher productivity among women.

Solvatten has demonstrated that the need for firewood can be reduced, leaving more trees untouched, which prevents soil erosion. With less wood burning, carbon emissions are also reduced. The reduction in black carbon emission has health benefits as well.

8.1.4 CONCLUSION: CONSTRUCTING A NEW NARRATIVE FOR FOOD SECURITY

The initiatives discussed in this text represent just a fraction of the amazing and diverse work of people all over the world. These efforts come together to create a new narrative that combines food and water security; it calls upon all of us to rethink our relationships with the land, each other, and our domestic animals, while rediscovering the power of dignity and of our communities to create health, peace, and prosperity.

REFERENCES

Food and Agriculture Organization of the United Nations (FAO), 2011. "Energy-Smart" food for people and climate. FAO, Rome.

Food and Agriculture Organization of the United Nations (FAO), 2008. Products: Sandalwood oil deal with Lush (UK). Non-wood forest products digest, 3 http://www.fao.org/forestry/49987/en/#P321_36484.

Galloway, J. N., Townsend, A. R., Willem Erisman, J., Bekunda, M., Cai, Z., Freney, J. R., 2008.

Growing Power. (2014). http://www.growingpower.org/about/.

Hopkins, R., Thurstain-Goodwin, M., Fairlie, S., 2009. Can totnes and district feed Itself? Exploring the practicalities of food re-localisation. Transition Network and Totnes Working Paper.

Incredible Edible, Todmorden, England, 2015. http://www.incredible-edible-todmorden.co.uk.

Jones, A., Pimbert, M., Jiggins, J., 2011. Virtuous circles: Values, systems, and sustainability. International Institute for Environment and Development (IEED) and the IUCN Commision on Environmental, Economic and Social Policy (CEESP). 34–40.

Martinelli, L.A., Seitzinger, S.P., Sutton, M.A., 2008. Transformation of the nitrogen cycle: Recent trends, questions, and potential solutions. Science 320 (5878), 889–892.

Nattrass, B., 2013. Figure 2.2. The natural step for business: Wealth, ecology and the evolutionary corporation. New Society Publishers, Gabriola Island, BC, Canada.

Porter, M.E., Kramer, M.R., 2011. Creating shared value. Harvard Bus. Rev. 89, 62–77.

Prem Rawat Foundation. (2015). The Food For People Story. http://www.tprf.org/programs/ffp.

Reisch, L., Eberle, U., Lorek, S., 2013. Sustainable food consumption: An overview of contemporary issues and policies. Sustain. Sci. Prac. Pol. 9, 7–25.

Savory, A., 1999. Holistic management: A new framework for decision making. Island Press, Washington, DC.

Savory Institute. (2012). Framework for holistic Management. http://savory.global/assets/docs/evidence-papers/framework-for-holistic-management.pdf.

Savory Institute. (2015). http://savory.global/.

Solvatten. (2015). http://www.solvatten.se.

The Humanitarian Water and Food Award, WAF (2014). http://www.wafaward.org/#!finalists2014/c1ikp.

The Humanitarian Water and Food Award, WAF (2015). http://www.wafaward.org.

Wackernagel, M., et al., 1999. National natural capital accounting with the ecological footprint concept. Ecol. Econ. 29 (3), 375–390.

Wildcat, D., 2013. Introduction: Climate change and indigenous peoples of the USA. Clim. Change 120 (3), 509–515.

LAND RESTORATION AND COMMUNITY TRUST: KEYS TO COMBATING POVERTY

8.2

A CASE STUDY FROM RURAL MAHARASHTRA, INDIA

Jared Buono[1], Jayashree Rao[2]

Ecohydrologist, Mumbai, India[1] Grampari, Maharashtra, India[2]

8.2.1 INTRODUCTION

Landscapes around the world are being degraded as demand rises for natural resources such as water, fuel, and food. The loss of biodiversity and vegetation cover and the mining of soil and water resources have damaged many ecosystems and impaired their ability to support future human needs (MEA, 2005). For the world's poor, most of whom still reside in rural areas, this is particularly concerning, as the local landscape is the main source of sustenance and income (IFAD, 2010). Further, ecosystem pressure will increase as the demand for resources grows with population and improved standards of living. Include the emerging impacts of a changing climate, and this has the potential to make poor, land-dependent communities more vulnerable and further complicate efforts to eradicate extreme poverty. Meeting the resource needs of a growing and changing world while ensuring environmental sustainability is one of the defining challenges of our time.

This is the domain of many land restoration efforts—reversing ecological degradation to reduce the vulnerability of local communities. While there have been many successful restoration initiatives, one of the biggest impediments to success remains the comprehensive inclusion of local communities. Ecologically, stakeholder buy-in can be essential for addressing the root causes of ecological decline, as local resource demand is what drives many negative ecosystem pressures. This is also crucial programmatically because poor communities lack institutional and regulatory support, making the local people the sole stewards of the landscape. Despite these imperatives, community inclusion in the restoration processes tends to be hasty or overlooked altogether because it is daunting and difficult. It is a long and complex process that can increase the time frame and budget and can take what seemed like a simple tree-planting project into areas of livelihoods, gender, behavior change, and local governance—issues that confound and deter funders, practitioners, and communities alike. But community empowerment and local decision making are often exactly what are needed to inspire behavior change and the adoption of best practices to sustainably restore ecological function within the world's most vulnerable communities and ecosystems.

Land Restoration. http://dx.doi.org/10.1016/B978-0-12-801231-4.00033-1

Rural India provides an excellent case-study environment for balancing resource demand and ecosystem health and the challenges of community inclusion. Ecosystems there are under immense pressure as the population grows and demand for resources increases along with the standard of living.

8.2.2 LOCAL ECOSYSTEMS AND ECONOMIC SETTING

India's ecosystems are under intense pressure from a high population density and rapid economic growth. In many ways, the country exists in the crowded future that we all face, where it is becoming increasingly difficult to balance social development and improved standards of living while maintaining the ecosystems that sustain us. Already leading the way and overcoming some of these issues, India has made significant progress toward some Millennium Development Goals. For example, 543 million people have gained improved access to safe drinking water since 1990 (WHO/UNICEF, 2014). This is likely one of the largest and most rapid improvements in human conditions in history and deserves to be lauded as being due to the hard work of many people—but at the same time, it is not without its drawbacks. In the case of improved water access, many of these gains have come from sinking of over 20 million wells, leading to overdependence on rapidly diminishing groundwater resources (Economist, 2009). Solving these types of challenges makes India a model for the rest of the world. Can we provide equitable access to ecosystem goods and services while increasing economic growth?

Nowhere is the interplay between growth and sustainability starker than in the Western Ghats, a mountain range spanning much of India's west coast. Home to both affluence and poverty, it is a study in extremes, with burgeoning cities such as Mumbai, which drive global commerce, to rural farm villages subsisting on rainfed agriculture as they have for centuries. Recent economic growth is apparent in cities and villages alike. Hours from Mumbai's financial centers, the countryside is dotted with billboards for mobile phones and Internet providers and villages with satellite dishes sprouting within them. This reflects the nascent purchasing power of rural households—a huge success for a country with hundreds of millions of people living in poverty. But again, there have been costs and tradeoffs. Recent growth and historic demand for ecosystem goods and services have led to widespread ecological degradation. Intensive agriculture, heavy grazing of domestic animals, harvesting of fuel wood, annual anthropogenic fire, construction, and groundwater exploitation have resulted in deforestation, biodiversity loss, erosion, and a declining water table. What was once an unbroken, closed-canopy forest covering much of the mountain range is now open grassland with sparse trees and isolated sections of forest; only 6% of the primary vegetation in the Western Ghats remains intact (Molur et al., 2011).

This represents a loss in ecological capacity that has the potential to affect millions. A world biodiversity hot spot, these mountains directly support 120 million people with water, food, and energy and over 400 million people indirectly as the headwaters for nearly every river basin in the lower half of India originate here (Molur et al., 2011).

This scenario is playing out all over the world. Ecosystems are in decline due to the escalating demand for resources. Not only does this diminish the amount of goods and services the ecosystem can provide to dependent communities, but the real danger is crossing an ecological threshold where the land becomes permanently diminished. While healthy ecosystems are resistant to change and resilient in terms of recovery, sustained human pressure, such as collection of fuel or the intensive grazing of domestic animals, disrupts the dynamic processes that maintain healthy landscapes. In the Western Ghats, the nutrient cycle, in which resources vital to plants and animals (such as nitrogen, phosphorus, and carbon) are circulated through biological and physical processes, is disrupted by the grazing of domestic animals and the collection of cooking fuel. Cattle eat grass and the nitrogen it contains, but rather than being integrated back into the soil after being consumed and excreted by the animals, the dung is collected, dried, and used as cooking fuel. Nitrogen is lost to the atmosphere during combustion, and nitrogen levels in the soil decrease, which leads to less grass for future cattle. A positive feedback loop is created, pushing the landscape toward vegetation loss, soil erosion, and reduced biocapacity. Beyond a certain threshold, ecosystems enter a new normal state and cannot return to historic conditions. This type of pressure has pushed many ecosystems to the breaking point, where the landscape is no longer capable of providing historic levels of ecosystem services such as fuel, fodder, food, and water.

It is estimated that half of the world's population is affected by this type of degradation of mountain watersheds (Hassan et al., 2005). This is dire news, particularly for many of the world's poorest communities. Despite rapid urbanization, 70% of the world's poor continue to reside in rural areas (IFAD, 2010). Rural dwellers are more likely to depend directly on the land to meet their most basic everyday needs, deriving their sustenance and any income from the local landscape. This makes these communities highly vulnerable to any decreases in ecosystem services or any disturbances, such as drought, conflict, or changing climate, that can result in food insecurity and reduced livelihood options. This is considered one of the largest barriers to global development and meeting Millennium Development Goals (MEA, 2005). Thus, halting and reversing ecological degradation are vital to eradicating extreme poverty.

This type of challenge is apparent in the Western Ghats today. Economic growth and development efforts have had a positive impact in the lives of many, and strides are being made in the standard of living, but new barriers to development are being generated at the same time. The rise in commercial, industrial, and leisure development is driving up demand for land and water throughout the mountain range. This is causing some communities to lose hard-won gains in access to water and land for livelihoods. New hotels, commercial farms, and factories provide jobs on the one hand, but competition for water on the other. With each new construction project, wells are sunk into the same aquifer shared by adjacent villages, and competition is on the rise as groundwater drops and once-perennial wells run dry. Even within villages, economic development has presented new challenges. Wells used to be the domain of the village, financed and managed as a common resource, but as farmer incomes climb, private wells are now affordable and have become more common. The problem is that there is not that much water in the ground. Overpumping is depleting aquifers and many communities now face severe water shortages during the dry months. Competition for and conflict over water are constant issues and are likely to rise under increased climate change. On the one hand, people have more income, but on the other, this has opened up a race for water in which everyone suffers as a result. This is how many communities are losing recent gains in access to safe drinking water.

8.2.3 A RESTORATION PROGRAM IN THE WESTERN GHATS

Water has always been a challenge for communities in the Western Ghats. Despite receiving significant rainfall, communities struggle to find water in the dry season. Open dug wells often go dry within months of the end of the monsoon, and many bore wells never produce any appreciable water. Villages that do not reside adjacent to a large river or downstream of a dam often must rely on tanker trucks provided by the government bringing water from other areas. Water rationing is common. Much of this is due to local geology. Created by massive volcanic eruptions 60 million years ago, the entire mountain range consists of horizontal layers of fine-grained, impervious basalt, which prevents the flow of groundwater. This means that there is a natural limit to the amount of water stored during season monsoons (Naik et al., 2002).

Compounding the natural limitations on water, degradation of the ecosystem by human activities has further reduced the ability of the landscape to capture rain. The deforestation from annual burning, land clearing, overgrazing, and tree cutting for construction and fuel, as mentioned previously, has reduced vegetation cover and depleted the soils necessary to slow and absorb the intense rainstorms. Land restoration that addresses the loss of forest and soil resources has the potential to improve watershed function and increase the amount of water available for local villages and regional cities alike.

This is why we started a land restoration initiative. Most communities in this part of Maharashtra prioritize water as their greatest challenge, in terms of both quantity and quality, for household and agricultural use. We therefore embarked to develop a program that restores land to provide more water for villages. We partner with willing communities to identify and protect water sources through community-based, participatory approaches. This mainly translates to spring protection because that is what most communities in the area rely on (Figure 8.2.1). So we endeavor to map geology and aquifers, in conjunction with the community, to locate the most likely source areas where groundwater recharge is occurring (Figure 8.2.2). Once the sites are identified, we help provide options for focused ecological restoration in those areas. This can be anything from protecting an existing forest from encroachment, fire, or grazing to a complete restoration of a springshed that has been completely denuded, including revegetating, soil strengthening, and installation of watershed and recharge practices (Figure 8.2.3). We

FIGURE 8.2.1

A typical village setting with a central water tank and livestock guzzler. Water is collected from a spring in the hills in the background and fed by gravity down 2 km of pipe.

Credit: Jared Buono

FIGURE 8.2.2

Community members surveying an unimproved springbox—a plaster box placed near the outlet of a spring used to collect water. We work with communities to upgrade boxes like this to better protect the spring and prevent contamination of the water.

Credit: Jared Buono

present technologies as options for community-led decision making and facilitate the entire process, from problem identification to planning, implementation, and long-term maintenance. As explained previously, this can be a lengthy and involved process, as we ask communities to invest a lot of time in the participatory activities. We also ask that they contribute significant amounts in terms of labor and funds—often, all funding comes from the communities because most of the impact comes from changing land use practices and behaviors (Figure 8.2.4). And from water, we usually expand into other areas, including governance, livelihoods, and sanitation and hygiene. Our village partnering processes usually takes 1–2 years, and thus far, our small team has worked with about 25 communities.

The response over the past 8 years has been good. Most communities are already demanding this type of work since water is perceived as a serious issue. That makes mobilization relatively easier, but there are still significant challenges, even in the most drought-prone areas, when it comes to changing land use practices and longstanding behaviors. Annual burning of the landscape continues in most places despite awareness of how this practice reduces infiltration and groundwater. We have had some success in that communities are protecting springsheds from fires and tree cutting. Grazing remains a

FIGURE 8.2.3

Contour trenches like this one are used to restore a degraded springshed. The trench harvests rainwater, prevents soil loss, and collects seeds and nutrients to inspire the revegetation and infiltration of water.

Credit: Jared Buono

FIGURE 8.2.4

Community members building a new springbox for their community. This village volunteered a week of labor, carrying building materials up a mountain to improve their water supply. The village funded the project independently through household donations.

Credit: Jared Buono

stress on the landscape, including spring recharge areas, probably because it is so closely tied to income from livestock. But we will continue to work closely with communities to find alternatives and solutions. Next, we share some of the practices that we found useful.

8.2.4 COMMUNITY PARTICIPATION FOR RESTORATION

Our community programs are participatory in nature, meaning that they seek to put the intended beneficiaries at the heart of all aspects of development—from conceptualization and planning to implementation, monitoring, and maintenance. This is done for a variety of reasons. While there are many successful examples of large-scale land restoration, such as the great green wall in China or the greening of parts of the Sahel (Reij et al., 2009), one of the greatest challenges remains the inclusion of and investment by locally dependent stakeholders and communities, an essential component of the restoration process, from both ecological and programmatic standpoints.

From an ecological perspective, stakeholder involvement is critical because many ecological pressures originate locally. In contrast to global or regional impacts such as drought or climate change, ecosystems suffer from a variety of pressures from human residents, such as fuel wood collection, annual burning and land clearing, intensive grazing of domestic animals, and groundwater exploitation. These are some of the most common causes of ecosystem decline, all of which can be tied to local sustenance and livelihood activities. It sounds simple and obvious, but this factor is often overlooked.

In Maharashtra, for example, most watershed projects include tree planting and digging of contour trenches—shallow pits perpendicular to the slope that capture water, soil, nutrients, and seeds during heavy monsoon rains. While these may be effective practices required to prevent erosion, promote reforestation, and restore watershed function, they do not inherently address the root cause of the degradation (namely, the overgrazing of domestic animals, tree cutting, and annual burning). These activities, all tied to local livelihoods, are ubiquitous in Maharashtra's mountainous areas and are the cause of widespread deforestation and erosion. They directly reduce the amount of water captured and stored during monsoons by exposing soil and allowing flood runoff. These activities must be addressed, and alternative livelihood options must be identified, in order to restore watershed function. Any intervention that does not identify the root cause of degradation is simply a stopgap, temporary measure. This point cannot be overemphasized, as it causes many restoration initiatives to fail.

Programmatically, it is crucial to incorporate local stakeholders, particularly in impoverished areas that tend to lack public services. In these areas, regulatory frameworks and on-the-ground institutions, such as extension agents or natural resource managers, can be nonexistent. Therefore, environmental stewardship and land management falls entirely on local individuals and grassroots institutions.

Participatory approaches that aim to include the intended beneficiaries in the planning, implementation, and maintenance of development programs have been around since at least the 1970s, and they have been largely incorporated into development philosophies. Their aim is to help ensure program sustainability by transferring ownership to community members through engendering participation in program planning, implementation, monitoring, and long-term maintenance, using tools such as participatory rural appraisals that are designed to engage and mobilize communities around an issue.

But engaging communities does not automatically lead to the adoption of new practices. Asking communities to rethink and change their historical practices is an intensive process that requires considerable investment by everyone involved, including the stakeholders themselves. Inspiring a behavioral change at the programmatic level, such as the sustained uptake of alternative practices, is still an emerging field in restoration and conservation efforts. For example, building awareness of an environmental issue does not typically translate to a change in behavior, and human thinking tends toward shortsighted responses because we often separate ourselves from nature—and the problem (Schultz, 2011). While the public health field has embraced behavior change challenges in the development context, the environment sector has not. We need a focus that takes us beyond participation, to building the science of grassroots environmental stewardship.

Inspiring behavior change, therefore, represents a significant process. It takes time to generate demand, build awareness, empower decision making, and inspire lasting change. These activities can add months or even years to a restoration initiative. From a programmatic point of view, you are frontloading costs, with very little concrete evidence to show for your work. And what started as a simple reforestation project has become a holistic program that includes issues of livelihoods, gender, local governance, and institutional capacity building. This is precisely why many restoration initiatives fall short of their goals, as it is far easier and faster to implement infrastructure than to fully address and eliminate the root causes of ecological decline through this process. But we call this treating the symptoms, not the disease.

However, agencies, donors, and implementers are not the only ones resistant to this level of investment. What are we asking of the stakeholders themselves? We are asking them to change the way they live their lives in often fundamental ways, and in many cases, this may intersect with the traditional way they have always done things—to learn new ways of living, to adopt new behaviors, to make time for learning new concepts—all in the context of very busy and very hard lives. It cannot be overstated how big of a commitment we are asking people to make.

Therefore, communities must buy into this process. That means going well beyond building awareness around an issue; it means that we have to build demand. Why? Because individuals must be willing to take time from their busy lives to partake in numerous meetings and activities. It takes effort to learn basic hydrogeology to be able to make decisions about village groundwater supply. Meetings often have to be held at night, starting as late as 9 p.m., to ensure the participation of women who are otherwise occupied throughout the day with cooking, washing, child rearing, and collecting water and wood. Asking people to do this over the period of months or even longer is a significant investment on the part of all shareholders (Figure 8.2.5). Yet, that is often what is required to halt degradation of the very ecosystems on which people depend. That is why building demand for and trust in this process is essential.

8.2.5 OPPORTUNITIES FOR BUILDING TRUST

We often find that while communities may understand the need for, and are willing to undertake, activities, such as tree planting or installation of check dams to restore watershed function, there is little desire to examine, let alone alter, long-held traditional activities that necessitated these interventions in the first place. For example, mobilizing a community to dig contour trenches (shallow pits to harvest water, nutrients, and seeds) is often easy compared to asking them to undertake managed grazing or adopt new cropping patterns—this is construction versus behavior change. One must first build faith and trust in the process, which then makes this often lengthy and involved journey more feasible. Next, we highlight some of our experiences with this process of building demand and trust and inspiring

FIGURE 8.2.5

A group of farmers learning about hydrogeology and ecology to better manage the springs that supply their village. This community participated in 18 months of meetings and surveys to learn, plan, and implement an improved water supply.

Credit: Milind Dange

behavior change, particularly those elements that are not often presented in community development literature and guides.

Although availability of water during the dry season is the primary concern for many villages in the area, there is often little demand for awareness-generating and capacity-building initiatives. Misconceptions about groundwater (that there is an unlimited supply, for example) must first be dispelled. There is little to no demand for ecological understanding. So our programs often start with trust-building projects. These are small, manageable initiatives conducted early in the process to demonstrate commitment and show that something will come of the process. One example is making small improvements to a village's drinking water systems, such as fixing leaks, replacing pipes, or building springboxes. This engages the community and helps establish trust early in the process. Trust projects need to be undertaken with care, as a failure at this stage can harm relations and set the process back. It is best to start with small, feasible projects; this often requires resisting pressure from communities that want to move quickly. In Maharashtra, government watershed restoration programs have initial infrastructure components such as libraries, solar streetlights, and gyms that villages can opt to use. These act as incentives, but they also are "proof of concept" that the process will yield results.

Community immersion is very important. The more time one spends in a community, the more we can hope to understand the issues, challenges, and dynamics at play. This includes living and working with the people in the communities. Often it takes spending a few nights in a village to see activities that only occur outside working hours. One example is open defecation. Most communities are not initially willing to share information about potentially embarrassing activities with development workers, such as open defecation, so many will tell you that it does not occur, even if it is a common practice. When we spend nights in the community, we can see these things firsthand during the peak hours of twilight and dawn. At one community meeting, where we were mapping activities across a village and we asked, "Where does open defecation occur?" the initial response was, "That does not happen in this village." But one of the villagers then stood up and said, "These people have stayed in our place and have seen what happens, so you must tell the truth!" It was a lighthearted exchange but a serious issue—to gain people's acceptance and trust, immersion will accomplish a great deal.

Local leadership is essential. Whether an elected member of a village council, a traditional elder, or simply a person ready to take responsibility, sharing a vision is the basis for restoration activities. If there is no strong leadership, or if you don't share common goals, then you should consider putting your efforts and resources elsewhere until these constraints are rectified. And it is important to keep in mind that leadership can come in many forms and from many quarters. That is why it is important to spend time in the community to identify its real leaders. While it is possible to assist in the development of local leadership, keep in mind that it will take time and energy from your other goals. We have built entire community programs around motivated local leaders. One example came from the husband of the elected village head, who, after attending one meeting with us on community responsibility, went back to his village and worked with the village council to improve water supply, sanitation, solid waste management, and substance abuse (by banning the sale of alcohol). The response from the community was so positive that we joined and supported his work for the next 2 years.

Be firm, but also have compassion. In some parts of the world, like India, rural communities are bombarded by plans, schemes, and promises. Do not assume that your goodwill and honest intentions will be taken at face value. Sometimes, in order to generate demand, it is best not to be too accessible. For example, we insist that women attend all meetings. In this part of India, they are usually busy all day with cooking, firewood collection, washing, caring for children, and farm work until after dinner. That means meetings have to be held at 9 p.m. or later, as already mentioned. However, if there is enough demand for whatever is happening at the meeting, then people will make time to come. Having meetings that are relevant, inclusive, and entertaining encourages wider participation. So there is a balance between adjusting to the community's needs and asking them to make sacrifices to improve their community. We have found that sometimes it is better to be firm and ask people to come during the day. The same holds true for communities that promise to attend a meeting and then don't show up. We usually do not go back until they can demonstrate a sincere desire to participate.

Cast a wide net. For a variety of reasons, communities work at their own pace. This doesn't often fit with things like grant work plans and fiscal-year reporting. Our year, in terms of work planning, is usually determined by the farming calendar and rain season. Few farmers are available to work during much of the farming season (from September to January), and many are not free until much later, such as March. That means we have March through mid-June to do most of the community work. If all your time and effort are focused on one or two communities and, for whatever reason, those communities fail to come together during the crucial time of year, then the organization has lost an entire year, and meanwhile, other villages that might have been ready to work have missed an opportunity. Therefore, it is always best to be working with multiple communities, at various stages of the projects, in case you need to make a last-minute change in plans. This can also mean doing other community development work in parallel with restoration efforts. Not only does this help promote a robust, holistic development, but also it gives an opportunity to maintain progress and stay connected with a community during a long break in seasonal work.

We have also found that sharing stories of where we have made personal decisions for change and resolved conflicts or ill feelings by apology or forgiveness are very powerful motivators for change, as are moments of silence. During community meetings, we often include 5 min of silence. Some of the issues we have seen tackled are those of ego, the acceptance of (or at least willingness to listen to) another person's point of view, or sometimes even finding innovative solutions to difficult problems. For example, we have seen that land disputes sometimes result in land degradation, as no one is able to overcome the conflict to manage the land under dispute. Holding quiet time and asking people to conduct inner listening have provided insight into solving some of these issues. This practice often takes participants onto another level that opens up space for dialogue. For many villagers, this is their first time holding a microphone and speaking out in public, so quiet time devoted to inner listening helps them focus their thoughts and center on the issues at hand. Another means for doing this is to literally use another space, such as a meeting hall in another community. The act of moving from familiar surroundings can break down barriers between people and groups, as well as changing the power dynamic and allowing others the platform to share. Because of this, we invite communities to our center for meetings or even overnight workshops—and the results are overwhelmingly positive.

Another powerful motivator is modern media, such as the screening of short videos of successful villages in the area. These videos show the ease with which high-quality, effective work can be done. Making this an achievable reality by showing videos from nearby villages and then interspersing this with clips of their own village can be an immensely motivating tool. One example is seeing the impact of effective groundwater recharge by contour trenching or tree plantation and water harvesting and then showing a film clip of the current condition of their own village. Showing negative images can be just as powerful—while villagers will walk past open defecation sites and plastic dumps every day, seeing photos of these places projected on a big screen at a village meeting always brings about motivation to work on it.

VOICE FROM THE VILLAGE – A COMMUNITY MEMBER COMMENTS ON THIS WORK

"My name is Ashok More from Patchputewadi village. It is a nice village and very beautiful in the hills of this part of Maharashtra. But we struggle in many ways. Our houses have little light. The floors are made of mud and there is smoke from the cook stoves. Our biggest problem is jobs and water. Our young men migrate to Pune and Mumbai for jobs. And we spend time collecting water for our houses. Especially the women and girls."

"I had to migrate many times to other places for work. I worked as a painter in Gujarat. But I would have to leave my family for many months and return back. I was able to save enough money to buy bullocks and I now rent them for carts and plowing the fields. I also have land and farm rice and wheat. But water is less and I cannot grow after the rains end."

"Everyone is worried about water. So when the village had a meeting about our springs and talked about increasing the water, I was excited to help. After the first meeting, we met in the morning to walk up the mountain. Our springs, where all our water comes from, are in another village's land on top of a mountain. We made an agreement many years ago to pipe the water (2 km) down to our village tank. That day we walked up the pipeline. We saw it was old and the pipes were leaking. At the springs, the springboxes we built many years ago were also leaking and old. We learned a lot that day. And when it was hot and we didn't have water (on the trail), I climbed a tree to get us honey from wild bees."

"The people were not interested to fix leaks. We wanted to find more water at first. But when we measured, it was almost 200 liters a day in lost water. So we made a plan to fix the leaks in the pipes and we built new springboxes. We had many meetings to do this."

"The people from my village, women and men and boys, worked for many hours and built new springboxes. We collected money from each house and had each house send workers. We hired a mason from nearby and we carried the sand, cement and bricks up the mountain on our heads. In the first year we build three new springboxes and made three more in the next year. The water was clean and enough. For the first time in many years, we did not need to get a water tanker truck to come for our Jatra Festival (a post-harvest festival held early in the new year). Everyone was happy we did this work."

"We also learned about where our water comes from. The rocks on top of the mountain are laterite. They take in the rain and goes down inside until it comes to the basalt. The water cannot pass the basalt, so it comes out of the springs. If we put the forest back on the mountaintop, we will get more water. All this learned over many meetings and walks up the mountain."

"Some months after the end of the work, we found less water reaching our village tank. So we trekked back up the mountain to see what was happening. Just near our springs we saw construction, many digging machines and new bore wells being made. We were surprised because the land belonged to a small farmer. Now there were many large excavations. Some of our springboxes were damaged and workers were using the place to defecate."

"The village heads tried to stop this by appealing to higher authorities, but there was nothing they could do. The springs are now drying. Water finishes early after the rains. We now get water from tanker trucks sent by the government. And a pipe pumps water up from a nearby dam when needed but it costs more money. Our water is lost."

Ashok More is now convinced that the protection of springsheds can help all villages in the area. He considers that his sharing his knowledge of geology and springwater is important work. He began meeting with other local communities to help them find and protect the sources of their water. So two years ago, he was hired by Grampari and now works full time on spring protection and building awareness of this issue. He works hard climbing mountains to map geology and build springboxes around the region. And he recently sold his bullocks to concentrate on springs full time.

8.2.6 CONCLUSIONS

Land restoration encompasses so much more than implementation of improved environmental practices. The United Nations Environmental Programme (UNEP) has recognized this and recently launched what it calls "environmental governance," which includes raising awareness, empowerment, and governance (UNEP, 2012). Similarly, in India, watershed management approaches have evolved to include livelihood considerations under a program called "Watershed Plus." And while there is a growing list of tools and approaches for ensuring stakeholder buy-in, this remains one of the largest impediments to environmental sustainability; even more than improved technologies, ecological understanding, or innovative best practices, we need an increased focus on overcoming our challenges in human behavior and relationships.

So while we often know what is needed to restore ecosystem balance at a landscape level (i.e., reforestation, wildlife corridors, or ecological flows in rivers), and we have developed and tested many best practices (such as controlled grazing, integrated fire management, and watershed practices), where we continually seem to fall short is in the sustained uptake of alternative behaviors.

REFERENCES

Economist 2009. India's water crisis: When the rains fail. The Economist, September 10, 2009.

Hassan, R., Scholes, R., Ash, N. (Eds.), 2005. Ecosystems and Human Well-Being: Current State and Trends Vol. 1. Island Press, Washington, DC.

International Fund for Agricultural Development (IFAD), 2010. New realities, new challenges: New opportunities for tomorrow's generation. Rural Poverty Report 2011, International Fund for Agricultural Development (IFAD), Rome, Italy.

Millennium Ecosystem Assessment (MEA), 2005. Ecosystems and human well-being: General synthesis. Island Press, Washington, DC.

Molur, S., Smith, K.G., Daniel, B.A., Darwall, W.R.T., Compilers, 2011. The status and distribution of freshwater biodiversity in the Western Ghats, India. Zoo Outreach Organisation, Cambridge, UK, and Gland, Switzerland: IUCN, and Coimbatore, India.

Naik, P.K., Awasthi, A.K., Mohan, P.C., 2002. Springs in a headwater basin in the deccan trap country of the Western Ghats. India. Hydrogeol. J. 10, 533–565.

Reij, C., Tappan, G., Smale, M., 2009. Agroenvironmental transformations in the Sahel: Another kind of green revolution. International Food Policy Research Institute (IFPRI). Discussion Paper 00914.

Schultz, P.W., 2011. Conservation means behavior. Conserv. Biol. 25(6).

United Nations Environment Programme (UNEP), 2012. Relationships and resources: Environmental governance for peacebuilding and resilient livelihoods in Sudan. Nairobi, Kenya.

World Health Organization/United Nations Children's Fund (WHO/UNICEF), 2014. Progress on drinking water and sanitation. In: Joint Monitoring Programme—Update 2014. WHO Press, Geneva, Switzerland.

SHIFTING FROM INDIVIDUAL TO COLLECTIVE ACTION: LIVING LAND'S EXPERIENCE IN THE BAVIAANSKLOOF, SOUTH AFRICA

8.3

Maura Talbot, Dieter van den Broeck

Living Lands, Patensie, South Africa

8.3.1 LAND DEGRADATION AND THREE DISCONNECTS

This section explores the way in which land degradation in the landscapes that Living Lands[1] is working on in South Africa is a symptom of what Scharmer and Kaufer (2013) have termed "the three disconnects." This phrase refers to the way in which people are disconnected from themselves, from each other, and from the natural environment. Scharmer and Kaufer (2013) refer to these as the spiritual/cultural divide, the social divide, and the ecological divide. After outlining how these operate on the landscapes where Living Lands and Commonland are working, we describe the ways in which we are attempting to reconnect people with themselves, with each other, and with the natural environment, as well as our achievements and discoveries.

8.3.2 ECOLOGICAL DIVIDE

The research undertaken in the Baviaanskloof indicates that much of the environmental degradation that has occurred there occurred as a result of human activities and interventions in the landscape over the past 200 years or so (Jansen, 2008; Bobbins, 2011; Rebelo, 2012; Stokhof de Jong, 2012). Some of these were intentional interventions for economic and other strategic social gains like the alteration of the river and alluvial fans during the 1980s after a major flood.

Other livelihood activities had unintended degradation outcomes as well. One of these was the degradation of the hill slopes as a result of overgrazing by goats (Mills et al., 2005; Mills and Cowling, 2006; Powell, 2009; Stokhof de Jong, 2012). Another has been the introduction and spread of invasive

---

[1]Living Lands is a South African izaNGO working to facilitate collaboration on living landscapes through design and facilitation of social learning processes and building leadership capacity. For more information, see www.livinglands.co.za.

Land Restoration. http://dx.doi.org/10.1016/B978-0-12-801231-4.00008-2

alien plants (MacDonald et al., 1986; Van Wilgen et al., 2012). This degradation has reduced rainwater penetration into the soil and the replenishment of groundwater resources and exacerbated the intensity of flooding events (Van Luik et al., 2013). This has reduced the capacity to store and use water downstream and increased damages from floods (Van Luik et al., 2013; Jansen, 2008).

Due to the complex nature of ecological systems, the people's livelihood activities and the changes these make in the land interact with other elements of the ecological system and often change the system in unexpected ways, producing many unintended consequences (Cilliers, 2000). So these effects arise out of a spatial, temporal, physical, and intellectual disconnection between people and their environment. Finding solutions to these problems requires a reconnection with the socioecological system that helps us to understand each of its components and, more important, the relationships and dynamic interactions between them.

8.3.3 SOCIAL DIVIDE

The second generally recognized disconnect is the one between people. While our society is a globally interdependent ecosystem, our individualistic and self-centered way of behaving is essentially a war of the parts against each other and the whole (system) (Scharmer and Kaufer, 2013). Although many of us are aware that our individual activities are having unintended negative effects on others in society, we all seem to be trapped in inherited ways of thinking, behaving, and relating to one another that keep us locked into our habitual behaviors (Ariely, 2010; Kahneman, 2011). Living Lands has found this general situation is well reflected in the attitudes and behavior of the farmers in the Baviaanskloof area, the South African government's Working for Water Programme, the conservation and park authorities, and the competing water users in the Kouga catchment.

This social divide stems from people's usual myopic focus of attention on their own lives or responsibilities and their immediate and short-term needs and wants (Kahneman, 2011). This limited perspective results in their not taking into consideration the temporal and spatial impacts of their activities on other people. This perspective is created through our socialization process and maintained by the social institutions[2] within which we live and work and interact with others. One of the key institutional characteristics of our society that keeps us stuck on this destructive path is its individualistic orientation and compartmentalization of activities into many separate and different organizations focusing on specific activities or issues (Scharmer and Kaufer, 2013). Therefore, everyone is separated from one another and focused on only a small part of the whole system. So long as we continue to see and act in this competitive, compartmentalized way, we cannot comprehend and deal with the challenges that the whole system is facing. There is an urgent need to critically evaluate the system; coordinate, align, and integrate activities; and find strategic interventions that could put us on a more sustainable path.

Most attempts to deal with land degradation processes focus on these two divides, the ecological and the social, and attempt to restore natural and social capital. Living Lands, however, has found that dealing with the third divide—the disconnect with the self and the lack of inspiration, hope, and

---

[2]The term *institutions* here refers to all the rules, norms, and habits of thinking and behaving and relating to one another that are formalized in some way in society. It cannot, therefore, be equated with organizations that only form a small part of what is understood as social institutions.

complacency and the disempowerment that is associated with it—is the key to successfully dealing with the other two divides—the social and ecological.

8.3.4 THE DISCONNECT FROM SELF

Scharmer and Kaufer (2013) refer to this spiritual-cultural divide as "a disconnect between the self and Self—that is, between one's current 'self' and the emerging future 'Self' that represents one's greatest potential" (Scharmer and Kaufer, 2013, pp. 4–5). Deep down, our highest Self wants to be heard, understood, and appreciated and to love and care for others. Despite this, most people feel unloved or not respected, understood, or appreciated most of the time and are very sparing with their love for others. Most people also spend most of their time doing things they do not really want to do but that they feel obligated or forced to do. They also tend to be very judgmental, cynical, and negative about the potential to change things. It is not surprising, therefore, that there are such high rates of suicide, depression, and illness in our society (Krug et al., 2002).

On the landscapes that Living Lands is working on, we see this disconnection from the Self among all the stakeholders. The farmers in the Baviaanskloof and Kouga know that their farming activities are degrading the environment and that their relationships with their workers do not make their staff feel happy, inspired, motivated, or productive, but they feel unable to behave otherwise. They are concerned about their children's future and want to secure their land and lifestyle, but they find themselves unable to do so. The South African government officials in the departments responsible for water, the environment, and agriculture want to have collaborative relationships with the land and water users, but the way they are doing things does not create opportunities for collaboration and often tends to antagonize those they want to work with and makes it more difficult for the government to achieve its goals. The government officials often see these problems with the way government does things, but they feel powerless to do anything about it. At the same time, everyone is blaming everyone else for the problems and seeing each other as the enemy.

8.3.5 THE LIVING LANDS EXPERIENCE AND APPROACH

Living Lands has found that using the U Theory collective social learning approach (see figure 8.3.1), developed by the Presencing Institute at the Massachusetts Institute of Technology (MIT), directly addresses these divides from the self and one another and empowers and inspires the participants to change the way they see and do things and collaborate with others toward achieving their goals and deepest desires. Social learning is increasingly recognized as a crucial component of participatory processes aimed at nurturing collective action around common environmental concerns like land degradation (Cundill and Rodela, 2012). Social learning has been found to improve decision making by increasing awareness of the social system and the relationships and interactions among the various stakeholders, and through the collective learning process building relationships and the problem-solving capacity of stakeholders (Schusler et al., 2003; Brown et al., 2008). Social learning has also been found to support the development of systems thinking among the stakeholders (Nerbonne and Lentz, 2003), and in so doing enhancing the ability of socioecological systems (i.e., society) to respond to change (Pahl-Wostl and Hare, 2004; Pahl-Wostl et al., 2007). Social learning has also been shown to

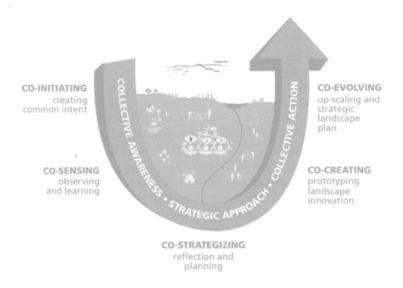

FIGURE 8.3.1

The U as one process with five movements (Scharmer, 2009).

promote collective action within social networks (Maurel et al., 2007; Lebel et al., 2010). Collins et al. (2007, p. 565) have argued that social learning promotes concerted action, which "involves multiple stakeholders working together in a purposeful way to achieve some common end that emerges during the process."

Living Lands adopted this approach after initially trying to impose its own restoration agenda on the stakeholders in the Baviaanskloof 6 years earlier. Living Lands was already involved with and inspired by the subtropical thicket restoration project,[3] as well as the potential to expand this over the 1.4 million ha of degraded spekboom veld in the Western and Eastern Cape Provinces of South Africa (Mills et al., 2010). When we began undertaking research and engaging with the farmers in this region, we thought that all we needed to do was educate the farmers and get them to revegetate their land with spekboom. We were surprised, however, to find that the farmers were not interested in our solution and wanted to have nothing to do with us. This forced us to reconsider our approach and begin using the key principals of the Theory U collective social learning approach (Scharmer, 2009). This approach (illustrated in Figure 8.3.1) required us to engage in deep listening and perceiving the system with an open mind and to look to the local stakeholders about their concerns and insights for direction on the challenges and potential innovations. This approach created the conditions through which the solutions could emerge from the engagement process. In the process of using this approach, Living Lands workers had to learn how to listen and empathize with the stakeholders, which made a huge difference to the willingness of the stakeholders to participate and consider alternative ways of doing things. Listening helped both Living Lands and the stakeholders understand and appreciate each other's

---

[3]The Subtropical Thicket Restoration Programme was initiated and managed by the Natural Resource Management Unit of the Department of Environmental Affairs and the Working for Water Programme. This project undertook considerable research to assess the potential to restore degraded spekboom veld through revegetation efforts and to sequester carbon and earn carbon credits on the global carbon markets.

perspectives and in the process broaden their understanding of the nature and dynamics of the system. To achieve this, Living Lands had to create a safe space in which all the different stakeholders could express their views, be heard, and interact. This built a great amount of trust.

Getting stakeholders to participate and to broaden their perspectives and become willing to change their behavior, is only possible if engagement and listening are done without the voices of judgment, cynicism, or fear toward one another. One needs to start from where the stakeholders are and to avoid generating fear and resistance. To achieve this, it is necessary to begin with a bilateral stakeholder engagement process between the landscape mobilizers (Living Lands) and each stakeholder group individually, before moving to a multilateral engagement processes. The initial bilateral engagement needs to build sufficient trust and (self) empathy for the stakeholders to become confident and willing enough to engage with other different or potentially competing or threatening groups. Living Lands experienced how powerful this process was in overcoming the tendency to act individualistically among the landowners in the Baviaanskloof and how it opened up the possibility of collaboration among all the landowners and among the landowners and other groups such as conservation authorities.

Once all the stakeholder groups feel that they have been heard and valued and are committed to participate, the next step is to scale up to multilateral stakeholder engagements. The interaction of all the stakeholder groups creates the opportunity to challenge existing narrow perspectives and broaden these to a more holistic and integrated perspective. These multilateral engagements are particularly important in situations where stakeholders are feeling insecure and threatened by one another's behavior. Such insecurity is likely to be prevalent in any situation where there are significant policy changes being proposed and implemented, as is the case in South Africa.

The Living Lands process with stakeholders was valuable in showing how most of the problems around land and water use are a direct consequence of the attitudes, thinking, and behavior of the stakeholders toward themselves, each other, and the environment. All of these groups, including the external stakeholders, were found to be acting separately and individualistically to protect and maintain their own interests and meet their needs and responsibilities. This effectively amounted to the "institutionalised irresponsibility" that Scharmer and Kaufer (2013) draws attention to in his recent book on the U theory processes and approach. Overcoming this problem depends on breaking down the social barriers among people, groups, and institutions, generating mutual understanding and respect, and building a more holistic awareness of the system and the interrelationships with all its elements among the participants. Once this has been achieved, it will then start becoming possible for people to collaborate, experiment, and create new types of relationships and partnerships that are considerate of the needs of all the elements of the system (social and environmental).

Living Lands has intentionally focused its efforts on facilitating the development of large collaborative efforts to deal with the complex social and ecological crises on the landscapes, rather than directly making practical interventions on the landscape. Living Lands found that it was easier for such collaborations to be created by concerned, external, and independent landscape mobilizers like its own organization, which was more willing and able to listen to and hear the voices and concerns of all the stakeholders without judgment, and in doing so, gain the trust and willingness of the stakeholders to participate (See figure 8.3.2 Living Lands listening to the stakeholders). External agencies that have vested interests in the landscape, such as government organizations, are more likely to be viewed with suspicion and as a threat by local stakeholders. Collective social learning processes also require skilled landscape mobilizers who can remain independent and mediate conflicts and tensions among the stakeholders. The networks of experts and other organizations that Living Lands has brought in to support the social change process on the landscape have also proved to be critical to the success of social learning processes.

FIGURE 8.3.2

Living Lands facilitating collective social learning in the Baviaanskloof.

The first phases of the U process are essentially research processes that Living Lands has undertaken using an action-oriented, transdisciplinary, and systems-thinking approach. This socioecological systems perspective is a holistic and proactive strategy for the integrated management of land, water, and living resources, which promotes equitable conservation and sustainable use as part of living landscapes. This approach is focused on the balancing of the on- and off-site services that an intact ecosystem can deliver to both upstream and downstream users, and therefore goes a step further than traditional restoration projects by balancing the effects at different landscape and land use scales.

This integrated approach helped us to understand the system from different perspectives, dimensions, and levels. It gave us important insights into people's future perspectives and interests, priorities, needs, expectations, and challenges. This was important because the priorities of external researchers and implementers like ourselves usually did not match those of local stakeholders. Expect to find considerable differences among stakeholders regarding their perspectives on history, conflicts and resolutions, organizational networks, stakeholder relationships, institutional priorities, and what has (and has not) worked in the past. This is valuable tacit knowledge about the elements of the socioecological system and the relationships among these elements that is embedded in the stakeholders and usually not consciously recognized as knowledge and not readily available from other sources (Craps, 2003; Pahl-Wostl et al., 2007). Accessing this knowledge is critical to gaining an understanding of the problematic relationships in the system that need to change in order to address the key challenges facing the people and their landscape. They can also provide insights into potential innovations, which are essentially new types of relationships that could be considered and piloted. Living Lands looks for solutions that support sustainable livelihoods and seek to be obtainable, location-specific approaches that are suited to the historic, cultural, and political context of the landscape.

As a result of applying this approach in the Baviaanskloof catchment, Living Lands has managed to secure an agreement with all the farmers in the area to collaborate and set up a conservancy, looking at sustainable collective land management, and enter into a stewardship agreement with the neighboring park authorities that includes setting aside a large portion of degraded spekboom veld for revegetation and alternative, more sustainable land uses being investigated and piloted. Figure 8.3.3 provides a map indicating the areas that the landowners are willing to make available for restoration. This marks a significant change from the situation 6 years ago, when the farmers were not working together and were not willing to consider restoration innovations or enter into any agreements with their surrounding park authorities.

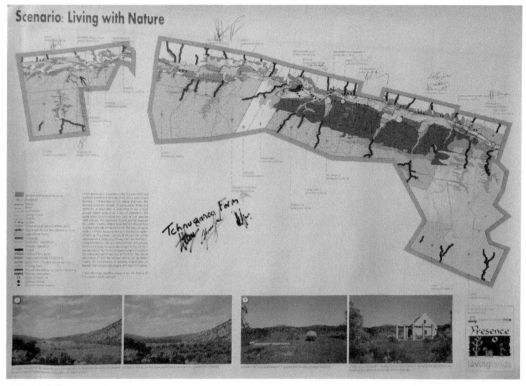

FIGURE 8.3.3

Signed map of Baviaanskloof land uses indicating areas that landowners are willing to set aside for spekboom rehabilitation.

(Source: Stokhof de Jong (2012))

Bottom-up approach with top-down support and guidance, and creating landscape business opportunities

Living Lands began this work very much as a bottom-up approach, recognizing the need for innovations that emerge from the local context and are aligned very directly with the local needs, opportunities, and constraints (illustrated in top half of figure 8.3.4). This stands in opposition to conventional, top-down approaches that identify the solutions as coming from the outside and maintain relationships of dependence and disempowerment. Living Lands recognizes that while such top-down approaches can sometimes affect change, they often have many negative and unintended outcomes that arise out of the complex dynamics and responses of all the different elements and agents within the socioecological system. To minimize these negative and unintended consequences and to build

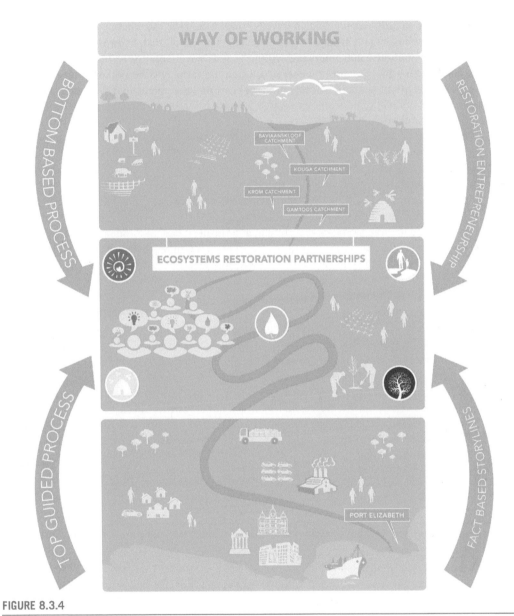

FIGURE 8.3.4

Living Land's approach to working on the landscape.

collaboration and effectiveness, it is critical to adopt a bottom-up approach that facilitates social learning and change.

The objective of the bottom-up-based process is to inspire, create a landscape strategy, and enable collective action to address the needs and dreams of the people—by and for the people. This process creates innovation that is sustainable, locally owned, applicable, and collaborative and addresses the local needs. Despite its impressive achievements with this bottom-up approach, Living Lands has, over time, come to appreciate that the innovations emerging at the local level needed to be aligned with and supported by more regional and national initiatives and has consequently begun integrating this bottom-up approach with a top-down-guided process that creates an enabling environment to prototype and upscale the local innovations and collaborations (as illustrated in Figure 8.3.4). This approach aims to build ownership and willingness among local stakeholders that is supported by an integrated effort by the government and private sector to implement and mainstream sustainable local policies and programs. The ultimate objective is to foster business partnerships and collaborations that integrate and catalyze resources and funding for sustainable innovations that facilitate adaptation to climate change and other social-ecological crises. This is to support the landscape and local stakeholders in their process of prototyping and mainstreaming their solutions.

To accomplish this, Living Lands has now teamed up with the Commonlands (based in the Netherlands) to implement their Four Returns model. This approach looks for ways to inspire, as well as produce returns on natural capital, social capital, and financial capital through collaborative investments in restoration and sustainable agriculture (see Chapter 4.6 in this book). This approach has created a language for engaging with business and investors to upscale investment in restoration and agriculture and build business partnerships.

Currently, Living Lands and Commonland are also in the process of expanding their social learning processes to the two neighboring catchments (the Kouga and Kromme), which, together with the Baviaanskloof, supply around 70% of the existing water supplies for Port Elizabeth City, as well as all the water for the downstream Gamtoos Irrigation Scheme. In partnership with Commonland, the Four Returns Project is currently being initiated in all three catchments and the downstream areas to form partnerships among business, government, and civil society in and around the Nelson Mandela Bay catchment areas to strategically and collectively explore investments in restoring the resilience of the degraded ecosystems. More specifically, we hope to increase local water and food security, improve livelihoods, and create a green and sustainable economy. We hope to encourage participants to work together to inspire people in other landscapes to take similar action. To support these processes, we are in the process of designing and initiating participatory water model development processes and water monitoring programs.

8.3.6 CONCLUSION

It is we—all of humanity—who are the cause of the ecological and social problems we are facing today. More particularly, it is the limited and disconnected way that we see ourselves and each other and the socio-ecological system that is the core of our problem. In this section, we have shown that we can change this situation. Living Lands has found that the key is to create safe spaces and opportunities for people to reconnect to themselves, one another, and their environment. We have found that we can do this through the collective social learning process called the U Theory process. This is a nonlinear (indirect) way that allows us to shift our focus back to ourselves (individually and collectively) so that we can begin to see ourselves, each other, and the system as a whole. Once we have made this shift in perspective, the solutions will become evident to us, and the mutual understanding and collaboration that has developed during the process of bending our perspective back onto ourselves will give us the collaborative power to implement the solutions.

We believe that integrating the living landscape approach and the Four Returns model creates the inspiration and social capital that we need to deal with the social and ecological divides. As we are experiencing in our Four Returns project, a return of inspiration (reconnection with Self) enables people to form collaborative ventures to deal with the other two divides. This reconnection process also involves reconnecting the bottom-based processes with key enabling agents such as business and government at the top (macro scale). In doing so, Living Lands and Commonland are enabling the first two reconnections that provide the means through which the third ecological divide is addressed and reconnected. However, we recognize that this is just our own experience and that each situation will be different.

REFERENCES

Ariely, D., 2010. The upside of irrationality: The unexpected benefits of defying logic at work and at home. Harper Collins, London.

Bobbins, K., 2011. Alluvial fan geomorphology and restoration. Master's thesis with Geography Department of Rhodes University, Grahamstown, South Africa.

Brown, H.C., Buck, L., Lassoie, J., 2008. Governance and social learning in the management of non-wood forest products in community forests in Cameroon. Intl. J. Agr. Res. Gov. Ecol. 7, 256–275.

Cilliers, P., 2000. What can we learn from a theory of complexity? Emergence 2 (1), 23–33.

Collins, K., Blackmore, C., Morris, D., Watson, D., 2007. A systemic approach to managing multiple perspectives and stakeholders in water catchments: Three UK studies. Environ. Sci. Pol. 10, 64–574.

Craps, M., 2003. Social learning in river basin management: HarmoniCOP WP2 reference document. European Commission ReportKU Leuven—Centre for Organisational and Personal Psychology, Leuven, Belgium.

Cundill, G., Rodela, R., 2012. A review of assertions about the processes and outcomes of social learning in natural resource management. J. Environ. Manage. 113, 7–14.

Jansen, H.C., 2008. Water for Food and Ecosystems in the Baviaanskloof Mega Reserve: Land and Water Resources Assessment in the Baviaanskloof, Eastern Cape Province, South Africa. Alterra Report 1812, Wageningen University, the Netherlands.

Kahneman, D., 2011. Thinking fast and slow. Penguin Books, London, UK.

Krug, E.G., et al., 2002. World report on violence and health. World Health Organization, Geneva, Switzerland, p. 185.

Lebel, L., Grothmann, T., Siebenhuner, B., 2010. The role of social learning in adaptiveness: Insights from water management. Intl. Environ. Agree.: Politics, Law Econ. 10, 333–353.

MacDonald, L.A.W., Kruger, F.J., Ferrar, A.A., 1986. The ecology and management of biological invasives in southern Africa. Oxford University Press, Cape Town, South Africa.

Maurel, P., Craps, M., Cernesson, F., Raymond, R., Valkering, P., Ferrand, N., 2007. Concepts and methods for analysing the role of information and communication tools (IC-tools) in social learning processes for river basin management. Environ. Model. Software 22, 630–639.

McConnachie, M.M., Cowling, R.M., Shackleton, C.M., Knight, A.T., 2013. The challenges of alleviating poverty through ecological restoration: Insights from South Africa's "Working for Water" Programme. Restor. Ecol. 21 (5), 544–550.

Mills, A.J., Cowling, R.M., 2006. Rate of carbon sequestation at two thicket sites in the eastern Cape. South Africa. Restor. Ecol. 14 (1), 38–49.

Mills, A.J., et al., 2005. The effects of goat pastoralism on ecosystem carbon storage in semi-arid thicket, Eastern Cape. South Africa. Austral Ecol. 30, 797–804.

Mills, A., et al., 2010. Investing in sustainability; restoring degraded thicket, creating jobs, capturing carbon and earning green credit. Report on Subtropical Restoration Project, In: Working for Water Programme. Department of Water Affairs, Cape Town, South Africa.

Nerbonne, J.F., Lentz, R., 2003. Rooted in grass: Challenging patterns of knowledge exchange as a means of fostering social change in a southeast Minnesota farm community. Agri. Human Values 20, 65–78.

Pahl-Wostl, C., Hare, M., 2004. Processes of social learning in integrated resources management. J. Comm. Appl. Soc. Psych. 14, 193–206.

Pahl-Wostl, C., Sendzimer, J., Jeffrey, P., Aerts, J., Berkamp, G., Cross, K., 2007. Managing change towards adaptive water management through social learning. Ecol. Soc. 12 (2), 1–18. Article 30, http://www.ecologyandsociety.org/vol12/iss2/art30/.

Powell, M.J., 2009. The restoration of degraded sub-tropical thicket in the Baviaanskloof Mega-Reserve, South Africa: The role of carbon stocks and Portulacaria afra survivorship. M.Sc. thesis, Rhodes University, Grahamstown, South Africa.

Rebelo, A., 2012. Hydrological impacts of wetland destruction, alien invasion, and subsequent restoration in the Kromme Catchment, Eastern Cape, SA. M.Sc. thesis, Conservation Ecology Department, Stellenbosch University, South Africa.

Scharmer, O., 2009. Theory U: Leading from the Future as it Emerges. Berrett-Koehler Publishers, San Francisco.

Scharmer, O., Kaufer, K., 2013. Leading from the Emerging Future: From Ego-System to Eco-System Economies. Applying the Theory U to Transforming Business, Society and Self. Berrett-Koehler Publishers, San Francisco.

Schusler, M.T., Decker, J.D., Pfeffer, J.M., 2003. Social learning for collaborative natural resource management. Soc. Nat. Res. 15, 309–326.

Stokhof de Jong, J., 2012. Living landscape restoration: The common vision for the Hartland in the Baviaanskloof. MSc's thesis, Landscape Architecture, Wageningen University, the Netherlands.

Van Luik, G., et al., 2013. Hydrological implications of desertification: Degradation of South African semi-arid subtropical thicket. J. Arid Environ. 91, 14–21.

Van Wilgen, B.M., et al., 2012. An assessment of the effectiveness of a large, national-scale invasive alien plant control strategy in South Africa. Biolog. Conserv. 148 (1), 28–38.

DEVELOPMENT AND SUCCESS, FOR WHOM AND WHERE: THE CENTRAL ANATOLIAN CASE

Erhan Akça[1], **Kume Takashi**[2], **Tetsu Sato**[3]

Adiyaman University, Technical Sciences, Vocational School, Adiyaman, Turkey[1]
Ehime University, Department of Rural Engineering, Faculty of Agriculture, Matsuyama City, Japan[2]
Research Institute for Humanity and Nature, Motoyama, Kamigamo, Kita-ku, Kyoto, Japan[3]

8.4.1 AGRICULTURAL DEVELOPMENT, PAST AND PRESENT

Human progression and development is embedded within human genes and directly affects all our decisions. Prior to the 20th century, development was not considered dangerous or negative; land was utilized within its natural capacity boundaries, water sources were not canalized to agriculture and industry, soils were not sealed under megacities, and greenhouse gases were below harmful levels. Prior to the 21st century, meeting population demands and natural resource consumption were not considered threats to the environment or human health. Rather, they were acknowledged as successes. In comparison, the 21st century may be known as the century of the "start of the end of human race" (McGuire, 2002), as activities such as agriculture have had catastrophic consequences around the globe. Integrating the needs of society and scientific endeavors may be the biggest challenge we will have in this century (Zalasiewicz et al., 2008).

World research and development (R&D) investment, excluding defense, has reach more than $1 trillion, and the accumulated information is beyond comparison to any previous decade. Global agricultural R&D spending increased by 22% during the 2000–2008 period, from $26.1 to $31.7 billion in 2005 purchasing power parity (PPP) prices (Beintema et al., 2012). Thus, agriculture is one of the sectors that benefited most from scientific developments; indeed, any land can be cultivated and any water source at any depth can be extracted thanks to the latest technologies. Nowadays, irrigated agriculture is a common practice even in deserts. In Saudi Arabia, drilling wells extract water from up to 1 km beneath the desert sands to irrigate fields of fruits, vegetables, and wheat (Figure 8.4.1) (Hussain et al., 2008; NASA Earth Observatory, 2012).

The arid highlands in Central Anatolia, Turkey, are no exception. For instance, agriculture in the Great Konya closed basin in Central Anatolia, spanning 55,000 km^2 and supported by a poor river network, was rainfed until the 1980s, when groundwater pumping for irrigation was introduced as a technological development. Whereas farmers argued that their drip or sprinkle irrigation systems consumed up to 50% less water than furrow irrigation, the practice itself allowed more land to be brought into

Land Restoration. http://dx.doi.org/10.1016/B978-0-12-801231-4.00034-3

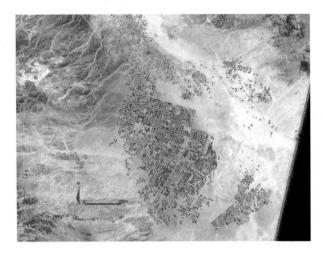

FIGURE 8.4.1

Irrigated area in Saudi Arabia (NASA Earth Observatory, 2012).

cultivation than previously, which did little to ease the pressure on water resources, despite the technological options.

Contrary to present development activities, the neolithic development in Central Anatolia did not compete with nature and climate. The majority of the crops were rainfed cereals, and the domesticated small ruminants were already adapted to natural fodders in wide grasslands. Indigenous knowledge (IK) and traditional ecological knowledge (TEK) systems (Colding and Folke, 2000) had been refined and reflected the production of crops that required a smaller amount of soil nutrients and were less water intensive, thereby supporting important ecosystem services to maintain soil quality and prevent desertification. This type of sustainable socioeconomic system can be considered successful for Central Anatolia in terms of both human livelihoods and environment.

Today's mismanagement of natural resources by agricultural activities inevitably affects all members of society. Large-scale commercial agricultural systems rely on pesticides, fertilizers, and tillage, all of which induce environmental degradation such as groundwater pollution and secondary soil salinization (Kume et al., 2010). Unfortunately, stakeholders frequently overlook the devastating outcomes of these practices since farmers, traders, laborers, government agencies, and politicians are focused on the economic gains. Few complain about environmental issues, as farmers are placated by their increased income, politicians garner support from farmers because of the legislations backing them, governmental officers do not pay much attention to mismanagement, and poorly educated local researchers are dismissive of alternative measures suggested by other stakeholders. At the same time, in order to protect the land from further deterioration, scientists often neglect local demands by instituting pure nature conservation without considering the needs of the human population dependent on its output. As a consequence, locals do not support this approach and generally trespass into protected areas.

However, the catastrophic results of land misuse around the globe triggered the need to take up environmental responsibility by all types of stakeholders for mitigating degradation problems. The initial step has been already undertaken by key players by shifting from a development-oriented aim to sustainable oriented goals. However, backing up excess resource-consuming technological

FIGURE 8.4.2

The marble mining on the karstic Taurus Mountains, Turkey (Erhan Akça).

developments of current industries created major conflicts between society and science. The short-term targeted development plans caused the invasion of natural resources by profit-oriented investments such as the case in marble mining. The high-production-capacity equipment of marble mining led to deterioration of freshwater sources in karstic areas (Figure 8.4.2).

In contrast, recent actions for better land management have been inspired by traditional sustainable land and marine management concepts, such as the Japanese practices of satoyama and satoumi. Duraiappah et al. (2012) described satoyama as "the mosaic landscapes of different types of ecosystems—secondary forests, farm lands, irrigation ponds, and grasslands, along with human settlements, managed to produce several ecosystem services for human well-being."

The Millennium Development Goal to ensure environmental sustainability calls for the integration of "principles of sustainable development into country policies and programmes and reverse the loss of environmental resources". However, the UN Report (UN, 2013) "A New Global Partnership: Eradicate Poverty and Transform Economies Through Sustainable Development" pointed out that although the international community has aspired to integrate the social, economic, and environmental dimensions of sustainability, no country has yet achieved it since the announcement of Millennium Development Goals in 2000 (UN, 2013).

8.4.2 WHAT DEVELOPMENT BROUGHT TO AND TOOK FROM CENTRAL ANATOLIA

Agriculture is the main income source in Central Anatolia, Turkey, employing nearly 85% of its 11.8 million households (MEVKA, 2014). Economic factors rather than environmental concerns primarily motivate this choice of employment. The primary way that agriculture is conducted demands a high input of agrochemicals and technologies, such as chemical fertilizers, pesticides, large-capacity irrigation pumps, irrigation equipment, and tractors, to produce cash crops like maize, sugar beets, and potatoes. The new agro-techniques, also a questionable topic in terms of sustainability, enabled semiarid and arid Central Anatolian farmers to triple, or even quadruple, their agricultural production. For example, the wheat yield increased from 1500 kg/ha in rain-fed areas to 7000 kg/ha in irrigated areas since the late 1980s by the introduction of water pumps capable of deep water extraction (MEVKA, 2014). On the other hand, this production increase led to excessive water extraction, where farmers drill down to 350 m to obtain irrigation water (Akça et al., 2012). The number of groundwater wells, both registered by state hydraulic works and unregistered ones, exceeded 100,000 in 2010 (Topak and Acar, 2010) and was estimated to be well over 110,000 in 2014. Approximately 6000 m^3 of water (almost equivalent to the discharge rate of the Niagara River on the border of the United States and Canada) was pumped from 110,000 groundwater wells from May to September 2014 in Central Anatolia to irrigate every crop of maize, sugar beets, wheat, alfalfa, and potatoes (Yılmaz, 2010). In addition to falling groundwater levels, excessive irrigation caused the formation of sinkholes in the Karapınar region of Central Anatolia. More than 20 sinkholes were formed, with varying depths from 2 to 30 m, since 1980 (Figure 8.4.3) (Yılmaz, 2010). Although the locals are aware of the relation between excessive water use and sinkhole formation, farming practices have remained unchanged in order to maintain yields.

Additionally, since irrigation allows the creation of new farmlands in the area, grasslands are being converted to irrigated lands (Figure 8.4.4). As a consequence, sheep herders complain about shrinkage of natural grazing areas, which decreases the sheep-breeding capacity of the region. Moreover, this shift also threatens the natural grasslands, where more than 120 fodder plants, adapted to arid conditions, can be found (Güner and Ekim, 2014). These endangered drought-resistant species are not under conservation or cultivation management, although they could serve as genetic sources for the anticipated, more arid future.

Furthermore, the excessive use of agrotools has induced nutrient loss, erosion, salinity increase, and, ultimately, desertification in Central Anatolia.

Agricultural practices in Central Anatolia have been implemented from developed countries, disregarding local hydrological conditions. Farmers in Central Anatolia continue to aim for high yields through intensive irrigation, despite limited water sources. As a result, agricultural activities in Central

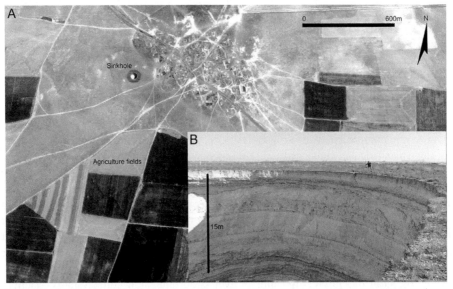

FIGURE 8.4.3

The İnobası sinkhole in Karapınar, Central Anatolia, Turkey (Erhan Akça). (A) Google Earth Image, (B) Erhan Akça.

FIGURE 8.4.4

Irrigated agriculture land expansion to grasslands (Erhan Akça).

Anatolia have become individually competitive rather than cooperatively progressive. In other words, this metamorphosis cannot be defined to be as successful as that achieved by previous ancestors during the "green revolution" in 10,000 Before Present (BP), whose production domains (that is, anthroscapes) were in harmony with the environment.

8.4.3 INDIGENOUS ANATOLIAN AGRICULTURE MANAGEMENT

The Central Anatolian anthroscape is adapted to less than 350 mm of precipitation per year through rainfed agriculture and small ruminant breeding. Barley, wheat, chickpeas, and lentils were the main cultivated crops, as they all can grow in arid conditions. Vegetables, such as green beans, melons, and green peppers, are mostly harvested before maturation, both to save water and so they can be used for pickle preparation for the winter. The Central Anatolian traditional land use and animal husbandry practices, which have been abandoned recently, actually may meet the economic expectations of locals and sustain environmental resources. For example, supporting sheep-breeding grasslands based on a carrying capacity of 1.5 sheep/ha is more profitable than irrigated maize and cereal production, since meat prices have exceeded cereal prices since 2013 (FAO, 2014; also see Figure 8.4.5). Another example is that harvesting melon before maturation requires only two or three irrigation episodes, while maize needs to be irrigated 10 to 12 times during the summer. In Karapınar, with 10–12 irrigation episodes, 13.5 tons of maize can be produced. The net income of maize per hectare in 2014 is $3030, whereas 15 tons/ha of melon for pickle making (Figure 8.4.6) may be produced with two or three irrigation episodes, generating a net income of $20,300/ha.

All these traditional agricultural practices, which had been abandoned since the introduction of intensive irrigation, should be merged with modern scientific analysis of their economic benefits and sociocultural potential to support livelihoods. Such integrated knowledge systems on environment, land use, and human livelihoods have emerged in Central Anatolia and elsewhere in the world, effectively promoting societal transformation based on decision making and actions by stakeholders

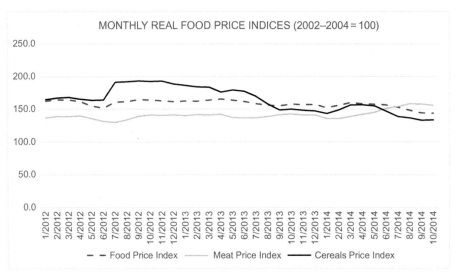

FIGURE 8.4.5

FAO Food Price Index of 2013–2014 (FAO, 2014).

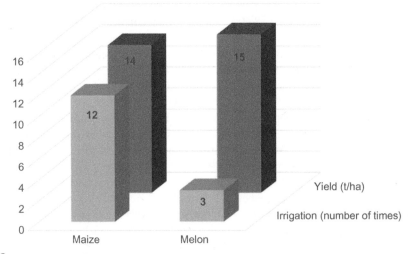

FIGURE 8.4.6

The water requirement (number of times) and yield (tons/ha) of traditional melon and recently introduced maize in Karapınar, Central Anatolia, Turkey (Erhan Akça).

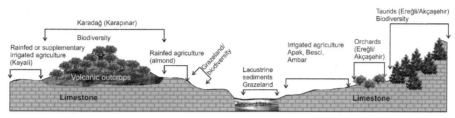

FIGURE 8.4.7

The Karapınar anthroscape (Akça et al., 2012).

(Sato, 2014). Scientists do not need to warn farmers about environmental dangers, but they need to grab farmers' attention with positive economic tools and incentives.

8.4.4 KARAPINAR ANTHROSCAPE MODEL

Karapınar's rich fodder vegetation can also provide an alternative means for sustaining farmers' high income through animal husbandry. However, in the 1960s, overgrazing nearly converted Karapınar into a desert due to the disregard shown to the grassland-carrying capacities. Nonetheless, following 50 years of restoration projects, initiated by the Ministry of Agriculture in cooperation with Wageningen University in the Netherlands in the 1960s (Groneman, 1968), cultivation and sheep breeding are again feasible in the region (Akça et al., 2009). The natural grasslands host more than 120 fodder plant species, which can sustain a high amount of meat demanded by the national market. These regional opportunities need to be assessed with the aim to build long-term systematic resilience to respond to present demands, as well as future demands. Using this traditional knowledge model, this text proposes that the Karapınar anthroscape (Figure 8.4.7; also see Akça et al., 2014) will enhance the strengths of the geographic properties and overcome their shortcomings by balancing sustainable production and natural assets. For instance, for future management of the region and maintenance of the carrying capacity of these grasslands, as well as for increasing vegetation cover and plant diversity of overgrazed grasslands, seeding with local plant species should be undertaken. In this way, even lands in an arid climate can provide ecosystem services to humans and living organisms without leading to degradation (Akça et al., 2012). The buffer zones in intensive agricultural areas and rich biodiversity regions in the highlands, along with patches of indigenous rainfed agriculture management, may also secure natural environmental conditions and diminish drastic fluctuations in farmers' income due to unstable climate conditions and market prices.

8.4.5 CONCLUSION

In order to find a way forward and to minimize conflicts between the governance of common resources and ecosystem services, a new type of solution-oriented approach that integrates traditional and scientific knowledge is required. This strategy needs, as a starting point, a paradigm shift from capitalizing all sources for individual benefits to establishing a balanced relation between society and resources. By including the target society along with natural properties, cultural and traditional values that have a positive impact on the environment should be sustained by stakeholders. The environmental capacity, particularly for irrigated agricultural activities, is unknown and investments on irrigation are often undertaken without long-term evaluation. By applying proven land use and management practices, rather than only short-term, profit-oriented solutions, crop production can become more environmentally friendly, as well as economically beneficial.

Through in-depth interactions with local stakeholders and transdisciplinary learning processes to produce integrated local and environmental knowledge, we found a potential approach to mitigate the drawbacks of excessive water use; that is, to promote highly valued crops traditionally cultivated in the region that are less demanding of water supplies. Farmlands can thereby function as integrated sustainable production domains and provide multiple ecosystem services without threatening the economic capital of farmers.

Recent approaches that integrate the human dimension into conservation by considering local values are producing promising results. Both the satoyama and satoumi concepts in Japan and the anthroscape in Anatolia integrate ancient knowledge and time-honored practices by including human livelihood needs into their treatment of surrounding natural environments (Kume et al., 2010; Sato, 2014; Eswaran et al., 2011). The ultimate aim is to improve the welfare of the local community through environmentally friendly methods derived from a blend of traditional approaches and modern sustainable technologies. As such, development that creates competition among individuals, regions, and nations is likely to produce losers as well as winners. On the other hand, sustainable development that integrates environmental protections ensures that humanity wins.

REFERENCES

Akça, E., et al., 2009. The Long-Term Land Conservation Effect on Soil Organic Sequestration. In: 1st National Drought and Desertification Sympoisum, June 6–18, 2009, Konya, Turkey, pp. 136–143.

Akça, E., et al., 2012. A decade of change in land use and development of sinkholes in Karapınar, C. Anatolia. In: Kapur, S. et al. (Eds.), Land Degradation and Challenges in Sustainable Soil Management, 8th Intl. Soil Science Congress, Vol. 1. DEMFO Press, İzmir, Turkey, pp. 222–226.

Akça, E., Kapur, S., Miavagni, S.R., 2014. Sustainable Land Management in Turkey. CIHEAM Watch Letter No. 28.

Beintema, N., Stads, G.-J., Fuglie, K., Heisey, P., 2012. ASTI Global Assessment of Agricultural R&D Spending Developing Countries Accelerate Investment. International Food Policy Research Institute, Washington, DC.

Colding, F.J., Folke, C., 2000. Rediscovery of traditional ecological knowledge as adaptive management. Ecol. Appl. 10 (5), 1251–1262.

Duraiappah, A.K., et al., 2012. Satoyama–Satoumi Ecosystems and Human Well-Being: Socio-Ecological Production Landscapes of Japan. UN University Press, Tokyo.

Eswaran, H., et al., 2011. The Anthroscape Approach in Sustainable Land Use. In: Kapur, S., et al. (Eds.), Sustainable Land Management. Springer, Berlin, pp. 1–50.

Food and Agricultural Organization (FAO), 2014. FAO Food Price Index. Accessed November 14, 2014, from: http://www.fao.org/worldfoodsituation/foodpricesindex/en/.

Groneman, A.F., 1968. The Soils of the Wind Erosion Control Cam Area, Karapınar, Turkey. Wageningen Agricultural University, Wageningen 161 p.

Güner, A., Ekim, T., 2014. Illustrated Flora of Turkey. İş Bank of Turkey. Cultural Series, İstanbul. Vol. I. Türkiye İş Bankası Pub.

Hussain, M., Ahmed, S.M., Abderrahman, W., 2008. Cluster analysis and quality assessment of logged water at an irrigation project, eastern Saudi Arabia. J. Environ. Manage. 86 (1), 297–307.

Kume, T., et al., 2010. Seasonal changes of fertilizer impacts on agricultural drainage in a salinized area in Adana. Turkey. Sci. Total Environ. 408 (16), 3319–3326.

McGuire, W.J., 2002. A Guide to the End of the World: Everything You Never Wanted to Know. Oxford University Press, New York.

MEVKA (Mevlana Kalkınma Ajansı), 2014. Karapınar Town Report. Mevlana Development Agency, Konya (in Turkish).

NASA Earth Observatory, 2012. Agricultural Fields, Wadi As-Sirhan Basin, Saudi Arabia. March 5, 2012. Thematic Mapper and Enhanced Thematic Mapper Plus, Landsat Satellites. 4, 5, and 7.

Sato, T., 2014. Integrated Local Environmental Knowledge Supporting Adaptive Governance of Local Communities. In: Alvares, C. (Ed.), Multicultural Knowledge and the University. Multiversity India, Mapusa, India, pp. 268–273.

Topak, R., Acar, B., 2010. Sustainable irrigation and importance of technological irrigation systems for Konya Basin. Tarım Bilimleri Araştırma Dergisi 3 (2), 65–70.

United Nations (UN), 2013. A New Global Partnership: Eradicate Poverty and Transform Economies Through Sustainable Development. United Nations Publications, New York.

Yılmaz, M., 2010. Environmental problems caused by ground water level changes around Karapinar. Ankara Univ. J. Environ. Sci. 2 (2), 145–163.

Zalasiewicz, J., et al., 2008. Are we now living in the Anthropocene? GSA Today 18 (2), 4–8. http://dx.doi.org/10.1130/GSAT01802A.1.

SHARING KNOWLEDGE TO SPREAD SUSTAINABLE LAND MANAGEMENT (SLM)

8.5

Rima Mekdaschi Studer, Isabelle Providoli, Hanspeter Liniger

Centre for Development and Environment, University of Bern, Bern, Switzerland

On the one hand, ensuring sustainable use of natural resources is crucial for maintaining the basis for our livelihoods. On the other hand, the prediction is that by 2050, global food production will have to increase by 60% from its 2005–2007 levels to satisfy consumer demand (according to Alexandratos and Bruinsma, cited in FAO, 2014, page 3). The huge challenge of being more productive while still protecting the environment and ensuring diversified and resilient livelihoods is further exacerbated by many global issues. Population growth, food security, climate change, land and water scarcity, natural disasters, loss of biodiversity, competing claims on land, changing markets, and migration are increasing worldwide. But this is not the full story. A large share of the problem of land/natural resource degradation is a result of poor land and water management at all levels and a lack of appropriate governance, including legal and regulatory frameworks.

For years, research and various national and international organizations have been working on issues of land degradation and alternative forms of land management. Numerous land users worldwide have been testing, adapting, and refining new and better ways of managing land more sustainably. The key principles of sustainable land management (SLM) are the productivity and protection of natural resources, coupled with economic viability and social acceptability (Schwilch et al., 2013). However, there has been relatively little documentation or evidence of the range of benefits generated by SLM practices in different farming systems and on different scales, which in turn is essential to persuade decision makers to invest in transitioning to more sustainable practices. Proper knowledge management is crucial in order for SLM to reach its full potential. Without it, land degradation will continue to be addressed in an ad hoc manner, all too often ignoring or only selectively using the knowledge and experience gained elsewhere.

Based on the premise that SLM experiences are not sufficiently or comprehensively documented, evaluated, and shared, the global World Overview of Conservation Approaches and Technologies initiative (WOCAT; www.wocat.net) and its network partners have developed standardized tools and methods for documenting, evaluating, and assessing the impact of SLM practices. Included in these methods are knowledge sharing and decision support in the field, at the planning level, and in the scaling up and out of good practices.

WOCAT was founded in 1992 as an informal global network of soil and water conservation (SWC) specialists and was one of the first programs to promote SWC/SLM in response to land degradation. The WOCAT network was reorganized in 2014 to become more formalized.

Land Restoration. http://dx.doi.org/10.1016/B978-0-12-801231-4.00035-5

Over the years, WOCAT experienced a shift in focus, as follows:

- From pure data collection to data evaluation, monitoring, training, and research
- From a narrower focus on soil erosion and fertility decline to good practices for SLM (covering soil, water, vegetation, and animals)
- From a set of questionnaires to a modular and flexible methodology
- From pure knowledge generation to the use of the knowledge for evidence-based decision making
- From on-site assessment to on- and off-site benefits of SLM, including watershed and landscape approaches. Joint and participatory development of the program with national and international partner institutions and organizations has permitted continuous improvements and adaptations to users' needs, while maintaining the standardization

The most frequently used WOCAT tools and methods are case study documentation questionnaires. Each case study comprises an SLM Approach and one or more SLM Technologies and can cover any area, ranging from as small as a single farmer's field to entire catchments or districts. All this information feeds into the WOCAT SLM Technology and SLM Approach database. Over the past 15 years, the global WOCAT database has grown to about 500 technologies and 250 approaches from all continents, with many case studies from Africa and Asia in particular (https://www.wocat.net/en/knowledge-base/technologiesapproaches.html). Over the past 2 years, WOCAT has made an important addition by producing videos of land users showing how SLM works, what problems it solves, how challenges can be overcome and what benefits can be achieved locally, regionally, and globally (https://www.wocat.net/en/knowledge-base/slm-videos.html).

Once documented, SLM experiences need to be made widely available and accessible in a form that allows land users, advisors, and planners to review a "basket" of alternative options, setting out the advantages and disadvantages of each, thereby enabling them to make informed choices rather than following set prescriptions of what to do. The development of innovations and implementation of new SLM efforts should build on existing knowledge from within a location itself or, alternatively, from similar conditions and environments elsewhere, as well as on a close interaction and exchange between land user and researchers/advisory services (Liniger et al., 2011; Schwilch et al., 2012).

Based on the WOCAT tools and methods, various knowledge products such as global, regional, and national overview books, inventories of practices, guidelines, and books covering/synthesizing different aspects of SLM have been produced and will continue to be, depending on need and actual circumstances (https://www.wocat.net/en/knowledge-base/documentation-analysis.html). For example in *Where the Land is Greener* (WOCAT, 2007) and *Desire for Greener Land* (Schwilch et al., 2012), case studies from around the world are analyzed and policy points for decision makers are drawn. *Sustainable Land Management in Practice* (Liniger et al., 2011) highlights the main principles of SLM and is a practical guide for investment and operation design. *Water Harvesting—Guidelines to Good Practice* (Mekdaschi Studer and Liniger, 2013) introduces the concepts behind water harvesting and proposes a harmonized classification system, followed by an assessment of suitability, adoption, and scaling up of practices. Several partner countries, such as Afghanistan, Bangladesh, China, Ethiopia, Mongolia, Nepal, Senegal, Tajikistan, and Tunisia, have produced national overviews on SLM good practices.

In 2014, WOCAT was officially recognized by the UN Convention to Combat Desertification (UNCCD) as the primary recommended database/platform for SLM best practices, including measures of climate change adaptation. UNCCD gave WOCAT a mandate to support the 196 signatory countries in recording their SLM technologies and approaches and using the SLM knowledge of stakeholders worldwide—from land users to decision makers—to improve local and regional land management.

The constant need for adapting land management to a rapidly changing and complex world with continued and newly emerging demands calls for continuous knowledge management, adjustment to upcoming users' needs, and the search for best SLM options.

REFERENCES

Food and Agriculture Organization (FAO), 2014. The State of Food and Agriculture—Innovation in Family Farming. Food and Agriculture Organization of the United Nations, Rome.

Liniger, H., Mekdaschi Studer, R., Hauert, C., Gurtner, M. et al., (Eds.), 2011. Sustainable Land Management in Practice. Guidelines and Best Practices for Sub-Saharan Africa. TerrAfrica, Swiss Agency for Development and Cooperation (SDC), World Overview of Conservation Approaches and Technologies (WOCAT), Food and Agriculture Organization of the United Nations (FAO).

Mekdaschi Studer, R., Liniger, H., 2013. Water Harvesting: Guidelines to Good Practice. Centre for Development and Environment (CDE), Bern; Rainwater Harvesting Implementation Network (RAIN), Amsterdam; Meta-Meta, Wageningen; International Fund for Agricultural Development (IFAD), Rome.

Schwilch, G., Hessel, R., Verzandvoort, S. (Eds.), 2012. Desire for Greener Land. Options for Sustainable Land Management in Drylands. Bern, Switzerland, and Wageningen, the Netherlands: University of Bern—Centre for Development and the Environment (CDE), Alterra-Wageningen UR, ISRIC–World Soil Information, and CTA–Technical Centre for Agricultural and Rural Cooperation.

Schwilch, G., Liniger, H., Hurni, H., 2013. Sustainable land management (SLM) practices in drylands: How do they address desertification threats? Environ. Manage. http://dx.doi.org/10.1007/s00267-013-0071-3. May 1–22.

World Overview of Conservation Approaches and Technologies (WOCAT), 2007. Where the Land is Greener - Case Studies and Analysis of Soil and Water Conservation Initiatives Worldwide. Liniger, H., Critchley, W. (Eds.), Bern, Switzerland.

SUGGESTIONS FOR WAYS TO USE THIS BOOK

BUFFETS, CAFES, OR A MULTICOURSE MEAL: ON THE MANY POSSIBLE WAYS TO USE THIS BOOK

9.1

Ilan Chabay

Professor and Senior Fellow, Institute for Advanced Sustainability Studies, Potsdam, Germany and Chair of the Knowledge, Learning, and Societal Change Alliance (www.KLaSiCA.org)

This book was written by a marvelous collection of authors who have widely different backgrounds and experiences; thus, they approached the crucial issue of land restoration from very different perspectives. Of course, you can use this volume as a buffet table of intellectually nutritious items from which to pick out only what best fits your own immediate needs. However, you can also use it as a collection of appetizers—provocations to stimulate your appetite for thinking anew about land restoration in a multitude of ways and contexts that are outside your usual palate. And in the same more exploratory fashion, you can use the ideas and case studies as a starting point for discussions with local experts or interested laypersons alike. Or you can invent new formats to engage children and adults to awaken their awareness and interest in the crucial issues raised and discussed in this book.

What resonates in the book as being particularly relevant to the conditions in which you live? What seems irrelevant to conditions or requirements that you face or know about? Which cases or approaches are set in very different conditions or locations, but nevertheless are really relevant in your context? What questions would you like to ask of the author or authors of a particular chapter?

One point of these questions is to suggest ideas for your own reflection upon reading some or all of the book. Another and more important point is to suggest that a very effective way to engage people with important ideas (whether in the book or encountered elsewhere) is through stimulating and engaging with their questions and raising your own for discussion. This is often an unfamiliar approach in some more authoritative and hierarchical learning environments. Yet it can be extremely effective once the barriers to open discussion are overcome. In many different cultural and educational settings, we have found that initial hesitation to question the speaker or teacher or authority can be overcome with patience, humor, and examples of familiar or surprising concrete experiences that provoke curiosity.

You might take a walk that allows the group to comment on the conditions and raise questions about what they see on the land around them. Or, in the meeting space, set up a simple experiment that provides fun, concrete examples to stimulate curiosity. For example, you could use a block of dry earth on a table tilted at an angle. Cover half of the earth with a cloth or some small leaves and leave half bare, and then pour water over both sides of it. Ask the group what they expect to happen before pouring the water and ask them to describe and explain what happens as the water hits the earth or leaves. A bin or bucket to collect the runoff can be used to show how much "soil" is carried away by the "rain." How much water drains out immediately compared to how much was poured on the dirt? Is this like something the group sees happening around them? Would the effects be different if the table were tilted more or less? Would it be different if the slab of earth were different—if it had a different composition or came from a different place? What other experiments with simple materials come to mind that might stimulate questions to begin thinking with others?

So how might you use this book? Here are some suggestions, which you can mix and match as appropriate to fit your interests and needs:

1. **Text**: The entire volume, or sections of it, could serve as a primary or supplementary text for a class, depending on the level, scope, and subject of the class. But rather than just assigning it as reading, ask students to pick out and discuss two chapters that, for example, offer contrasting cases under similar conditions; or one chapter that is similar to or in contrast to the local or regional conditions.

   Additional text resources, among many, include the *Soil Atlas,* published in 2015 by the Heinrich Böll Foundation and the Institute for Advanced Sustainability Studies (http://globalsoilweek.org/soilatlas-2015) and the DryNet initiative's web site at http://www.dry-net.org.

2. **Seminar**: Any one of the individual chapters could be a good common starting point for an informal or formal seminar in a corporate, nongovernmental organization (NGO) or academic setting. A prospective participant might choose to read a chapter prior to the seminar and then summarize the content of the chapter at the start of the seminar, perhaps beginning or ending with what he or she sees as a crucial question or issue of local relevance that is touched upon in the chapter. Then that participant, or another person acting as moderator, can start and lead a discussion among all present around the thesis or implications of the chapter content.

   In many countries, there are specific resource centers, either in academia or in agricultural or geophysical research agencies, that in many cases can suggest local experts available to participate in such seminars or discussions.

3. **Event**: Key ideas from sections or chapters of the book could form the core of an event that features a moderated discussion in a cafe, tavern, museum, school cafeteria, gathering place in a garden or field, or other informal and culturally familiar venue. Science centers, museums, and zoos often have staff who organize such events on a regular basis—see the web site of the Association of Science Technology Centers (ASTC; http://www.ASTC.org) to find a center near you and some ideas for events. A moderator could start the event by introducing a person with policy, practice, historical, or legal experience and knowledge regarding some aspect of soil as a resource or land restoration to give a brief, broadly accessible informal talk, followed by an informal question-and-answer session and general discussion. Or the moderator could conduct an open interview with one or more people to focus on a particular set of questions of general interest to the visitors. An important opportunity that this type of event provides is for children and adults to relate not only to the content per se but also to the personal stories and lives of people with expertise and significant experience.

The Global Soil Week (http://globalsoilweek.org/) and other events related to soil and land restoration are also noted on the Institute for Advanced Sustainability Studies (IASS; http://www .iass-potsdam.de) web site.

The Caux Dialogues on Land and Security (http://www.landlivespeace.org) is another event that not only provides a spectacularly beautiful venue conducive to dialogues and brings together a truly diverse group of people each year, but it was the inspiration for this book, as noted in other chapters herein.

4. **Games, films, and simulations**: How about creating a game based on the choices that farmers, policy makers, or agribusiness have to make to deal with the issues covered in the book on land, soil, water, and human security? It could be a simulation updated and more specific to land and soil than the classic SimCity or SimEarth game, or it might be an adventure game featuring a farmer and her journey of discovery after inheriting a degraded farm in a hardscabble region. Two examples of interactive games dealing with farming in Africa are available online at http://globalsoilweek.org/category/resources/interactive.

An animated video on soil as a critical resource was developed by IASS for the first Global Soil Week in 2012; it is available at http://globalsoilweek.org/resources/video-lets-talk-about-soil.

The main point here is that each chapter, as well as the book as a whole, is a resource that provides a rich picture of the successes, failures, and challenges still before us. We hope they are also a starting point to stimulate your curiosity and lead to opportunities for mutual learning and thinking about the issues of land restoration that are so vital in the transition to a more sustainable future.

CONCLUDING REMARKS AND A WAY FORWARD

CONCLUDING REMARKS

10.1

Luc Gnacadja

Executive Secretary of the UN Convention to Combat Desertification (2007–2013)

It's time to enter the restoration age.

In the face of the complex challenges of sustainability, nothing connects humanity more fundamentally than the air we breathe and the soil we walk on. However, we often fail to value sufficiently the land and soils that support our lives.

Everything we have ever possessed, currently possess, or will possess stems from soil, the precious skin of the land. It is our most valuable geo-resource, and the mother of all other human resources; but it is finite and in acute danger.

When we deplete soil or degrade land, it is done at our own expense. Land degradation and soil nutrient depletion can be ugly side effects of human civilization. Its consequences distort the stream of ecosystem services and places heavy burdens on generations to come. The area of abandoned land since the beginning of farming (some 2 billion ha) has now surpassed the total area of cropland currently under cultivation around the world (nearly 1.5 billion ha); also see Pimentel and Burgess (2013).

Population dynamics, globalization, and global warming exacerbate the soil crisis. But despite its central importance, soil is still the forgotten capital of humanity: Its significance has been poorly reflected in international discourses on climate change and sustainability. Land degradation and restoration have only recently become issues debated during discussions on human security.

Coming to Caux, Switzerland, in July 2008 for the first time to participate in the first Caux Forum on Human Security, I experienced the fertility and amazing ability of the conferences at Caux, which have a reputation as unique incubators for reconciliation. However, in this particular conference, which focused on human security and the root causes of insecurity, I noticed the lack of discussion about the essential role that soil plays in sustaining human security as a crucial asset for ensuring the freedom from want and from fear and the freedom to live in dignity.

So, when I came back to Caux two years later to deliver the keynote speech of the third Caux Forum on Human Security in July 2010, I emphasized that "Our security as a global community depends on the extent to which we preserve our soils, because enhancing soil anywhere enhances life everywhere—and grounds security."

The following year, a new conference was created to further explore these thoughts: the Caux Dialogue on Land and Security. The distinguished role of being the keynote speaker for the first session was given to Yacouba Sawadogo, an illiterate peasant from Burkina Faso, who left his audience speechless about his amazing achievements. Sawadogo earned the nickname "the man who stopped the desert" for his work transforming the severely desertified, imminently doomed land of his community during the terrible drought conditions of the Sahel in the 1980s into a thriving agroforestry

Land Restoration. http://dx.doi.org/10.1016/B978-0-12-801231-4.00010-0

parkland some two decades later. His story is also featured also in the award-winning documentary *The Man Who Stopped the Desert*. His exemplary grassroots leadership on land stewardship was an inspiration to all attendees.

I have learnt lessons from Yacouba's story and from many other community level ground-breaking land stewardship achievements I have encountered around the world:

- Governments need to continually capitalize on grassroots-level success stories and good practices in order to forge effective policies and governance for sustainable development.
- We must balance our narrative about the negatives surrounding land degradation by highlighting success stories in sustainable land management and assessing their socioeconomic impacts, including their return on investment. We need to emphasize the positive trends that we can build through scaling up and dissemination.
- Land degradation is all about the consecutive tradeoffs that we overlook in making land use decisions that end up transforming productive landscapes into "manmade deserts." Keeping those tradeoffs in close check and continually addressing them in a given landscape is the challenge all concerned actors and stakeholders have to face together.
- We must debunk the misconceptions about land degradation: too costly, too little return on investment, and if anything is gained, it is only in the long term. These claims are far from the truth. Land degradation has far-reaching impacts, but prevention and restoration can reach even further. We need to improve our understanding of the overall cost of inaction versus the benefits of sustainable land management; among them is how restoration contributes to addressing or mitigating present and potential resource-based conflicts around livelihoods supporting resources such as water and productive land. Such land stewardship contributes to ground security.

Reversing the tide of land degradation is at the core of solving a central 21st-century's conundrum: feeding sustainably 9 billion people on a "thirsty planet" while eradicating extreme poverty. "If just 12% of the world's degraded lands were restored to production, we could feed another 200 million people and farmers' incomes would be raised by US$40 billion a year" (Global Commission on the Economy and Climate, 2014, p. 31).

It was encouraging to see that at the United Nations Conference on Sustainable Development (i.e., the Rio + 20 Summit) in June 2012, the imperative of achieving land-degradation neutrality as soon as possible made it into the summit's outcome document—"The Future We Want"—and is reflected in the post-2015 development agenda as an operational target for Sustainable Development Goals.

But that is just one (albeit good) step in the direction of building an effective framework for managing our soils as a global common.

The international community has set 2015 as the year to conclude its global agenda for achieving sustainability, including a more comprehensive and effective climate deal; 2015 is also the International Year of Soils. This year offers a promising context and good ground for releasing this book as a seed that will contribute to building the sustainable future for which we aim; a seed that will add to the growing momentum about landscape restoration in the context of the overall land stewardship and sustainability.

No doubt it shall bear much fruit and new seeds for an age of increased land restoration and protection.

Luc Gnacadja

REFERENCES

Global Commission on the Economy and Climate, 2014. Better Growth, Better Climate. The New Climate Economy Report. New Climate Economy, London, UK.

Pimentel, D., Burgess, M., 2013. Soil erosion threatens food production. Agri. August 2013.

Index

Note: Page numbers followed by *f* indicate figures, *b* indicate boxes, *t* indicate tables and *np* indicate footnotes.